T0362107

STABILITY OF STRUCTURES

STABILITY OF STRUCTURES

Principles and Applications

CHAI H. YOO
Auburn University

SUNG C. LEE
Dongguk University

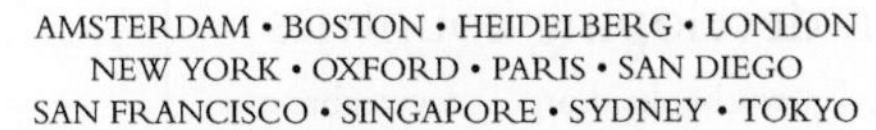

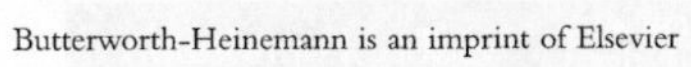

Butterworth-Heinemann is an imprint of Elsevier
30 Corporate Drive, Suite 400, Burlington, MA 01803, USA
The Boulevard, Langford Lane, Kidlington, Oxford OX5 1GB, UK

Library of Congress Cataloging-in-Publication Data
Yoo, Chai Hong.
Stability of structures: principles and applications/Chai H. Yoo, Sung C. Lee.
p. cm.
Summary: "The current trend of building more streamlined structures has made stability analysis a subject of extreme importance. It is mostly a safety issue because Stability loss could result in an unimaginable catastrophe. Written by two authors with a combined 80 years of professional and academic experience, the objective of Stability of Structures: Principles and Applications is to provide engineers and architects with a firm grasp of the fundamentals and principles that are essential to performing effective stability analysts"– Provided by publisher.
Includes bibliographical references and index.
ISBN 978-0-12-385122-2 (hardback)
1. Structural stability. 2. Safety factor in engineering. I. Lee, Sung Chul, 1957- II. Title.
TA656.Y66 2011
624.1'71–dc22
2010048927

British Library Cataloguing-in-Publication Data
A catalogue record for this book is available from the British Library.

ISBN: 978-0-12-385122-2

CONTENTS

PREFACE

The subject of this book is the stability of structures subjected to external loading that induces compressive stresses in the body of the structures. The structural elements examined are beams, columns, beam-columns, frames, rectangular plates, circular plates, cylindrical shells, and general shells. Emphasis is on understanding the behavior of structures in terms of load-displacement characteristics; on formulation of the governing equations; and on calculation of the critical load.

Buckling is essentially flexural behavior. Therefore, it is imperative to examine the condition of equilibrium in a flexurally deformed configuration (adjacent equilibrium position). The governing stability equations are derived by both the equilibrium method and the energy method based on the calculus of variations invoking the Trefftz criterion.

Stability analysis is a topic that fundamentally belongs to nonlinear analysis. The fact that the eigenvalue procedure in modern matrix and/or finite element analysis is a fortuitous by-product of incremental nonlinear analysis is a reaffirming testimony. The modern emphasis on fast-track education designed to limit the number of required credit hours for core courses in curriculums left many budding practicing structural analysts with gaping gaps in their understanding of the theory of elastic stability. Many advanced works on structural stability describe clearly the fundamental aspects of general nonlinear structural analysis. We believe there is a need for an introductory textbook such as this, which will present the fundamentals of structural stability analysis within the context of elementary nonlinear flexural analysis. It is believed that a firm grasp of these fundamentals and principles is essential to performing the important interpretation required of analysts when computer solutions are adopted.

The book has been planned for a two-semester course. The first chapter introduces the buckling of columns. It begins with the linear elastic theory and proceeds to include the effects of large deformations and inelastic behavior. In Chapter 2 various approximate methods are illustrated along with the fundamentals of energy methods. The chapter concludes by introducing several special topics, some of them advanced, that are useful in understanding the physical resistance mechanisms and consistent and rigorous mathematical analysis. Chapters 3 and 4 cover buckling of beam-columns. Chapter 5 presents torsion in structures in some detail, which is one of the least-well-understood subjects in the entire spectrum of structural mechanics. Strictly speaking, torsion itself does not belong to a work on structural stability, but it needs to be covered to some extent if one is to have a better understanding of buckling accompanied with torsional behavior. Chapters 6 and 7 consider stability of framed structures in conjunction with torsional behavior of structures. Chapters 8 to 10 consider buckling of plate elements, cylindrical shells, and general shells. Although the book is devoted primarily to analysis, rudimentary design aspects are also discussed.

The reader is assumed to have a good foundation in elementary mechanics of deformable bodies, college-level calculus, and analytic geometry, and some exposure to differential equations. The book is designed to be a textbook for advanced seniors and/or first-year graduate students in aerospace, civil, mechanical, engineering mechanics, and possibly naval architects and shipbuilding fields and as a reference book for practicing structural engineers.

Needless to say, we have relied heavily on previously published work. Consequently, we have tried to be meticulous in citing the works and hope that we have not erred on the side of omission.

AUTHORS BIOGRAPHY

Chai H. Yoo, PhD., PE., F.ASCE, is professor emeritus at Auburn University. He has over 40 years of teaching, research, and consulting experience. He received his BS from Seoul National University in Korea and his MS and PhD degrees from the University of Maryland. He is the recipient of the 2008 ASCE Shortridge Hardesty Award for his outstanding research and practical engineering efforts dealing with the strength and stability of thin-walled sections, especially as it applies to bridge girders. Author of over 75 refereed journal papers and a book on curved steel girder bridges, Dr. Yoo has been active on numerous ASCE and SSRC technical committee activities. He was the principal investigator/project director of NCHRP Project 12-38 that produced AASHTO Guide Specifications for Horizontally Curved Steel Girder Highway Bridges with Design Examples for I-Girder and Box-Girder Bridges.

Sung C. Lee, PhD., M.ASCE, is professor and head of the Department of Civil and Environmental Engineering at Dongguk University in Seoul, Korea. He received his BS from Seoul National University in Korea, his MS from Oregon State University, and his PhD from Auburn University. He is the author of over 100 refereed journal articles, technical papers, and proceedings. Dr. Lee's breakthrough research on the strength and stability of web panels in plate- and box-girders renders a correct understanding on a topic that has been controversial over the last five decades.

CHAPTER 1

Buckling of Columns

Contents

1.1. INTRODUCTION

A physical phenomenon of a reasonably straight, slender member (or body) bending laterally (usually abruptly) from its longitudinal position due to compression is referred to as buckling. The term *buckling* is used by engineers as well as laypeople without thinking too deeply. A careful examination reveals that there are two kinds of buckling: (1) bifurcation-type buckling; and (2) deflection-amplification-type buckling. In fact, most, if not all, buckling phenomena in the real-life situation are the deflection-amplification type. A bifurcation-type buckling is a purely conceptual one that occurs in a perfectly straight (geometry) homogeneous (material) member subjected to a compressive loading of which the resultant must pass

Stability of Structures
ISBN 978-0-12-385122-2, doi:10.1016/B978-0-12-385122-2.10001-6

though the centroidal axis of the member (concentric loading). It is highly unlikely that any ordinary column will meet these three conditions perfectly. Hence, it is highly unlikely that anyone has ever witnessed a bifurcation-type buckling phenomenon. Although, in a laboratory setting, one could demonstrate setting a deflection-amplification-type buckling action that is extremely close to the bifurcation-type buckling. Simulating those three conditions perfectly even in a laboratory environment is not probable.

Structural members resisting tension, shear, torsion, or even short stocky columns fail when the stress in the member reaches a certain limiting strength of the material. Therefore, once the limiting strength of material is known, it is a relatively simple matter to determine the load-carrying capacity of the member. Buckling, both the bifurcation and the deflection-amplification type, does not take place as a result of the resisting stress reaching a limiting strength of the material. The stress at which buckling occurs depends on a variety of factors ranging from the dimensions of the member to the boundary conditions to the properties of the material of the member. Determining the buckling stress is a fairly complex undertaking.

If buckling does not take place because certain strength of the material is exceeded, then, why, one may ask, does a compression member buckle? Chajes (1974) gives credit to Salvadori and Heller (1963) for clearly elucidating the phenomenon of buckling, a question not so easily and directly explainable, by quoting the following from *Structure in Architecture*:

> *A slender column shortens when compressed by a weight applied to its top, and, in so doing, lowers the weight's position. The tendency of all weights to lower their position is a basic law of nature. It is another basic law of nature that, whenever there is a choice between different paths, a physical phenomenon will follow the easiest path. Confronted with the choice of bending out or shortening, the column finds it easier to shorten for relatively small loads and to bend out for relatively large loads. In other words, when the load reaches its buckling value the column finds it easier to lower the load by bending than by shortening.*

Although these remarks will seem excellent to most laypeople, they do contain nontechnical terms such as choice, easier, and easiest, flavoring the subjective nature. It will be proved later that buckling is a phenomenon that can be explained with fundamental natural principles.

If bifurcation-type buckling does not take place because the aforementioned three conditions are not likely to be simulated, then why, one may ask, has so much research effort been devoted to study of this phenomenon? The bifurcation-type buckling load, the critical load, gives the upper-bound solution for practical columns that hardly satisfies any one of the three conditions. This will be shown later by examining the behavior of an eccentrically loaded cantilever column.

1.2. NEUTRAL EQUILIBRIUM

The concept of the stability of various forms of equilibrium of a compressed bar is frequently explained by considering the equilibrium of a ball (rigid-body) in various positions, as shown in Fig. 1-1 (Timoshenko and Gere 1961; Hoff 1956).

Although the ball is in equilibrium in each position shown, a close examination reveals that there are important differences among the three cases. If the ball in part (a) is displaced slightly from its original position of equilibrium, it will return to that position upon the removal of the disturbing force. A body that behaves in this manner is said to be in a state of stable equilibrium. In part (a), any slight displacement of the ball from its position of equilibrium will raise the center of gravity. A certain amount of work is required to produce such a displacement. The ball in part (b), if it is disturbed slightly from its position of equilibrium, does not return but continues to move down from the original equilibrium position. The equilibrium of the ball in part (b) is called unstable equilibrium. In part (b), any slight displacement from the position of equilibrium will lower the center of gravity of the ball and consequently will decrease the potential energy of the ball. Thus in the case of stable equilibrium, the energy of the system is a minimum (local), and in the case of unstable equilibrium it is a maximum (local). The ball in part (c), after being displaced slightly, neither returns to its original equilibrium position nor continues to move away upon removal of the disturbing force. This type of equilibrium is called neutral equilibrium. If the equilibrium is neutral, there is no change in energy during a displacement in the conservative force system. The response of the column is very similar to that of the ball in Fig. 1-1. The straight configuration of the column is stable at small loads, but it is unstable at large loads. It is assumed that a state of neutral equilibrium exists at the

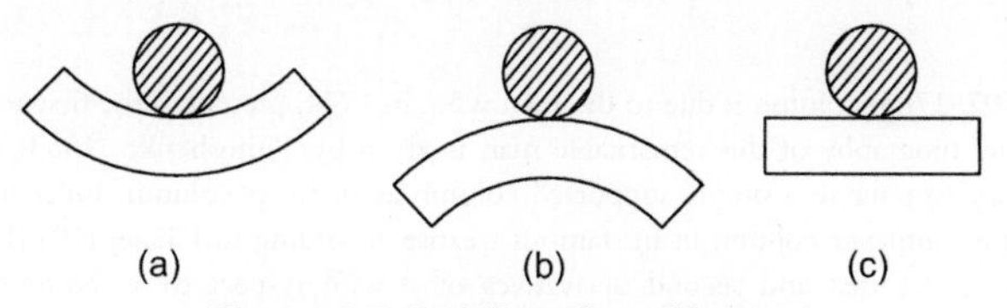

Figure 1-1 Stability of equilibrium

transition from stable to unstable equilibrium in the column. Then the load at which the straight configuration of the column ceases to be stable is the load at which neutral equilibrium is possible. This load is usually referred to as the critical load.

To determine the critical load, eigenvalue, of a column, one must find the load under which the member can be in equilibrium, both in the straight and in a slightly bent configuration. How slightly? The magnitude of the slightly bent configuration is indeterminate. It is conceptual. This is why the free body of a column must be drawn in a slightly bent configuration. The method that bases this slightly bent configuration for evaluating the critical loads is called the method of neutral equilibrium (neighboring equilibrium, or adjacent equilibrium).

At critical loads, the primary equilibrium path (stable equilibrium, vertical) reaches a bifurcation point and branches into neutral equilibrium paths (horizontal). This type of behavior is called the buckling of bifurcation type.

1.3. EULER LOAD

It is informative to begin the formulation of the column equation with a much idealized model, the Euler[1] column. The axially loaded member shown in Fig. 1-2 is assumed to be prismatic (constant cross-sectional area) and to be made of homogeneous material. In addition, the following further assumptions are made:

1. The member's ends are pinned. The lower end is attached to an immovable hinge, and the upper end is supported in such a way that it can rotate freely and move vertically, but not horizontally.
2. The member is perfectly straight, and the load P, considered positive when it causes compression, is concentric.
3. The material obeys Hooke's law.
4. The deformations of the member are small so that the term $(y')^2$ is negligible compared to unity in the expression for the curvature, $y''/[1+(y')^2]^{3/2}$. Therefore, the curvature can be approximated by y''.[2]

[1] The Euler (1707–1783) column is due to the man who, in 1744, presented the first accurate column analysis. A brief biography of this remarkable man is given by Timoshenko (1953). Although it is customary today to refer to a simply supported column as an Euler column, Euler in fact analyzed a flag-pole-type cantilever column in his famous treatise according to Chajes (1974).

[2] y' and y'' denote the first and second derivatives of y with respect to x. Note: $|y''| < |y'|$ but $|y'| \approx$ thousandths of a radian in elastic columns.

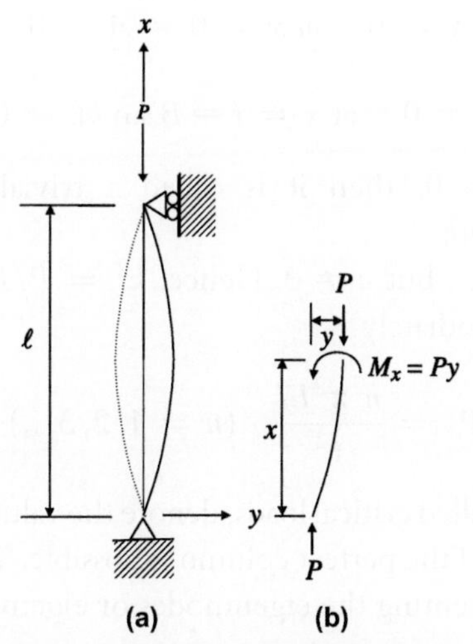

Figure 1-2 Pin-ended simple column

$$-\frac{M}{EI} = \frac{y''}{[1 + (y')^2]^{3/2}} \simeq y'' \tag{1.3.1}$$

From the free body, part (b) in Fig. 1-2, the following becomes immediately obvious:

$$EIy'' = -M(x) = -Py \quad \text{or} \quad EIy'' + Py = 0 \tag{1.3.2}$$

Equation (1.3.2) is a second-order linear differential equation with constant coefficients. Its boundary conditions are

$$y = 0 \quad \text{at } x = 0 \quad \text{and} \quad x = \ell \tag{1.3.3}$$

Equations (1.3.2) and (1.3.3) define a linear eigenvalue problem. The solution of Eq. (1.3.2) will now be obtained. Let $k^2 = P/EI$, then $y'' + k^2 y = 0$. Assume the solution to be of a form $y = \alpha e^{mx}$ for which $y' = \alpha m e^{mx}$ *and* $y'' = \alpha m^2 e^{mx}$. Substituting these into Eq. (1.3.2) yields $(m^2 + k^2)\alpha e^{mx} = 0$.

Since αe^{mx} cannot be equal to zero for a nontrivial solution, $m^2 + k^2 = 0$, $m = \pm ki$. Substituting gives

$$y = C_1 \alpha e^{kix} + C_2 \alpha e^{-kix} = A \cos kx + B \sin kx$$

A and B are integral constants, and they can be determined by boundary conditions.

$$y = 0 \quad at\ x = 0 \Rightarrow A = 0$$

$$y = 0 \quad \text{at } x = \ell \Rightarrow B \sin k\ell = 0$$

As $B \neq 0$ (if $B = 0$, then it is called a trivial solution; $0 = 0$), $\sin k\ell = 0 \Rightarrow k\ell = n\pi$
where $n = 1, 2, 3, \ldots$ but $n \neq 0$. Hence, $k^2 = P/EI = n^2\pi^2/\ell^2$, from which it follows immediately

$$P_{cr} = \frac{n^2\pi^2 EI}{\ell^2} \quad (n = 1, 2, 3, ..) \tag{1.3.4}$$

The eigenvalues P_{cr}, called critical loads, denote the values of load P for which a nonzero deflection of the perfect column is possible. The deflection shapes at critical loads, representing the eigenmodes or eigenvectors, are given by

$$y = B \sin \frac{n\pi x}{\ell} \tag{1.3.5}$$

Note that B is undetermined, including its sign; that is, the column may buckle in any direction. Hence, the magnitude of the buckling mode shape cannot be determined, which is said to be immaterial.

The smallest buckling load for a pinned prismatic column corresponding to $n=1$ is

$$P_E = \frac{\pi^2 EI}{\ell^2} \tag{1.3.6}$$

If a pinned prismatic column of length ℓ is going to buckle, it will buckle at $n = 1$ unless external bracings are provided in between the two ends.

A curve of the applied load versus the deflection at a point in a structure such as that shown in part (a) of Fig. 1-3 is called the equilibrium path. Points along the primary (initial) path (vertical) represent configurations of the column in the compressed but straight shape; those along the secondary path (horizontal) represent bent configurations. Equation (1.3.4) determines a periodic bifurcation point, and Eq. (1.3.5) represents a secondary (adjacent or neighboring) equilibrium path for each value of n. On the basis of Eq. (1.3.5), the secondary path extends indefinitely in the horizontal direction. In reality, however, the deflection cannot be so large and yet satisfies the assumption of rotations to be negligibly small. As P in Eq. (1.3.4) is not a function of y, the secondary path is horizontal. A finite displacement formulation to be discussed later shows that the secondary equilibrium path for the column curves upward and has a horizontal tangent at the critical load.

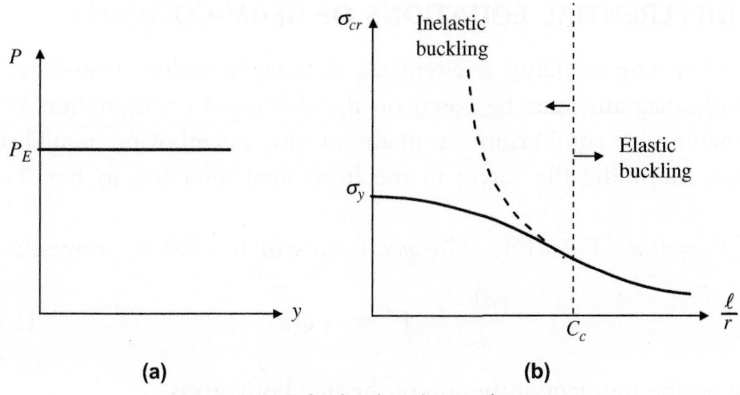

Figure 1-3 Euler load and critical stresses

Note that at P_{cr} the solution is not unique. This appears to be at odds with the well-known notion that the solutions to problems of classical linear elasticity are unique. It will be recalled that the equilibrium condition is determined based on the deformed geometry of the structure in part (b) of Fig. 1-2. The theory that takes into account the effect of deflection on the equilibrium conditions is called the *second-order* theory. The governing equation, Eq. (1.3.2), is an ordinary linear differential equation. It describes neither linear nor nonlinear responses of a structure. It describes an eigenvalue problem. Any nonzero loading term on the right-hand side of Eq. (1.3.2) will induce a second-order (nonlinear) response of the structure.

Dividing Eq. (1.3.4) by the cross-sectional area A gives the critical stress

$$\sigma_{cr} = \frac{P_{cr}}{A} = \frac{\pi^2 EI}{\ell^2 A} = \frac{\pi^2 EAr^2}{\ell^2 A} = \frac{\pi^2 E}{(\ell/r)^2} \tag{1.3.7}$$

where ℓ/r is called the slenderness ratio and $r = \sqrt{I/A}$ is the radius of gyration of the cross section. Note that the critical load and hence, the critical buckling stress is independent of the yield stress of the material. They are only the function of modulus of elasticity and the column geometry. In Fig. 1-3(b), C_c is the threshold value of the slenderness ratio from which elastic buckling commences.

$$\text{eigen pair} \begin{cases} \text{eigenvalue} = P_{cr} = \dfrac{n^2\pi^2 EI}{\ell^2} \\ \text{eigenvector} = y = B\sin\dfrac{n\pi x}{\ell} \end{cases}$$

1.4. DIFFERENTIAL EQUATIONS OF BEAM-COLUMNS

Bifurcation-type buckling is essentially flexural behavior. Therefore, the free-body diagram must be based on the deformed configuration as the examination of equilibrium is made in the neighboring equilibrium position. Summing the forces in the horizontal direction in Fig. 1-4(a) gives

$\sum F_y = 0 = (V + dV) - V + qdx$, from which it follows immediately

$$\frac{dV}{dx} = V' = -q(x) \tag{1.4.1}$$

Summing the moment at the top of the free body gives

$$\sum M_{\text{top}} = 0 = (M + dM) - M + Vdx + Pdy - q(dx)\frac{dx}{2}$$

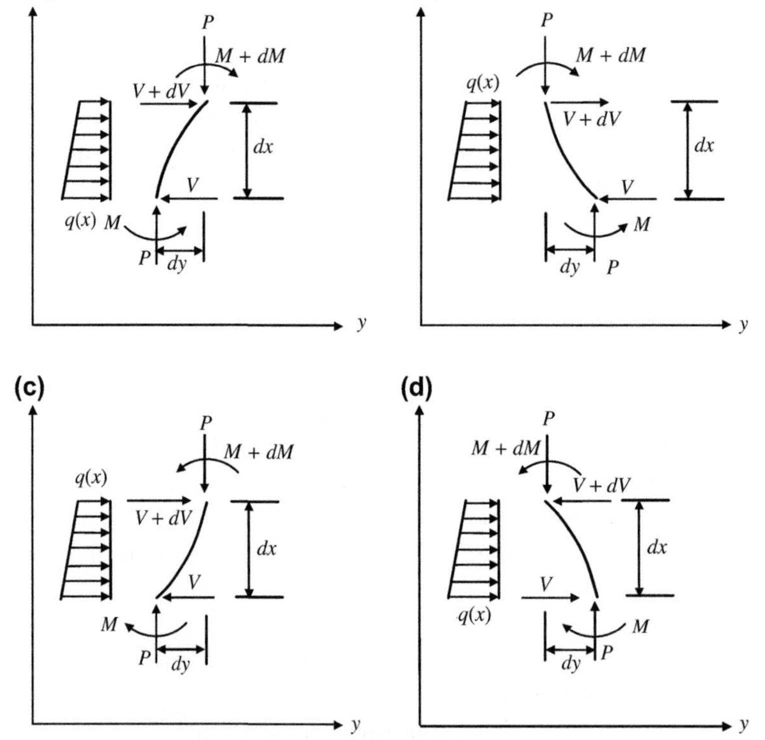

Figure 1-4 Free-body diagrams of a beam-column

Neglecting the second-order term leads to

$$\frac{dM}{dx} + P\frac{dy}{dx} = -V \tag{1.4.2}$$

Taking derivatives on both sides of Eq. (1.4.2) gives

$$M'' + (Py')' = -V' \tag{1.4.3}$$

Since the convex side of the curve (buckled shape) is opposite from the positive y axis, $M = EIy''$. From Eq. (1.4.1), $V' = -q(x)$. Hence, $(EIy'')'' + (Py')' = q(x)$. For a prismatic (EI $=$ const) beam-column subjected to a constant compressive force P, the equation is simplified to

$$EIy^{iv} + Py'' = q(x) \tag{1.4.4}$$

Equation (1.4.4) is the fundamental beam-column governing differential equation.

Consider the free-body diagram shown in Fig. 1-4(d). Summing forces in the y direction gives

$$\sum F_y = 0 = -(V + dV) + V + qdx \Rightarrow \frac{dV}{dx} = V' = q(x) \tag{1.4.5}$$

Summing moments about the top of the free body yields

$$\sum M_{top} = 0$$

$$= -(M + dM) + M - Vdx - Pdy - \cancel{qdxdx/2} \Rightarrow$$

$$-\frac{dM}{dx} - P\frac{dy}{dx} = V \tag{1.4.6}$$

For the coordinate system shown in Fig. 1-4(d), the curve represents a decreasing function (negative slope) with the convex side to the positive y direction. Hence, $-EIy'' = M(x)$. Thus,

$$-(-EIy'')' - (-Py') = V \tag{1.4.7}$$

which leads to

$$EIy''' + Py' = V \quad \text{or} \quad EIy^{iv} + Py'' = q(x) \tag{1.4.8}$$

It can be shown that the free-body diagrams shown in Figs. 1-4(b) and 1-4(c) will lead to Eq. (1.4.4). Hence, the governing differential equation is independent of the shape of the free-body diagram assumed.

The homogeneous solution of Eq. (1.4.4) governs the bifurcation buckling of a column (characteristic behavior). The concept of geometric imperfection (initial crookedness), material heterogeneity, and an eccentricity is equivalent to having nonvanishing $q(x)$ terms.

Rearranging Eq. (1.4.4) gives

$$EIy^{iv} + Py'' = 0 \Rightarrow y^{iv} + k^2 y'' = 0, \quad \text{where } k^2 = \frac{P}{EI}$$

Assuming the solution to be of a form $y = \alpha e^{mx}$, then $y' = \alpha m e^{mx}$, $y'' = \alpha m^2 e^{mx}$, $y''' = \alpha m^3 e^{mx}$, and $y^{iv} = \alpha m^4 e^{x}$. Substituting these derivatives back to the simplified homogeneous differential equation yields

$$\alpha m^4 e^{mx} + \alpha k^2 m^2 e^{mx} = 0 \Rightarrow \alpha e^{mx}(m^4 + k^2 m^2) = 0$$

Since $\alpha \neq 0$ *and* $e^{mx} \neq 0 \Rightarrow m^2(m^2 + k^2) = 0 \Rightarrow m = \pm 0, \ \pm ki$. Hence,

$$y_h = c_1 e^{kix} + c_2 e^{-kix} + c_3 x e^0 + c_4 e^0$$

Know the mathematical identities $\begin{cases} e^0 = 1 \\ e^{ikx} = \cos kx + i \sin kx \\ e^{-ikx} = \cos kx - i \sin kx \end{cases}$

Hence, $y_h = A \sin kx + B \cos kx + Cx + D$ where integral constants A, B, C, and D can be determined uniquely by applying proper boundary conditions of the structure.

Example 1 Consider a both-ends-fixed column shown in Fig. 1-5.

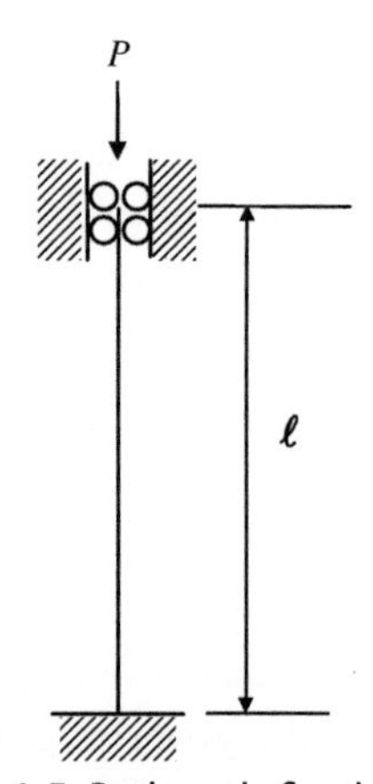

Figure 1-5 Both-ends-fixed column

$$y' = Ak\cos kx - Bk\sin kx + C$$

$$y'' = -Ak^2\sin kx - Bk^2 kx$$

$$y = 0 \quad \text{at } x = 0 \Rightarrow B + D = 0$$

$$y' = 0 \quad \text{at } x = 0 \Rightarrow Ak + C = 0$$

$$y = 0 \quad \text{at } x = \ell \Rightarrow A\sin k\ell + B\cos k\ell + C\ell + D = 0$$

$$y' = 0 \quad \text{at } x = \ell \Rightarrow Ak\cos k\ell - Bk\sin k\ell + C = 0$$

For a nontrivial solution for A, B, C, and D (or the stability condition equation), the determinant of coefficients must vanish. Hence,

$$Det = \begin{vmatrix} 0 & 1 & 0 & 1 \\ k & 0 & 1 & 0 \\ \sin k\ell & \cos k\ell & \ell & 1 \\ k\cos k\ell & -k\sin k\ell & 1 & 0 \end{vmatrix} = 0$$

Expanding the determinant (**Maple**®) gives

$$2(\cos k\ell - 1) + k\ell\sin k\ell = 0$$

Know the following mathematical identities:

$$\begin{cases} \sin k\ell = \sin\left(\dfrac{k\ell}{2} + \dfrac{k\ell}{2}\right) = \sin\dfrac{k\ell}{2}\cos\dfrac{k\ell}{2} + \cos\dfrac{k\ell}{2}\sin\dfrac{k\ell}{2} = 2\sin\dfrac{k\ell}{2}\cos\dfrac{k\ell}{2} \\ \cos k\ell = \cos\left(\dfrac{k\ell}{2} + \dfrac{k\ell}{2}\right) = \cos\dfrac{k\ell}{2}\cos\dfrac{k\ell}{2} - \sin\dfrac{k\ell}{2}\sin\dfrac{k\ell}{2} = 1 - 2\sin^2\dfrac{k\ell}{2} \\ \Rightarrow \cos k\ell - 1 = -2\sin^2\dfrac{k\ell}{2} \end{cases}$$

Rearranging the determinant given above yields:

$$2\left(-2\sin^2\frac{k\ell}{2}\right) + k\ell\left(2\sin\frac{k\ell}{2}\cos\frac{k\ell}{2}\right) = 0$$

$$\Rightarrow \sin\frac{k\ell}{2}\left(\frac{k\ell}{2}\cos\frac{k\ell}{2} - \sin\frac{k\ell}{2}\right) = 0$$

Let $u = k\ell/2$, then the solution becomes $\sin u = 0$ *or* $\tan u = u$. For $\sin u = 0 \Rightarrow u = n\pi$ *or* $k\ell = 2n\pi \Rightarrow P_{cr} = 4n^2\pi^2 EI/\ell^2$. Substituting the eigenvalue $k = 2n\pi/\ell$ into the buckling mode shape yields

$$y = c_1 \sin\frac{2n\pi x}{\ell} + c_2 \cos\frac{2n\pi x}{\ell} + c_3 x + c_4$$

$y = 0$ at $x = 0 \Rightarrow 0 = c_2 + c_4 \Rightarrow c_4 = -c_2$ Hence, $y = c_1 \sin(2n\pi x/\ell) + c_2\left(\cos(2n\pi x/\ell) - 1\right) + c_3 x$

$$y = 0 \text{ at } x = \ell \Rightarrow 0 = c_1 \sin 2n\pi + c_2(\cos 2n\pi - 1) + c_3\ell \Rightarrow c_3 = 0$$

$$y' = -\frac{2n\pi}{\ell} c_2 \sin\frac{2n\pi x}{\ell} + \frac{2n\pi}{\ell} c_1 \cos\frac{2n\pi x}{\ell}$$

$$y' = 0 \quad \text{at } x = 0 \Rightarrow y' = 0 + \frac{2n\pi}{\ell} c_1 \Rightarrow c_1 = 0$$

Hence, $y = c_2\left(\cos(2n\pi x/\ell) - 1\right) \Leftarrow$ eigenvector or mode shape as shown in Fig. 1-6.

$$\text{If } n = 1, \quad p_{cr} = \frac{\pi^2 EI}{\left(\frac{\ell}{2}\right)^2} = \frac{\pi^2 EI}{(\ell_e)^2}$$

where $\ell_e = \ell/2$ is called the effective buckling length of the column. For $\tan u = u$, the smallest nonzero root can be readily computed using **Maple®**. In the old days, it was a formidable task to solve such a simple transcendental equation. Hence, a graphical solution method was frequently employed, as shown in Fig. 1-7.

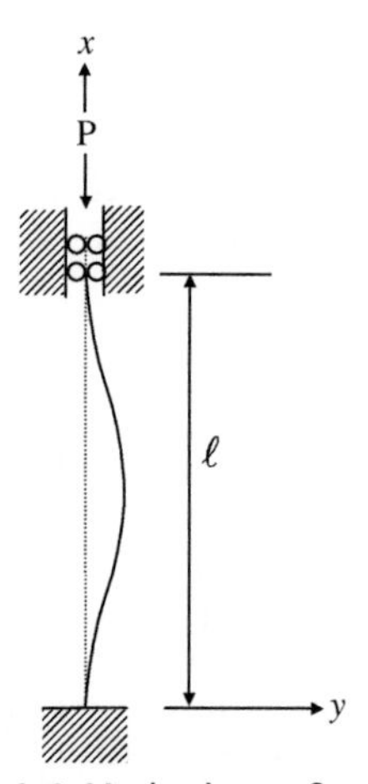

Figure 1-6 Mode shape, first mode

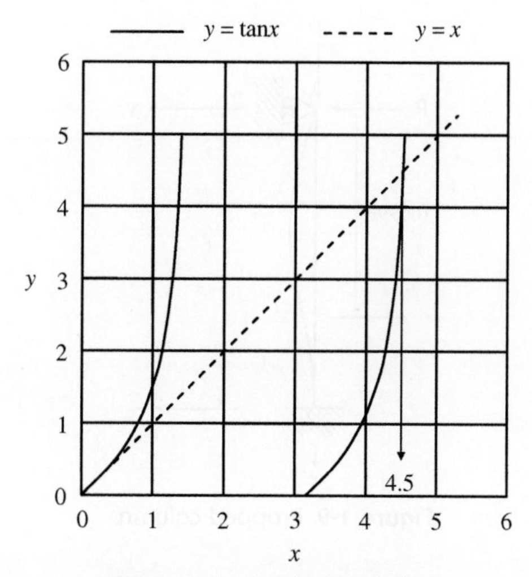

Figure 1-7 Graphical solution

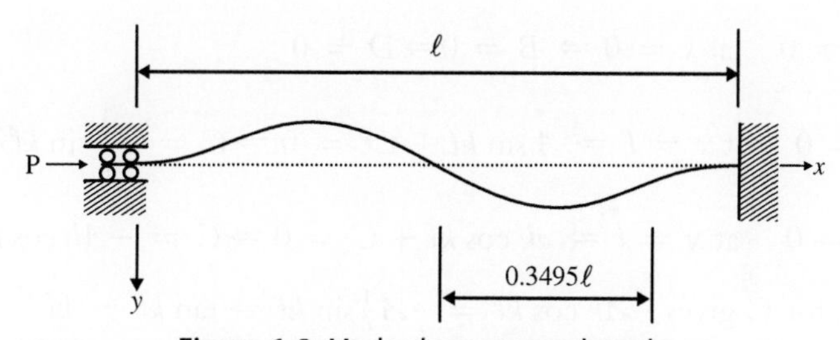

Figure 1-8 Mode shape, second mode

From **Maple®** output, the smallest nonzero root is

$$u = 4.4934094 \Rightarrow \frac{k\ell}{2} = 4.493 \Rightarrow k\ell = 8.9868 \Rightarrow k^2\ell^2 = 80.763$$

$$P_{cr} = \frac{80.763EI}{\ell^2} = \frac{8.183\pi^2 EI}{\ell^2} = \frac{\pi^2 EI}{(0.349578\ell)^2} = \frac{\pi^2 EI}{[0.699156(0.5\ell)]^2}.$$

The corresponding mode shape is shown in Fig. 1-8.

Example 2 Consider propped column as shown in Fig. 1-9.

$$y = A \sin kx + B \cos kx + Cx + D$$

$$y' = Ak \cos kx - Bk \sin kx + C$$

$$y'' = -Ak^2 \sin kx - Bk^2 \cos kx$$

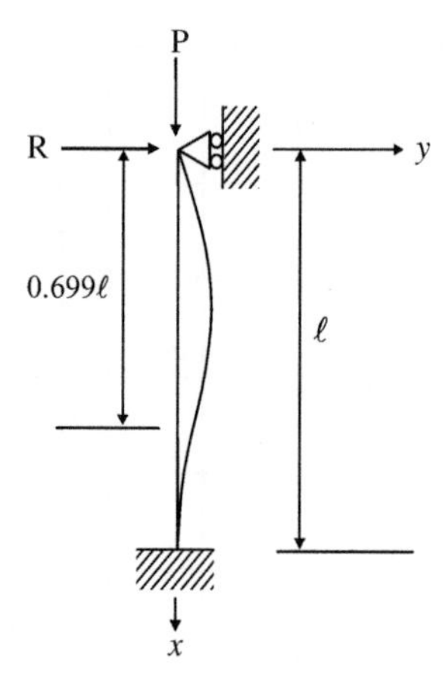

Figure 1-9 Propped column

$$y = 0 \quad \text{at } x = 0 \Rightarrow B + D = 0$$

$$y'' = 0 \quad \text{at } x = 0 \Rightarrow B = 0 \Rightarrow D = 0$$

$$y = 0 \quad \text{at } x = \ell \Rightarrow A \sin k\ell + C\ell = 0 \Rightarrow C = -\frac{1}{\ell} \sin k\ell A$$

$$y' = 0 \quad \text{at } x = \ell \Rightarrow ak \cos k\ell + C = 0 \Rightarrow C = -Ak \cos k\ell$$

Equating for C gives $-\cancel{A}k \cos k\ell = -\cancel{A}\frac{1}{\ell} \sin k\ell \Rightarrow \tan k\ell = k\ell$
Let $u = k\ell \Rightarrow \tan u = u$, then from the previous example, $u = 4.9340945$

$$k\ell = 4.934 = \sqrt{\frac{P}{EI}}\ell$$

$$P_{cr} = \frac{20.19EI}{\ell^2} = \frac{2.04575\pi^2 EI}{\ell^2} = \frac{\pi^2 EI}{(0.699155\ell)^2}$$

Substituting the eigenvalue of $k = 4.934/\ell$ into the eigenvector gives

$$y = A \sin kx - \left(\frac{A}{\ell} \sin k\ell\right)x = A\left[\sin\left(\frac{4.934x}{\ell}\right) - \left(\frac{1}{\ell} \sin 4.934\right)x\right]$$

$$\begin{aligned} y_i|_{x=0.699\ell} &= A[-0.30246 - (-9.7755 \times 10^{-1} \times 0.699155)] \\ &= A(0.3796) > 0 \end{aligned}$$

Summing the moment at the inflection point yields

$$\sum M|_{x=0.699\ell} = 0 = \frac{20.19EI}{\ell^2}A(0.3796) - R(0.699155\ell) \Rightarrow$$

$$R = 11\frac{EI}{\ell^3}A \neq 0$$

For $W10 \times 49$, $I_y = 93.4\ \text{in}^4$, $r_y = 2.54$ in, *say* $\ell = 25$ ft $= 300$ in, $Area = 14.4\ \text{in}^2$

If it is assumed that this column has initial imperfection of $\ell/250$ at the inflection point, then

$$y_{x=0.699155\ell} = \ell/250 = 300/250 = 1.2\ \text{in} \Rightarrow A = 1.758$$

Then, $R = 11 \times \left((29 \times 10^3 \times 93.4)/300^3\right) \times 1.758 = 1.94$ kips

$$\frac{k\ell}{r} = \frac{1 \times 0.699155 \times 300}{2.54} = 82.6 \Rightarrow F_{cr} = 15.6\ \text{ksi} \Rightarrow P_{cr} = 224.6\ \text{kips}$$

$$R = 1.94/224.6 \times 100 = 0.86\% < 2\% \Leftarrow \text{rule of thumb}$$

1.5. EFFECTS OF BOUNDARY CONDITIONS ON THE COLUMN STRENGTH

The critical column buckling load on the same column can be increased in two ways.

1. Change the boundary conditions such that the new boundary condition will make the effective length shorter.
   (a) pinned-pinned $\Rightarrow \ell_e = \ell$
   (b) pinned-fixed $\Rightarrow \ell_e = 0.7\ \ell$
   (c) fixed-fixed $\Rightarrow \ell_e = 0.5\ \ell$
   (d) flag pole (cantilever) $\Rightarrow \ell_e = 2.0\ \ell$, etc.
2. Provide intermediate bracing to make the column buckle in higher modes $\Rightarrow$ achieve shorter effective length.

Consider an elastically constrained column *AB* shown in Fig. 1-10.

The two members, *AB* and *BC*, are assumed to have identical member length and flexural rigidity for simplicity. The moments, *m* and *M*, are due to the rotation at point *B* and possibly due to the axial shortening of member *AB*.

Since $Q = (M + m)/\ell <<< p_{cr}$, Q is set equal to zero and the effect of any axial shortening is neglected.

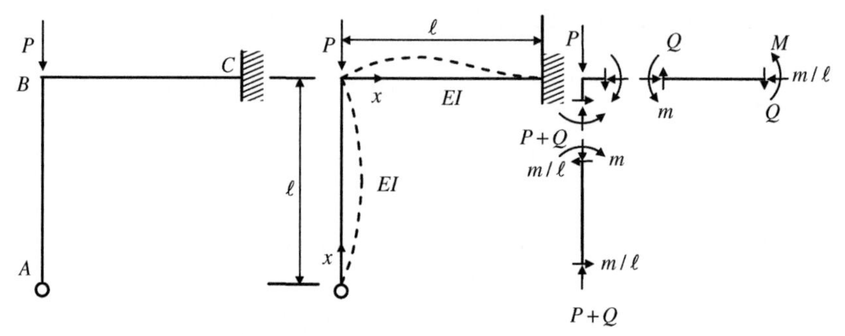

Figure 1-10 Buckling of simple frame

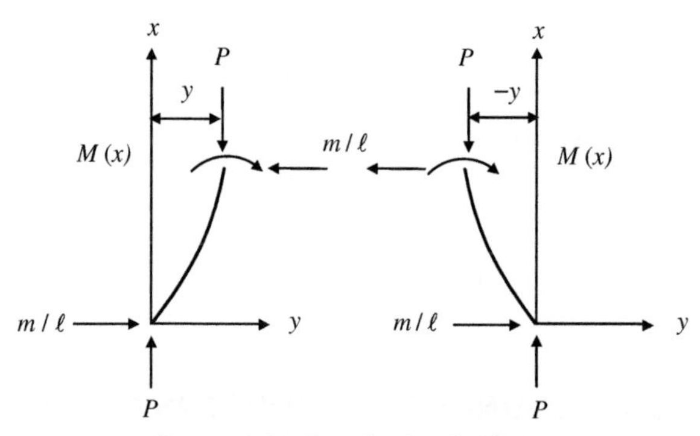

Figure 1-11 Free body of column

Summing moment at the top of the free body gives

(from the left free body) (from the right free body)

$$M(x) + Py - \frac{mx}{\ell} = 0 \qquad M(x) - P(-y) - \frac{mx}{\ell} = 0$$

$$EIy'' = -M(x) = -\left(Py - \frac{mx}{\ell}\right) \qquad EIy'' = M(x) = -Py + \frac{mx}{\ell}$$

As expected, the assumed deformed shape does not affect the Governing Differential Equation (GDE) of the behavior of member AB.

$$EIy'' + Py = \frac{mx}{\ell}$$

Let $k^2 = P/EI \Rightarrow y'' + k^2 y = (mx/\ell P)\, k^2$

The general solution to this DE is given

$$y = A \sin kx + B \cos kx + \frac{m}{\ell P} x$$

$$y = 0 \quad \text{at } x = 0 \Rightarrow B = 0$$

$$y = 0 \quad \text{at } x = \ell \Rightarrow A = -\frac{m}{P \sin k\ell}$$

$$y = \frac{m}{P}\left(\frac{x}{\ell} - \frac{\sin kx}{\sin k\ell}\right) \Leftarrow \text{buckling mode shape}$$

Since joint B is assumed to be rigid, continuity must be preserved. That is

$$\left.\frac{dy}{dx}\right|_{col} = \left.\frac{dy}{dx}\right|_{bm}$$

$$\text{for col } \left.\frac{dy}{dx}\right|_{x=\ell} = \frac{m}{P}\left(\frac{1}{\ell} - \frac{k \cos kx}{\sin k\ell}\right) = \frac{m}{P}\left(\frac{1}{\ell} - \frac{k}{\tan k\ell}\right)$$

$$= \frac{m}{kEI}\left(\frac{1}{k\ell} - \frac{1}{\tan k\ell}\right)$$

$$\text{for beam } \left.\frac{dy}{dx}\right|_{x=0} = \theta_N = \frac{m\ell}{4EI}$$

Recall the slope deflection equation: $m = (2EI/\ell)(2\theta_N + \cancel{\theta_F} - \cancel{3\varphi}) \Rightarrow$

$\theta_N = m\ell/4EI$

Equating the two slopes at joint B gives

$$\frac{m\ell}{4EI} = -\frac{m}{kEI}\left(\frac{1}{k\ell} - \frac{1}{\tan k\ell}\right) \Leftarrow \text{Note the direction of rotation at joint } B!$$

If the frame is made of the same material, then

$$\frac{\ell}{4I_b} = -\frac{1}{kI_c}\left(\frac{1}{k\ell} - \frac{1}{\tan k\ell}\right) \quad \text{or}$$

$$\frac{k\ell}{4} = -\frac{I_b}{I_c}\left(\frac{1}{k\ell} - \frac{1}{\tan k\ell}\right) \Leftarrow \text{stability condition equation}$$

Rearranging the stability condition equation gives

$$\frac{k\ell I_c}{4I_b} = -\frac{1}{k\ell} + \frac{1}{\tan k\ell} \Rightarrow \frac{1}{\tan k\ell} = \frac{1}{k\ell} + \frac{k\ell I_c}{4I_b} = \frac{4I_b + (k\ell)^2 I_c}{k\ell 4I_b} \Rightarrow$$

$$\tan k\ell = \frac{4k\ell I_b}{4I_b + (k\ell)^2 I_c}$$

If $I_b = 0$, then $P_{cr} = \dfrac{\pi^2 EI_c}{\ell^2}$

If $I_b = \infty$, then $P_{cr} = \dfrac{2\pi^2 EI_c}{\ell^2}$

For $I_b = I_c$, then $\tan k\ell = 4k\ell/(4 + (k\ell)^2)$, the smallest root of this equation is $k\ell = 3.8289$.

$$P_{cr} = \frac{14.66EI_c}{\ell^2} = \frac{1.485\pi^2 EI_c}{\ell^2} \Rightarrow \text{as expected } 1 < 1.485 < 2.$$

1.6. INTRODUCTION TO CALCULUS OF VARIATIONS

The calculus of variations is a generalization of the minimum and maximum problem of ordinary calculus. It seeks to determine a function, $y = f(x)$, that minimizes/maximizes a definite integral

$$I = \int_{x_1}^{x_2} F(x, y, y', y'', \ldots\ldots) \tag{1.6.1}$$

which is called a functional (function of functions) and whose integrand contains y and its derivatives and the independent variable x.

Although the calculus of variations is similar to the maximum and minimum problems of ordinary calculus, it does differ in one important aspect. In ordinary calculus, one obtains the actual value of a variable for which a given function has an extreme point. In the calculus of variations, one does not obtain a function that provides extreme value for a given

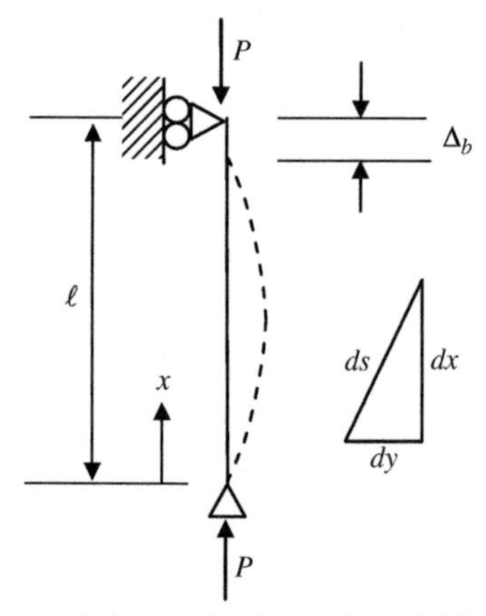

Figure 1-12 Deformed shape of column (in neighboring equilibrium)

integral (functional). Instead, one only obtains the governing differential equation that the function must satisfy to make the given function have a stationary value. Hence, the calculus of variations is not a computational tool, but it is only a device for obtaining the governing differential equation of the physical stationary value problem.

The bifurcation buckling behavior of a both-end-pinned column shown in Fig. 1-12 may be examined in two different perspectives. Consider first that the static deformation prior to buckling has taken place and the examination is being conducted in the neighboring equilibrium position where the axial compressive load has reached the critical value and the column bifurcates (is disturbed) without any further increase of the load. The strain energy stored in the elastic body due to this flexural action is

$$U = \frac{1}{2}\int_v \underset{\sim}{\sigma}^T \underset{\sim}{\varepsilon}\, dv = \frac{1}{2}\int_v \left(\frac{EIy''}{I}y\right)(y''y)dv$$

$$= \frac{E}{2}\int_\ell (y'')^2 \int_A (y)^2 dA d\ell = \frac{EI}{2}\int_0^\ell (y'')^2 dx \tag{1.6.2}$$

In calculating the strain energy, the contributions from the shear strains are generally neglected as they are very small compared to those from normal strains.[3]

Neglecting the small axial shortening prior to buckling ($\Delta_s < \varepsilon\ell$ where $\varepsilon < 0.0005''/''$, hence, $\Delta_s < 0.05\ \%$ of ℓ), the vertical distance, Δ_b, due to the flexural action can be computed as

$$\Delta_b = \int_0^\ell ds - \ell = \int_0^\ell \sqrt{dx^2 + dy^2} - \ell = \int_0^\ell \sqrt{1 + (y')^2}dx - \ell$$

$$= \int_0^\ell \left[1 + \frac{1}{2}(y')^2\right]dx - \ell = \frac{1}{2}\int_0^\ell (y')^2 dx$$

Hence, the change (loss) in potential energy of the critical load is

$$\mathrm{V} = -\frac{1}{2}P\int_0^\ell (y')^2 dx \tag{1.6.3}$$

[3] Of course, the shear strains can be included in the formulation. The resulting equation is called the differential equation, considering the effect of shear deformations.

and the total potential energy functional becomes

$$I = \Pi = U + V = \frac{EI}{2}\int_0^{\ell} (y'')^2 dx - \frac{1}{2}P\int_0^{\ell} (y')^2 dx \qquad (1.6.4)$$

Now the task is to find a function, $y = f(x)$ which will make the total functional, π, have a stationary value.

$$\delta \Pi = \delta(U + V)$$
$$= 0 \Leftarrow \text{necessary condition for equilibrium or stationary value}$$

$$\delta^2 \Pi \begin{Bmatrix} > 0 \Leftarrow \text{minimum value or stable equilbrium} \\ < 0 \Leftarrow \text{maximum value or unstable equilbrium} \\ = 0 \Leftarrow \text{neutral or neutral equilibrium} \end{Bmatrix}$$

$$\Leftarrow \text{sufficient condition}$$

If one chooses an arbitrary function, $\bar{y}(x)$, which only satisfies the boundary conditions (geometric) and lets $y(x)$ be the real exact function, then

$$\bar{y}(x) = y(x) + \varepsilon\eta(x) \qquad (1.6.5)$$

where ε = small number and $\eta(x)$ = twice differentiable function satisfying the geometric boundary conditions. A graphical representation of the above statement is as follows:

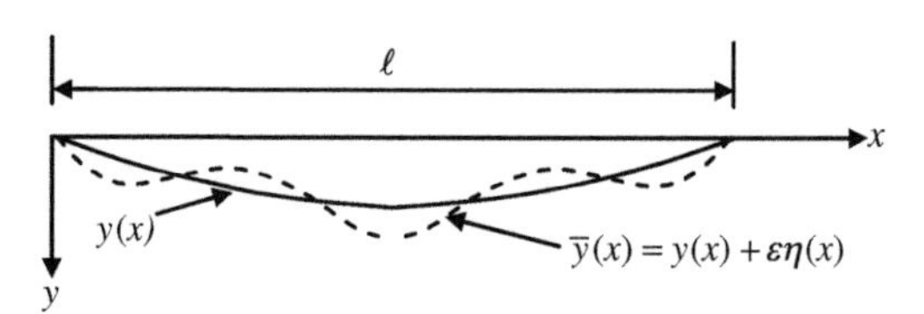

Figure 1-13 Varied path

If one expresses the total potential energy functional in terms of the generalized (arbitrarily chosen) displacement, $\bar{y}(x)$, then

$$\Pi = U + V = \int_0^{\ell} \left[\frac{EI}{2}(y'' + \varepsilon\eta'')^2 - \frac{P}{2}(y' + \varepsilon y')^2\right] dx \qquad (1.6.6)$$

Note that π is a function of ε for a given $\eta(x)$. If $\varepsilon = 0$, then $\bar{y}(x) = y(x)$, which is the curve that provides a stationary value to π. For this to happen

$$\left.\frac{d(U+V)}{d\varepsilon}\right|_{\varepsilon=0} = 0 \tag{1.6.7}$$

Differentiating Eq. (1.6.6) under the integral sign leads to

$$\frac{d(U+V)}{d\varepsilon} = \int_0^{\ell} [EI(y'' + \varepsilon\eta'')\eta'' - P(y' + \varepsilon\eta')\eta']dx$$

Making use of Eq. (1.6.7) yields

$$\int_0^{\ell} (EIy''\eta'' - Py'\eta')dx = 0 \tag{1.6.8}$$

To simplify Eq. (1.6.8) further, use integration by parts. Consider the second term in Eq. (1.6.8).
Let $u = y'$, $du = y''$, $dv = \eta' dx, \mathrm{v} = \eta$ ($\int udv = uv - \int vdu$)

$$\int_0^{\ell} y'\eta' dx = y'\eta\Big|_0^{\ell} - \int_0^{\ell} \eta y'' dx$$

$$= -\int_0^{\ell} \eta y'' dx \quad (\eta \text{ satisfies the geometric bc's}) \tag{a}$$

Similarly,

$$\int_0^{\ell} y''\eta'' dx = y''\eta'\Big|_0^{\ell} - \int_0^{\ell} \eta' y''' dx = y''\eta'\Big|_0^{\ell} - y'''\eta\Big|_0^{\ell} + \int_0^{\ell} y^{iv}\eta dx \tag{b}$$

Equations (a) and (b) lead to

$$\int_0^{\ell} (EIy^{iv} + Py'')\eta dx + (EIy''\eta')\Big|_0^{\ell} = 0 \tag{1.6.9}$$

Except $\eta(0) = \eta(\ell) = 0$, $\eta(x)$ is completely arbitrary and therefore nonzero; hence, the only way to hold Eq. (1.6.9) to be true is that each part of Eq. (1.6.9) must vanish simultaneously. That is

$$\int_0^{\ell} (EIy^{iv} + Py'')\eta dx = 0 \quad \text{and} \quad (EIy''\eta')\Big|_0^{\ell} = 0$$

Since $\eta'(0)$, $\eta'(1)$, and $\eta(x)$ are not zero and $\eta'(0) \neq \eta'(\ell)$, it follows that $y(x)$ must satisfy

$$EIy^{iv} + Py'' = 0 \Leftarrow \text{Euler-Lagrange differential equation} \tag{1.6.10}$$

$$EIy''|_{x=0} = 0 \Leftarrow \text{natural boundary condition} \tag{1.6.11}$$

$$EIy''|_{x=\ell} = 0 \Leftarrow \text{natural boundary condition} \tag{1.6.12}$$

It is recalled that one imposed the geometric boundary conditions, $y(0) = y(\ell) = 0$ at the beginning; however, it can be shown that these conditions are not necessarily required. Shames and Dym (1985) elegantly explain the case for the problem that has the properties of being self-adjoint and positive definite.

The governing differential equation can be obtained either by (1) considering the equilibrium of deformed elements of the system or (2) using the principle of stationary potential energy and the calculus of variations. For a simple system such as a simply supported column buckling, method (1) is much easier to apply, but for a complex system such as cylindrical or spherical shell or plate buckling, method (2) is preferred as the concept is almost automatic although the mathematical manipulations involved are fairly complex. In dealing with the total potential energy, the kinematic (or geometric) boundary conditions involve displacement conditions (deflection or slope) of the boundary, while natural boundary conditions involve internal force conditions (moment or shear) at the boundary.

Example 1 Derive the Euler-Lagrange differential equation and the necessary kinematic (geometric) and natural boundary conditions for the prismatic cantilever column with a linear spring (spring constant α) attached to its free end shown in Fig. 1-14.

The strain energy stored in the deformed body is

$$U = \frac{EI}{2}\int_0^\ell (y'')^2\,dx + \frac{\alpha}{2}(y_\ell)^2 \tag{1.6.13}$$

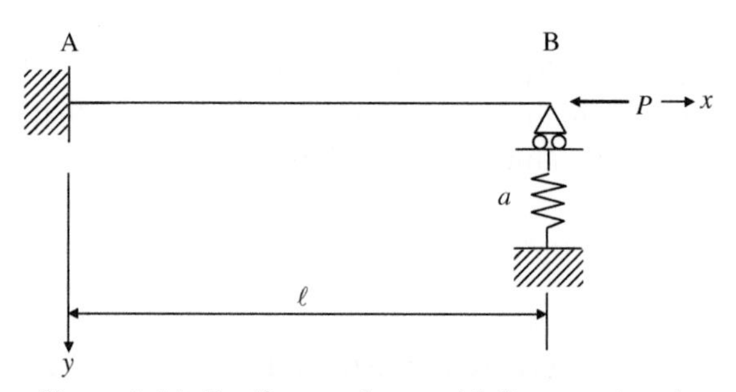

Figure 1-14 Cantilever column with linear spring tip

The loss of potential energy of the external load due to the deformation to the neighboring equilibrium position is

$$V = -\frac{P}{2}\int_0^\ell (y')^2 dx \tag{1.6.14}$$

Hence, the total potential energy functional becomes

$$\Pi = U + V = \frac{EI}{2}\int_0^\ell (y'')^2 dx + \frac{\alpha}{2}(y_\ell)^2 - \frac{P}{2}\int_0^\ell (y')^2 dx$$

or

$$\Pi = \int_0^\ell \left[\frac{EI}{2}(y'')^2 - \frac{P}{2}(y')^2\right] dx + \frac{\alpha}{2}(y_\ell)^2 \tag{1.6.15}$$

The total potential energy functional must be stationary if the first variation $\delta\Pi = 0$. Since the differential operator and the variational operator are interchangeable, one obtains

$$\delta\Pi = \int_0^\ell (EIy''\delta y'' - Py'\delta y')dx + \alpha y_\ell \delta y_\ell = 0 \tag{1.6.16}$$

Integrating by parts each term in the parenthesis of Eq. (1.6.16) yields

$$\int_0^\ell EIy''\delta y'' dx = [EIy''\delta y']_0^\ell - [EIy'''\delta y]_0^\ell + \int_0^\ell EIy^{IV}\delta y dx \tag{1.6.17}$$

$$-\int_0^\ell Py'\delta y' dx = -[Py'\delta y]_0^\ell + \int_0^\ell Py''\delta y dx \tag{1.6.18}$$

It becomes obvious by inspection of the sketch that (1) the deflection and slope must be equal to zero due to the unyielding support at A ($x = 0$) and the variation will also be equal to zero, that is, $y_0 = 0$, $y_0' = 0$ and $\delta y_0 = 0$, $\delta y_0' = 0$, and (2) the moment and its variation must also be equal to zero due to the roller support at B ($x = \ell$), that is, $y_\ell'' = 0$ and $\delta y_\ell'' = 0$ where the subscripts 0 and ℓ represent the values at A ($x = 0$) and B ($x = \ell$), respectively. The first and second term of Eq. (1.6.17) can be written, respectively, as

$$[EIy''\delta y']_0^\ell = EIy_\ell''\delta y_\ell' - EIy_0''\delta y_0' = 0$$

and

$$-\left[EIy'''\delta y\right]_0^\ell = -EIy_\ell'''\delta y_\ell + EIy_0'''\delta y_0 = -EIy_\ell'''\delta y_\ell.$$

The first term of Eq. (1.6.18) can be written as

$$-\left[Py'\delta y\right]_0^\ell = -Py_\ell'\delta y_\ell + Py_0'\delta y_0 = -Py_\ell'\delta y_\ell$$

Equation (1.6.16) may now be rearranged

$$\delta\pi = (\alpha y_\ell - EIy_\ell''' - Py_\ell')\delta y_\ell + \int_0^\ell (EIy^{IV} + Py'')\delta y dx = 0 \quad (1.6.19)$$

It is noted here in Eq. (1.6.19) that δy_ℓ is not zero. In order for Eq. (1.6.19) to be equal to zero for all values of δy between $x = 0$ and $x = \ell$, it is required that the function y must satisfy the Euler-Lagrange differential equation (the integrand inside the parenthesis)

$$EIy^{IV} + Py'' = 0 \quad (1.6.20)$$

and additional condition

$$\alpha y_\ell - EIy_\ell''' - Py_\ell' = 0 \quad (1.6.21)$$

must be met.

Equation (1.6.21), along with the condition $y_\ell'' = 0$, are the natural boundary conditions of the problem, and y_0 (and/or δy_0) $= 0$ y_0' (and/or $\delta y_0'$) $= 0$ are the geometric boundary conditions of the problem. Hence, four boundary conditions are available as required for a fourth-order differential equation. The sum of all of the expanded integral terms at the end points consisting of a multiple of the geometric boundary conditions and/or the natural boundary conditions is collectively called a conjunct or a concomitant and is equal to zero for all positive definite and self-adjoint problems.

1.7. DERIVATION OF BEAM-COLUMN GDE USING FINITE STRAIN

Recall the following Green-Lagrange finite strain:

$$e_{ij} = \frac{1}{2}(u_{i,j} + u_{j,i} + u_{k,i}u_{k,j}) \quad (1.7.1)$$

$$e_{xx} = \frac{du_x}{dx} + \frac{1}{2}\left[\left(\frac{du_x}{dx}\right)^2 + \left(\frac{du_y}{dx}\right)^2 + \left(\frac{du_z}{dx}\right)^2\right] \Leftarrow \text{axial strain} \quad (1.7.2)$$

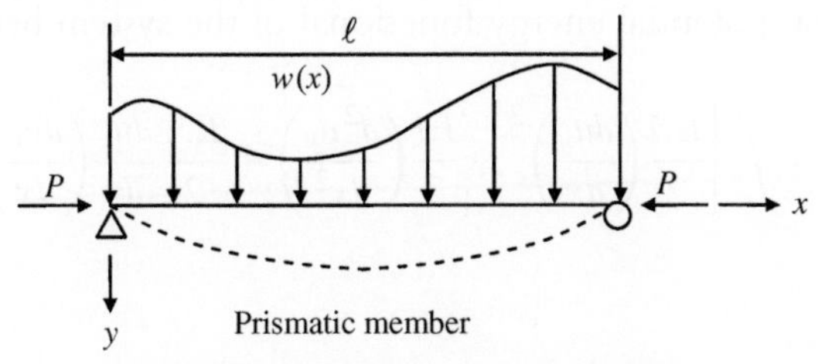

Figure 1-15 Beam-column model

where $(du_x/dx)^2 \doteq 0$ (considered to be a higher order term) and $(du_z/dx)^2 = 0$ (only uniaxial bending is considered here). For the given coordinate system in the sketch, the axial strain due to bending is

$$e_b = -\frac{d^2u_y}{dx^2}y \tag{1.7.3}$$

where $d^2u_y/dx^2 = 1/\rho$ is the curvature of the elastic curve. The sum of axial strains due to axial force and flexure constitutes the total normal strain. Hence,

$$\varepsilon_{xx} = e_a + e_b = \frac{du_x}{dx} + \frac{1}{2}\left(\frac{du_y}{dx}\right)^2 - \frac{d^2u_y}{dx^2}y \tag{1.7.4}$$

The strain energy stored in the elastic body becomes

$$\begin{aligned} U &= \frac{1}{2}\int_v \sigma^T \varepsilon dv = \frac{E}{2}\int_v \varepsilon_{xx}^2 dv = \frac{E}{2}\int_v \left[\frac{du_x}{dx} - y\frac{d^2u_y}{dx^2} + \frac{1}{2}\left(\frac{du_y}{dx}\right)^2\right]^2 dv \\ &= \frac{E}{2}\int_0^\ell \int_A \left[\left(\frac{du_x}{dx}\right)^2 + \left(\frac{d^2u_y}{dx^2}\right)^2 y^2 + \frac{1}{4}\left(\frac{du_y}{dx}\right)^4 - 2\frac{du_x}{dx}\frac{d^2u_y}{dx^2}y \right. \\ &\quad \left. - \frac{d^2u_y}{dx^2}\left(\frac{du_y}{dx}\right)^2 y + \frac{du_x}{dx}\left(\frac{du_y}{dx}\right)^2\right] dAdx \end{aligned} \tag{1.7.5}$$

Neglecting the higher order term and integrating over the cross-sectional area A while noting all integrals of the form $\int ydA$ to be zero as y is measured from the centroidal axis, one gets

$$U = \int_\ell \left[\frac{EA}{2}\left(\frac{du_x}{dx}\right)^2 + \frac{EI}{2}\left(\frac{d^2u_y}{dx^2}\right) + \frac{EA}{2}\frac{du_x}{dx}\left(\frac{du_y}{dx}\right)^2\right] dx \tag{1.7.6}$$

The loss of potential energy of the applied transverse load is

$$V = -\int_\ell w\, u_y dx \tag{1.7.7}$$

Hence, the total potential energy functional of the system becomes

$$\Pi = U + V = \int_\ell \left[\frac{EA}{2}\left(\frac{du_x}{dx}\right)^2 + \frac{EI}{2}\left(\frac{d^2u_y}{dx^2}\right) + \frac{EA}{2}\frac{du_x}{dx}\left(\frac{du_y}{dx}\right)^2 - w\,u_y \right] dx \tag{1.7.8}$$

or

$$\Pi = U + V = \int_\ell \left[\frac{EA}{2}\left(\frac{du_x}{dx}\right)^2 + \frac{EI}{2}\left(\frac{d^2u_y}{dx^2}\right) - \frac{P}{2}\left(\frac{du_y}{dx}\right)^2 - w\,u_y \right] dx \tag{1.7.9}$$

Note that $P = \sigma A = EAe_a = EA(du_x/dx)$, which is called the stress resultant. The negative sign corresponds to the fact that P is in compression. The quantity inside the square bracket, the integrand, is denoted by F. Applying the principle of the minimum potential energy (or applying the Euler-Lagrange differential equation), one obtains

$$F = \frac{EA}{2}(u')^2 + \frac{EI}{2}(y'')^2 - \frac{P}{2}(y')^2 - wy \tag{1.7.10}$$

where $u = u_x$, $y = u_y$.

Recall the Euler-Lagrange DE (see Bleich 1952, pp. 91–103):

$$F_u - \frac{d}{dx}F_{u'} + \frac{d^2}{dx^2}F_{u''} - \ \ldots = 0 \tag{1.7.11}$$

$$F_y - \frac{d}{dx}F_{y'} + \frac{d^2}{dx^2}F_{y''} - \ \ldots = 0 \tag{1.7.12}$$

$$F_u = 0,\ F_{u'} = EAu' \Rightarrow -\frac{d}{dx}F_{u'} = -EAu'',\ F_{u''} = 0$$

$$EAu'' = 0 \tag{1.7.13}$$

$$F_y = -w,\ F_{y'} = -Py' \Rightarrow -\frac{d}{dx} = Py'',$$

$$F_{y''} = EIy'' \Rightarrow \frac{d^2}{dx^2}F_{y''} = EIy^{iv}$$

$$EIy^{iv} + Py'' = w \tag{1.7.14}$$

It should be noted that the concept of finite axial strain implicitly implies the buckled shape (lateral displacement) and any prebuckling state is ignored.

1.8. GALERKIN METHOD

The requirement that the total potential energy of a hinged column has a stationary value is shown in the following equation:

$$\int_0^{\ell} (EIy^{iv} + Py'')\delta y dx + (EIy'')\delta y' \Big|_0^{\ell} = 0 \tag{1.8.1}$$

where δy is a virtual displacement.

Assume that it is possible to approximate the deflection of the column by a series of independent functions, $g_i(x)$, multiplied by undetermined coefficients, a_i.

$$y_{approx} \doteq a_1 g_1(x) + a_2 g_2(x) + \ \ldots\ldots + a_n g_n(x) \tag{1.8.2}$$

If each $g_i(x)$ satisfies the geometric and natural boundary conditions, then the second term in Eq. (1.8.1) vanishes when it substitutes y_{approx} to y. Also, the coefficients, a_i, must be chosen such that y_{approx} will satisfy the first term. Let the operator be

$$Q = EI\frac{d^4}{dx^4} + P\frac{d^2}{dx^2} \tag{1.8.3}$$

and

$$\phi = \sum_{i=1}^{n} a_i g_i(x) \tag{1.8.4}$$

From Eqs. (1.8.3) and (1.8.4), the first term of Eq. (1.8.1.) becomes:

$$\int_0^{\ell} Q(\phi)\delta\phi \, dx = 0 \tag{1.8.5}$$

Since ϕ is a function of n parameters, a_i,

$$\begin{aligned} \delta\phi &= \frac{\partial\phi}{\partial a_1}\delta a_1 + \frac{\partial\phi}{\partial a_2}\delta a_2 + \ \ldots . + \frac{\partial\phi}{\partial a_n}\delta a_n \\ &= g_1\delta a_1 + g_2\delta a_2 + \ \ldots\ldots + g_n\delta a_n = \sum_{i=1}^{n} g_i\delta a_i \end{aligned} \tag{1.8.6}$$

$$\int_0^{\ell} Q(\phi)\sum_{i=1}^{n} g_i(x)\delta a_i \, dx = 0 \tag{1.8.7}$$

Since it has been assumed that $g_i(x)$ are independent of each other, the only way to hold Eq. (1.8.7) is that each integral of Eq. (1.8.7) must vanish, that is

$$\int_0^\ell Q(\phi)g_i(x)\delta a_i\, dx = 0 \quad i = 1, 2, \;....., n$$

a_i are arbitrary; hence $\delta a_i \neq 0$.

$$\int_0^\ell Q(\phi)g_i(x)dx = 0 \quad i = 1, 2, \;....., n \tag{1.8.8}$$

Equation (1.8.8) is somewhat similar to the weighted integral process in the finite element method.

Example 1 Consider the axial buckling of a propped column.
The Galerkin method is to be applied. For y_{approx}, use the lateral displacement function of a propped beam subjected to a uniformly distributed load. Hence,

$$y_{approx} = \phi = A(x\ell^3 - 3x^3\ell + 2x^4)$$

$$Q(\phi) = EI\frac{d^4\phi}{dx^4} + P\frac{d^2\phi}{dx^2} = A[48EI + P(24x^2 - 18\ell x)]$$

$$g(x) = (\ell^3 x - 3\ell x^3 + 2x^4)$$

$$\int_0^\ell A[48EI + P(24x^2 - 18\ell x)](\ell^3 x - 3\ell x^3 + 2x^4)dx = 0$$

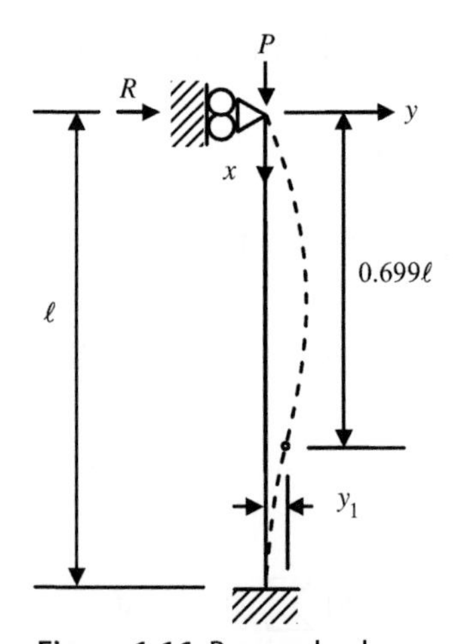

Figure 1-16 Propped column

Carrying out the integration gives

$$A\Big((36EI\ell^5/5) - (12P\ell^7/35)\Big) = 0 \Rightarrow A \neq 0 \text{ for a nontrivial solution}$$

$$P_{cr} = 21EI/\ell^2 \Leftarrow 3.96\% \text{ greater than the exact value, } P_{cr\,exact} = 20.2EI/\ell^2$$

1.9. CONTINUOUS BEAM-COLUMNS RESTING ON ELASTIC SUPPORTS

A general method to evaluate the minimum required spring constants of a beam-column resting on an elastic support is to apply the slope-deflection equations with axial compression. In order to simplify the illustration, all beam-columns are assumed to be rigid and equal spans.

1.9.1. One Span

Assume that a small displacement occurs at b, so that the bar becomes inclined to the horizontal by a small angle, α. As the stability of a system is examined in the neighboring equilibrium position, free body for equilibrium must be extracted from a deformed state. Owing to this displacement, the load P moves to the left by the amount

$$L(1 - \cos\alpha) \doteq \frac{L\alpha^2}{2} \tag{1.9.1}$$

and the decrease in the potential energy of the load P, equal to the work done by P, is

$$\frac{PL\alpha^2}{2} \tag{1.9.2}$$

At the same time the spring deforms by the amount αL, and the increase in strain energy of the spring is

$$\frac{k(\alpha L)^2}{2} \tag{1.9.3}$$

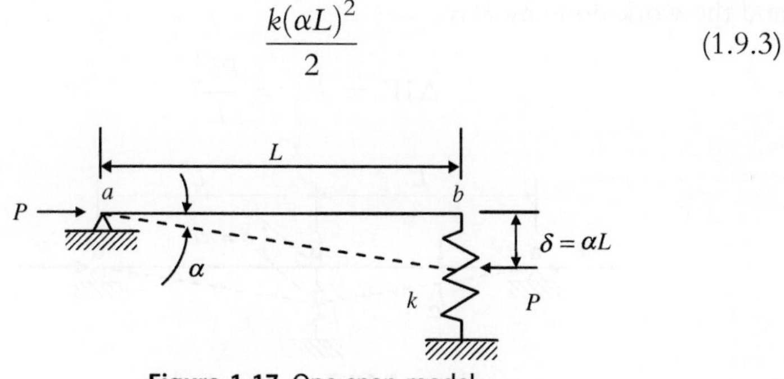

Figure 1-17 One-span model

where k denotes the spring constant. The system will be stable if

$$\frac{k(\alpha L)^2}{2} > \frac{PL\alpha^2}{2} \tag{1.9.4}$$

and will be unstable if

$$\frac{k(\alpha L)^2}{2} < \frac{PL\alpha^2}{2} \tag{1.9.5}$$

Therefore the critical value of the load P is found from the condition that

$$\frac{k(\alpha L)^2}{2} = \frac{PL\alpha^2}{2} \tag{1.9.6}$$

from which

$$k = \frac{\beta P_{cr}}{L} \Rightarrow \beta = 1 \tag{1.9.7}$$

The same conclusion can be reached by considering the equilibrium of the forces acting on the bar. However, if the system has three or more springs, simple statics may not be sufficient to determine the small displacement associated with each spring. Hence, the energy method appears to be better suited.

1.9.2. Two Span

For small deflection δ, the angle of inclination of the bar ab is δ/L, and the distance λ moved by the force P is found to be

$$\lambda = 2\left[\frac{1}{2}L\left(\frac{\delta}{L}\right)^2\right] = \frac{1}{L}\delta^2 \tag{1.9.8}$$

and the work done by P is

$$\Delta W = P\lambda = \frac{P\delta^2}{L} \tag{1.9.9}$$

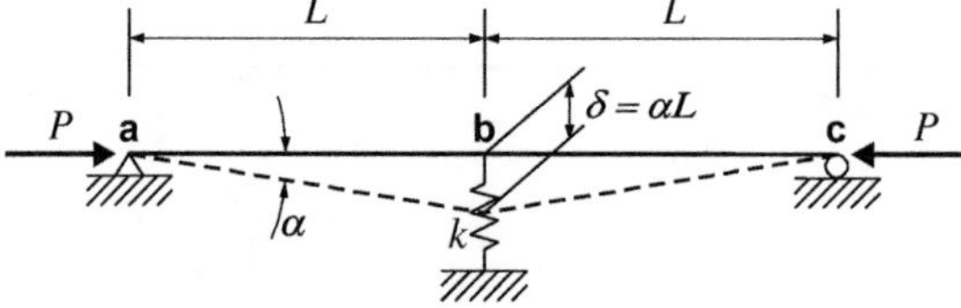

Figure 1-18 Two-span model

The strain energy stored in the spring is

$$\Delta U = \frac{k\delta^2}{2} \tag{1.9.10}$$

The critical value of the load P is found from the equation

$$\Delta U = \Delta W \tag{1.9.11}$$

which represents the condition when the equilibrium configuration changes from stable to unstable. Hence,

$$k = \frac{\beta P_{cr}}{L} = \frac{2P_{cr}}{L} \Rightarrow \beta = 2 \tag{1.9.12}$$

1.9.3. Three Span

For small displacements, the rotation of bars *ab* and *cd* may be expressed as

$$\alpha_1 = \frac{\delta_1}{L} \quad \text{and} \quad \alpha_2 = \frac{\delta_2}{L} \tag{1.9.13}$$

and the rotation of bar *bc* is

$$\frac{\delta_2 - \delta_1}{L} \tag{1.9.14}$$

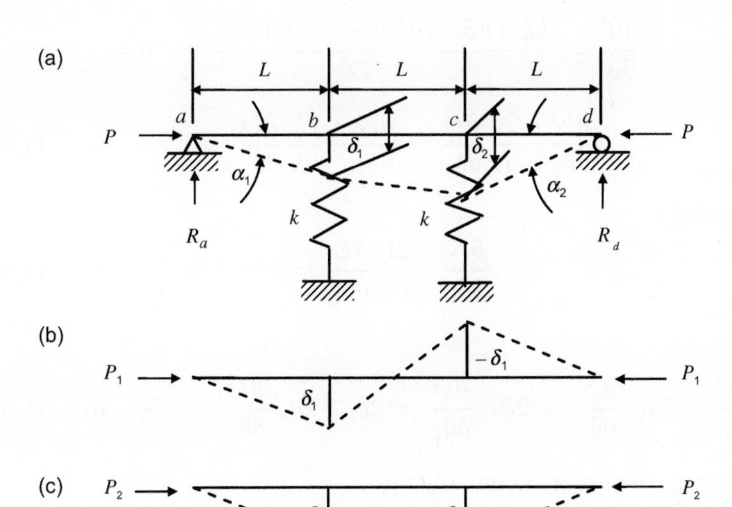

Figure 1-19 Three-span model

The distance λ moved by the force P is found to be

$$\lambda = \frac{1}{2}L\left[\left(\frac{\delta_1}{L}\right)^2 + \left(\frac{\delta_2 - \delta_1}{L}\right)^2 + \left(\frac{\delta_2}{L}\right)^2\right]$$

$$= \frac{1}{2L}(\delta_1^2 + \delta_2^2 + \delta_1^2 - 2\delta_1\delta_2 + \delta_2^2) = \frac{1}{L}(\delta_1^2 - \delta_1\delta_2 + \delta_2^2) \quad (1.9.15)$$

and the work done by the force P is

$$\Delta W = P\lambda = \frac{P}{L}(\delta_1^2 - \delta_1\delta_2 + \delta_2^2) \quad (1.9.16)$$

The strain energy stored in the elastic supports during buckling is

$$\Delta U = \frac{k}{2}(\delta_1^2 + \delta_2^2) \quad (1.9.17)$$

The critical condition is found by equating these two expressions

$$\frac{P}{L}(\delta_1^2 - \delta_1\delta_2 + \delta_2^2) = \frac{k}{2}(\delta_1^2 + \delta_2^2) \Rightarrow P = \frac{kL}{2}\frac{\delta_1^2 + \delta_2^2}{\delta_1^2 - \delta_1\delta_2 + \delta_2^2} = \frac{kL}{2}\frac{N}{D} \quad (1.9.18)$$

where N and D represent the numerator and denominator of the fraction. To find the critical value of P, one must adjust the deflections δ_1 and δ_2, which are unknown, so as to make P a minimum value. This is accomplished by setting $\partial P/\partial \delta_1 = 0$ and $\partial P/\partial \delta_2 = 0$.

$$\frac{\partial P}{\partial \delta_1} = \frac{kL}{2}\frac{D(\partial N/\partial \delta_1) - N(\partial D/\partial \delta_1)}{D^2} = 0 \Rightarrow$$

$$\frac{\partial N}{\partial \delta_1} - \frac{N}{D}\frac{\partial D}{\partial \delta_1} = \frac{\partial N}{\partial \delta_1} - \frac{2P}{kL}\frac{\partial D}{\partial \delta_1} = 0 \quad (1.9.19)$$

Similarly,

$$\frac{\partial N}{\partial \delta_2} - \frac{2P}{kL}\frac{\partial D}{\partial \delta_2} = 0 \quad (1.9.20)$$

and

$$\frac{\partial N}{\partial \delta_1} = 2\delta_1, \ \frac{\partial N}{\partial \delta_2} = 2\delta_2, \ \frac{\partial D}{\partial \delta_1} = 2\delta_1 - \delta_2, \ \frac{\partial D}{\partial \delta_2} = 2\delta_2 - \delta_1 \quad (1.9.21)$$

Substituting these values, one obtains

$$2\delta_1 - \frac{2P}{kL}(2\delta_1 - \delta_2) = \delta_1\left(1 - \frac{2P}{kL}\right) + \delta_2\frac{P}{kL} = 0 \quad (1.9.22)$$

$$2\delta_2 - \frac{2P}{kL}(2\delta_2 - \delta_1) = \delta_1 \frac{P}{kL} + \delta_2\left(1 - \frac{2P}{kL}\right) = 0 \tag{1.9.23}$$

For nontrivial solutions, the coefficient determinant must vanish. Hence,

$$\begin{vmatrix} 1 - \frac{2P}{kL} & \frac{P}{kL} \\ \frac{P}{kL} & 1 - \frac{2P}{kL} \end{vmatrix} = 0 \Rightarrow \left(1 - \frac{2P}{kL}\right)^2 - \left(\frac{P}{kL}\right)^2 = 0 \Rightarrow P_1 = \frac{kL}{3},\ P_2 = kL \tag{1.9.24}$$

The critical load P_1 corresponds to the buckling mode shape shown in Fig. 1-19(b), and the critical load P_2 corresponds to the buckling mode shape shown in Fig. 1-19(c). For a given system, the critical load is the small one. Hence, P_1 is the correct solution. Hence,

$$k = \frac{\beta P_{cr}}{L} = \frac{3P_{cr}}{L} \Rightarrow \beta = 3 \tag{1.9.25}$$

The same problem can be solved readily by using equations of equilibrium. Noting that the reactive force of the spring is given by $k\delta$, the end reactions are

$$R_a = \frac{2}{3}k\delta_1 + \frac{1}{3}k\delta_2 \tag{1.9.26}$$

$$R_d = \frac{1}{3}k\delta_1 + \frac{2}{3}k\delta_2 \tag{1.9.27}$$

Another equation for R_a is found by taking the moment about point B for bar ab, which gives

$$P\delta_1 = R_a L \tag{1.9.28}$$

and similarly, for ad

$$P\delta_2 = R_d L \tag{1.9.29}$$

Combining these four equations yields

$$\frac{P}{L}\delta_1 = \frac{2}{3}k\delta_1 + \frac{1}{3}k\delta_2 \Rightarrow \delta_1\left(2 - \frac{3P}{kL}\right) + \delta_2 = 0 \tag{1.9.30}$$

$$\frac{P}{L}\delta_2 = \frac{1}{3}k\delta_1 + \frac{2}{3}k\delta_2 \Rightarrow \delta_1 + \delta_2\left(2 - \frac{3P}{kL}\right) = 0 \tag{1.9.31}$$

Setting the determinant equal to zero yields"

$$\begin{vmatrix} 2-\dfrac{3P}{kL} & 1 \\ 1 & 2-\dfrac{3P}{kL} \end{vmatrix} = \left(2-\frac{3P}{kL}\right)^2 - 1 = 0 \Rightarrow P_1 = \frac{kL}{3} \text{ and } P_2 = kL \tag{1.9.32}$$

By definition, P_1 is the correct solution.

1.9.4. Four Span

For small displacements, the rotation of bars *ab* and *de* may be expressed as

$$\alpha_1 = \frac{\delta_1}{L} \quad \text{and} \quad \alpha_2 = \frac{\delta_3}{L} \tag{1.9.33}$$

and the angles of rotation of bar *bc* and *cd* are

$$\frac{\delta_2 - \delta_1}{L} \quad \text{and} \quad \frac{\delta_3 - \delta_2}{L} \tag{1.9.34}$$

The distance λ moved by the force P is found to be

$$\begin{aligned} \lambda &= \frac{1}{2}L\left[\left(\frac{\delta_1}{L}\right)^2 + \left(\frac{\delta_2-\delta_1}{L}\right)^2 + \left(\frac{\delta_3-\delta_2}{L}\right)^2 + \left(\frac{\delta_3}{L}\right)^2\right] \\ &= \frac{1}{2L}(\delta_1^2 + \delta_2^2 + \delta_1^2 - 2\delta_1\delta_2 + \delta_3^2 + \delta_2^2 - 2\delta_2\delta_3 + \delta_3^2) \\ &= \frac{1}{L}(\delta_1^2 - \delta_1\delta_2 + \delta_2^2 - \delta_2\delta_3 + \delta_3^2) \end{aligned} \tag{1.9.35}$$

and the work done by the force P is

$$\Delta W = P\lambda = \frac{P}{L}(\delta_1^2 - \delta_1\delta_2 + \delta_2^2 - \delta_2\delta_3 + \delta_3^2) \tag{1.9.36}$$

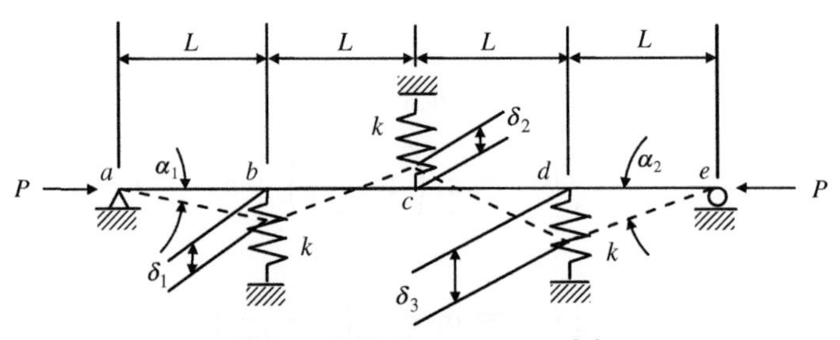

Figure 1-20 Four-span model

The strain energy stored in the elastic supports during buckling is

$$\Delta U = \frac{k}{2}(\delta_1^2 + \delta_2^2 + \delta_3^2) \tag{1.9.37}$$

The critical condition is found by equating these two expressions

$$\frac{P}{L}(\delta_1^2 - \delta_1\delta_2 + \delta_2^2 - \delta_2\delta_3 + \delta_3^2) = \frac{k}{2}(\delta_1^2 + \delta_2^2 + \delta_3^2) \Rightarrow$$

$$P = \frac{kL}{2}\frac{\delta_1^2 + \delta_2^2 + \delta_3^2}{\delta_1^2 - \delta_1\delta_2 + \delta_2^2 - \delta_2\delta_3 + \delta_3^2} = \frac{kL}{2}\frac{N}{D} \tag{1.9.38}$$

where N and D represent the numerator and denominator of the fraction. To find the critical value of P, one must adjust the deflections δ_1, δ_2 and δ_3, which are unknown, so as to make P a minimum value. This is accomplished by setting $\partial P/\partial \delta_1 = 0, \partial P/\partial \delta_2$ and $\partial P/\partial \delta_3 = 0$.

$$\frac{\partial P}{\partial \delta_1} = \frac{kL}{2}\frac{D(\partial N/\partial \delta_1) - N(\partial D/\partial \delta_1)}{D^2} = 0 \Rightarrow$$

$$\frac{\partial N}{\partial \delta_1} - \frac{N}{D}\frac{\partial D}{\partial \delta_1} = \frac{\partial N}{\partial \delta_1} - \frac{2P}{kL}\frac{\partial D}{\partial \delta_1} = 0 \tag{1.9.39}$$

Similarly,

$$\frac{\partial N}{\partial \delta_2} - \frac{2P}{kL}\frac{\partial D}{\partial \delta_2} = 0 \tag{1.9.40}$$

$$\frac{\partial N}{\partial \delta_3} - \frac{2P}{kL}\frac{\partial D}{\partial \delta_3} = 0 \tag{1.9.41}$$

and

$$\frac{\partial N}{\partial \delta_1} = 2\delta_1, \quad \frac{\partial N}{\partial \delta_2} = 2\delta_2, \quad \frac{\partial N}{\partial \delta_3} = 2\delta_3, \quad \frac{\partial D}{\partial \delta_1} = 2\delta_1 - \delta_2,$$

$$\frac{\partial D}{\partial \delta_2} = 2\delta_2 - \delta_1 - \delta_3, \quad \frac{\partial D}{\partial \delta_3} = 2\delta_3 - \delta_2 \tag{1.9.42}$$

Substituting these values, one obtains

$$2\delta_1 - \frac{2P}{kL}(2\delta_1 - \delta_2) = \delta_1\left(1 - \frac{2P}{kL}\right) + \delta_2\frac{P}{kL} + 0\delta_3 = 0 \tag{1.9.43}$$

$$2\delta_2 - \frac{2P}{kL}(2\delta_2 - \delta_1 - \delta_3) = \delta_1\frac{P}{kL} + \delta_2\left(1 - \frac{2P}{kL}\right) + \delta_3\frac{P}{kL} = 0 \quad (1.9.44)$$

$$2\delta_3 - \frac{2P}{kL}(2\delta_3 - \delta_2) = 0 \quad \delta_1 + \delta_2\frac{P}{kL} + \delta_3\left(1 - \frac{2P}{kL}\right) = 0 \quad (1.9.45)$$

For nontrivial solutions, the coefficient determinant must vanish. Hence,

$$\begin{vmatrix} 1-\frac{2P}{kL} & \frac{P}{kL} & 0 \\ \frac{P}{kL} & 1-\frac{2P}{kL} & \frac{P}{kL} \\ 0 & \frac{P}{kL} & 1-\frac{2P}{kL} \end{vmatrix} = \left(1-\frac{2P}{kL}\right)^3 - 2\left(1-\frac{2P}{kL}\right)\left(\frac{P}{kL}\right)^2 \quad (1.9.46)$$

$$= \left(1-\frac{2P}{kL}\right)\left[\left(1-\frac{2P}{kL}\right)^2 - 2\left(\frac{P}{kL}\right)^2\right] = 0$$

The smallest critical load $P_1 = 0.29289kL$ corresponds to the buckling mode shape shown in sketch.

$$k = \frac{\beta P_{cr}}{L} = \frac{P_{cr}}{0.29289L} = \frac{3.414P_{cr}}{L} \Rightarrow \beta = 3.414 \quad (1.9.47)$$

The equilibrium method cannot be applied to problems with three or more elastic supports as there are only two equations of equilibrium available, that is, $\sum$ moment $= 0$ and $\sum$ vertical force $= 0$. It is further noted that β varies from 1 for one span to 4 for infinite equal spans. Since β equals 3.414 for four equal spans, the use of $\beta = 4$ for multistory frames would seem justified.

Compression members in real structures are not perfectly straight (sweep, camber), perfectly aligned, or concentrically loaded as is assumed in design calculations; there is always an initial imperfection. Examining the single-story column of Fig. 1-17 assuming there is an initial deflection δ_0 reveals that the following equilibrium equation is required:

$$(k\delta)L = P(\delta + \delta_0) \quad (1.9.48)$$

for $P = P_{cr}$

$$k_{\text{reqd}} = \frac{P_{cr}}{L}\left(1 + \frac{\delta_0}{\delta}\right) \quad (1.9.49)$$

Since $k_{\text{ideal}} = P_{cr}/L$, Eq. (1.9.48) becomes

$$k_{\text{reqd}} = k_{\text{ideal}}\left(1 + \frac{\delta_0}{\delta}\right) \tag{1.9.50}$$

which is the stiffness requirement for compression members having initial imperfection δ_0. The stiffness requirement is

$$Q = k_{\text{reqd}}\delta = k_{\text{ideal}}\left(1 + \frac{\delta_0}{\delta}\right)\delta = k_{\text{ideal}}(\delta + \delta_0) \tag{1.9.51}$$

Winter (1960) has suggested $\delta = \delta_0 = L/500$. Substitution of this into Eqs. (1.9.49) and (1.9.50) gives the following design equations:

$$\text{For stiffness, } k_{\text{reqd}} = 2k_{\text{ideal}} \tag{1.9.52}$$

For nominal strength

$$Q_n = k_{\text{ideal}}(2\delta_0) = k_{\text{ideal}}(0.004L) = \frac{\beta P_{cr}}{L}(0.004L) \tag{1.9.53}$$

Example 1 Turn-buckled threaded rods ($F_y = 50$ ksi, $F_u = 70$ ksi) are to be provided for the bracing system for a single-story frame shown in Fig. 1-21. The typical loading on each girder consists of three concentrated loads. The factored loads are: $P_1 = 200$ kips and $P_2 = 100$ kips. Determine the diameter of the rod by the AISC (2005) *Specification for Structural Steel Building*, 13th edition.

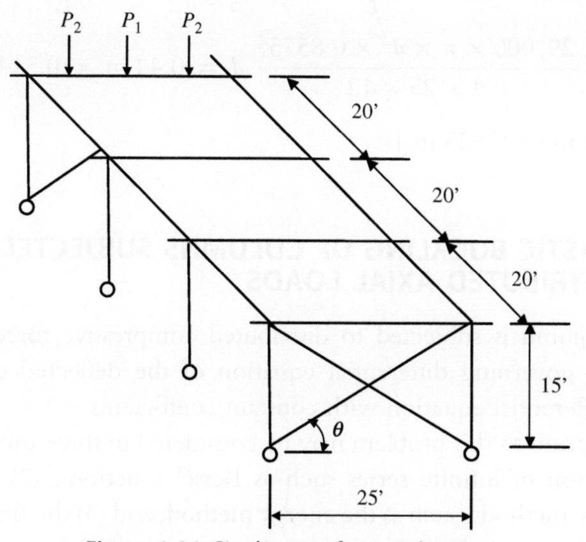

Figure 1-21 Single-story frame X-bracing

$$\sum P = 4 \times (200 + 2 \times 100) = 1,600 \text{ kips}, \beta = 1, A_e = UA_n$$
$$= 1 \times A_n$$

$$Q_u = 1 \times 1,600 \times 0.004 = 6.4 \text{ kips}, \cos\theta = \frac{25}{\sqrt{(25^2 + 15^2)}} = 0.8575$$

Design for strength:

$$Q_n \text{ for yielding, } Q_n = Q_u/0.9 = 6.4/0.9 = 7.11 \text{ kips}$$

$$Q_n \text{ for fracture, } Q_n = Q_u/0.75 = 6.4/0.75 = 8.53 \text{ kips}$$

The required diameter of the rod against yielding is

$$7.11 = \frac{\pi}{4} \times d^2 \times 50 \times 0.8575, d = 0.46 \text{ in.}$$

The required diameter of the rod against fracture is

$$8.53 = \frac{\pi}{4} \times \left(d - \frac{0.9743}{11}\right)^2 \times 70 \times 0.8575,$$

$$d = 0.154 \text{ in.} (11 \text{ threads per inch is justified})$$

Design for stiffness:

$$k_{\text{reqd}} = 2k_{\text{ideal}} = \frac{2\beta P_{cr}}{L} = \frac{2 \times 1 \times 1,600}{25 \times 12} = \frac{EA}{L}\cos^2\theta$$

$$10.67 = \frac{29,000 \times \pi \times d^2 \times 0.8575^2}{4 \times 25 \times 12}, d = 0.44 \text{ in.} < 0.514 \text{ in., use}$$

$$d = 5/8 \text{ in.} (= 0.625 \text{ in.})$$

1.10. ELASTIC BUCKLING OF COLUMNS SUBJECTED TO DISTRIBUTED AXIAL LOADS

When a column is subjected to distributed compressive forces along its length, the governing differential equation of the deflected curve is no longer a differential equation with constant coefficients.

The solution to this problem may be considered in three different ways: (1) application of infinite series such as Bessel functions, (2) one of the approximate methods, such as the energy method, and (3) the finite element method (the solution converges to the exact one following the grid

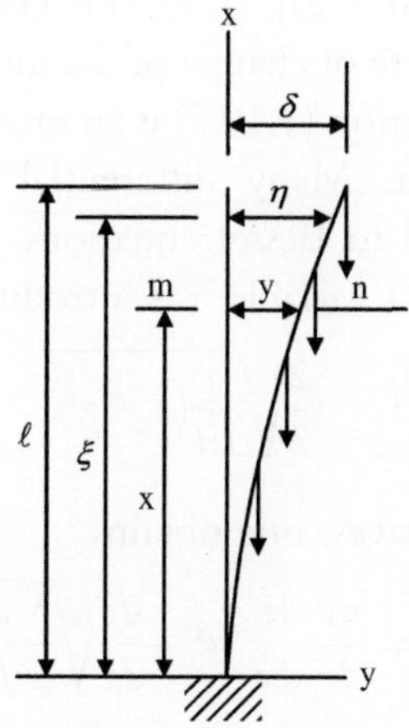

Figure 1-22 Cantilever column subjected to distributed axial load

refinement). The energy methods and the finite element analysis will be illustrated in the next chapter.

Consider the problem of elastic buckling of a prismatic column subjected to its own weight.[4,5] Figure 1-22 shows a flagpole-type cantilever column. The lower end of the column is built in, the upper end is free, and the weight is uniformly distributed along the column length. Assuming the buckled shape of the column as shown in Fig. 1-22, the differential equation of the deflected curve can be shown as:

$$EI\frac{d^2y}{dx^2} = \int_x^{\ell} q(\eta - y)\, d\xi \tag{1.10.1}$$

where the integral on the right-hand side of the equation represents the bending moment at any cross section *mn* produced by the uniformly distributed load of intensity *q*. Likewise, the shearing force at any cross section *mn* can be expressed as

$$EI\frac{d^3y}{dx^3} = -q(\ell - x)\frac{dy}{dx} \tag{1.10.2}$$

[4] This problem was first discussed by L. Euler (1707–1783), but Euler did not succeed in obtaining a satisfactory solution according to I. Todhunter, *A History* of Elasticity and of the Strength of Materials, edited and completed by K. Pearson, Vol. I (Cambridge: 1886; Dover edition, 1960), pp. 45–50.

[5] According to S. Timoshenko and J. Gere, *Theory* of Elastic Stability (New York: McGraw-Hill, 1961), 2nd ed., pp. 100–103, the problem was solved by A.G. Greenhill (1847–1927) using Bessel functions.

Note that the moment given in Eq. (1.10.1) is a decreasing function against the x-axis, and hence, the rate of change of the moment must be negative as shown in Eq. (1.10.2). Equation (1.10.2) is an ordinary differential equation with a variable coefficient. Many differential equations with variable coefficients can be reduced to Bessel equations. In order to facilitate the solution, a new independent variable z is introduced such that

$$z = \frac{2}{3}\sqrt{\frac{q}{EI}(\ell - x)^3} \tag{1.10.3}$$

By taking successive derivatives, one obtains

$$\frac{dy}{dx} = \frac{dy}{dz}\frac{dz}{dx} = -\frac{dy}{dz}\sqrt[3]{\frac{3}{2}\frac{qz}{EI}} \tag{1.10.4}$$

$$\frac{d^2y}{dx^2} = \left(\frac{3}{2}\frac{q}{EI}\right)^{\frac{2}{3}}\left(\frac{1}{3}z^{-\frac{1}{3}}\frac{dy}{dz} + z^{\frac{2}{3}}\frac{d^2y}{dz^2}\right) \tag{1.10.5}$$

$$\frac{d^3y}{dx^3} = \frac{3}{2}\frac{q}{EI}\left(\frac{1}{9}z^{-1}\frac{dy}{dz} - \frac{d^2y}{dz^2} - z\frac{d^3y}{dz^3}\right) \tag{1.10.6}$$

Substituting Eqs. (1.10.4) and (1.10.5) into Eq. (1.10.2) and letting

$$\frac{dy}{dz} = u \tag{1.10.7}$$

One obtains

$$\frac{d^2u}{dz^2} + \frac{1}{z}\frac{du}{dz} + \left(1 - \frac{1}{9z^2}\right)u = \frac{d^2u}{dz^2} + \frac{1}{z}\frac{du}{dz} + \left(1 - \frac{p^2}{z^2}\right)u = 0 \tag{1.10.8}$$

Equation (1.10.8) is a Bessel equation, and its solution can be expressed in terms of Bessel functions.

Invoking the method of Frobenius,[6] it is assumed that a solution of the form

$$u(z) = \sum_{n=0}^{\infty} c_n z^{r+n} \tag{1.10.9}$$

exists for Bessel's equation, Eq. (1.10.8) of[7] order p ($\pm 1/3$ in this case). Substituting Eq. (1.10.9) into Eq. (1.10.8), one obtains:

6 Frobenius (1848–1917) was a German mathematician.

7 See, for example, S.I. Grossman and W.R. Derrick, *Advanced Engineering Mathematics* (New York: Harper & Row, 1988), pp. 272–274.

$$\sum_{n=o}^{\infty} c_n(r+n)(r+n-1)z^{r+n-2} + \sum_{n=o}^{\infty} c_n(r+n)z^{r+n-2} + \sum_{n=o}^{\infty}(-p^2)c_n z^{r+n-2}$$

$$+\sum_{n=2}^{\infty} c_{n-2}z^{r+n-2} = 0$$

or

$$c_0(r^2-p^2)z^{r-2} + c_1[(r+1)^2-p^2]z^{r-1}$$
$$+\sum_{n=2}^{\infty}\{c_n[(n+r)^2-p^2]+c_{n-2}\}z^{r+n-2} = 0 \qquad (1.10.10)$$

The indicial equation is $r^2 - p^2 = 0$ with roots $r_1 = p = 1/3$ and $r_2 = -p = -1/3$. Setting $r = p$ in Eq. (1.10.10) yields

$$(1+2p)c_1 z^{p-1} + \sum_{n=2}^{\infty}[n(n+2p)c_n + c_{n-2}]z^{n+p-2} = 0$$

indicating that $c_1 = 0$ and $c_n = \dfrac{-c_{n-2}}{n(n+2p)}$, for $n \geq 2$. (1.10.11)

Hence, all the coefficients with odd-numbered subscripts equal to zero. Letting $n = 2j + 2$ one sees that the coefficients with even-numbered subscripts satisfy

$$c_{2(j+1)} = \frac{-c_{2j}}{2^2(j+1)(p+j+1)}, \quad \text{for } j \geq 0,$$

which yields

$$c_2 = \frac{-c_0}{2^2(p+1)}, \quad c_4 = \frac{-c_2}{2^2(2)(p+2)} = \frac{c_0}{2^4(2!)(p+1)(p+2)},$$

$$c_6 = \frac{-c_4}{2^2(3)(p+3)} = \frac{-c_0}{2^6(3!)(p+1)(p+2)(p+3)}, \ldots..$$

Hence, the series of Eq. (1.10.9) becomes

$$u_1 = z^p\left[c_0 - \frac{c_0}{2^2(p+1)}z^2 + \frac{c_0}{2^4 2!(p+1)(p+2)}z^4 - \ldots\right]$$
$$= c_0 z^p \sum_{n=0}^{\infty}(-1)^n \frac{z^{2n}}{2^{2n}n!(p+1)(p+2)\ldots(p+n)} \qquad (1.10.12)$$

It is customary in Eq. (1.10.12) to let the integral constant, $c_0 = [2^p\Gamma(p+1)]^{-1}$ in which $\Gamma(p+1)$ is the gamma function. Then, Eq. (1.10.12) becomes

$$J_p(z) = (z/2)^p \sum_{n=0}^{\infty} (-1)^n \frac{(z/2)^{2n}}{n!\Gamma(p+n+1)}$$

which is known as the Bessel function of the first kind of order p. Thus $J_p(z)$ is the first solution of Eq. (1.10.8). One will again be able to apply the method of Frobenius with $r = -p$ to find the second solution. From Eq. (1.10.10), one immediately obtains

$$(1-2p)c_1 z^{-p-1} + \sum_{n=2}^{\infty} [n(n-2p)c_n + c_{n-2}] z^{n-p-2} = 0 \qquad (1.10.13)$$

indicating $c_1 = 0$ as before and

$$c_n = \frac{-c_{n-2}}{n(n-2p)} \qquad (1.10.14)$$

With algebraic operations similar to those done earlier, one obtains the second solution of Eq. (1.10.8)

$$J_{-p}(z) = (z/2)^{-p} \sum_{n=0}^{\infty} (-1)^n \frac{(z/2)^{2n}}{n!\Gamma(n-p+1)} \qquad (1.10.15)$$

Hence, the complete solution of Eq. (1.10.8) is

$$u(z) = u_1(z) + u_2(z) = AJ_p(z) + BJ_{-p}(z) \qquad (1.10.16)$$

In Eq. (1.10.16), A and B are constants of integration, and they must be determined from the boundary conditions of the column. Since the upper end of the column is free, the condition yields

$$\left(\frac{d^2y}{dx^2}\right)_{x=\ell} = 0$$

Observing that $z = 0$ at $x = \ell$ and using Eqs. (1.10.5) and (1.10.7), one can express this condition as

$$\left(\frac{1}{3} z^{-\frac{1}{3}} u + z^{\frac{2}{3}} \frac{du}{dz}\right)_{z=0} = 0$$

Substituting Eq. (1.10.16) into this equation, one obtains $A = 0$ and hence

$$u(z) = BJ_{-p}(z) \qquad (1.10.17)$$

At the lower end of the column the condition is

$$\left(\frac{dy}{dx}\right)_{x=0} = 0$$

With the use of Eqs. (1.10.3), (1.10.4), and (1.10.7), this condition is expressed in the form

$$u = 0 \quad \text{when } z = \frac{2}{3}\sqrt{\frac{q\ell^3}{EI}}.$$

The value of z which makes $u = 0$ can be found from Eq. (1.10.17) by trial and error, from a table of the Bessel function of order $-(1/3)$, or from a computerized symbolic algebraic code such as **Maple®**. The lowest value of z which makes $u = 0$, corresponding to the lowest buckling load, is found from **Maple®** to be $z = 1.866350859$, and hence

$$z = \frac{2}{3}\sqrt{\frac{q\ell^3}{EI}} = 1.866$$

or

$$(ql)_{cr} = \frac{7.837EI}{\ell^2}. \tag{1.10.18}$$

This is the critical value of the uniform load for the column shown in Fig. 1-22.

Equation (1.10.2) above is differentiated once more to derive the governing equation of the buckling of the column under its own weight as

$$EI\frac{d^2}{dx^2}\left(\frac{d^2y}{dx^2}\right) + q\frac{d}{dx}\left[(\ell - x)\frac{dy}{dx}\right] = 0 \tag{1.10.19}$$

Equation (1.10.19) is accompanied by appropriate boundary conditions. For the column that is pinned, clamped, and free at its end, the boundary conditions are, respectively

$$y = 0, \quad \frac{d^2y}{dx^2} = 0 \tag{1.10.20a}$$

$$y = 0, \quad \frac{dy}{dx} = 0 \tag{1.10.20b}$$

$$\frac{d^2y}{dx^2} = 0, \quad \frac{d^3y}{dx^3} = 0 \tag{1.10.20c}$$

As the differential equation is an ordinary homogeneous equation with a variable constant, the power series method, or a combination of Bessel and Lommel functions, are used after a clever transformation. Elishakoff (2005) gives[8] credit to Dinnik (1912) for the solution of the pin-ended column as

$$(q\ell)_{cr} = \frac{18.6EI}{\ell^2} \tag{1.10.21}$$

and to Engelhardt (1954) for the solution of the column that is clamped at one end (bottom) and pinned at the other (top) as

$$(q\ell)_{cr} = \frac{52.5EI}{\ell^2} \tag{1.10.22}$$

as well as for the column that is clamped at both ends as

$$(q\ell)_{cr} = \frac{74.6EI}{\ell^2} \tag{1.10.23}$$

Structural Stability (STSTB)[9] computes critical load for the column that is clamped at one end (top) and pinned at the other (bottom) as

$$(q\ell)_{cr} = \frac{30.0EI}{\ell^2} \tag{1.10.24}$$

Solutions given by Eqs. (1.10.18), (1.10.21), (1.10.22), (1.10.23), and (1.10.24) can be duplicated closely (within the desired accuracy) by most present-day computer programs, for example, STSTB. Wang et al. (2005) present exact solutions for columns with other boundary conditions. A case of considerable practical importance, in which the moment of inertia of the column section varies along its length, has been investigated. However, these problems can be effectively treated by the present-day computer programs, and efforts associated with the complex mathematical manipulations can now be diverted into other endeavors.

1.11. LARGE DEFLECTION THEORY (THE ELASTICA)

Although it is not likely to be encountered in the construction of buildings and bridges, a very slender compression member may exhibit a nonlinear elastic large deformation so that a simplifying assumption of the small

[8] I. Elishakoff, *Eigenvalues of Inhomogeneous Structures* (Boca Raton, FL: CRC Press, 2005), p. 75.

[9] C.H. Yoo, "Bimoment Contribution to Stability of Thin-Walled Assemblages," *Computers and Structures*, 11, No. 5 (May 1980), pp. 465–471. Fortran source code is available at the senior author's Website.

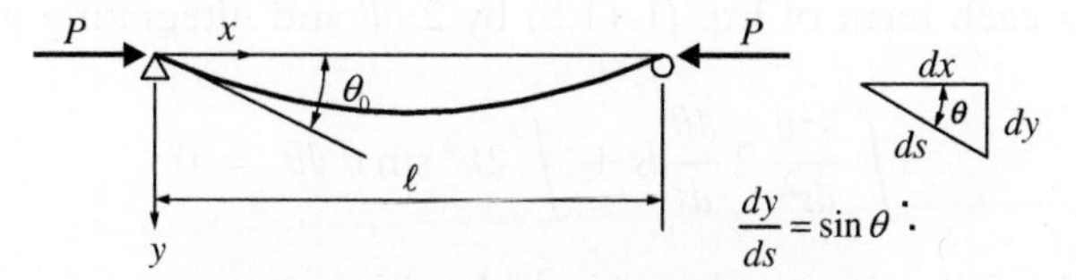

Figure 1-23 Large deflection model

displacement theory may not be valid, as illustrated by Timoshenko and Gere (1961) and Chajes (1974). Consider the simply supported wiry column shown in Fig. 1-23. Aside from the assumption of small deflections, all the other idealizations made for the Euler column are assumed valid. The member is assumed perfectly straight initially and loaded along its centroidal axis, and the material is assumed to obey Hooke's law.

From an isolated free body of the deformed configuration of the column, it can be readily observed that the external moment, Py, at any section is equal to the internal moment, $-EI/\rho$.

Thus

$$Py = -\frac{EI}{\rho} \tag{1.11.1}$$

where $1/\rho$ is the curvature. Since the curvature is defined by the rate of change of the unit tangent vector of the curve with respect to the arc length of the curve, the curvature and slope relationship is established.

$$\frac{1}{\rho} = \frac{d\theta}{ds} \tag{1.11.2}$$

Substituting Eq. (1.11.1) into Eq. (1.11.2) yields

$$EI\frac{d\theta}{ds} + Py = 0 \tag{1.11.3}$$

Introducing $k^2 = P/EI$, Eq. (1.11.3) transforms into

$$\frac{d\theta}{ds} + k^2 y = 0 \tag{1.11.4}$$

Differentiating Eq. (1.11.4) with respect to s and replacing dy/ds by $\sin\theta$ yields

$$\frac{d^2\theta}{ds^2} + k^2 \sin\theta = 0 \tag{1.11.5}$$

Multiplying each term of Eq. (1.11.5) by 2 $d\theta$ and integrating gives

$$\int \frac{d^2\theta}{ds^2} 2 \frac{d\theta}{ds} ds + \int 2k^2 \sin\theta \, d\theta = 0 \tag{1.11.6}$$

Recalling the following mathematical identities

$$\frac{d}{ds}\left(\frac{d\theta}{ds}\right)^2 = 2\left(\frac{d\theta}{ds}\right)\left(\frac{d^2\theta}{ds^2}\right) \quad \text{and} \quad \sin\theta \, d\theta = -d(\cos\theta),$$

it follows immediately that

$$\int d\left(\frac{d\theta}{ds}\right)^2 - 2k^2 \int d(\cos\theta) = 0 \tag{1.11.7}$$

Carrying out the integration gives

$$\left(\frac{d\theta}{ds}\right)^2 - 2k^2 \cos\theta = C \tag{1.11.8}$$

The integral constant C can be determined from the proper boundary condition. That is

$$\frac{d\theta}{ds} = 0 \quad \text{at } x = 0,$$

$$\left(\text{moment} = 0 \Rightarrow \frac{1}{\rho} = 0 \text{ or } \rho = \infty, \text{straight line}\right) \text{ and } \theta = \theta_0$$

Hence,

$$C = -2k^2 \cos\theta_0$$

and Eq. (1.11.8) becomes

$$\left(\frac{d\theta}{ds}\right)^2 - 2k^2(\cos\theta - \cos\theta_0) = 0 \tag{1.11.9}$$

Taking the square root of Eq. (1.11.9) and rearranging gives

$$ds = -\frac{d\theta}{\sqrt{2}k\sqrt{\cos\theta - \cos\theta_0}} \tag{1.11.10}$$

Notice the negative sign in Eq. (1.11.10), which implies that θ decreases as s increases. Carrying out the integral of Eq. (1.11.10) gives

$$\int_0^{\ell/2} ds = -\frac{1}{\sqrt{2}k}\int_{\theta_0}^{0} \frac{d\theta}{\sqrt{\cos\theta - \cos\theta_0}} \text{ or } \frac{\ell}{2} = \frac{1}{\sqrt{2}k}\int_0^{\theta_0} \frac{d\theta}{\sqrt{\cos\theta - \cos\theta_0}}$$

or

$$\ell = \frac{2}{k}\int_0^{\theta_0} \frac{d\theta}{\sqrt{2\cos\theta - 2\cos\theta_0}} \tag{1.11.11}$$

Notice the negative sign is eliminated by reversing the limits of integration. Making use of mathematical identities

$$\cos\theta = 1 - 2\sin^2\frac{\theta}{2} \quad \text{and} \quad \cos\theta_0 = 1 - 2\sin^2\frac{\theta_0}{2}$$

in Eq. (1.11.11) yields:

$$\ell = \frac{1}{k}\int_0^{\theta_0} \frac{d\theta}{\sqrt{\sin^2\frac{\theta_0}{2} - \sin^2\frac{\theta}{2}}} \tag{1.11.12}$$

In order to simplify Eq. (1.11.12) further, let

$$\sin\frac{\theta_0}{2} = \alpha \tag{1.11.13}$$

and introduce a new variable ϕ such that

$$\sin\frac{\theta}{2} = \alpha\sin\phi \tag{1.11.14}$$

Then $\theta = 0 \Rightarrow \phi = 0$ and $\theta = \theta_0 \Rightarrow \sin\phi = 1 \Rightarrow \phi = \pi/2$. Differentiating Eq. (1.11.14) yields

$$\frac{1}{2}\cos\frac{\theta}{2}\,d\theta = \alpha\cos\phi\,d\phi \tag{1.11.15}$$

which can be rearranged to show

$$d\theta = \frac{2\alpha\cos\phi\,d\phi}{\sqrt{1 - \sin^2\frac{\theta}{2}}} = \frac{2\alpha\cos\phi\,d\phi}{\sqrt{1 - \alpha^2\sin^2\phi}} \tag{1.11.16}$$

Substituting Eqs. (1.11.13), (1.11.14), (1.11.15), and (1.11.16) into Eq. (1.11.12) yields

$$\ell = \frac{1}{k}\int_0^{\theta_0} \frac{d\theta}{\sqrt{\sin^2\frac{\theta_0}{2} - \sin^2\frac{\theta}{2}}} = \frac{1}{k}\int_0^{\pi/2} \frac{1}{\sqrt{\alpha^2 - \alpha^2\sin^2\phi}} \frac{2\alpha\cos\phi\, d\phi}{\sqrt{1-\alpha^2\sin^2\phi}}$$

$$= \frac{2}{k}\int_0^{\pi/2} \frac{1}{\alpha\cos\phi} \frac{\alpha\cos\phi\, d\phi}{\sqrt{1-\alpha^2\sin^2\phi}}$$

$$\ell = \frac{2}{k}\int_0^{\pi/2} \frac{d\phi}{\sqrt{1-\alpha^2\sin^2\phi}} = \frac{2K}{k} \tag{1.11.17}$$

where:

$$K = \int_0^{\pi/2} \frac{d\phi}{\sqrt{1-\alpha^2\sin^2\phi}} \tag{1.11.18}$$

Equation (1.11.18) is known as the complete elliptic integral of the first kind. Its value can be readily evaluated from a computerized symbolic algebraic code such as **Maple**®. Equation (1.11.17) can be rewritten in the form

$$\ell = \frac{2K}{k} = \frac{2K}{\sqrt{P/EI}} \quad \text{as } k^2 = \frac{P}{EI}$$

or

$$\frac{P}{P_{cr}} = \frac{4K^2}{\pi^2} \tag{1.11.19}$$

as

$$P = \frac{4K^2}{\ell^2/EI} = \frac{4EIK}{\ell^2} \quad \text{and} \quad P_{cr} = \frac{\pi^2 EI}{\ell^2}$$

If the lateral deflection of the member is very small (just after the initial bulge), then θ_0 is small and consequently $\alpha^2\sin^2\phi$ in the denominator of K becomes negligible. The value of K approaches $\pi/2$ and from Eq. (1.11.19) $P = P_{cr} = \pi^2 EI/\ell^2$.

The midheight deflection, y_m (or δ), can be determined from $dy = ds\sin\theta$.

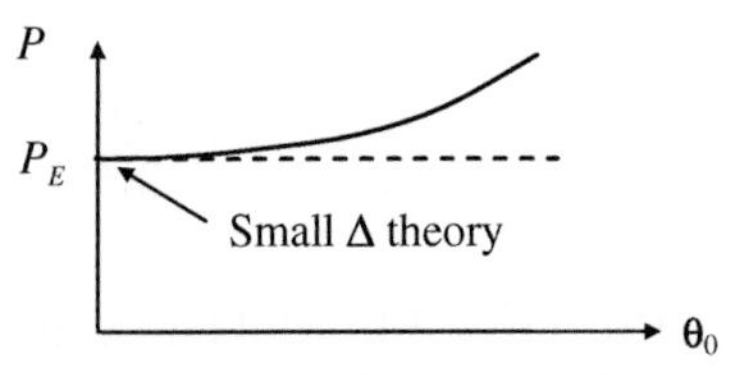

Figure 1-24 Postbuckling behavior

Substituting Eq. (1.11.10) into the above equation yields

$$dy = -\frac{\sin\theta\, d\theta}{\sqrt{2}k\sqrt{\cos\theta - \cos\theta_0}}$$

Integrating the above equation gives

$$\int_0^{y_m} dy = -\frac{1}{2k}\int_{\theta_0}^{0} \frac{\sin\theta\, d\theta}{\sqrt{\cos\theta - \cos\theta_0}} \quad \text{or} \quad y_m = \frac{1}{2k}\int_0^{\theta_0} \frac{\sin\theta\, d\theta}{\sqrt{\sin^2\dfrac{\theta_0}{2} - \sin^2\dfrac{\theta}{2}}}$$

Recall $\sin(\theta/2) = \alpha\sin\phi$ and $d\theta = 2\alpha\cos\phi\, d\phi/\sqrt{1-\alpha^2\sin^2\phi}$
Hence,

$$\sin\theta = 2\sin\frac{\theta}{2}\cos\frac{\theta}{2} = 2\sin\frac{\theta}{2}\sqrt{1-\sin^2\frac{\theta}{2}} = 2\alpha\sin\phi\sqrt{1-\alpha^2\sin^2\phi}$$

$$y_m = \frac{1}{2k}\int_0^{\theta_0} \frac{\sin\theta\, d\theta}{\sqrt{\sin^2\dfrac{\theta_0}{2} - \sin^2\dfrac{\theta}{2}}}$$

$$= \frac{1}{2k}\int_0^{\pi/2} \frac{2\alpha\sin\phi\sqrt{1-\alpha^2\sin^2\phi}\, 2\alpha\cos\phi\, d\phi}{\sqrt{\alpha^2 - \alpha^2\sin^2\phi}\sqrt{1-\alpha^2\sin^2\phi}}$$

$$y_m = \delta = \frac{2\alpha}{k}\int_0^{\pi/2} \sin\phi\, d\phi = \frac{2\alpha}{k} \quad \text{or} \quad \frac{y_m}{\ell} = \frac{2\alpha}{\pi\sqrt{\dfrac{P}{P_E}}}$$

The distance between the two load points (x-coordinates) can be determined from

$$dx = ds\cos\theta$$

Substituting Eq. (1.11.10) into the above equation yields

$$dx = -\frac{\cos\theta\, d\theta}{\sqrt{2}k\sqrt{\cos\theta - \cos\theta_0}}$$

Integrating (x_m is the x-coordinate at the midheight) the above equation gives

$$\int_0^{x_m} dx = -\frac{1}{\sqrt{2}k}\int_{\theta_0}^{0} \frac{\cos\theta\, d\theta}{\sqrt{\cos\theta - \cos\theta_0}} = -\frac{1}{\sqrt{k}}\int_{\theta_0}^{0} \frac{\cos\theta\, d\theta}{\sqrt{2\cos\theta - 2\cos\theta_0}} \text{ or}$$

$$x_m = \frac{1}{2k}\int_0^{\theta_0} \frac{\cos\theta\, d\theta}{\sqrt{\sin^2\dfrac{\theta_0}{2} - \sin^2\dfrac{\theta}{2}}}$$

Recall $\sin(\theta/2) = \alpha \sin\phi$ and $d\theta = 2\alpha \cos\phi \, d\phi / \sqrt{1-\alpha^2 \sin^2\phi}$

and $\cos\theta = \cos^2(\theta/2) - \sin^2(\theta/2) = 1 - 2\sin^2(\theta/2) = 1 - 2\alpha^2 \sin^2\phi$

$$
\begin{aligned}
x_m &= \frac{1}{2k}\int_0^{\theta_0} \frac{\cos\theta \, d\theta}{\sqrt{\sin^2\frac{\theta_0}{2} - \sin^2\frac{\theta}{2}}} \\
&= \frac{1}{2k}\int_0^{\pi/2} \frac{(1-2\alpha^2\sin^2\phi)2\alpha\cos\phi\, d\phi}{\sqrt{\alpha^2-\alpha^2\sin^2\phi}\sqrt{1-\alpha^2\sin^2\phi}} \\
&= \frac{1}{k}\int_0^{\pi/2} \frac{(1-2\alpha^2\sin^2\phi)d\phi}{\sqrt{1-\alpha^2\sin^2\phi}}
\end{aligned}
$$

$$
\begin{aligned}
x_0 &= 2x_m = \frac{2}{k}\int_0^{\pi/2} \frac{[2(1-\alpha^2\sin^2\phi)-1]d\phi}{\sqrt{1-\alpha^2\sin^2\phi}} \\
&= \frac{4}{k}\int_0^{\pi/2} \sqrt{1-\alpha^2\sin^2\phi}\, d\phi \\
&\quad - \frac{2}{k}\int_0^{\pi/2} \frac{d\phi}{\sqrt{1-\alpha^2\sin^2\phi}} = \frac{4}{k}E(\alpha) - \ell
\end{aligned}
$$

where $E(\alpha)$ is the complete elliptic integral of the second kind

$$
\frac{x_0}{\ell} = \frac{4E(\alpha)}{\ell\sqrt{\frac{P}{EI}}} - 1 = \frac{4E(\alpha)}{\pi\sqrt{\frac{P}{P_E}}} - 1
$$

The complete elliptic integral of the first kind can be evaluated by an infinite series given by

$$
\begin{aligned}
K &= \int_0^{\pi/2} \frac{d\phi}{\sqrt{1-\alpha^2\sin^2\phi}} \\
&= \frac{\pi}{2}\left[1 + \left(\frac{1}{2}\right)^2\alpha^2 + \left(\frac{1\cdot 3}{2\cdot 4}\right)^2\alpha^4 + \left(\frac{1\cdot 3\cdot 5}{2\cdot 4\cdot 6}\right)^2\alpha^6 + \cdots\right] \text{ with } \alpha^2 < 1
\end{aligned}
$$

Summing the first four terms of the above infinite series for $\alpha = 0.5$ yields $K = 1.685174$.

Likewise, the complete elliptic integral of the second kind can be evaluated by an infinite series given by

$$E = \int_0^{\pi/2} \sqrt{1 - \alpha^2 \sin^2 \phi}\, d\phi$$
$$= \frac{\pi}{2}\left[1 - \left(\frac{1}{2}\right)^2 \alpha^2 - \left(\frac{1\cdot 3}{2\cdot 4}\right)^2 \frac{\alpha^4}{3} - \left(\frac{1\cdot 3\cdot 5}{2\cdot 4\cdot 6}\right)^2 \frac{\alpha^6}{5} - \cdots\right] \text{ with } \alpha^2 < 1$$

Summing the first four terms of the above infinite series for $\alpha = 0.5$ yields $E = 1.46746$. These two infinite series can be programmed as shown or can be evaluated by commercially available symbolic algebraic codes such as **Maple**®, **Matlab**®, and/or **MathCAD**®.

```
C      THIS IS TO EXPAND THE COMPLETE ELLIPTIC INTEGRAL OF THE FIRST KIND
C      AND THE COMPLETE ELLIPTIC INTEGRAL OF THE SECOND KIND
C      USING AN INFINITE SERIES
       IMPLICIT DOUBLE PRECISION (A-H,O-Z)
       DIMENSION VAL(50),VALK(50),VALE(50)
C      ALPHA=ONE OF THE ARGUMENTS
C      N=THE NUMBER OF TERMS DESIRED TO BE SUMMED, GENERALLY LESS THAN 10
    99 READ (5,*,END=98) ALPHA,N
       J=N-1
       IF (J.EQ.0) GO TO 2
       DO 3 K=1,J
       IF (K.GT.1) GO TO 6
       VAL(1)=1./4.
       VALK(1)=VAL(1)
       VALE(1)=VAL(1)
       GO TO 8
     6 VAL(K)=VAL(K-1)*((2.*K-1.)/(2.*K))**2
       VALK(K)=VAL(K)
       VALE(K)=VAL(K)
     8 VALE(K)=VALE(K)*ALPHA**(2.*K)/(2.*K-1.)
       VALK(K)=VALK(K)*ALPHA**(2.*K)
     3 CONTINUE
       GO TO 4
     2 SUM=1.0
       SUB=1.0
       GO TO 7
     4 SUM=0.0
       SUB=0.0
       DO 5 L=1,J
       SUB=SUB-VALE(L)
     5 SUM=SUM+VALK(L)
       SUM=SUM+1.0
       SUB=SUB+1.0
     7 SUM=SUM*DATAN(1.0D0)*2.
       SUB=SUB*DATAN(1.0D0)*2.
       WRITE(6,600) SUM,SUB,ALPHA,N
   600 FORMAT (' ','FIRST=',E11.5,' SECOND=',E11.5,' FOR ALPHA=',E11.5,
     1 ' BY SUMMING ',I2, ' TERMS')
       GO TO 99
    98 STOP
       END
```

Table 1-1 Load vs. deflection data for large deflection theory

θ_0 / rad	K	E	α	P/P_E	y_m/ℓ	x_0/ℓ
0/0	$\pi/2$	$\pi/2$	.0	1.	.0	1.
20/.349	1.583	1.5588	.174	1.015	.110	0.9700
40/.698	1.620	1.5238	.342	1.063	.211	0.8818
60/1.047	1.686	1.4675	.500	1.152	.296	0.7408
90/1.5708	1.8539	1.3507	.707	1.3929	.3814	0.4572
120/2.0944	2.1564	1.2111	.866	1.8846	.4016	0.1233
150/2.618	2.7677	1.0764	.9659	3.1045	.349	−0.2222
170/2.967	4.4956	1.0040	.999	8.1910	.2222	−0.5533
179.996/π	12.55264	1.0000	0.9999999999	63.86	.07966	−0.8407

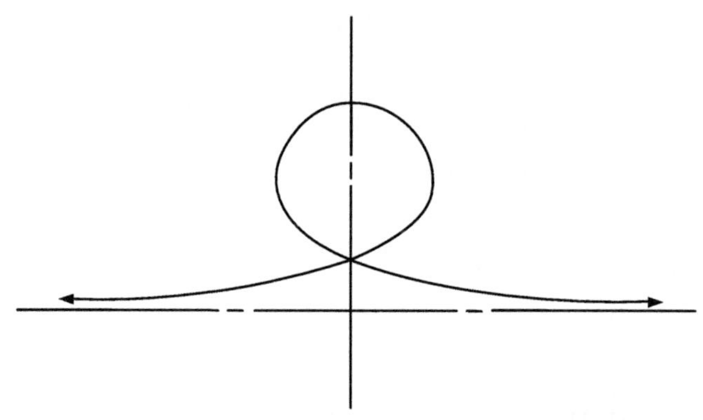

Figure 1-25 Postbuckling shape of wiry column

Consider the postbuckling shape of the wiry column. This type of postbuckling behavior may only be imagined for a very thin high-strength wire. Notice that the two end support positions are reversed. The θ_0 to make the two end points contact (x_0) is found to be 130.6 degrees by trial and error. Many ordinary materials may not be able to withstand the high-stress level required to develop a shape similar to that shown in Fig. 1-25 in an elastic manner, and the stresses in the critical column sections are likely to be extended well into the plastic region. Therefore, the practical value of the large deflection theory at large deflections is questionable.

1.12. ECCENTRICALLY LOADED COLUMNS—SECANT FORMULA

In the derivation of the Euler model, a both-end pinned column, it is assumed that the member is perfectly straight and homogeneous, and that the loading is assumed to be concentric at every cross section so that the structure and loading are symmetric. These idealizations are made to simplify the problem. In real life, however, a perfect column that satisfies all three conditions does not exist. It is, therefore, interesting to study the behavior of an imperfect column and compare it with the behavior predicted by the Euler theory. The imperfection of a monolithic slender column is predominantly affected by the geometry and eccentricity of loading. As an imperfect column begins to bend as soon as the initial amount of the incremental load is applied, the behavior of an imperfect column can be investigated successfully by considering either an initial imperfection or an eccentricity of loading.

Consider the eccentrically loaded slender column shown in Fig. 1-26. From equilibrium of the isolated free body of the deformed configuration, Eq. (1.12.1) becomes obvious

$$EIy'' + P(e + y) = 0 \tag{1.12.1}$$

or

$$y'' + k^2 y = -k^2 e \quad \text{with } k^2 = P/EI \tag{1.12.2}$$

It should be noted in Eq. (1.12.2) that the system (both-end pinned prismatic column of length ℓ with constant EI) eigenvalue remains unchanged from the Euler critical load as it is evaluated from the homogeneous differential equation.

The general solution of Eq. (1.12.2) is

$$y = y_h + y_p = A \sin kx + B \cos kx - e \tag{1.12.3}$$

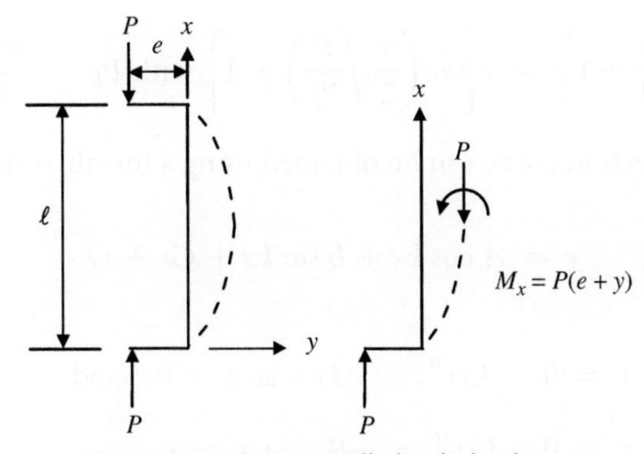

Figure 1-26 Eccentrically loaded column

The integral constants are evaluated from the boundary conditions. (The notion of solving an nth order ordinary differential equation implies that a direct or an indirect integral process is applied n times and hence there should be n integral constants in the solution of an nth order equation.) Thus the condition

$$y = 0 \quad \text{at } x = 0$$

leads to

$$B = e$$

and the condition

$$y = 0 \quad \text{at } x = \ell$$

gives

$$A = e\frac{1 - \cos k\ell}{\sin k\ell}$$

Substituting A and B into Eq. (1.12.3) yields

$$y = e\left(\cos kx + \frac{1 - \cos k\ell}{\sin k\ell}\sin kx - 1\right) \tag{1.12.4}$$

Letting $x = \ell/2$ in Eq. (1.12.4) for the midheight deflection, δ, gives

$$y\Big|_{x=\ell/2} = \delta = e\left(\cos\frac{k\ell}{2} + \frac{1 - \cos k\ell}{\sin k\ell}\sin\frac{k\ell}{2} - 1\right)$$

$$= e\left(\cos\frac{k\ell}{2} + \frac{1 - 1 + 2\sin^2\dfrac{k\ell}{2}}{2\sin\dfrac{k\ell}{2}\cos\dfrac{k\ell}{2}}\sin\frac{k\ell}{2} - 1\right) \tag{1.12.5}$$

$$\delta = e\left(\sec\frac{k\ell}{2} - 1\right) = e\left[\sec\left(\frac{\pi}{2}\sqrt{\frac{P}{P_E}}\right) - 1\right] \text{ with } P_E = \frac{\pi^2 EI}{\ell^2}$$

The same deflection curve can be obtained using a fourth-order differential equation,

$$y = A\cos kx + B\sin kx + Cx + D$$

with

$$y = 0, \quad EIy'' = -Pe \quad \text{at } x = 0 \quad \text{and}$$

$$y = 0, \quad EIy'' = -Pe \quad \text{at } x = \ell.$$

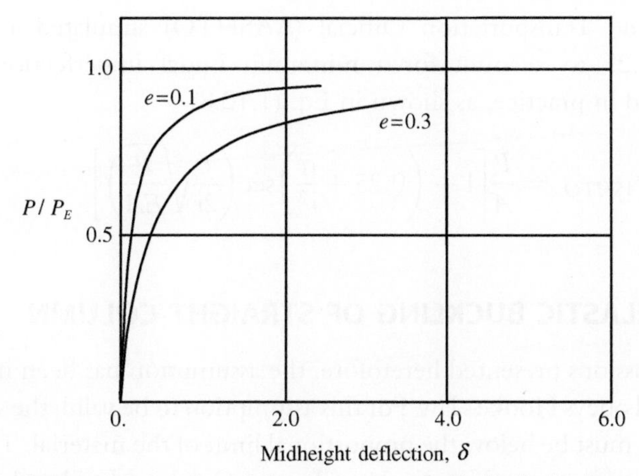

Figure 1-27 Load vs. deflection, eccentrically loaded column

Figure 1-27 shows the variation of the midheight deflection for two values of eccentricity, e.

The behavior of an eccentrically loaded column is essentially the same as that of an initially bent column except there will be the nonzero initial deflection at the no-load condition in the case of a column initially bent. A slightly imperfect column begins to bend as soon as the load is applied. The bending remains small until the load approaches the critical load, after which the bending increases very rapidly. Hence, the Euler theory provides a reasonable design criterion for real imperfect columns if the imperfections are small.

The maximum stress in the extreme fiber is due to the combination of the axial stress and the bending stress. Hence,

$$\begin{aligned}\sigma_{\max} &= \frac{P}{A} + \frac{M_{\max}c}{I} = \frac{P}{A} + \frac{P(\delta + e)c}{I} = \frac{P}{A} + \frac{ceP\sec\left(\dfrac{\ell}{2}\sqrt{\dfrac{P}{EI}}\right)}{I} \\ &= \frac{P}{A}\left[1 + \frac{ecA}{I}\sec\left(\frac{\ell}{2}\sqrt{\frac{P}{EI}}\right)\right] \qquad (1.12.6)\end{aligned}$$

$$\sigma_{\max} = \frac{P}{A}\left[1 + \frac{ec}{r^2}\sec\left(\frac{\ell}{2r}\sqrt{\frac{P}{EA}}\right)\right] \qquad (1.12.7)$$

Equation (1.12.7) is known as the secant formula. In an old edition of *Standard Specification of Highway Bridges*, American Association of State

Highway and Transportation Official (AASHTO) stipulated a constant value of 0.25 to account for a minimum initial imperfection usually encountered in practice, as shown in Eq. (1.12.8)

$$\sigma_{AASHTO} = \frac{P}{A}\left[1 + \left(0.25 + \frac{ec}{r^2}\right)\sec\left(\frac{\ell}{2r}\sqrt{\frac{P}{EA}}\right)\right] \tag{1.12.8}$$

1.13. INELASTIC BUCKLING OF STRAIGHT COLUMN

In the discussions presented heretofore, the assumption has been made that the material obeys Hooke's law. For this assumption to be valid, the stresses in the column must be below the proportional limit of the material. The linear elastic analysis is correct for slender columns. On the other hand, the axial stress in a shot column will exceed the proportional limit. Consequently, the elastic analysis is not valid for short columns, and the limiting load for short columns must be determined by taking inelastic behavior into account. Before proceeding to consider the development of the theory of inelastic column behavior, it would be informative to review its historic perspective. The Euler hyperbola was derived by Leonhard Euler in 1744. It was believed at the time that the formula applied to all columns, short and slender. It was soon discovered that the formula was grossly unconservative for short columns; the Euler formula was considered to be completely erroneous and was discarded for a lengthy period of time, approximately 150 years. An anecdotal story reveals that people ridiculed Euler when he could not adequately explain why a coin (a compression member with an extremely small slenderness ratio) on an anvil smashed by a hammer yielded (flattened) instead of carrying an infinitely large stress. It is of interest to note that the concept of flexural rigidity, *EI*, was not clearly defined at the time, and the modulus of elasticity of steel was determined by Thomas Young in 1807.[10]

However, Theodore von Kármán developed the double-modulus theory in 1910 in his doctoral dissertation at Göttingen University under Ludwig Prandtl direction. It gained widespread acceptance and the validity of Euler's work reestablished if the constant modulus E is replaced by an effective modulus for short columns. Later in 1947, F.R. Shanley[11] demonstrated that the tangent modulus and not the double modulus is the correct effective modulus, which leads to lower buckling load than the double-modulus

[10] S.P. Timoshenko, *History* of Strength of Materials (New York: Dover Edition, 1983), p. 92.

[11] A. Chajes, *Principles of Structural Stability* Theory (Englewood Cliffs, NJ: Prentice-Hall, 1974), p. 37.

theory and agrees better than the double-modulus theory with test results. These inelastic buckling analyses using effective modulus are just academic history today. The present-day finite element codes capable of conducting incremental analyses of the geometric and material nonlinearities, as refined in their final form in the 1980s, can correctly evaluate the inelastic column strengths, including the effects of initial imperfections, inelastic material properties including strain hardening, and residual stresses.

1.13.1. Double-Modulus (Reduced Modulus) Theory

Assumptions

1) Small displacement theory holds.
2) Plane sections remain plane. This assumption is called Bernoulli, or Euler, or Navier hypothesis.
3) The relationship between the stress and strain in any longitudinal fiber is given by the stress-strain diagram of the material (compression and tension, the same relationship).
4) The column section is at least singly symmetric, and the plane of bending is the plane of symmetry.
5) The axial load remains constant as the member moves from the straight to the deformed position.

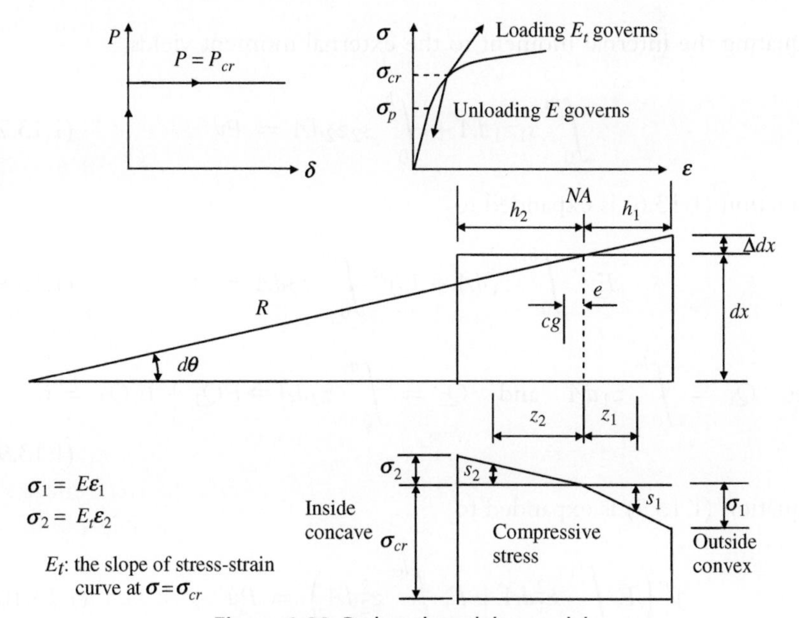

Figure 1-28 Reduced modulus model

In small displacement theory, the curvature of the bent column is

$$\frac{1}{R} \doteq \frac{d^2y}{dx^2} = \frac{d\phi}{dx} \tag{1.13.1}$$

From a similar triangle relationship, the flexural strains are computed

$$\varepsilon_1 = z_1 y'' \tag{1.13.2}$$

$$\varepsilon_2 = z_2 y'' \tag{1.13.3}$$

and the corresponding stresses are

$$\sigma_1 = E h_1 y'' \tag{1.13.4}$$

$$\sigma_2 = E_t h_2 y'' \tag{1.13.5}$$

where E_t = tangent modulus, s_1 (tension) $= E z_1 y''$ and s_2 (compression) $=$ $E_t z_2 y''$.
The pure bending portion (no net axial force) requires

$$\int_0^{h_1} s_1 dA + \int_0^{h_2} s_2 dA = 0 \tag{1.13.6}$$

Equating the internal moment to the external moment yields

$$\int_0^{h_1} s_1 z_1 dA + \int_0^{h_2} s_2 z_2 dA = Py \tag{1.13.7}$$

Equation (1.13.6) is expanded to

$$Ey'' \int_0^{h_1} z_1 dA + E_t y'' \int_0^{h_2} z_2 dA = 0 \tag{1.13.8}$$

$$\text{Let} \quad Q_1 = \int_0^{h_1} z_1 dA \quad \text{and} \quad Q_2 = \int_0^{h_2} z_2 dA \Rightarrow EQ_1 + E_t Q_2 = 0 \tag{1.13.9}$$

Equation (1.13.7) is expanded to

$$y''\left(E \int_0^{h_1} z_1^2 dA + E_t \int_0^{h_2} z_2^2 dA\right) = Py \tag{1.13.10}$$

$$\text{Let} \quad \overline{E} = \frac{EI_1 + E_t I_2}{I} \tag{1.13.11}$$

which is called the reduced modulus that depends on the stress-strain relationship of the material and the shape of the cross section. I_1 is the moment of inertia of the tension side cross section about the neutral axis and I_2 is the moment of inertia of the compression side cross section such that

$$I_1 = \int_0^{h_1} z_1^2 dA \quad \text{and} \quad I_2 = \int_0^{h_2} z_2^2 dA \tag{1.13.12}$$

Equation (1.13.10) takes the form

$$\overline{E}Iy'' + Py = 0 \tag{1.13.13}$$

Equation (1.13.13) is the differential equation of a column stressed into the inelastic range identical to Eq. (1.3.3) except that E has been replaced by $\overline{E}$, the reduced modulus. If it can be assumed that $\overline{E}$ is constant, then Eq. (1.13.13) is a linear differential equation with constant coefficients, and its solution is identical to that of Eq. (1.3.3), except that E is replaced by $\overline{E}$. Corresponding critical load and critical stress based on the reduced modulus are

$$P_{r,cr} = \frac{\pi^2 \overline{E} I}{\ell^2} \tag{1.13.14}$$

and

$$\sigma_{r,cr} = \frac{\pi^2 \overline{E}}{\left(\frac{\ell}{r}\right)^2} \tag{1.13.15}$$

Introducing

$$\tau_r = \frac{\overline{E}}{E} = \frac{E_t}{E}\frac{I_2}{I} + \frac{I_1}{I} < 1.0 \quad \text{and} \quad \tau = \frac{E_t}{E} < 1.0 \tag{1.13.16}$$

the differential equation based on the reduced modulus becomes

$$EI\tau_r y'' + Py = 0 \tag{1.13.17}$$

and

$$\tau_r = \tau\frac{I_2}{I} + \frac{I_1}{I} \quad \text{and} \quad \sigma_{r,cr} = \frac{P_{r,cr}}{A} = \frac{\pi^2 E\tau_r}{\left(\frac{\ell}{r}\right)^2} \tag{1.13.18}$$

The procedure for determining $\sigma_{r,cr}$ may be summarized as follows:
1) For $\sigma - \varepsilon$ diagram, prepare $\sigma - \tau$ diagram.
2) From the result of step 1, prepare $\tau_r - \sigma$ curve.
3) From the result of step 2, prepare $\sigma_r - (\ell/r)$ curve.

1.13.2. Tangent-Modulus Theory

Assumptions

The assumptions are the same as those used in the double-modulus theory, except assumption 5. The axial load increases during the transition from the straight to slightly bent position, such that the increase in average stress in compression is greater than the decrease in stress due to bending at the extreme fiber on the convex side. The compressive stress increases at all points; the tangent modulus governs the entire cross section.
If the load increment is assumed to be negligibly small such that

$$\Delta P <<< P \tag{1.13.19}$$

then

$$E_t I y'' + P y = 0 \tag{1.13.20}$$

and the corresponding critical stress is

$$\sigma_{t,cr} = \frac{P_{t,cr}}{A} = \frac{\pi^2 : E\tau}{\left(\frac{\ell}{r}\right)^2} \quad \text{with } \tau = \frac{E_t}{E} \tag{1.13.21}$$

Hence, σ_t vs ℓ/r curve is not affected by the shape of the cross section.
The procedure for determining the $\sigma_t - (\ell/r)$ curve may be summarized as follows:
1) From $\sigma - \varepsilon$ diagram, establish $\sigma - \tau$ curve.
2) From the result of step 1, prepare $\sigma_t - (\ell/r)$.

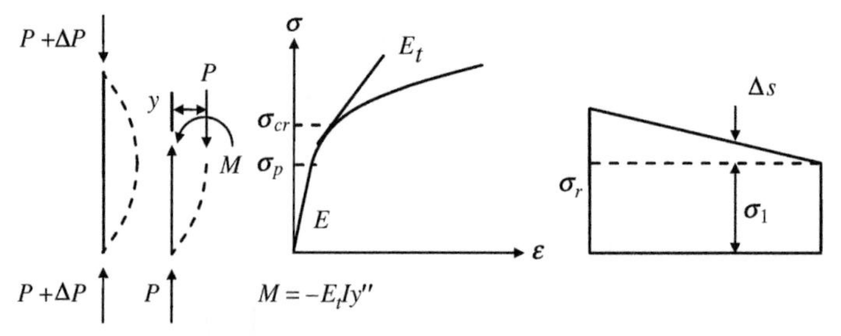

Figure 1-29 Tangent-modulus model

Example 1

An axially loaded, simply supported column is made of structural steel with the following mechanical properties: $E = 30 \times 10^3$ ksi, $\sigma_p = 28.0$ ksi, $\sigma_y = 36$ ksi, and tangent moduli given in Table 1-2.

Determine the following:

1) The value of ℓ/r, which divides the elastic buckling range and the inelastic buckling range
2) The value of τ_r and ℓ/r for $P/A = 28, 30, 32, 34, 35, 35.5$ ksi using the double-modulus theory and assuming that the cross section of the column is a square of side h.
3) The critical average stress $P/$A for $\ell/r = 20$, 40, 60, 80, 100, 120, 140, 160, 180, and 200 using the tangent-modulus theory in the inelastic range.

From the results of 1), 2), and 3), plot

4) The " $(P/A) - \tau_r$" curve for the double-modulus theory.
5) The " $(P/A) - (\ell/r)$" curves, distinguishing the portion of the curve derived by the tangent-modulus theory from that derived by the double-modulus theory. Present short discussions.
6) The current AISC LRFD Specification specifies (Chapter E) that the critical value of P/A for axially loaded column shall not exceed the following:

   (i) For $\lambda_c \leq 1.5$ $\quad F_{cr} = (0.658^{\lambda_c^2})F_y$

   (ii) For $\lambda_c > 1.5$ $\quad F_{cr} = \left[0.877/\lambda_c^2\right]F_y$

   Plot these curves and superimpose them on the graph in 5 using double arguments (ℓ/r and λ_c) on the horizontal axis.

Table 1-2 Tangent moduli measured

σ_t or σ_r (ksi)	$\tau = E_t/E$
28.0	1.00
29.0	0.98
30.0	0.96
31.0	0.93
32.0	0.88
33.0	0.77
34.0	0.55
35.0	0.31
35.5	0.16
36.0	0.00

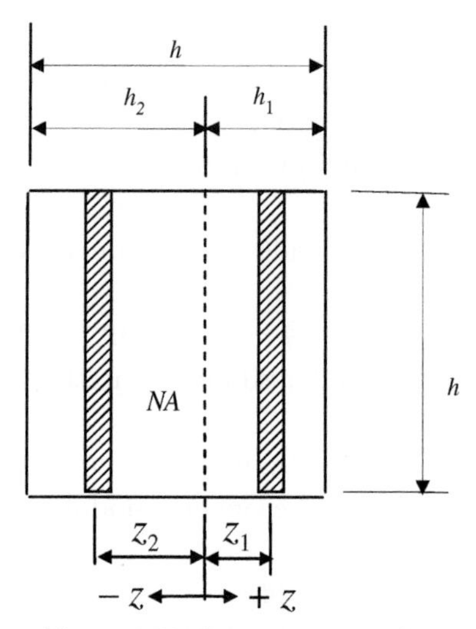

Figure 1-30 Square cross section

Locations of NA at various stages

$$EQ_1 + E_t Q_2 = 0 \tag{1.13.9}$$

$$h_1 + h_2 = h$$

$$Q_1 = \int_0^{h_1} z_1 dA$$

$$Q_2 = \int_0^{h_2} z_2 dA$$

$$Q_1 = \int_0^{h_1} z_1 h dz_1 = \frac{h}{2} z_1^2 \bigg|_0^{h_1} = \frac{hh_1^2}{2}$$

Likewise $Q_2 = -(hh_2^2/2)$

$$Q_1 + \frac{E_t}{E} Q_2 = \frac{\not{h}h_1^2}{2} - \tau \frac{\not{h}h_2^2}{2} = h_1^2 - \tau(h - h_1)^2 = 0$$

$$h_1^2 + 2\tau hh_1 - \tau h^2 - \tau h_1^2 = 0$$

$$(1 - \tau)h_1^2 + 2\tau hh_1 - \tau h^2 = 0$$

Table 1-3 Cross-sectional properties vs. shifting neutral axis

τ (1)	$h_1(/h)$ (2)	$I_1(/I)$ (3)	$I_2(/I)$ (4)	(1) × (4) (5)	τ_r [(3) + (5)] (6)	σ, σ_t, σ_r (7)	σ/τ [(7)/(1)] (8)
1.00	.5000	.5000	.5000	.5000	1.0000	28.0	28.00
0.98	.4975	.4925	.5076	.4975	0.9899	29.0	29.60
0.96	.4950	.4848	.5155	.4948	0.9797	30.0	31.25
0.93	.4910	.4733	.5277	.4908	0.9640	31.0	33.33
0.88	.4840	.4536	.5495	.4835	0.9371	32.0	36.36
0.77	.4674	.4084	.6044	.4654	0.8738	33.0	42.86
0.55	.4258	.3088	.7572	.4165	0.7253	34.0	61.82
0.31	.3576	.1830	1.0602	.3287	0.5116	35.0	112.90
0.16	.2857	.0933	1.4577	.2332	0.3265	35.5	221.88
0.00	.0000	.0000	4.0000	.0000	0.0000	36.0	∞

$$h_1 = \frac{-\tau h + \sqrt{\tau^2 h^2 + (1-\tau)(\tau h^2)}}{(1-\tau)} = \frac{h(-\tau + \sqrt{\tau})}{(1-\tau)}$$

$$\sigma_r = \frac{\pi^2 E \tau_r}{\left(\frac{\ell}{r}\right)^2} = \frac{\pi^2 \times 30 \times 10^3 \tau_r}{\left(\frac{\ell}{r}\right)^2} \Rightarrow$$

$$\left(\frac{\ell}{r}\right) = \sqrt{\frac{30 \times 10^3 \pi^2 \tau_r}{\sigma_r}} = 544.14\sqrt{\frac{\tau_r}{\sigma_r}}$$

1)

$$\sigma_P = \frac{\pi^2 E}{\left(\frac{\ell}{r}\right)^2} \Rightarrow \frac{\ell}{r} = \pi\sqrt{\frac{E}{\sigma_p}} = 102.83$$

2) and 3)

Table 1-4 Slenderness ratio vs. critical stress

ℓ/r	σ_r	ℓ/r	λ_c	$F_{cr\ AISC}$	σ_t/τ	σ_t	Remarks
102.83	28.0	200	2.21	6.49	7.402	7.402	elastic
100.53	29.0	180	1.98	8.01	9.138	9.138	
98.33	30.0	160	1.76	10.14	11.566	11.566	
95.96	31.0	140	1.54	13.25	15.107	15.107	

(Continued)

Table 1-4 Slenderness ratio vs. critical stress—cont'd

ℓ/r	σ_r	ℓ/r	λ_c	$F_{cr\ AISC}$	σ_t/τ	σ_t	Remarks
93.12	32.0	120	1.32	18.03	20.562	20.562	
88.54	33.0	100	1.10	21.64	29.609	29.000	
79.47	34.0	80	0.88	25.99	46.264	33.200	
65.79	35.0	60	0.66	29.97	82.247	34.200	σ_t, from graph
52.18	35.5	40	0.44	33.18	185.055	35.300	
0.00	36.0	20	0.22	35.27	740.220	35.990	
		0	0.00	36.00	∞	36.000	

$$\sigma_t = \frac{\pi^2 E\tau}{\left(\frac{\ell}{r}\right)^2} \Rightarrow \frac{\sigma_t}{\tau} = \frac{\pi^2 E}{\left(\frac{\ell}{r}\right)^2} \Rightarrow \text{in the elastic range,}$$

$$\tau = 1.0\left(\sigma_t = \sigma_r = \sigma_E\right)$$

4) and 5)

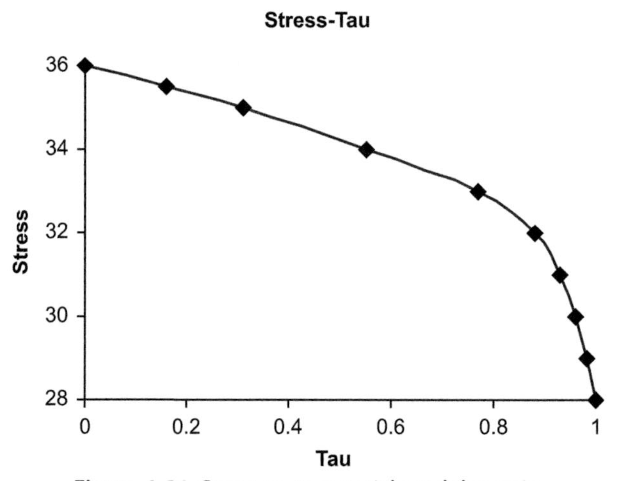

Figure 1-31 Stress vs. tangential-modulus ratio

6) (i) For $\lambda_c \leq 1.5 \Rightarrow F_{cr} = (0.658^{\lambda_c^2})F_y$ (ii) For $\lambda_c > 1.5 \Rightarrow F_{cr} = \left[0.877/\lambda_c^2\right]F_y$ where $\lambda_c = (k\ell/r\pi)\sqrt{(F_y/E)}$

Compression members (or elements) may be classified into three different regions depending on their slenderness ratios (or width-to-thickness ratios): yield zone, inelastic transition zone, and elastic buckling

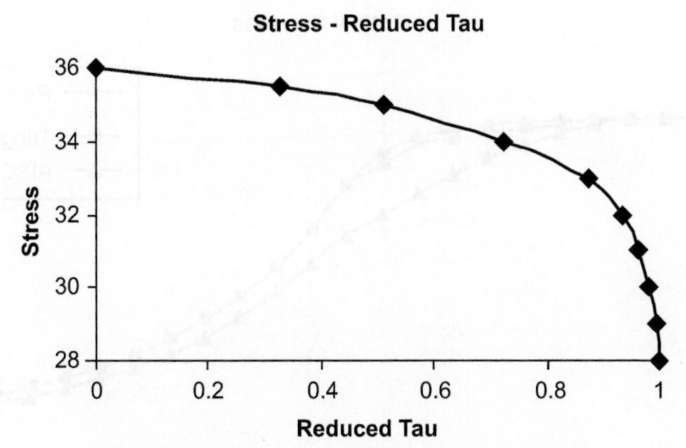

Figure 1-32 Stress vs. reduced modulus ratio

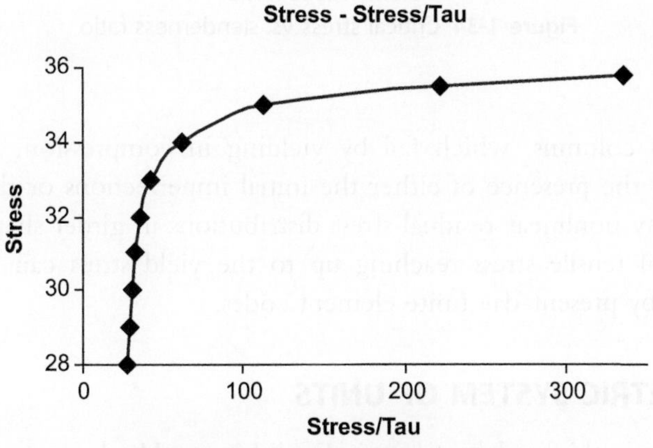

Figure 1-33 Stress vs. modular ratio

zone (or compact, noncompact, and slender). As can be seen from Fig. 1-34, the tangent-modulus theory reduces the critical compressive stress only slightly compared to that by the reduced modulus theory in the inelastic transition zone. Furthermore, both theories give the inelastic critical stresses much higher (unconservative) for a solid square cross section considered herein than those computed from the AISC LRFD formulas that are considered to be representative (Salmon and Johnson 1996) of many test data scattered over the world reported by Hall (1981). Experience (Yoo et al. 2001; Choi and Yoo 2005) has shown that the effect of the initial imperfections is significant in columns of intermediate slenderness, whereas the presence of residual stresses reduces the elastic buckling strength. The lowest

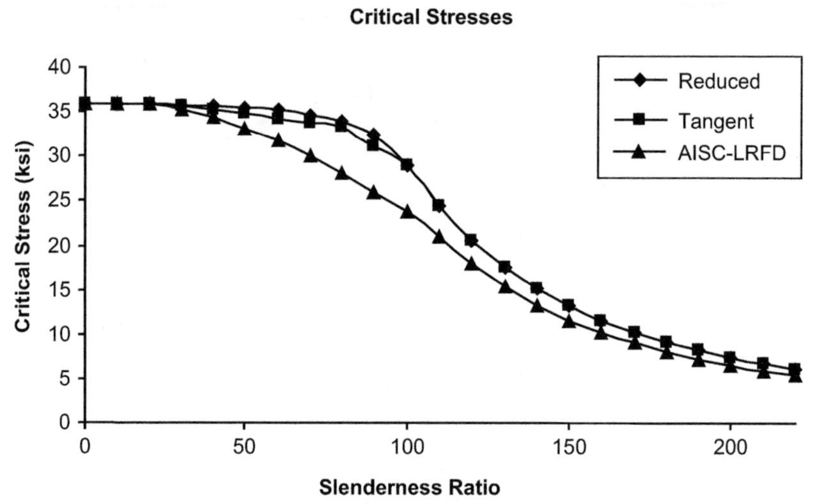

Figure 1-34 Critical stress vs. slenderness ratio

slenderness columns, which fail by yielding in compression, are hardly affected by the presence of either the initial imperfections or the residual stresses. Any nonlinear residual stress distributions in girder shapes having the residual tensile stress reaching up to the yield stress can readily be examined by present-day finite element codes.

1.14. METRIC SYSTEM OF UNITS

Dimensions in this book are given in English units. Hard conversion factors to the metric system are given in Table 1-5. The unit of force in the International System of units (Systéme International) is the Newton (N). In European countries and Japan, however, the commonly used unit is kilogram-force (kgf). Both units are included in the table. Metrication is the process of converting from the various other systems of units used throughout the world to the metric or SI (Systéme International) system. Although the process was begun in France in the 1790s and is currently converted 95% throughout the world, it is confronting stubborn resistance in a handful of countries. The main large-scale popular opposition to metrication appears to be based on tradition, aesthetics, cost, and distaste for a foreign system. Even in some countries where the international system is officially adopted, some sectors of the industry. or in a special product line, old tradition units are still being practiced.

Table 1-5 Conversion Factors

SI to English	English to SI
Length	
1 mm = 0.03937 in	1 in = 25.4 mm
1 m = 3.281 ft	1 ft = 0.3048 m
1 km = 0.6214 mi	1 mi = 1.609 km
Area	
1 mm^2 = 1.55 × 10^{-3} in^2	1 in^2 = 0.6452 × 10^3 mm^3
1 cm^2 = 1.55 × 10^{-1} in^2	1 in^2 = 6.452 cm^2
1 m^2 = 10.76 ft^2	1 ft^2 = 0.0929 m^2
1 m^2 = 1.196 yd^2	1 yd^2 = 0.836 m^2
Volume	
1 mm^3 = 6.102 × 10^{-5} in^3	1 in^3 = 16.387 × 10^3 mm^3
1 cm^3 = 6.102 × 10^{-2} in^3	1 in^3 = 16.387 cm^3
1 m^3 = 35.3 ft^3	1 ft^3 = 0.0283 m^3
1 m^3 = 1.308 yd^3	1 yd^3 = 0.765 m^3
Moment of inertia	
1 in^4 = 41.62 × 10^4 mm^4	1 mm^4 = 0.024 × 10^{-4} in^4
1 in^4 = 41.62 cm^4	1 cm^4 = 0.024 in^4
1 in^4 = 41.62 × 10^{-8} m^4	1 m^4 = 0.024 × 10^8 in^4
Mass	
1 kg = 2.205 lb	1 lb = 0.454 kg
1 kg = 1.102 × 10^{-3} ton	1 ton (2000 lb) = 907 kg
1 Mg = 1.102 ton	1 tonne (metric) = 1000 kg
Force	
1 N = 0.2248 lbf	1 lbf = 4.448 N
1 kgf = 2.205 lbf	1 kip = 4.448 kN
Stress	
1 kgf/cm^2 = 14.22 psi	1 psi = 0.0703 kgf/cm^2
1 kN/m^2 = 0.145 psi	1 psi = 6.895 kPa (kN/m^2)
1 MN/m^2 = 0.145 ksi	1 ksi = 6.895 MN/m^2 (MPa)

GENERAL REFERENCES

Some of the more general references on the stability of structures are collected in this section for convenience. References cited in the text are listed at the ends of the respective chapters. References requiring further details are given in the footnotes. Relatively recent textbooks and reference books include those by Bleich (1952), Timoshenko and Gere (1961), Ziegler (1968), Britvec (1973), Chajes (1974), Brush and Almroth (1975), Allen and Bulson (1980), Chen and Lui (1991), Bazant and Cedolin (1991), Godoy (2000), Simitses and Hodges (2006), and Galambos and Surovek (2008). Some of these books address only the elastic stability of framed structures, while others extend the coverage into the stability of plates and shells, including dynamic stability and stability of nonconservative force systems.

The design of structural elements and components is beyond the scope of this book. For stability design criteria for columns and plates, *Guide to Stability Design Criteria* (Galambos, 1998) is an excellent reference. The design of highway bridge structures is to be carried out based on AASHTO (2007) specifications, and steel building frames are to follow AISC (2005) specifications. In the case of ship structures, separate design rules are stipulated for different vessel types such as IACS (2005) and IACS (2006). A variety of organizations and authorities are claiming jurisdiction over the certificates of airworthiness of civil aviation aircrafts.

REFERENCES

AASHTO. (2007). *LRFD Bridge Design Specifications* (4th ed.). Washington, DC: American Association of State Highway and Transportation Officials.

AASHTO. (1973). *Standard Specifications for Highway Bridges* (11th ed.). Washington, DC: American Association of State Highway and Transportation Officials.

AISC. (2005). *Specification for Structural Steel Building* (13th ed.). Chicago, IL: American Institute of Steel Construction.

Allen, H. G., & Bulson, P. S. (1980). *Background to Buckling*. London: McGraw-Hill (UK).

Bazant, Z. P., & Cedolin, L. (1991). *Stability of Structures, Elastic, Inelastic, Fracture, and Damage Theories*. Oxford: Oxford University Press.

Bleich, F. (1952). *Buckling Strength of Metal Structures*. New York: McGraw-Hill.

Britvec, S. J. (1973). *The Stability of Elastic Systems*. Elmsford, NY: Pergamon Press.

Brush, D. O., & Almroth, B. O. (1975). *Buckling of Bars, Plates, and Shells*. New York: McGraw-Hill.

Chajes, A. (1974). *Principles of Structural Stability Theory*. Englewood Cliffs, NJ: Prentice-Hall.

Chen, W. F., & Lui, E. M. (1991). *Stability Design of Steel Frames*. Boca Raton, FL: CRC Press.

Choi, B. H., & Yoo, C. H. (2005). Strength of Stiffened Flanges in Horizontally Curved Box Girders. *J. Engineering Mechanics, ASCE, Vol. 131*(No. 2), 167–176, February 2005.

Dinnik, A.N. (1912). Buckling Under Own Weight, Proceedings of Don Polytechnic Institute, 1, Part 2, pp. 12 (in Russian).

Elishakoff, I. (2005). *Eigenvalues of Inhomogeneous Structures, Unusual Closed-Form Solutions*. Boca Raton, FL: CRC Press.

Engelhardt., H. (1954). *Die einheitliche Behandlung der Stabknickung mit Berücksichtung des Stabeiggengewichte in den Eulerfällen 1 bis 4 als Eigenwertproblem*. Der Stahlbau. 23(4), 80–84 (in German).

Euler, L. (1744). *Methodus Inveniendi Lineas Curvas Maximi Minimive Propreietate Gaudentes* (Appendix, De Curvis Elasticis). Lausanne and Geneva: Marcum Michaelem Bousquet.

Galambos, T. V. (Ed.). (1998). *Guide to Stability Design Criteria for Metal Structures* (5th ed.). New York: Structural Stability Research Council, John Wiley and Sons.

Galambos, T. V., & Surovek, A. E. (2008). *Structural Stability of Steel, Concepts and Applications for Structural Engineers*. Hoboken, NJ: John Wiley and Sons.

Godoy, L. A. (2000). *Theory of Elastic Stability: Analysis and Sensitivity*. Philadelphia, PA: Taylor and Francis.

Grossman, S. I., & Derrick, W. R. (1988). *Advanced Engineering Mathematics*. New York: Harper & Row.

Hall, D. H. (1981). Proposed Steel Column Strength Criteria. *J. Structural Div., ASCE, Vol. 107*(No. ST4). April 1981, pp. 649–670. Discussed by B.G. Johnston, Vol. 108, No. ST4 (April 1982), pp. 956–957; by Z. Shen and L.W. Lu, Vol. 108, No.ST7 (July 1982), pp. 1680–1681; by author, Vol. 108, No. ST12 (December 1982), pp. 2853–2855.

Hoff, N. F. (1956). *The Analysis of Structures*. New York: John Wiley and Sons.

IACS. (2005). *Common Structural Rules for Bulk Carriers*. London: International Association of Classification Societies.

IACS. (2006). *Common Structural Rules for Double Hull Oil Tankers*. London: International Association of Classification Societies.

Salmon, C. G., & Johnson, J. E. (1996). *Steel Structures, Design and Behavior* (4th ed.). New York: Haper Collins.

Salvadori, M., & Heller, R. (1963). *Structure in Architecture*. Englewood Cliffs, NJ: Prentice-Hall.

Shames, I. H., & Dym, C. L. (1985). *Energy and Finite Element Methods in Structural Mechanics*. New York: Hemisphere Publishing Corporation.

Simitses, G. J., & Hodges, D. H. (2006). *Fundamentals of Structural Stability*. Burlington, MA: Elsevier.

Timoshenko, S. P. (1953). *History of Strength of Materials*. New York: McGraw-Hill.

Timoshenko, S. P., & Gere, J. M. (1961). *Theory of Elastic Stability* (2nd ed.). New York: McGraw-Hill.

Todhunter, I. (1886). In K. Pearson (Ed.) (Dover ed). *A History of Elasticity and of the Strength of Materials, Vol. I.* New York: Cambridge. 1960.

Wang, C. M., Wang, C. Y., & Reddy, J. N. (2005). *Exact Solutions for Buckling of Structural Members*. Boca Raton, FL: CRC Press.

Winter, G. (1960). Lateral Bracing of Columns and Beams. *Transactions, ASCE, Vol. 125*, 807–845.

Yoo, C. H. (1980). Bimoment Contribution to Stability of Thin-Walled Assemblages. *Computers and Structures, Vol. 11*(No. 5), 465–471.

Yoo, C. H., Choi, B. H., & Ford, E. M. (2001). Stiffness Requirements for Longitudinally Stiffened Box Girder Flanges. *J. Structural Engineering, ASCE, Vol. 127*(No. 6), 705–711, June 2001.

Ziegler, H. (1968). *Principles of Structural Stability*. Waltham, MA: Blaisdell Publishing Co.

PROBLEMS

1.1 For structures shown in Fig. P1-1, determine the following:

(a) Using fourth-order DE, determine the lowest three critical loads.

(b) Determine the lowest two critical loads.

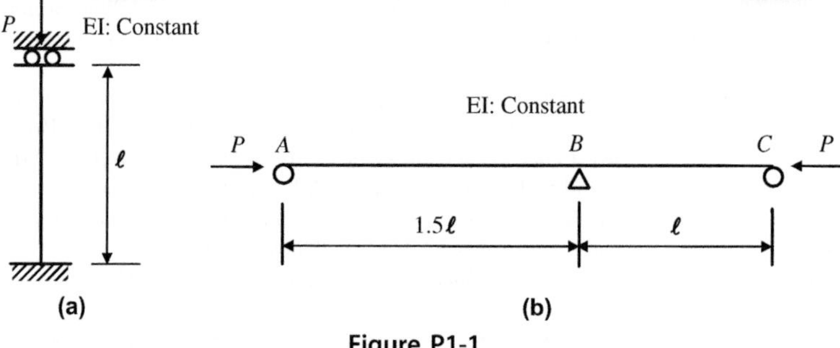

Figure P1-1

1.2 Two rigid bars are connected with a linear rotational spring of stiffness $C = M/\theta$ as shown in Fig. P1-2. Determine the critical load of the structure in terms of the spring constant and the bar length.

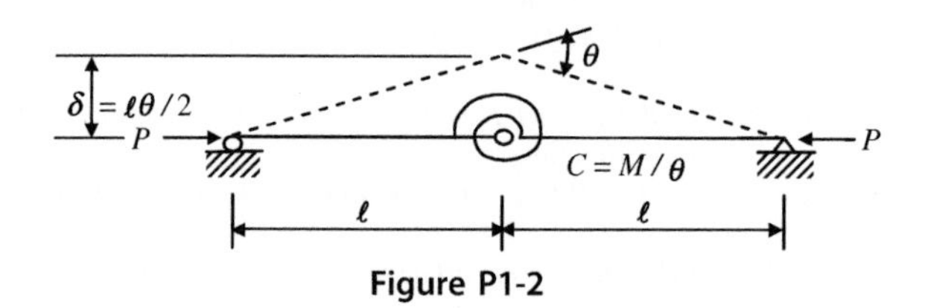

Figure P1-2

1.3 For the structure shown in Fig. P1-2, plot the load versus transverse deflection in a qualitative sense when:

(a) the transverse deflections are large,

(b) the load is applied eccentrically, and

(c) the model has an initial transverse deflection d_0.

1.4 Determine the critical load of the structure shown in Fig. P1-4.

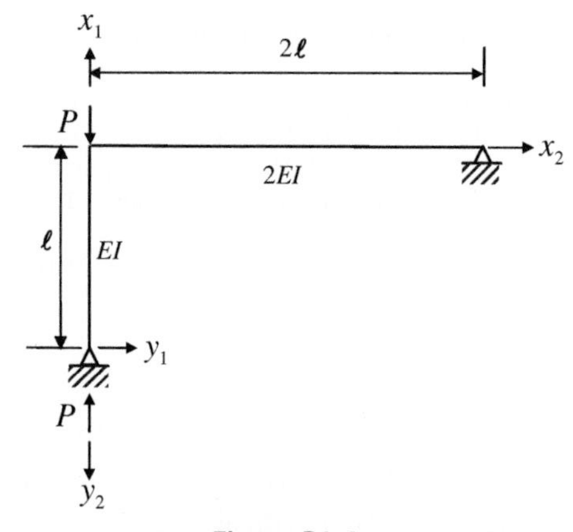

Figure P1-4

1.5 Derive the Euler-Lagrange differential equation and the necessary geometric and natural boundary conditions for a prismatic column of length ℓ and elastically supported by a rotational spring of constant β at A and a linear spring of constant α at B as shown in Fig. P1-5. Determine the critical load, P_{cr}.

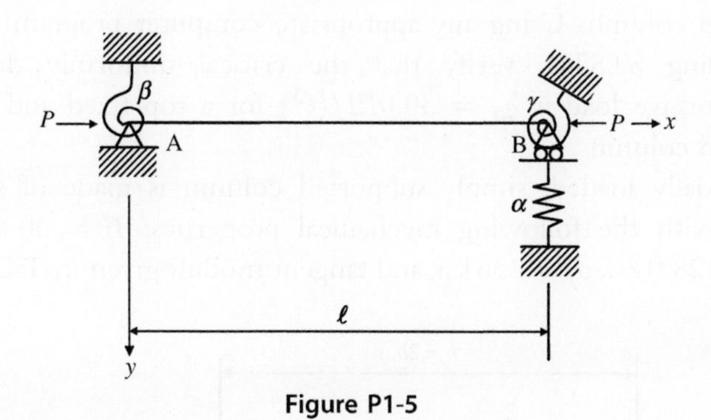

Figure P1-5

1.6 Turn-buckled threaded rods ($F_y = 50$ ksi, $F_u = 70$ ksi) are to be provided for the bracing system for a single-story frame shown in Fig. P1-6. Determine the diameter of the rod by the AISC Specifications, 13th edition, for each loading,

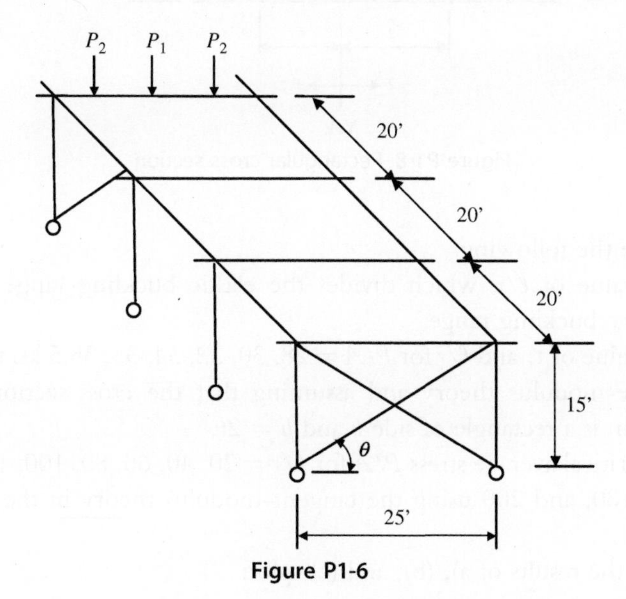

Figure P1-6

(a) when the typical factored loads on each girder are $P_1 = 250$ kips and $P_2 = 150$ kips, and

(b) when the frame is subjected to a horizontal wind load of intensity 20 psf on the vertical projected area.

1.7 Equation (1.10.22) gives the critical uniformly distributed axial compressive load as $q_{cr} = 52.5EI/(\ell^3)$ for a bottom fixed and top pinned column. Using any appropriate computer program available, including STSTB, verify that the critical uniformly distributed compressive load is $q_{cr} = 30.0EI/(\ell^3)$ for a top-fixed and bottom-pinned column.

1.8 An axially loaded, simply supported column is made of structural steel with the following mechanical properties: $E = 30 \times 10^3$ ksi, $\sigma_p = 28.0$ ksi, $\sigma_y = 36$ ksi, and tangent moduli given in Table 1-2.

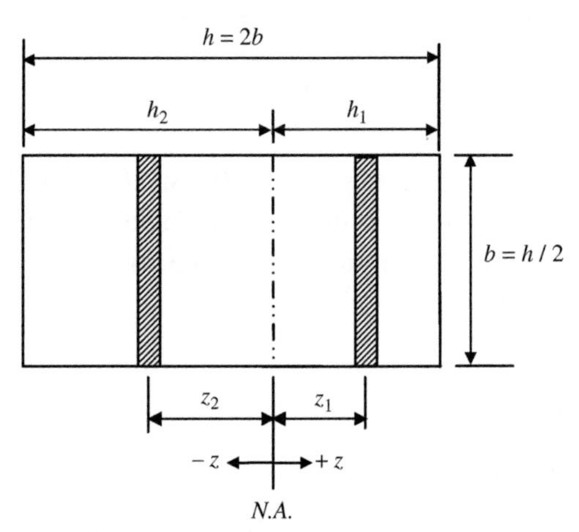

Figure P1-8 Rectangular cross section

Determine the following:

(a) The value of ℓ/r, which divides the elastic buckling range and the inelastic buckling range.

(b) The value of τ_r and ℓ/r for $P/A = 28, 30, 32, 34, 35, 35.5$ ksi using the double-modulus theory and assuming that the cross section of the column is a rectangle of side b and $h = 2b$.

(c) The critical average stress P/A for $\ell/r = 20, 40, 60, 80, 100, 120, 140, 160, 180$, and 200 using the tangent-modulus theory in the inelastic range.

From the results of a), (b), and (c), plot:

(d) The " $(P/A) - \tau_r$," curve for the double-modulus theory.

(e) The " $(P/A) - (\ell/r)$" curves, distinguishing the portion of the curve derived by the tangent-modulus theory from that derived by the double-modulus theory. Present short discussions.

(f) The current AISC LRFD Specification specifies (Chapter E) that the critical value of P/A for axially loaded column shall not exceed the following:

(i) For $\lambda_c \leq 1.5$ $\quad F_{cr} = (0.658^{\lambda_c^2})F_y$

(ii) For $\lambda_c > 1.5$ $\quad F_{cr} = \left[0.877/\lambda_c^2\right]F_y$

Plot these curves and superimpose them on the graph in (e) using double arguments (ℓ/r and λ_c) on the horizontal axis.

CHAPTER 2

Special Topics in Elastic Stability of Columns

Contents

Stability of Structures
ISBN 978-0-12-385122-2, doi:10.1016/B978-0-12-385122-2.10002-8

2.1. ENERGY METHODS

It has been shown that energy methods provide a convenient means of formulating the governing differential equation and necessary natural boundary conditions. The solutions that are obtained by solving the governing equations are exact within the framework of the theory (for example, classical beam theory) computing unknown forces and displacements in elastic structures. Besides providing convenient methods for computing unknown displacements and forces in structures, the energy principles are fundamental to the study of structural stability and structural dynamics. However, one of the greatest advantages of the energy methods is its usefulness in obtaining approximate solutions (Washizu 1974) in situations where exact solutions are difficult or impossible to obtain (Tauchert 1974). Hence, thorough familiarity with the energy principles will be an invaluable asset in the study of structural mechanics. Additional references for a more detailed treatment of energy methods may be found in Hoff (1956), Langhaar (1962), Fung and Tong (2001), Sokolnikoff (1956), and Shames and Dym (1985).

2.1.1. Preliminaries

Consider an infinitesimal rectangular parallelepiped at a point in a stressed body and let the stress vectors (traction vectors) $\boldsymbol{T}_1$, $\boldsymbol{T}_2$, and $\boldsymbol{T}_3$ represent the stress vectors[1] on each face perpendicular to the coordinate axes x_1, x_2, and x_3, respectively, as shown in Fig. 2-1. The component of the stress vector

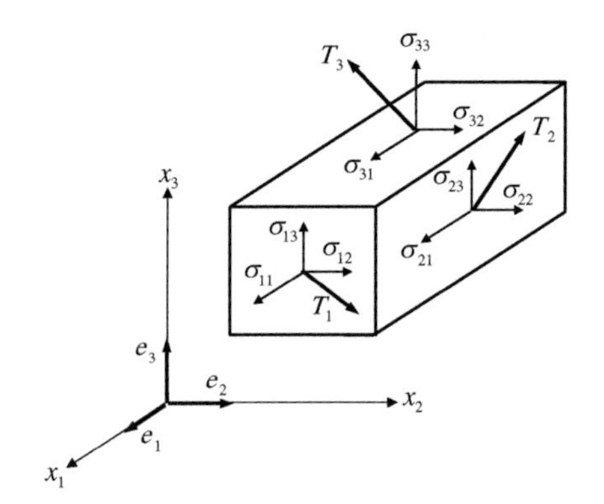

Figure 2-1 Stress vectors and their components

[1] Boldfaced-fonted quantities represent vectors.

$\boldsymbol{T}_i$ denoted by σ_{ij}[2] represents the projection of $\boldsymbol{T}_i$ on the face whose normal is x_j.

Hence,

$$\begin{aligned} \boldsymbol{T}_1 &= \sigma_{11}\boldsymbol{e}_1 + \sigma_{12}\boldsymbol{e}_2 + \sigma_{13}\boldsymbol{e}_3 \\ \boldsymbol{T}_2 &= \sigma_{21}\boldsymbol{e}_1 + \sigma_{22}\boldsymbol{e}_2 + \sigma_{23}\boldsymbol{e}_3 \\ \boldsymbol{T}_3 &= \sigma_{31}\boldsymbol{e}_1 + \sigma_{32}\boldsymbol{e}_2 + \sigma_{33}\boldsymbol{e}_3 \end{aligned} \tag{2.1.1}$$

Or in a compact form (index notation)

$$\boldsymbol{T}_i = \sigma_{ij}\boldsymbol{e}_j \tag{2.1.2}$$

Figure 2-2 shows the stress vector $\boldsymbol{T}$ acting on an arbitrary plane identified by $\boldsymbol{n}$ (unit outward normal to the plane), along with stress vectors $\boldsymbol{T}_i$ acting on the projected plane indicated by $\boldsymbol{e}_i$ and the body force per unit volume $\boldsymbol{f}$. The force acting on the arbitrary sloping plane ABC is $\boldsymbol{T}_n dA_n$, while the force on each projected plane is $-\boldsymbol{T}_i dA_i$ as each has a unit normal in the negative $\boldsymbol{e}_i$ direction.

Each projected area can be computed by

$$dA_i = dA_n \cos(\boldsymbol{n}, \boldsymbol{e}_i) = dA_n \boldsymbol{n} \cdot \boldsymbol{e}_i \tag{2.1.3}$$

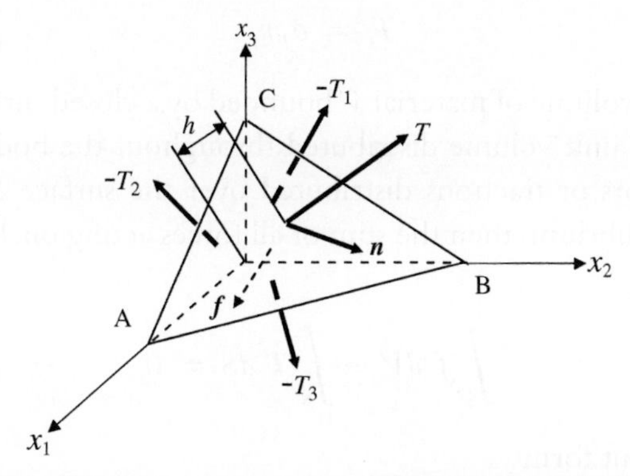

Figure 2-2 Stress vectors on an infinitesimal tetrahedron

[2] The first subscript i of σ_{ij} denotes the direction of the normal of the face on which the stress acts, and j indicates the direction of the stress itself. Denoting quantities with indices having a range of three is called an index notation (or indicial notation). The index notation is a mathematical agreement just to shorten the long write-ups adopted by Einstein in his general theory of relativity (Hjelmstad 2005, Wikipedia 2009). An index appearing once in a term is called a free index, and repeated subscripts are called dummy indices. The number of free indices determines how many quantities are represented by a symbol. Unless explicitly forbidden, a summation convention is executed on all dummy indexed quantities.

so that

$$dA_n = \frac{dA_i}{\boldsymbol{n} \cdot \boldsymbol{e}_i} = \frac{dA_i}{n_i} \tag{2.1.4}$$

where

$$n_i = \boldsymbol{n} \cdot \boldsymbol{e}_i = \cos(\boldsymbol{n}, \boldsymbol{e}_i) \tag{2.1.5}$$

is a direction cosine of $\boldsymbol{n}$.

Since the tetrahedron is in equilibrium, the resultant of all forces acting on it must vanish. Hence,

$$\left(\boldsymbol{T}_n - \boldsymbol{T}_i n_i + \frac{h}{3} \boldsymbol{f} \right) dA_n = 0 \tag{2.1.6}$$

Resolving $\boldsymbol{T}_n$ into Cartesian components ($\boldsymbol{T}_n = T_i \boldsymbol{e}_i$) and taking the limit as $h \to 0$, Eq. (2.1.6) reduces to

$$\boldsymbol{T}_n = T_i \boldsymbol{e}_i = \boldsymbol{T}_i n_i \tag{2.1.7}$$

Substituting Eq. (2.1.2) into Eq. (2.1.7) yields

$$T_i \boldsymbol{e}_i = \boldsymbol{T}_i n_i = \boldsymbol{T}_j n_j = \sigma_{ji} \boldsymbol{e}_i n_j \tag{2.1.8}$$

from which

$$T_i = \sigma_{ji} n_j \tag{2.1.9}$$

Consider a volume of material V bounded by a closed surface S. Let the body force per unit volume distributed throughout the body V be $\boldsymbol{f}$, and the stress vectors or tractions distributed over the surface S be $\boldsymbol{T}$. If the body is in equilibrium, then the sum of all forces acting on V must vanish; that is

$$\int_V \boldsymbol{f} \, dV + \int_S \boldsymbol{T} \, dS = 0 \tag{2.1.10}$$

or in component form

$$\int_V f_i \, dV + \int_S T_i \, dS = 0 \tag{2.1.11}$$

Equation (2.1.9) may be rewritten as

$$\int_S T_i \, dS = \int_S \sigma_{ji} n_j \, dS \tag{2.1.12}$$

Assuming that the components σ_{ji} and their first derivatives are continuous, the surface integral in Eq. (2.1.12) can be transformed into a volume integral using the divergence theorem, as

$$\int_S \sigma_{ji} n_j \, dS = \int_V \sigma_{ji,j} \, dV \tag{2.1.13}$$

From Eqs. (2.1.11), (2.1.12), and (2.1.13), it follows immediately that

$$\int_V \left(f_i + \sigma_{ji,j}\right) dV = 0 \tag{2.1.14}$$

Equation (2.1.14) can only be satisfied if the integrand is equal to zero at every point in the body. Hence,

$$f_i + \sigma_{ji,j} = 0 \tag{2.1.15}$$

Equation (2.1.15) presents three equations of equilibrium written in terms of stresses and body forces.

2.1.2. Principle of Virtual Work

If a structure is in equilibrium and remains in equilibrium while it is subjected to a virtual displacement, the external virtual work δW_E done by the external (real) forces acting on the structure is equal to the internal virtual work δW_I done by the internal stresses (due to real forces).

The external virtual work is

$$\delta W_E = \int_S T_i \delta u_i \, dS + \int_V f_i \delta u_i \, dV \tag{2.1.16}$$

Using Eq. (2.1.9) and the divergence theorem, the first term in Eq. (2.1.16) can be transformed into

$$\begin{aligned} \int_S T_i \delta u_i \, dS &= \int_S \sigma_{ij} n_j \delta u_i \, dS = \int_V \left(\sigma_{ij} \delta u_{i,j}\right) dV \\ &= \int_V \left(\sigma_{ij,j} \delta u_i + \sigma_{ij} \delta u_{i,j}\right) dV \end{aligned} \tag{2.1.17}$$

Substituting Eq. (2.1.17) into Eq. (2.1.16) yields

$$\delta W_E = \int_V \left[\left(\sigma_{ij,j} + f_i\right) \delta u_i + \sigma_{ij} \delta u_{i,j}\right] dV \tag{2.1.18}$$

Since the structure is in equilibrium, $f_i + \sigma_{ji,j} = 0$. Hence, Eq. (2.1.18) reduces to

$$\delta W_E = \int_V \sigma_{ij} \delta u_{i,j} \, dV \tag{2.1.19}$$

Recalling that $\delta e_{ij} = (\delta u_{i,j} + \delta u_{j,i})/2$ and $\delta u_{i,j} = \delta u_{j,i}$ leads to:

$$\sigma_{ij} \delta u_{i,j} = \sigma_{ij} \delta e_{ij} \tag{2.1.20}$$

This transforms Eq. (2.1.20) to

$$\int_V \sigma_{ij} \delta e_{ij} \, dV = \delta W_I = \delta U \tag{2.1.21}$$

Equation (2.1.21) describes the internal work done by the actual stresses (due to real forces) and virtual strains produced during the virtual displacement. The internal work is frequently referred to as the strain energy stored in the elastic body. From Eqs. (2.1.16), (2.1.20), and (2.1.21), one immediately obtains

$$\delta W_E = \int_S T_i \delta u_i \, dS + \int_V f_i \delta u_i \, dV = \int_V \sigma_{ij} \delta e_{ij} \, dV = \delta W_I = \delta U \tag{2.1.22}$$

Equation (2.1.22) is a mathematical statement of the principle of virtual work. The reverse of this principle is also true. That is, if $\delta W_E = \delta W_I$ for virtual displacement, then the body is in equilibrium (Tauchert 1974). The principle of virtual work is valid regardless of the material stress-strain relations as shown in the derivation.

2.1.3. Principle of Complementary Virtual Work

Figure 2-3 shows the stress-strain diagram of a nonlinearly elastic rod. The strain energy U represents the energy stored in a deformed elastic body; however, the physical interpretation of the complementary strain energy U^* is not clear.

The strain energy U in the rod is defined by

$$U = \int_V \left(\int_0^{e_{11}} \sigma_{11} de_{11} \right) dV = V \int_0^{e_{11}} \sigma_{11} de_{11} \tag{2.1.23}$$

The strain energy density or the strain energy per unit volume is equal to the area under the material's stress-strain curve (Fig. 2-3). The complementary strain energy U^* in the rod is defined by

$$U^* = \int_V \left(\int_0^{\sigma_{11}} e_{11} d\sigma_{11} \right) dV = V \int_0^{\sigma_{11}} e_{11} d\sigma_{11} \tag{2.1.24}$$

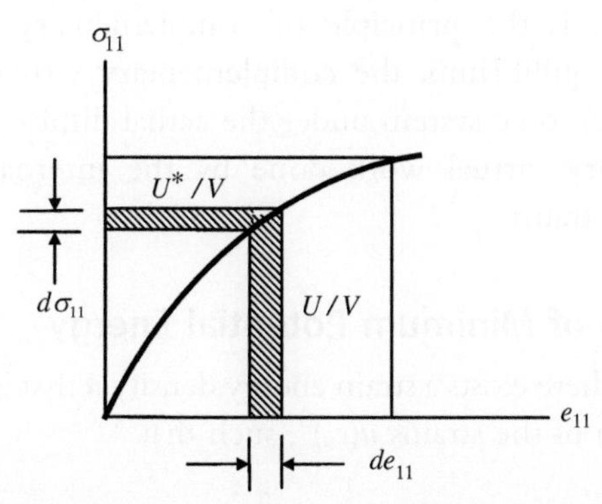

Figure 2-3 Stress-strain curve of a nonlinearly elastic rod

Therefore, the complementary strain energy density corresponds to the area above the stress-strain curve. For a linearly elastic material, the two areas are equal, and $U^* = U$. In order to maintain the generality, the structure under consideration is assumed to have arbitrary material properties. Consider an imaginary system of surface tractions δT_i and body forces δf_i that produce a state of stresses $\delta\sigma_{ij}$ inside the structure. If these quantities are in equilibrium, they must satisfy the equilibrium equations such that

$$\left(\delta\sigma_{ij}\right)_{,j} + \delta f_i = 0$$

The work done by these virtual forces during the actual displacements u_i is referred to as the complementary virtual work δW_E^* and is expressed as

$$\delta W_E^* = \int_S \delta T_i u_i \, dS + \int_V \delta f_i u_i \, dV \tag{2.1.25}$$

Proceeding in a manner similar to that used in the derivation of Eq. (2.1.22) with the roles of the actual and virtual quantities interchanged, one obtains the following:

$$\int_S \delta T_i u_i \, dS + \int_V \delta f_i u_i \, dV = \int_V \delta\sigma_{ij} e_{ij} \, dV \tag{2.1.26}$$

The right-hand side of Eq. (2.1.26) is denoted as

$$\delta U^* = \delta W_I^* = \int_V \delta\sigma_{ij} e_{ij} \, dV \tag{2.1.27}$$

From Eqs. (2.1.25) and (2.1.27), Eq. (2.1.26) is rewritten symbolically as

$$\delta W_E^* = \delta U^* = \delta W_I^* \tag{2.1.28}$$

Equation (2.1.28) is the principle of complementary virtual work. If a structure is in equilibrium, the complementary virtual work done by the external virtual force system under the actual displacement is equal to the complementary virtual work done by the internal virtual stresses under the actual strains.

2.1.4. Principle of Minimum Potential Energy

It is assumed that there exists a strain energy density u that is a homogeneous quadratic function of the strains $u(e_{ij})^3$, such that

$$\sigma_{ij} = \frac{\partial u}{\partial e_{ij}} \tag{2.1.29}$$

It is recalled that the virtual displacement field δu_i was a priori not related to the stress field σ_{ij} when applying the principle of virtual work. They are now related through a constitutive law expressed by Eq. (2.1.29). Substituting Eq. (2.1.29) into the principle of virtual work, Eq.(2.1.22), one obtains

$$\begin{aligned}\int_S T_i \delta u_i \, dS + \int_V f_i \delta u_i \, dV &= \int_V \frac{\partial u}{\partial e_{ij}} \delta e_{ij} \, dV = \int_V \delta^{(1)} u \, dV \\ &= \delta^{(1)} \int_V u \, dV = \delta^{(1)} U \end{aligned} \tag{2.1.30}$$

Notice that the variation and integration operations are interchanged. The (loss of) potential energy of the applied loads is now defined as a function of displacement field u_i and the applied loads.

$$V = -\int_V f_i u_i \, dV - \int_S T_i u_i \, dS \tag{2.1.31}$$

Taking the first variation of Eq. (2.1.31) gives

$$\delta^{(1)} V = -\int_V f_i \frac{\partial u_i}{\partial u_j} \delta u_j \, dV - \int_S T_i \frac{\partial u_i}{\partial u_j} \delta u_j \, dS$$

Noting that $\delta u_i / \delta u_j = \delta_{ij}$ and $\delta_{ij} = 1$ for $i = j$ and $\delta_{ij} = 0$ for $i \neq j$, the equation leads to

[3] This concept is attributed to George Green (1793–1841). It can be shown to be a positive definite quadratic function (Shames and Dym 1985; Sokolnikoff 1956).

$$\delta^{(1)}V = -\int_V f_i \delta u_j \, dV - \int_S T_i \delta u_j \, dS \tag{2.1.32}$$

From Eqs. (2.1.30) and (2.1.32), it follows immediately

$$\delta^{(1)}(U + V) = 0 \tag{2.1.33}$$

The quantity $(U + V)$ denoted by π is the total potential energy of the body and is given as

$$\Pi = \int_V \left(\int_0^{e_{ij}} \sigma_{ij} de_{ij} \right) dV - \int_V f_i u_i \, dV - \int_S T_i u_i \, dS \tag{2.1.34}$$

Equation (2.1.33) is known as the principle of minimum potential energy; it may be stated as follows (several variations are also used):

> *An elastic structure is in equilibrium if no change occurs in the total potential energy (stationary value) of the system when its displacement is changed by a small arbitrary amount.*

Equation (2.1.33) is the necessary condition for the stationary value of the total potential energy provided that (a) f_i and T_i are statically compatible and (b) the deformation field e_{ij}, to which the stress filed σ_{ij} is related through a constitutive law (not necessarily linear elastic) for elastic behavior, extremizes Π with respect to all other kinematically compatible, admissible deformation field (Shames and Dym 1985).

In the early days of the original development of the calculus of variations, the developers including Bernoulli (1654–1705), Euler (1707–1783), and Lagrange (1736–1813) did not consider the stationary value of the total potential energy as indeed a minimum until Legendre (1752–1833) postulated the so-called Legendre test seeking a mathematical rigor for a minimum (Forsyth 1960). A proof that Π actually assumes a minimum value in the case of stable equilibrium is illustrated below.

From Eqs. (2.1.33) and (2.1.34), it follows immediately that $\delta^{(1)}\Pi = \delta^{(1)}(U + V) = 0$. Hence

$$\delta^{(1)}\Pi = 0 = \int_V \frac{\partial u}{\partial e_{ij}} \delta e_{ij} \, dV - \int_V f_i \delta u_i \, dV - \int_S T_i \delta u_i \, dS \tag{2.1.35}$$

Using the constitutive relations of Eq. (2.1.29) and the strain-displacement relations for small displacement theory (Cauchy strain), the first integral of Eq. (2.1.35) is expanded to

$$\int_V \frac{\partial u}{\partial e_{ij}} \delta e_{ij}\, dV = \int_V \frac{1}{2} \sigma_{ij} \delta \left(u_{i,j} + u_{j,i} \right) dV$$

$$= \int_V \left[\frac{1}{2} \sigma_{ij} \delta \left(u_{i,j} \right) + \frac{1}{2} \sigma_{ij} \delta \left(u_{j,i} \right) \right] dV \tag{2.1.36}$$

Noting that $\sigma_{ij} = \sigma_{ji}$ and interchanging the dummy indices j and i, the right-hand side of Eq. (2.1.36) is expanded to

$$\int_V \frac{\partial u}{\partial e_{ij}} \delta e_{ij}\, dV = \int_V \left[\frac{1}{2} \sigma_{ij} \delta \left(u_{i,j} \right) + \frac{1}{2} \sigma_{ij} \delta \left(u_{j,i} \right) \right] dV = \int_V \sigma_{ij} \delta \left(u_{i,j} \right)\, dV$$

$$= \int_V \sigma_{ij} \delta \left(u_i \right)_{,j}\, dV = \int_V \left(\sigma_{ij} \delta u_i \right)_{,j}\, dV - \int_V \sigma_{ij,j} \delta u_i\, dV$$

$$= \int_S \sigma_{ij} \delta u_i n_j\, dS - \int_V \sigma_{ij,j} \delta u_i\, dV$$

Substituting this into Eq. (2.1.35) yields

$$\int_S \sigma_{ij} \delta u_i n_j\, dS - \int_V \sigma_{ij,j} \delta u_i\, dV - \int_V f_i \delta u_i\, dV - \int_S T_i \delta u_i\, dS = 0$$

or

$$\int_S \left(\sigma_{ij} n_j - T_i \right) \delta u_i\, dS - \int_V \left(\sigma_{ij,j} + f_i \right) \delta u_i\, dV = 0$$

This must be true for all δu_i. Then it follows that

$$\sigma_{ij,j} + f_i = 0 \tag{2.1.15}$$

and

$$\sigma_{ij} n_j = T_i \tag{2.1.9}$$

The Euler-Lagrange equations are the equations of equilibrium, and the necessary boundary conditions are embedded into the Cauchy formula Eq. (2.1.9). Hence it has been proved that $\delta^{(1)} \Pi = 0$ is a sufficient condition for equilibrium (Shames and Dym 1985). If it can be shown that the total potential energy of an admissible state having a displacement field $u_i + \delta u_i$ and a corresponding strain field $e_{ij} + \delta e_{ji}$ is always greater than that of the equilibrium state, then it suffices that the total potential energy Π is a local minimum for the equilibrium configuration.

$$\Pi_{e_{ij}+\delta e_{ij}} - \Pi_{e_{ij}} = \int_V \left[u\left(e_{ij}+\delta e_{ij}\right) - u\left(e_{ij}\right)\right] dV - \int_V f_i \delta u_i \, dV - \int_S T_i \delta u_i \, dS \tag{2.1.37}$$

Expanding $u\left(e_{ij}+\delta e_{ij}\right)$ by a Taylor series gives

$$u\left(e_{ij}+\delta e_{ij}\right) = u\left(e_{ij}\right) + \frac{\partial u}{\partial e_{ij}}\delta e_{ij} + \frac{1}{2}\frac{\partial^2 u}{\partial e_{ij}\partial e_{kl}}\delta e_{ij}\delta e_{kl} + \cdots \tag{2.1.38}$$

Substituting Eq. (2.1.38) into Eq. (2.1.37) gives

$$\begin{aligned}
\Pi e_{ij} + \delta e_{ij} - \Pi_{e_{ij}} &= \int_V \frac{\partial u}{\partial e_{ij}}\delta e_{ij} \, dV - \int_V f_i \delta u_i \, dV - \int_S T_i \delta u_i \, dS \\
&\quad + \int_V \frac{1}{2}\frac{\partial^2 u}{\partial e_{ij}\partial e_{kl}}\delta e_{ij}\delta e_{kl} \, dV + \cdots \\
&= \delta^{(1)} + \int_V \frac{1}{2}\frac{\partial^2 u}{\partial e_{ij}\partial e_{kl}}\delta e_{ij}\delta e_{kl} \, dV + \cdots \\
&= 0 + \delta^{(2)} + \cdots
\end{aligned}$$

$$\delta^{(2)}\Pi = \int_V \frac{1}{2}\frac{\partial^2 u}{\partial e_{ij}\partial e_{kl}}\delta e_{ij}\delta e_{kl} \, dV \tag{2.1.39}$$

It will be demonstrated that the integrand of Eq. (2.1.39) is $u(\delta_{\mathrm{eij}})$ for $e_{ij} = 0$. Examination of Eq. (2.1.38) in association with $e_{ij} = 0$ reveals that the first term is a constant throughout the body and is taken to be zero, so that the strain energy vanishes in the unrestrained body. By definition $\partial u/\partial e_{ij}$ in the second term is stress σ_{ij}. The stress in the unrestrained state must be equal to zero. Considering up to second-order terms, it gives

$$u\left(\delta e_{ij}\right) = \frac{1}{2}\left(\frac{\partial^2 u}{\partial e_{ij}\partial e_{kl}}\right)_{e_{ij}=0} \delta e_{ij}\delta e_{kl}$$

Hence, Eq. (2.1.39) can be written as

$$\delta^{(2)}\Pi = \int_V u\left(\delta e_{ij}\right) \, dV$$

Since u is a positive definite function, the second variation of the total potential energy is positive. Hence, the total potential energy is a minimum for the equilibrium state $e_{ij} = 0$ when compared to all other neighboring admissible deformation fields. Fung and Tong (2001), Love (1944), Saada (1974), Shames and Dym (1985), and Washizu (1974) use logic similar to that shown above in the proof of the nature of the total potential energy being a minimum. It appears that Sokolnikoff (1956) did not impose $e_{ij} = 0$ to show that Π actually assumes a minimum value.

2.1.5. Principle of Minimum Complementary Potential Energy

Parallel to the concept of the strain energy density introduced in Eq. (2.1.29), it is assumed that there exists the complementary energy density function u^* defined for elastic bodies as function of stress such that

$$\frac{\partial u^*}{\partial \sigma_{ij}} = e_{ij} \tag{2.1.40}$$

Substituting Eq. (2.1.40) into Eq. (2.1.26) gives

$$\int_S \delta T_i u_i \, dS + \int_V \delta f_i u_i \, dV = \int_V \delta \sigma_{ij} \frac{\partial u^*}{\partial \sigma_{ij}} \, dV \tag{2.1.41}$$

As per Eq. (2.1.27), the right-hand side of Eq. (2.1.41) is δU^*, the first variation of the complementary energy for the structure. A complementary potential energy function is defined by

$$V^* = -\int_V u_i f_i \, dV - \int_S u_i T_i \, dV$$

for which the first variation is given by

$$\delta V^* = -\int_V u_i \delta f_i \, dV - \int_S u_i \delta T_i \, dS \tag{2.1.42}$$

From Eqs. (2.1.41) and (2.1.42), it can be concluded that

$$\delta \Pi^* = \delta(U^* + V^*) = 0 \tag{2.1.43}$$

Equation (2.1.43) is the principle of total complementary energy, and Π^* is given by

$$\Pi^* = \int_V \sigma_{ij} e_{ij} \, dV - \int_V u_i f_i \, dV - \int_S u_i T_i \, dV \tag{2.1.46}$$

It may be shown that the total complementary energy is a minimum for the proper stress field following a procedure similar to that used in the principle of minimum total potential energy.

2.1.6. Castigliano Theorem, Part I

The principle of minimum total potential energy can be used to derive the Castigliano theorem, part I,[4] which is extremely useful in the analysis of elastic structures. For a structure in equilibrium under a set of discrete generalized forces Q_i $(i = 1, 2, ..., n)$, the total potential energy is given by

$$\Pi = U(\Delta_i) - \sum_{i=1}^{n} Q_i \Delta_i \tag{2.1.47}$$

For equilibrium the first variation of Π, found by varying Δ_i, must be equal to zero.

$$\begin{aligned} \delta\left[U(\Delta_i) - \sum_{i=1}^{n} Q_i \Delta_i\right] &= \sum_{i=1}^{n}\left(\frac{\partial U}{\partial \Delta_i}\delta\Delta_i - Q_i \delta\Delta_i\right) \\ &= \sum_{i=1}^{n}\left(\frac{\partial U}{\partial \Delta_i} - Q_i\right)\delta\Delta_i = 0 \end{aligned} \tag{2.1.48}$$

Since the variations $\delta\Delta_i$ are arbitrary, the quantities in each parenthesis must vanish; hence,

$$\frac{\partial U}{\partial \Delta_i} = Q_i \quad i = 1, 2, ..., n \tag{2.1.49}$$

Equation (2.1.49) is the Castigliano's theorem, part I. It states that if the strain energy U stored in an elastic structure is expressed as a function of the generalized displacements Δ_i, then the first partial derivative of U with respect to any one of the generalized displacements Δ_i is equal to the corresponding generalized force Q_i.

As the stiffness influence coefficient k_{ij} is defined as the generalized force required at i for a unit displacement at j while suppressing all other generalized displacements, k_{ij} can be expressed as

$$k_{ij} = \frac{\partial Q_i}{\partial \Delta_j} \tag{2.1.50}$$

Using Eq. (2.1.49), it can be rewritten as

$$k_{ij} = \frac{\partial^2 U}{\partial \Delta_i \partial \Delta_j} \tag{2.1.51}$$

[4] Carlo Alberto Castigliano (1847–1884) presented his famous theorem in 1873 in his thesis for the engineer's degree at Turin Polytechnical Institute.

2.1.7. Castigliano Theorem, Part II

For an elastic (not necessarily linearly elastic) structure that is in equilibrium under a system of applied generalized forces Q_i, the principle of minimum complementary energy states that

$$\delta\Pi^* = \delta(U^* + V^*) = 0 \tag{2.1.52}$$

Assuming that the complementary strain energy U^* is expressed as a function of Q_i, then Eq. (2.1.52) may be rewritten as

$$\begin{aligned}\delta\Pi^* = \delta(U^* + V^*) &= \sum_{i=1}^{n}\left(\frac{\partial U^*}{\partial Q_i}\delta Q_i - \Delta_i \delta Q_i\right) \\ &= \sum_{i=1}^{n}\left(\frac{\partial U^*}{\partial Q_i} - \Delta_i\right)\delta Q_i = 0\end{aligned} \tag{2.1.53}$$

Since δQ_i are arbitrary, Eq. (2.1.53) requires that

$$\frac{\partial U^*}{\partial Q_i} = \Delta_i \quad i = 1, 2, ..., n \tag{2.1.54}$$

Equation (2.1.54) is known as the Engesser[5] theorem, derived by Friedrich Engesser in 1889 (Tauchert 1974) and is valid for any elastic structure. If the structure is linearly elastic, the strain energy U and the complementary strain energy U^* are equal, and the Castigliano theorem, part II results.

$$\frac{\partial U}{\partial Q_i} = \Delta_i \quad i = 1, 2, ..., n \tag{2.1.55}$$

Equation (2.1.55) states that if the strain energy U in a linearly elastic structure is expressed as a function of the generalized forces Q_i, then the partial derivative of U with respect to the generalized force Q_i is equal to the corresponding displacement Δ_i. The flexibility coefficient of a linearly elastic structure is given by

$$f_{ij} = \frac{\partial U}{\partial Q_i \partial Q_j} \tag{2.1.56}$$

2.1.8. Summary of the Energy Theorems

Table 2-1 summarizes the energy theorem derived here. It is noted that a duality exists between those principles and theorems involving generalized

[5] Engesser (1848–1931) was a German engineer who introduced the concept of complementary energy (Fung and Tong 2001).

Table 2-1 Variational Principles (After Tauchert, *Energy Principles in Structural Mechanics*, McGraw-Hill, 1974). Reproduced by permission.

Displacement Methods	Force Methods
Principle of Virtual Work	**Principle of Complementary Virtual Work**
$\delta W_E = \delta U$	$\delta W_E^* = \delta U^*$
$\delta W_E = \sum_{i=1}^{n} Q_i \delta \Delta_i$	$\delta W_E^* = \sum_{i=1}^{n} \Delta_i \delta Q_i$
$\delta U = \int_V \sigma_{ij} \delta e_{ij}\, dV$	$\delta U^* = \int_V e_{ij} \delta \sigma_{ij}\, dV$
Principle of Minimum Potential Energy	**Principle of Minimum Complementary Energy**
$\delta \Pi = \delta(U + V) = 0$	$\delta \Pi^* = \delta(U^* + V^*) = 0$
$U = \int_V (\int_0^{e_{ij}} \sigma_{ij} de_{ij})\, dv$	$U^* = \int_V (\int_0^{\sigma_{ij}} e_{ij} d\sigma_{ij})\, dv$
$\boldsymbol{U = \int_V \left(v e_{ij} e_{ij} + \frac{\lambda}{2} e_{kk}^2 \right) dV = U^*}$	$\boldsymbol{U^* = \int_V \left(\frac{1+\mu}{2E} \sigma_{ij}\sigma_{ij} - \frac{\mu}{2E}\sigma_{kk}^2 \right) dV = U}$
$V = -\sum_{i=1}^{n} Q_i \Delta_i$	$V^* = -\sum_{i=1}^{n} Q_i \Delta_i$
Castigliano Theorem, Part I	**Castigliano Theorem, Part II**
$Q_i = \frac{\partial U}{\partial \Delta_i}$	$\Delta_i = \frac{\partial U^*}{\partial Q_i} = \frac{\partial U}{\partial Q_i}$
$\boldsymbol{k_{ij} = \frac{\partial^2 U}{\partial \Delta_i \partial \Delta_j}}$	$\boldsymbol{f_{ij} = \frac{\partial^2 U}{\partial Q_i \partial Q_j}}$

Notes: The Lamé constants λ and v in the table are given by

$$\lambda = \frac{\mu E}{(1+\mu)(1-2\mu)}$$

and

$$v = \frac{E}{2(1+\mu)}$$

Terms in "bold font" are valid for linearly elastic materials only.

displacements as the varied quantities (displacement methods) and those involving variations in the generalized forces (force methods). Principles and theorems related to the principle of virtual work are grouped as displacement methods, and those related to the principle of the complementary virtual work are grouped as force methods. These equations apply to nonlinear as well as linearly elastic materials, except where noted otherwise in Table 2-1.

2.2. STABILITY CRITERIA

The stability criteria must be established in order to answer the question of whether a structure is in stable equilibrium under a given set of loadings. If upon releasing the structure from its virtually displaced state the structure returns to its previous configuration, then the structure is said to be in stable equilibrium. On the other hand if the structure does not return to its undisturbed state following the release of the virtual displacements, the condition is either neutral equilibrium or unstable equilibrium. Stability can also be defined in terms of the total potential energy Π of the structure. Recall that Π is the sum of the strain energy U stored in the deformed elastic body and the loss of the potential of the generalized external forces V. If the total potential energy increases during a virtual displacement, then the equilibrium configuration is defined to be stable; if Π decreases or remains unchanged, the configuration is unstable.

The stability criteria can also be expressed in mathematical form. For simplicity it is assumed that the structure's deformation is characterized by a finite number of generalized displacements Δ_i.

$$\lambda = \frac{\mu E}{(1+\eta)(1-2\mu)}$$

$$\nu = \frac{E}{2(1+\mu)}$$

If the structure is given a virtual displacement $\delta\Delta_i$, then it is possible to write the total potential energy in a Taylor series expansion about Δ_i. Consider, for example, a two-degree-of-freedom system.

$$\begin{aligned}\Pi(\Delta_1+\delta\Delta_1,\Delta_2+\delta\Delta_2) &= \Pi(\Delta_1,\Delta_2)+\frac{\partial\Pi}{\partial\Delta_1}\delta\Delta_1+\frac{\partial\Pi}{\partial\Delta_2}\delta\Delta_2 \\ &+\frac{1}{2!}\left[\frac{\partial^2\Pi}{\partial\Delta_1^2}(\delta\Delta_1)^2+2\frac{\partial^2\Pi}{\partial\Delta_1\partial\Delta_2}\delta\Delta_1\delta\Delta_2+\frac{\partial^2\Pi}{\partial\Delta_2^2}(\delta\Delta_2)^2\right]+\cdots\end{aligned} \tag{2.2.1}$$

The change in potential energy is then

$$\Delta\Pi = \delta\Pi+\frac{1}{2!}\delta^2\Pi+\cdots \tag{2.2.2}$$

where the first variation is equal to zero by virtue of the principle of the minimum total potential energy.

$$\delta\Pi = \frac{\partial\Pi}{\partial\Delta_1}\delta\Delta_1+\frac{\partial\Pi}{\partial\Delta_2}\delta\Delta_2 = 0 \tag{2.2.3}$$

and the second variation is

$$\delta^2\Pi = \delta(\delta\Pi) = \frac{\partial^2\Pi}{\partial\Delta_1^2}(\delta\Delta_1)^2+2\frac{\partial^2\Pi}{\partial\Delta_1\partial\Delta_2}\delta\Delta_1\delta\Delta_2+\frac{\partial^2\Pi}{\partial\Delta_2^2}(\delta\Delta_2)^2 \tag{2.2.4}$$

Note that the sign of $\Delta\Pi$ in Eq. (2.2.2) is determined by the first nonvanishing term in the Taylor's expansion. Since $\delta\Pi = 0$, the second variation is the relevant term. If $\delta^2\Pi$ is positive, then $\Delta\Pi$ is positive, Π is a local minimum, and the equilibrium condition is stable. The special case in which the second variation is zero corresponds to a state known as neutral equilibrium. When a structure that is in neutral equilibrium is released from a virtual displacement, there is no net restoring force present, and the system remains in its virtual displaced state. Hence, by the first definition of stability, neutral equilibrium is a special case of unstable equilibrium. The criteria for stability are summarized as follows:

$$\begin{aligned} \Delta\Pi &> 0 \quad \text{stable equilibrium} \\ \Delta\Pi &= 0 \quad \text{neutral equilibrium} \\ \Delta\Pi &< 0 \quad \text{unstable equilibrium} \end{aligned} \tag{2.2.5}$$

If the potential energy Π is quadratic in the displacements Δ_i, which is the case when the structure is linearly elastic and the deformations are small, then all variations higher than the second are necessarily zero. In this case the type of equilibrium is governed by the following conditions:

$$\begin{aligned} \delta^2\Pi &> 0 \quad \text{stable equilibrium} \\ \delta^2\Pi &= 0 \quad \text{neutral equilibrium} \\ \delta^2\Pi &< 0 \quad \text{unstable equilibrium} \end{aligned} \tag{2.2.6}$$

Equation (2.2.6) is called the sufficient condition. A rigid body stability concept can be illustrated as follows:

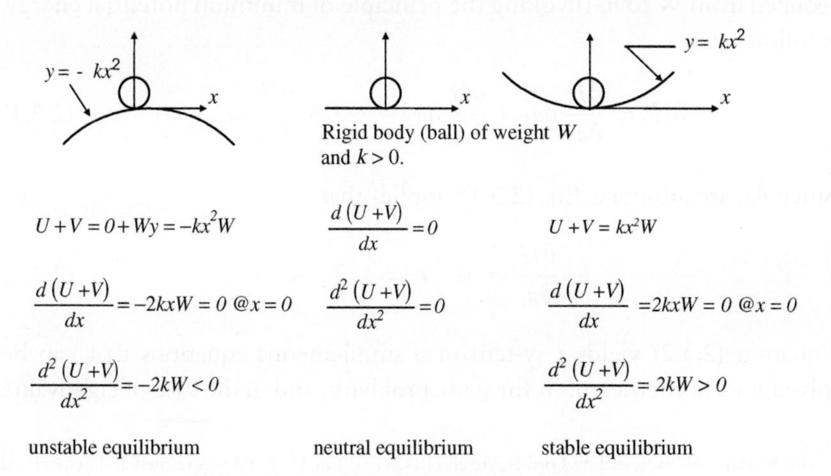

y = - kx²
x
x
x
y = kx²
Rigid body (ball) of weight W and k > 0.
U + V = 0 + Wy = −kx²W
d(U + V)/dx = 0
U + V = kx²W
d(U + V)/dx = −2kxW = 0 @x = 0
d²(U + V)/dx² = 0
d(U + V)/dx = 2kxW = 0 @x = 0
d²(U + V)/dx² = −2kW < 0
d²(U + V)/dx² = 2kW > 0
unstable equilibrium
neutral equilibrium
stable equilibrium

Figure 2-4 Concept of rigid body equilibrium

2.3. RAYLEIGH-RITZ METHOD

The energy methods introduced in Section 2.1 are a convenient means of computing unknown forces and displacements in elastic structures. They can be the basis of deriving the governing differential equations and required boundary conditions of the problem. They are also the starting point of many modern matrix/finite element methods. The solutions that are obtained using these methods are exact within the framework of the theory (for example, classical beam theory). Energy methods are also used to derive approximate solutions in situations where exact solutions are difficult or nearly impossible to obtain. The most widely known and used approximate procedure is the Rayleigh-Ritz method,[6] in which the structure's displacement field is approximated by functions that include a finite number of independent coefficients (or natural coordinates; one for the Rayleigh method and more than one for the Rayleigh-Ritz method). The assumed solution functions must satisfy the kinematic boundary conditions (otherwise, the convergence is not guaranteed, no matter how many functions are assumed), but they need not satisfy the natural boundary conditions (if they satisfy the natural boundary condition, a fairly good solution accuracy can be expected). The unknown constants in the assumed functions are determined by invoking the principle of minimum potential energy. Suppose, for example, the assumed function has n independent constants a_i ($i = 1, 2,\ldots,n$). Since the approximate state of deformation of the structure is characterized (amplitude as well as shape) by these n constants, the degrees of freedom of the structure have been reduced from ∞ to n. Invoking the principle of minimum potential energy, it follows that

$$\delta\Pi = \frac{\partial \Pi}{\partial a_1}\delta a_1 + \frac{\partial \Pi}{\partial a_2}\delta a_2 + \cdots + \frac{\partial \Pi}{\partial a_n}\delta a_n = 0 \qquad (2.3.1)$$

Since δa_i are arbitrary, Eq. (2.3.1) implies that

$$\frac{\partial \Pi}{\partial a_i} = 0 \quad i = 1, 2\ldots, n \qquad (2.3.2)$$

Equation (2.3.2) yields a system of n simultaneous equations that can be solved for the coefficients a_i for static problems, and in the case of eigenvalue

[6] This method was proposed by Lord Rayleigh (1842–1919) in 1877 and was refined and generalized by Walter Ritz (1878–1909) in 1908 (Tauchert 1974).

problems, the determinant (characteristic determinant) for the unknown constants is set equal to zero for the n eigenvalues.

Before illustrating detailed applications of the Rayleigh-Ritz method, a few general comments are in order. Although the accuracy is generally improved by increasing the number of independent functions, the computation efforts increase proportionally to the square of the number of independent functions. The type of functions to be selected for a particular problem is based on an intuitive idea of what the true deformation looks like. Trigonometric or polynomial functions are frequently used simply because of the ease of analysis involved. By virtue of using the principle of minimum potential energy, all approximate solutions make the structure stiffer than what it is. Consequently, the displacements predicted by the Rayleigh-Ritz method are always smaller than exact ones, and eigenvalues are greater than those predicted by exact solution methods.

Finally, if the approximate displacements are used to evaluate internal forces or stresses, the latter results should be viewed with caution because the stress components depend on the derivatives of displacements. Although displacements themselves may be reasonably accurate, their derivatives may not be the case. In fact, the higher the derivatives, the accuracy involved is further deteriorated. In a similar fashion, the accuracy of eigenvalues associated with higher mode eigenvectors deviates much more rapidly than those associated with lower mode eigenvectors.

Example 1 Consider a both-ends pinned column shown in Fig. 2-5. The strain energy stored in the deformed body is

$$U = \frac{1}{2}\int_0^\ell \frac{M^2}{EI}dx = \frac{1}{2}\int_0^\ell \frac{(-EIy'')^2}{EI}dx = \frac{EI}{2}\int_0^\ell (y'')^2 dx$$

The potential energy of the applied load is

$V = -P\Delta\ell$ (the reason for the negative sign: as $\Delta\ell$ increases, V decreases)

$$ds^2 = dx^2 + dy^2 = \left[1 + \left(\frac{dy}{dx}\right)^2\right] dx^2 \Rightarrow ds = \sqrt{1 + (y')^2}\,dx$$

It is noted that the static deformation has already taken place and the examination is being conducted on the neighboring equilibrium configuration. Hence, the shortening of the column, $\Delta\ell$, is entirely due to the flexural action.

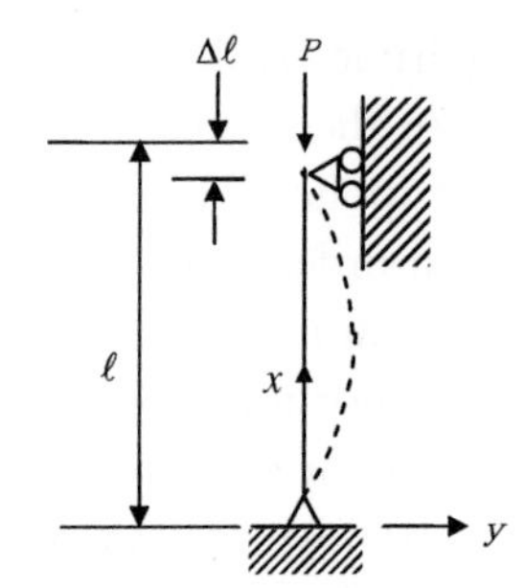

Figure 2-5 Simple column model

$$\Delta\ell = \int_0^\ell ds - \int_0^\ell dx = \int_0^\ell \sqrt{1+(y')^2}dx - \int_0^\ell dx$$

$$= \int_0^\ell \left[1+\frac{1}{2}(y')^2 + ..\right]dx - \int_0^\ell dx \doteq \frac{1}{2}\int_0^\ell (y')^2 dx$$

$$V = -\frac{P}{2}\int_0^\ell (y')^2 dx$$

Invoking the principle of minimum potential energy, it follows immediately that

$$\delta\Pi = \delta U + \delta V = \delta\left[\frac{EI}{2}\int_0^\ell (y'')^2 dx - \frac{P}{2}\int_0^\ell (y')^2 dx\right] = 0$$

In order to use the energy method, one must know the equation of the deformed shape of the structure. In general, the exact displacement function is not known at this stage of the solution. Experience has shown, however, that any assumed reasonable displacement shape function that satisfies at least the geometric boundary conditions leads a very fast-converging upper-bound solution.

It is assumed that the column shown in Fig. 2-5 is prismatic just for simplicity. An example having a nonprismatic member will be illustrated later. Assume the solution function to be of the form $y = \sum_{i=1}^{n} a_i\phi_i = \sum_{i=1}^{n} a_i \sin(i\pi x/\ell)$. This assumed y satisfies not only the *GBC* but also the *NBC*. Hence, it will lead to the exact solution or a fast-converging one.

$$\Pi = \frac{EI}{2}\int_0^{\ell}(y'')^2 dx - \frac{P}{2}\int_0^{\ell}(y')^2 dx$$

$$= \frac{EI}{2}\int_0^{\ell}\left[\sum_{i=1}^{n}(-1)\frac{i^2\pi^2}{\ell^2}a_i \sin\frac{i\pi x}{\ell}\right]^2 dx - \frac{P}{2}\int_0^{\ell}\left(-\sum_{i=1}^{n}\frac{i\pi}{\ell}a_i \cos\frac{i\pi x}{\ell}\right)^2 dx$$

$$= \frac{EI}{2}\left(\frac{\pi^4}{2\ell^3}\sum_{i=1}^{n}i^4 a_i^2\right) - \frac{P}{2}\left(\frac{\pi^2}{2\ell}\sum_{i=1}^{n}i^2 a_i^2\right)$$

Recall the following orthogonality of finite integrals of trigonometric functions: $\int_0^{\ell}(\sin^2 ax)\,dx = (\ell/2)$, $\int_0^{\ell}(\cos^2 ax)\,dx = (\ell/2)$, $\int_0^{\ell}(\sin ix)(\sin jx)\,dx = 0\ (i \neq j)$, and $\int_0^{\ell}(\cos ix)(\cos jx)\,dx = 0\ (i \neq j)$

$$\frac{\partial \Pi}{\partial a_i} = 0 = \frac{EI\pi^4}{4\ell^3}i^4(2a_i) - \frac{P}{2}\frac{\pi^2}{2\ell}i^2(2a_i) = \left(\frac{EI\pi^4}{\ell^2}i^2 - P\pi^2\right)a_i = 0$$

$$\text{As } a_i \neq 0,\ P_i = \frac{i^2\pi^2 EI}{\ell^2} \text{ or } (P_{cr})_{i=1} = \frac{\pi^2 EI}{\ell^2} \Leftarrow \text{exact solution}$$

2.4. THE RAYLEIGH QUOTIENT

Mikhlin (1964) proposes that the approximate solution of the eigenvalue problem usually reduces to the integration of a differential equation of the form

$$Lw - \lambda Mw = 0 \tag{2.4.1}$$

where w is the displacement that satisfies not only the differential equation, Eq. (2.4.2), but also certain homogeneous boundary conditions (this condition may preclude the cantilevered end condition), L and M are certain differential operators, and λ is an unknown numerical parameter. For the stability of a column, the governing differential equation is

$$\frac{d^2}{dx^2}\left(EI\frac{d^2 w}{dx^2}\right) = -P\frac{d^2 w}{dx^2} \tag{2.4.2}$$

For Eq. (2.4.2)

$$L \equiv \frac{d^2}{dx^2} EI \frac{d^2}{dx^2} \tag{2.4.3}$$

$$M \equiv -\frac{d^2}{dx^2} \tag{2.4.4}$$

$$\lambda = P \tag{2.4.5}$$

Equations (2.4.3) and (2.4.4) are self-adjoint (symmetric), positive definite operators for the usual end supports of columns. If a linear differential operator L has the following property, it is called a self-adjoint or symmetric operator:

$$(Lu, v) = (u, Lv) \tag{2.4.6}$$

The inner product of two functions g and h over the domain V is defined as

$$(g, h) \equiv \text{ inner product of } g \text{ and } h \equiv \int_V gh\, dv \tag{2.4.7}$$

An operator is said to be positive definite if the following inequality is valid for any function from its field of definition, $u(q) \neq 0$:

$$(Lu, u) > 0, \quad (Lu, u) \equiv 0 \quad \text{for } u(q) \equiv 0 \tag{2.4.8}$$

The reason why one is concerned whether or not a boundary-value problem has the properties of being self-adjoint (symmetric) and positive definite is that boundary-value problems having these properties are said to be properly posed, and there exists a unique solution to a properly posed boundary-value problem. An improperly posed boundary-value problem due to haphazardly or arbitrarily assigned boundary conditions is meaningless.

Multiplying both sides of Eq. (2.4.2) by w and integrating over the domain yields

$$\int_0^\ell w \frac{d^2}{dx^2}\left(EI \frac{d^2 w}{dx^2}\right) dx = -P \int_0^\ell w \frac{d^2 w}{dx^2} dx \tag{2.4.9}$$

Integrate the left-hand side of Eq. (2.4.9) by parts twice, as follows:

$$\int_0^\ell w \frac{d^2}{dx^2}\left(EI \frac{d^2 w}{dx^2}\right) dx = \int_0^\ell EI \left(\frac{d^2 w}{dx^2}\right)^2 dx + w \frac{d}{dx}\left(EI \frac{d^2 w}{dx^2}\right)\Bigg|_0^\ell - EI \frac{dw}{dx}\frac{d^2 w}{dx^2}\Bigg|_0^\ell$$

For simply supported, fixed, or cantilevered end conditions, the last two quantities are zero. Integrating the right-hand side of Eq. (2.4.9) gives

$$-P\int_0^\ell w\frac{d^2w}{dx^2}dx = P\int_0^\ell\left(\frac{dw}{dx}\right)^2 dx - Pw\frac{dw}{dx}\bigg|_0^\ell$$

The last expression vanishes for fixed and simple supports (not for the cantilevered end). Substituting the expanded integrals back into Eq. (2.4.2) gives

$$P = \frac{EI\int_0^\ell\left(d^2w/dx^2\right)^2 dx}{\int_0^\ell(dw/dx)^2 dx} \quad \text{(C1 method)} \tag{2.4.10}$$

It is noted that Eq. (2.4.10) works for cantilevered columns despite the fact that one of the concomitants is not zero.

As mentioned earlier, the error involved in the approximate solution propagates much faster in the higher order derivatives. In order to improve the critical value computed from the Rayleigh quotient, d^2w/dx^2 in the numerator is replaced by M/EI. Then

$$P_{cr} = \frac{\left(1/EI\right)\int_0^\ell M^2 dx}{\int_0^\ell(w')^2 dx} \quad \text{(C2 method)} \tag{2.4.11}$$

Example 1 Consider a pin-ended prismatic column shown in Fig. 2-6. Assume $w = ax(\ell - x)$, which satisfies the *GBC*.

$$w' = a(\ell - 2x), \quad w'' = -2a$$

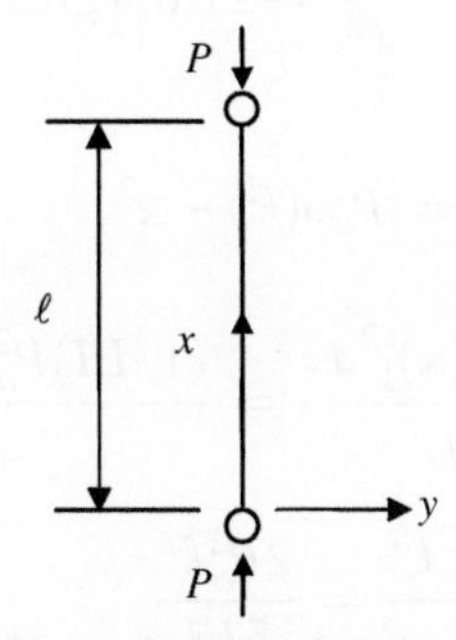

Figure 2-6 Pin-ended simple prismatic column

$$P_{cr} = \frac{4EIa^2 \int_0^\ell dx}{a^2 \int_0^\ell (\ell - 2x)^2 dx} = \frac{12EI}{\ell^2} \Leftarrow 21.6\% \text{ greater than Euler load}$$

For C2 method: $M = P_{cr}w = P_{cr}ax(\ell - x)$

$$P_{cr} = \frac{(1/EI) \int_0^\ell (P_{cr}w)^2 dx}{\int_0^\ell (w')^2 dx} = \frac{(P_{cr}^2 a^2/EI) \int_0^\ell (\ell x - x^2)^2 dx}{a^2 \int_0^\ell (\ell - 2x)^2 dx}$$

$$P_{cr} = \frac{(10EI)}{\ell^2} \Leftarrow \text{only 1.32\% greater than the exact solution,}$$

$$P_E = \frac{\pi^2 EI}{\ell^2} = \frac{9.8696EI}{\ell^2}$$

If the true deflection curve is used, both the C1 method and the C2 method lead to the same exact solution. However, if an approximate expression for the deflection curve is used for w, the error in w'' is considerably greater than the error in w or w'. Hence, C2 method gives a better solution than the C1 method does. In general, the energy method leads to the values of the critical load that are greater than the exact solution as a consequence of using the principle of minimum potential energy. Such greater values are called the upper-bound solution.

Example 2 Consider a prismatic cantilever column with the fixed support at $x = 0$.
Assume $w = ax^2$, which satisfies the *GBC*.

$$w' = 2ax, \quad w'' = 2a$$

$$P_{cr} = \frac{4EIa^2 \int_0^\ell dx}{4a^2 \int_0^\ell x^2 dx} = \frac{3EI}{\ell^2} \Leftarrow 21.6\% \text{ greater than exact load}$$

For the C2 method: $M(x) = P_{cr}a(\ell^2 - x^2)$

$$P_{cr} = \frac{(1/EI) \int_0^\ell [M(x)]^2 dx}{\int_0^\ell (w')^2 dx} = \frac{(1/EI)P_{cr}^2 a^2 \int_0^\ell (\ell^2 - x^2)^2 dx}{4a^2 \int_0^\ell x^2 dx}$$

$$= \frac{(8/15EI)P_{cr}^2 a^2 \ell^5}{4/3a^2\ell^3} = \frac{2P_{cr}^2 \ell^2}{5EI}$$

$$P_{cr} = 2.5\frac{EI}{\ell^2} \Leftarrow 1.32\% \text{ larger than exact load}$$

As can be seen here, Eq. (2.4.2) works equally well for a cantilever column, despite the fact that one of the concomitants, $w(dw/dx)$, does not vanish at the cantilevered end.

2.5. ENERGY METHOD APPLIED TO COLUMNS SUBJECTED TO DISTRIBUTED AXIAL LOADS

2.5.1. Cantilever Column

As illustrated in Section 1.10, this problem results in a governing differential equation with variable coefficients. In order to facilitate a closed-form solution, various ingenious schemes have been tried. Successful attempts reported include the application of power series, Bessel function, and Lommel function and their combination after a clever transformation. As demonstrated by Timoshenko and Gere (1961), the Rayleigh-Ritz method can effectively be applied to this problem with the desired accuracy of the solution by considering a number of independent functions.

Revisit the problem of buckling of a prismatic bar shown in Fig. 2-7 as considered in Section 1.10. The Rayleigh method can also be applied to the calculation of the critical value of the distributed compressive loads. As a first approximation of the deflection curve, the following equation may be tried:

$$y = \delta\left(1 - \cos\frac{\pi x}{2\ell}\right) \tag{2.5.1.1}$$

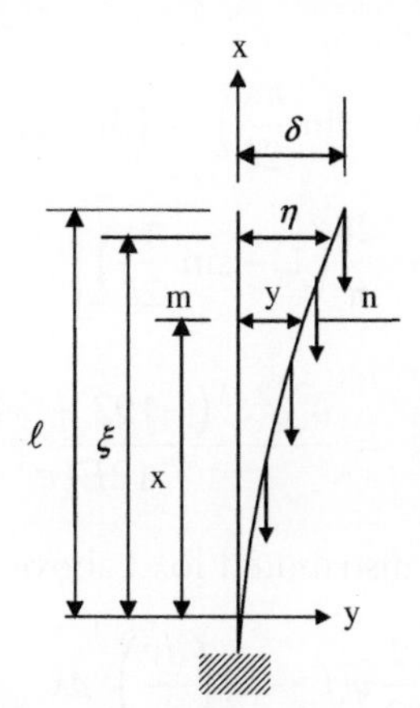

Figure 2-7 Cantilever column subjected to distributed axial load

Equation (2.5.1.1) is the exact solution curve for the case where buckling occurs under the concentrated load applied at the free end of the cantilever. In the case of a uniformly distributed axial load, the true curve is much more complicated as is shown in Section 1.10. Nevertheless, the curve of Eq. (2.5.1.1) satisfies the geometrical as well as the natural boundary conditions and, therefore, is expected to yield a fairly good approximated solution. The bending moment at any cross section *mn* is

$$M = \int_x^\ell q(\eta - y)d\xi \tag{2.5.1.2}$$

The deflection η is also expressed as

$$\eta = \delta\left(1 - \cos\frac{\pi\xi}{2\ell}\right) \tag{2.5.1.3}$$

Substituting Eqs. (2.5.1.1) and (2.5.1.3) into Eq. (2.5.1.2) gives

$$M = q\int_x^\ell (\eta - y)d\xi = q\left[\int_x^\ell \eta d\xi - y(\ell - x)\right] \tag{2.5.1.4}$$

The integral on the right-hand side of Eq. (2.5.1.4) is expanded to

$$\int_x^\ell \eta d\xi = \delta\int_x^\ell \left(1 - \cos\frac{\pi\xi}{2\ell}\right)d\xi = \delta\left[(\ell - x) - \frac{2\ell}{\pi}\sin\frac{\pi\xi}{2\ell}\Bigg|_x^\ell\right]$$

$$= \delta\left[(\ell - x) - \frac{2\ell}{\pi}\left(1 - \sin\frac{\pi x}{2\ell}\right)\right]$$

Hence

$$M = q\delta\left[(\ell - x) - \frac{2\ell}{\pi}\left(1 - \sin\frac{\pi x}{2\ell}\right) - \left(1 - \cos\frac{\pi x}{2\ell}\right)(\ell - x)\right]$$

$$= q\delta\left[(\ell - x)\cos\frac{\pi x}{2\ell} - \frac{2\ell}{\pi}\left(1 - \sin\frac{\pi x}{2\ell}\right)\right] \tag{2.5.1.5}$$

$$U = \frac{1}{2EI}\int_0^\ell M^2 dx = \frac{\delta^2 q^2 \ell^3\left(-192 + 54\pi + \pi^3\right)}{12EI\pi^3} \tag{2.5.1.6}$$

and the work done by the distributed load above the section *mn* is

$$\frac{1}{2}q(\ell - x)\left(\frac{dy}{dx}\right)^2 dx$$

The total loss of potential energy of the distributed load during buckling is

$$V = -\frac{1}{2}q\int_0^\ell (\ell - x)\left(\frac{dy}{dx}\right)^2 dx = -\frac{1}{2}q\delta^2\int_0^\ell (\ell - x)\left(\frac{\pi}{2\ell}\sin\frac{\pi x}{2\ell}\right)^2 dx$$
$$= -\frac{\delta^2 q}{32}(\pi^2 - 4)$$
$$\tag{2.5.1.7}$$

By virtue of the principle of minimum potential energy, it follows that

$$\frac{\partial \Pi}{\partial \delta} = \frac{\partial U}{\partial \delta} + \frac{\partial V}{\partial \delta} = 0 = \frac{\delta q^2 \ell^3(-192 + 54\pi + \pi^3)}{6EI\pi^3} - \frac{\delta q}{16}(\pi^2 - 4) = 0$$

$$q = \frac{(\pi^2 - 4)}{16}\frac{6EI\pi^3}{(-192 + 54\pi + \pi^3)\ell^3} = 7.888\frac{EI}{\ell^3} \tag{2.5.1.8}$$

Although Eq. (2.5.1.8) is only 0.65% greater than the exact solution, it would seem interesting to see how much the accuracy can be improved by taking one more term in the assumed displacement function. Consider the following function for the deflection of the cantilever shown in Fig. 2-7:

$$y = a\left(1 - \cos\frac{\pi x}{2\ell}\right) + b\left(1 - \cos\frac{3\pi x}{2\ell}\right) \tag{2.5.1.9}$$

Equation (2.5.1.9) also satisfied the geometric boundary conditions. As is done earlier, η is taken as

$$\eta = a\left(1 - \cos\frac{\pi\xi}{2\ell}\right) + b\left(1 - \cos\frac{3\pi\xi}{2\ell}\right) \tag{2.5.1.10}$$

The integral on the right-hand side of Eq. (2.5.1.4) is expanded to

$$\int_x^\ell \eta d\xi = \int_x^\ell \left[a\left(1 - \cos\frac{\pi\xi}{2\ell}\right) + b\left(1 - \cos\frac{3\pi\xi}{2\ell}\right)\right]d\xi$$
$$= a\left[(\ell - x) - \frac{2\ell}{\pi}\sin\frac{\pi\xi}{2\ell}\Big|_x^\ell\right] + b\left[(\ell - x) - \frac{2\ell}{3\pi}\sin\frac{3\pi\xi}{2\ell}\Big|_x^\ell\right]$$
$$= a\left[(\ell - x) - \frac{2\ell}{\pi}\left(1 - \sin\frac{\pi x}{2\ell}\right)\right] + b\left[(\ell - x)\right.$$
$$\left. + \frac{2\ell}{3\pi}\left(1 + \sin\frac{3\pi x}{2\ell}\right)\right]$$

Hence

$$M = q\left\{\begin{aligned} & a\left[(\ell - x) - \frac{2\ell}{\pi}\left(1 - \sin\frac{\pi x}{2\ell}\right) - \left(1 - \cos\frac{\pi x}{2\ell}\right)(\ell - x)\right] \\ & +b\left[(\ell - x) + \frac{2\ell}{3\pi}\left(1 + \sin\frac{3\pi x}{2\ell}\right) - \left(1 - \cos\frac{3\pi x}{2\ell}\right)(\ell - x)\right] \end{aligned}\right\}$$

$$= q\left\{a\left[(\ell - x)\cos\frac{\pi x}{2\ell} - \frac{2\ell}{\pi}\left(1 - \sin\frac{\pi x}{2\ell}\right)\right]\right.$$

$$\left. +b\left[(\ell - x)\cos\frac{3\pi x}{2\ell} + \frac{2\ell}{3\pi}\left(1 + \sin\frac{3\pi x}{2\ell}\right)\right]\right\} \tag{2.5.1.11}$$

$$\begin{aligned} U &= \frac{1}{2EI}\int_0^\ell M^2 dx \\ &= \frac{q^2\ell^3}{108\pi^3 EI}\left[\begin{aligned} & \left(-1728 + 486\pi + 9\pi^3\right)a^2 + \left(64 + 54\pi + 9\pi^3\right)b^2 \\ & + (384 - 9\pi)ab \end{aligned}\right] \end{aligned} \tag{2.5.1.12}$$

The total loss of potential energy of the distributed load during buckling is

$$\begin{aligned} V &= -\frac{1}{2}q\int_0^\ell (\ell - x)\left(\frac{dy}{dx}\right)^2 dx \\ &= -\frac{1}{2}q\int_0^\ell (\ell - x)\left(\frac{\pi}{2\ell}\sin\frac{\pi x}{2\ell}a + \frac{3\pi}{2\ell}\sin\frac{3\pi x}{2\ell}b\right)^2 dx \\ &= -\frac{1}{32}q\left[\left(\pi^2 - 4\right)a^2 + \left(9\pi^2 - 4\right)b^2 + 24ab\right] \end{aligned} \tag{2.5.1.13}$$

$$\Pi = U + V$$

$$\begin{aligned} \frac{\partial \Pi}{\partial a} = \frac{\partial U}{\partial a} + \frac{\partial V}{\partial a} &= \frac{q^2\ell^3}{108\pi^3 EI}\left[\left(-3456 + 972\pi + 18\pi^3\right)a + (384 - 9\pi)b\right] \\ & \quad - \frac{1}{32}q\left[\left(2\pi^2 - 8\right)a + 24b\right] = 0 \end{aligned}$$

$$\frac{\partial \Pi}{\partial b} = \frac{\partial U}{\partial b} + \frac{\partial V}{\partial b} = \frac{q^2\ell^3}{108\pi^3 EI}\left[\left(128 + 108\pi + 18\pi^3\right)b + (384 - 9\pi)a\right]$$

$$- \frac{1}{32}q\left[\left(18\pi^2 - 8\right)b + 24a\right] = 0$$

For a nontrivial solution (a and b cannot be equal to zero simultaneously), the determinant for the coefficient matrix for a and b must be equal to zero. Solving the resulting polynomial for the critical value yields

$$q_{cr} = 7.888 \frac{EI}{\ell^3} \tag{2.5.1.14}$$

In this case, the addition of an extra term in the assumed displacement function does not improve the accuracy up to the fourth effective digit. The numerical computation in the example has been carried out using **Maple®**

The uniform load $q\ell$ reduces the critical buckling load P applied at the cantilever tip. It is written in the form

$$P_{cr} = \frac{mEI}{\ell^2} \tag{2.5.1.15}$$

where the factor m is equal to $\pi^2/4$ when $q\ell$ is equal to zero and it approaches zero when $q\ell$ approaches the value given by Eq. (2.5.1.14). Using the notation

$$n = \frac{4q\ell^3}{\pi^2 EI}$$

the values of the coefficient m in Eq. (2.5.1.15) for values of n can be found in Timoshenko and Gere (1961). The following illustration is an example case of using the energy method to compute values of n and m interactively.

The moment due to the concentrated load P is

$$M_P = P(\delta - y) = \delta P \cos \frac{\pi x}{2\ell}$$

From Eq. (2.5.1.5).

$$M_q = q\delta\left[(\ell - x)\cos\frac{\pi x}{2\ell} - \frac{2\ell}{\pi}\left(1 - \sin\frac{\pi x}{2\ell}\right)\right]$$

$$M = M_P + M_q = \delta\left\{P\cos\frac{\pi x}{2\ell} + q\left[(\ell - x)\cos\frac{\pi x}{2\ell} - \frac{2\ell}{\pi}\left(1 - \sin\frac{\pi x}{2\ell}\right)\right]\right\}$$

$$\begin{aligned} U &= \frac{1}{2EI}\int_0^\ell M^2 dx \\ &= \frac{\delta^2}{2EI}\int_0^\ell \left\{P\cos\frac{\pi x}{2\ell} + q\left[(\ell - x)\cos\frac{\pi x}{2\ell} - \frac{2\ell}{\pi}\left(1 - \sin\frac{\pi x}{2\ell}\right)\right]\right\}^2 dx \\ &= \delta^2\ell\left(-12\pi\ell\pi qP + 54\ell^2\pi q^2 - 192\ell^2 q^2 + \ell^2\pi^3 q^2 + 3\pi^3 P^2\right. \\ &\quad \left. + 3\ell\pi^3 qP\right)/\left(12EI\pi^3\right) \\ &= \frac{\delta^2\ell}{12EI\pi^3}\left(93.01883P^2 + 55.3197182\ell qP + 8.65228\ell^2 q^2\right) \end{aligned}$$

$$V_P = -\frac{P}{2}\int_0^\ell \left(\frac{dy}{dx}\right)^2 dx = -\frac{\delta^2 P}{2}\left(\frac{\pi}{2\ell}\right)^2\int_0^\ell \left(\sin\frac{\pi x}{2\ell}\right)^2 dx = -\frac{\delta^2 P}{2}\left(\frac{\pi}{2\ell}\right)^2\frac{\ell}{2}$$

$$\begin{aligned} V_q &= -\frac{1}{2}q\int_0^\ell (\ell - x)\left(\frac{dy}{dx}\right)^2 dx = -\frac{1}{2}q\delta^2\int_0^\ell (\ell - x)\left(\frac{\pi}{2\ell}\sin\frac{\pi x}{2\ell}\right)^2 dx \\ &= -\frac{\delta^2 q}{32}\left(\pi^2 - 4\right) \end{aligned}$$

$$\begin{aligned} V &= V_P + V_q = -\frac{\delta^2 P}{2}\left(\frac{\pi}{2\ell}\right)^2\frac{\ell}{2} - \frac{\delta^2 q}{32}\left(\pi^2 - 4\right) \\ &= -\delta^2\left(0.6168503\frac{P}{\ell} + 0.183425138q\right) \end{aligned}$$

$$\begin{aligned} \frac{\partial U}{\partial\delta} + \frac{\partial V}{\partial\delta} &= 0.25\frac{\ell}{EI}P^2 + 0.148678816\frac{\ell^2}{EI}qP + 0.0232541088\frac{\ell^3}{EI}q^2 \\ &\quad - 0.6168503\frac{P}{\ell} - 0.183425138q = 0 \end{aligned}$$

If $P = 0$, then

$$q_{cr} = \frac{7.88786EI}{\ell^3} \tag{2.5.1.16}$$

The critical load given by Eq. (2.5.1.16) is only 0.65% greater than that given by Timoshenko and Gere (1961).

If $n = 1(q = \pi^2 EI/4\ell^3)$, then
$P_{cr} = 1.7223\ EI/\ell^2 \Leftarrow 0.13\%$ greater than the exact solution.

2.5.2. Simply Supported

Consider a both-end simply supported column subjected to a distributed axial load q and a concentrated axial load P at the top of the column shown in Fig. 2-8.

Assume a one-term trial deflection curve.

$$y = \delta \sin \frac{\pi x}{\ell} \tag{2.5.2.1}$$

The bending moment at any section mn in Fig. 2-8 is

$$M_q = \int_x^\ell q(y - \eta)d\xi \tag{2.5.2.2}$$

It is noted that the deflection η is also expressed as

$$\eta = \delta \sin \frac{\pi \xi}{\ell} \tag{2.5.2.3}$$

Substituting η into the moment equation and noting that y is not a function of ξ yields

$$M_q = \int_x^\ell q(y - \eta)d\xi = q\left[y(\ell - x) - \int_x^\ell \eta d\xi\right]$$

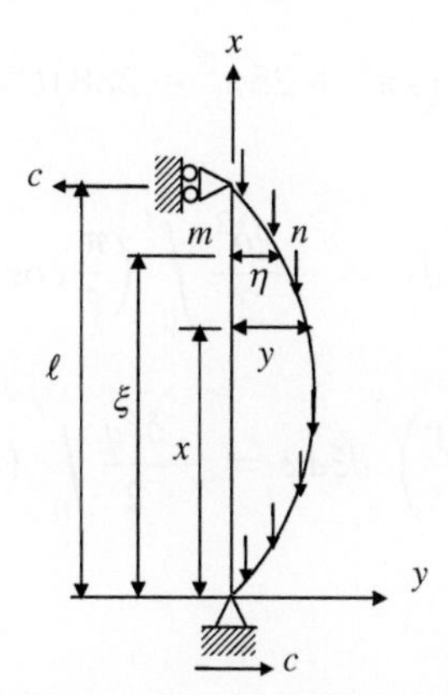

Figure 2-8 Simple column with distributed load

The integral on the right-hand side is expanded to

$$\int_x^\ell \eta d\xi = \delta \int_x^\ell \sin\frac{\pi\xi}{\ell} d\xi = -\frac{\delta\ell}{\pi}\cos\frac{\pi\xi}{\ell}\bigg|_x^\ell = \frac{\delta\ell}{\pi}\left(1+\cos\frac{\pi x}{\ell}\right)$$

Hence,

$$M_q = q\delta\left[(\ell - x)\sin\frac{\pi x}{\ell} - \frac{\ell}{\pi}\left(1+\cos\frac{\pi x}{\ell}\right)\right] \tag{2.5.2.4}$$

The moment at the bottom support (hinged end) must be equal. However, the moment equation shows a moment equal to $-2\ell/\pi$ upon substitution of $x = 0$. In order to maintain equilibrium, a correction couple force c (also known as the continuity shear) is required, as shown in Fig. 2-8. Hence, the corrected moment at any point along the column length is

$$M_q = q\delta\left[(\ell - x)\sin\frac{\pi x}{\ell} - \frac{\ell}{\pi}\left(1+\cos\frac{\pi x}{\ell}\right) + \frac{2}{\pi}(\ell - x)\right]$$

$$M_P = Py = P\delta\sin\frac{\pi x}{\ell}$$

$$M = M_P + M_q$$

$$\begin{aligned} U &= \frac{1}{2EI}\int_0^\ell M^2 dx = \frac{1}{2EI}\int_0^\ell \left\{P\delta\sin\frac{\pi x}{\ell} + q\delta\left[(\ell - x)\sin\frac{\pi x}{\ell}\right.\right. \\ &\quad \left.\left. - \frac{\ell}{\pi}\left(1+\cos\frac{\pi x}{\ell}\right) + \frac{2}{\pi}(\ell - x)\right]\right\}^2 dx \\ &= \frac{\delta^2\ell}{24\pi^4 EI}\left[6\pi^4P^2 + (2\pi^4 + 25\pi^2 - 288)\ell^2q^2 + 6\pi^4\ell Pq\right] \end{aligned} \tag{2.5.2.5}$$

$$V_P = -\frac{P}{2}\int_0^\ell \left(\frac{dy}{dx}\right)^2 dx = -\frac{P\delta^2}{2}\int_0^\ell \left(\frac{\pi}{\ell}\cos\frac{\pi x}{\ell}\right)^2 dx = -\frac{\delta^2 P\pi^2}{4\ell}$$

$$\begin{aligned} V_q &= -\frac{1}{2}\int_0^\ell \int_x^\ell q\left(\frac{dy}{dx}\right)^2 d\xi dx = -\frac{\delta^2 q}{2}\int_0^\ell (\ell - x)\left(\frac{\pi}{\ell}\cos\frac{\pi x}{\ell}\right)^2 dx \\ &= -\frac{\delta^2 q\pi^2}{8} \end{aligned} \tag{2.5.2.6}$$

$$V = V_P + V_q = -\frac{\delta^2 P\pi^2}{4\ell} - \frac{\delta^2 q\pi^2}{8}$$

$$\frac{\partial U}{\partial \delta} + \frac{\partial V}{\partial \delta} = \frac{\ell}{12\pi^4 EI}\left[6\pi^4 P^2 + \left(2\pi^4 + 25\pi^2 - 288\right)\ell^2 q^2 + 6\pi^4 \ell Pq\right]$$

$$-\frac{P\pi^2}{2\ell} - \frac{q\pi^2}{4} = 0 \tag{2.5.2.7}$$

If $q = 0$, then $P_{cr} = (\pi^2 EI/\ell^2) \Leftarrow$ As expected.
If $P = 0$, then $q_{cr} = (18.78EI/\ell^3) \Leftarrow$ this is only 0.98% greater than the exact value $\frac{18.6EI}{\ell^3}$.

For this example, there appears to be an opportunity to improve the solution accuracy by adding a second term in the assumed deflection curve.

$$y = a \sin\frac{\pi x}{\ell} + b \sin\frac{2\pi x}{\ell} \tag{2.5.2.8}$$

$$y' = a\frac{\pi}{\ell}\cos\frac{\pi x}{\ell} + b\frac{2\pi}{\ell}\cos\frac{2\pi x}{\ell}$$

$$y'' = -a\left(\frac{\pi}{\ell}\right)^2 \sin\frac{\pi x}{\ell} - b\left(\frac{2\pi}{\ell}\right)^2 \sin\frac{2\pi x}{\ell}$$

Only the uniformly distributed axial load is considered in this illustration. The bending moment at any section *mn* in Fig. 2-8 is

$$M_q = \int_x^\ell q(y - \eta)\,d\xi$$

The deflection η is also expressed as

$$\eta = a \sin\frac{\pi\xi}{\ell} + b \sin\frac{2\pi\xi}{\ell}$$

Substituting η into the moment equation and noting that y is not a function of ξ yields:

$$M_q = \int_x^\ell q(y - \eta)\,d\xi = q\left[y(\ell - x) - \int_x^\ell \eta\,d\xi\right]$$

The integral on the right-hand side is expanded to

$$\begin{aligned}\int_x^\ell \eta d\xi &= \int_x^\ell \left(a \sin\frac{\pi\xi}{\ell} + b \sin\frac{2\pi x}{\ell}\right) d\xi \\ &= \left(-\frac{a\ell}{\pi}\cos\frac{\pi\xi}{\ell} - \frac{b\ell}{2\pi}\cos\frac{2\pi\xi}{\ell}\right)\Bigg|_x^\ell \\ &= \left[\frac{a\ell}{\pi}\left(1+\cos\frac{\pi x}{\ell}\right) - \frac{b\ell}{2\pi}\left(1-\cos\frac{2\pi x}{\ell}\right)\right]\end{aligned}$$

Hence,

$$\begin{aligned}M_q = q\Bigg[&(\ell - x)\left(a \sin\frac{\pi x}{\ell} + b\sin\frac{2\pi x}{\ell}\right) - \frac{a\ell}{\pi}\left(1+\cos\frac{\pi x}{\ell}\right) \\ &+ \frac{b\ell}{2\pi}\left(1-\cos\frac{2\pi x}{\ell}\right)\Bigg]\end{aligned} \tag{2.5.2.9}$$

The moment at the bottom support (hinged end) must be equal. However, the moment equation shows a moment equal to $-2\ell a/\pi$ upon substitution of $x = 0$. In order to maintain equilibrium, a correction couple force c is required as shown in Fig. 2-8. Hence, the corrected moment at any point along the column length is

$$\begin{aligned}M_q = q\Bigg[&(\ell - x)\left(a \sin\frac{\pi x}{\ell} + b\sin\frac{2\pi x}{\ell}\right) - \frac{a\ell}{\pi}\left(1+\cos\frac{\pi x}{\ell}\right) \\ &+ \frac{b\ell}{2\pi}\left(1-\cos\frac{2\pi x}{\ell}\right) + \frac{2a}{\pi}(\ell - x)\Bigg]\end{aligned}$$

$$\begin{aligned}U = \frac{q^2}{2EI}\int_0^\ell \Bigg[&(\ell - x)\left(a \sin\frac{\pi x}{\ell} + b\sin\frac{2\pi x}{\ell}\right) - \frac{a\ell}{\pi}\left(1+\cos\frac{\pi x}{\ell}\right) \\ &+ \frac{b\ell}{2\pi}\left(1-\cos\frac{2\pi x}{\ell}\right) + \frac{2a}{\pi}(\ell - x)\Bigg]^2 dx \\ = \frac{q^2\ell^3}{288\pi^4 EI}\Big[&\left(24\pi^4 + 300\pi^2 - 3456\right)a^2 \\ &+ \left(24\pi^4 + 99\pi^2\right)b^2 - 400\pi^2 ab\Big]\end{aligned}$$

Let $s = \dfrac{q^2\ell^3}{288\pi^4 EI}$

$$V = -\frac{1}{2}\int_0^\ell \int_x^\ell q\left(\frac{dy}{dx}\right)^2 d\xi\, dx$$

$$= -\frac{q}{2}\int_0^\ell (\ell - x)\left(a\frac{\pi}{\ell}\cos\frac{\pi x}{\ell} + b\frac{2\pi}{\ell}\cos\frac{2\pi x}{\ell}\right)^2 dx$$

$$= -q\left(\frac{20}{9}ab + \frac{\pi^2}{8}a^2 + \frac{\pi^2}{2}b^2\right)$$

$$\frac{\partial U}{\partial a} + \frac{\partial V}{\partial a} = (3685.4s - 2.4674)a + (3947.84s - 2.22222)b = 0$$

$$\frac{\partial U}{\partial b} + \frac{\partial V}{\partial b} = (3947.84s - 2.222222)a + (6629.818s - 9.8696)b = 0$$

$$\begin{vmatrix} 3685.4s - 2.4674 & 3947.84s - 2.22222 \\ 3947.84s - 2.22222 & 6629.818s - 9.8696 \end{vmatrix} = 0$$

For a nontrivial solution (a and b cannot be equal to zero simultaneously), the determinant of the coefficient must be zero.
Solving for s gives

$$s = 0.000661938 = \frac{q^2\ell^3}{288\pi^4 EI} \Rightarrow q_{cr} = 18.57\frac{EI}{\ell^3}$$

2.5.3. Pinned-Clamped Column

A propped column with the top rotationally clamped and the bottom pinned is subjected to a uniformly distributed axial compression as shown in Fig. 2-9. Because of the boundary condition, a continuity shear or a correction couple force is expected for equilibrium.
Assume a one-term trial displacement function as

$$y = a(\ell^3 x - 3\ell x^3 + 2x^4) \tag{2.5.3.1}$$

Boundary conditions are

$$y = 0 \ @\ x = 0 \text{ and } y = 0 \ @\ x = \ell$$

$$y' = 0 \ @\ x = \ell \text{ and } y'' = 0 \ @\ x = 0$$

The function satisfies the geometric and natural boundary conditions at both ends.

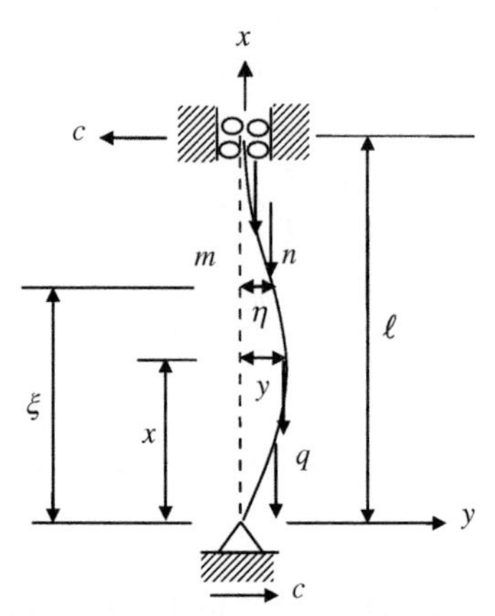

Figure 2-9 Clamped-pinned column

The bending moment at any section *mn* in Fig. 2-9 is:

$$M = \int_x^\ell q(y - \eta)d\xi$$

It is noted that the deflection η is also expressed as:

$$\eta = a\left(\ell^3\xi - 3\ell\xi^3 + 2\xi^4\right)$$

Substituting η into the moment equation and noting that y is not a function of ξ yields:

$$M = \int_x^\ell q(y - \eta)d\xi = q\left[y(\ell - x) - \int_x^\ell \eta d\xi\right]$$

The integral on the right-hand side is expanded to:

$$\int_x^\ell \eta d\xi = a\int_x^\ell \left(\ell^3\xi - 3\ell\xi^3 + 2\xi^4\right)d\xi$$

$$= a\left[\frac{\ell^3}{2}\left(\ell^2 - x^2\right) - \frac{3\ell}{4}\left(\ell^4 - x^4\right) + \frac{2}{5}\left(\ell^5 - x^5\right)\right]$$

Hence,

$$M = qa\left\{(\ell - x)\left(\ell^3 x - 3\ell x^3 + 2x^4\right) - \left[\frac{\ell^3}{2}\left(\ell^2 - x^2\right) - \frac{3\ell}{4}\left(\ell^4 - x^4\right) + \frac{2}{5}\left(\ell^5 - x^5\right)\right]\right\} \tag{2.5.3.2}$$

The moment at the bottom support (hinged end) must be equal to zero. However, the moment equation shows a moment equal to $-3\ell^5/20$ upon substitution of $x = 0$. In order to maintain equilibrium, a correction couple force c (also known as the continuity shear) is required as shown in Fig. 2-9. Hence, the corrected moment at any point along the column length is

$$M = qa\left\{(\ell - x)\left(\ell^3 x - 3\ell x^3 + 2x^4 + \frac{3\ell^4}{20}\right) - \left[\frac{\ell^3}{2}(\ell^2 - x^2) - \frac{3\ell}{4}(\ell^4 - x^4) + \frac{2}{5}(\ell^5 - x^5)\right]\right\}$$

The assumed deflection function has an inflection point at $x = 0.75\ell$. In order to ensure the moment to be equal to zero at the inflection point, the moment equation needs an additional adjustment.

$$M = qa\left\{\begin{array}{l}(\ell - x)\left(\ell^3 x - 3\ell x^3 + 2x^4 + \dfrac{3\ell^4}{20}\right) - 0.074211875\ell^4 x \\ -\left[\dfrac{\ell^3}{2}(\ell^2 - x^2) - \dfrac{3\ell}{4}(\ell^4 - x^4) + \dfrac{2}{5}(\ell^5 - x^5)\right]\end{array}\right\} \tag{2.5.3.3}$$

$$U = \frac{1}{2EI}\int_0^\ell M^2 dx$$

$$= \frac{q^2 a^2}{2EI}\int_0^\ell \left\{\begin{array}{l}(\ell - x)\left(\ell^3 x - 3\ell x^3 + 2x^4 + \dfrac{3\ell^4}{20}\right) - 0.07421875\ell^4 x \\ -\left[\dfrac{\ell^3}{2}(\ell^2 - x^2) - \dfrac{3\ell}{4}(\ell^4 - x^4) + \dfrac{2}{5}(\ell^5 - x^5)\right]\end{array}\right\}^2 dx$$

$$= 0.007519762734\, \ell^{11} q^2 a^2 / 2EI$$

$$V_q = -\frac{1}{2}\int_0^\ell \int_x^\ell q\left(\frac{dy}{dx}\right)^2 d\xi\, dx$$

$$= -\frac{a^2 q}{2}\int_0^\ell (\ell - x)(\ell^3 - 9\ell x^2 + 8x^3)^2 dx = \frac{3\ell^8 q a^2}{28}$$

$$\frac{\partial U}{\partial a} + \frac{\partial V}{\partial a} = \frac{\ell^{11} q}{EI}(0.0075197627534) - \frac{6\ell^8}{28} = 0 \Rightarrow$$

$$q = 28.5\frac{EI}{\ell^3} \text{ (5\% less than the exact solution)}$$

2.5.4. Clamped-Pinned Column

A propped column with the top rotationally clamped and the bottom pinned is subjected to a uniformly distributed axial compression as shown in Fig. 2-10.

Assume a one-term trial displacement function as

$$y = a(3\ell^2x^2 - 5\ell x^3 + 2x^4) \tag{2.5.4.1}$$

Boundary conditions are

$$y = 0 \quad @ \; x = 0 \text{ and } y = 0 \quad @ \; x = \ell$$

$$y' = 0 \quad @ \; x = 0 \text{ and } y'' = 0 \quad @ \; x = \ell$$

The function satisfies the geometric and natural boundary conditions at both ends.

The bending moment at any section *mn* in Fig. 2-10 is

$$M = \int_x^\ell q(y - \eta)d\xi$$

It is noted that the deflection η is also expressed as

$$\eta = a\left(3\ell^2\xi^2 - 5\ell\xi^3 + 2\xi^4\right)$$

Substituting η into the moment equation and noting that y is not a function of ξ yields

$$M = \int_x^\ell q(y - \eta)d\xi = q\left[y(\ell - x) - \int_x^\ell \eta d\xi\right]$$

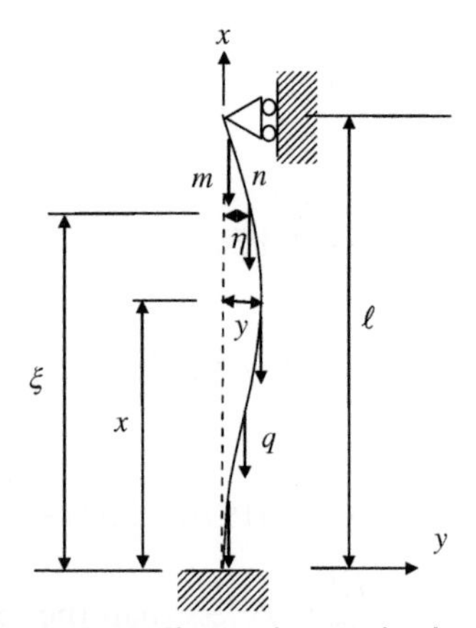

Figure 2-10 Clamped-pinned column

The integral on the right-hand side is expanded to

$$\int_x^{\ell} \eta d\xi = a\int_x^{\ell} \left(3\ell^2\xi^2 - 5\ell\xi^3 + 2\xi^4\right) d\xi$$

$$= a\left[\ell^2\left(\ell^3 - x^3\right) - \frac{5\ell}{4}\left(\ell^4 - x^4\right) + \frac{2}{5}\left(\ell^5 - x^5\right)\right]$$

Hence,

$$M = qa\left\{\left(\ell - x\right)\left(3\ell^2x^2 - 5\ell x^3 + 2x^4\right) - \left[\ell^2\left(\ell^3 - x^3\right) - \frac{5\ell}{4}\left(\ell^4 - x^4\right) + \frac{2}{5}\left(\ell^5 - x^5\right)\right]\right\}$$

The assumed deflection function has an inflection point at $x = 0.25\ell$. In order to ensure the moment to be equal to zero at the inflection point, the moment equation needs an adjustment.

$$M = qa\left\{\begin{array}{c}\left(\ell - x\right)\left(3\ell^2x^2 - 5\ell x^3 + 2x^4 + 0.067968747\ell^4\right) \\ -\left[\ell^2\left(\ell^3 - x^3\right) - \frac{5\ell}{4}\left(\ell^4 - x^4\right) + \frac{2}{5}\left(\ell^5 - x^5\right)\right]\end{array}\right\} \quad (2.5.4.2)$$

$$U = \frac{1}{2EI}\int_0^{\ell} M^2 dx$$

$$= \frac{q^2a^2}{2EI}\int_0^{\ell}\left\{\begin{array}{c}\left(\ell - x\right)\left(3\ell^2x^2 - 5\ell x^3 + 2x^4 + 0.067968747\ell^4\right) \\ -\left[\ell^2\left(\ell^3 - x^3\right) - \frac{5\ell}{4}\left(\ell^4 - x^4\right) + \frac{2}{5}\left(\ell^5 - x^5\right)\right]\end{array}\right\}^2 dx$$

$$= 0.002425669952\ell^{11}q^2a^2/2EI$$

$$V_q = -\frac{1}{2}\int_0^{\ell}\int_x^{\ell} q\left(\frac{dy}{dx}\right)^2 d\xi dx$$

$$= -\frac{a^2q}{2}\int_0^{\ell}\left(\ell - x\right)\left(6\ell^2x - 15\ell x^2 + 8x^3\right)^2 dx = -\frac{9\ell^8qa^2}{140}$$

$$\frac{\partial U}{\partial a} + \frac{\partial V}{\partial a} = \frac{\ell^{11}q}{EI}(0.002425669952) - \frac{18\ell^8}{140} = 0 \Rightarrow$$

$$q = 53\frac{EI}{\ell^3} \text{ (0.95\% greater than the exact solution)}$$

2.5.5. Both-Ends Clamped Column

A both-end clamped column is subjected to a uniformly distributed axial compression as shown in Fig. 2-11. Assume the deflection curve to be the form

$$y = a\left(1 - \cos\frac{2\pi x}{\ell}\right) \tag{2.5.5.1}$$

The bending moment at any section mn in Fig. 2-11 is

$$M = \int_x^\ell q(y - \eta)d\xi$$

It is noted that the deflection η is also expressed as

$$\eta = a\left(1 - \cos\frac{2\pi\xi}{\ell}\right)$$

Substituting η into the moment equation and noting that y is not a function of ζ yields

$$M = \int_x^q q(y - \eta)d\xi = q\left[y(\ell - x) - \int_x^\ell \eta d\xi\right]$$

The integral on the right-hand side is expanded to

$$\int_x^\ell \eta d\xi = a\int_x^\ell \left(1 - \cos\frac{2\pi\xi}{\ell}\right)d\xi = a\left[\ell - x) + \frac{\ell}{2\pi}\sin\frac{2\pi x}{\ell}\right]$$

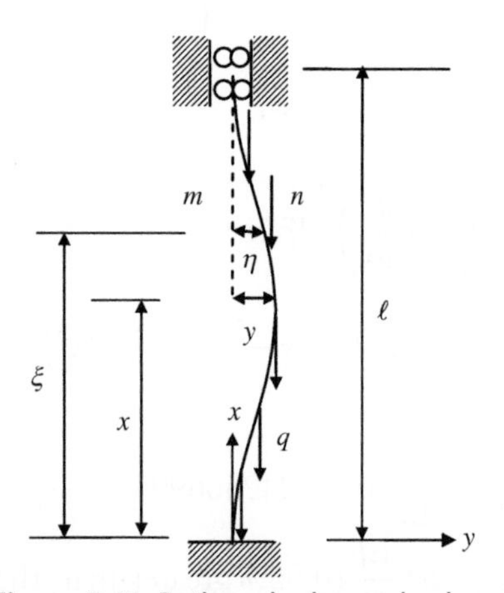

Figure 2-11 Both-ends clamped column

Hence,

$$M = qa\left\{(\ell - x)\left(1 - \cos\frac{2\pi x}{\ell}\right) - \left[(\ell - x) + \frac{\ell}{2\pi}\sin\frac{2\pi x}{\ell}\right]\right\} \quad (2.5.5.2)$$

The assumed deflection curve is to have two inflection points at $x = 0.25\ell$ and $x = 0.75\ell$.

$$M = qa\left\{(\ell - x)\left(1 - \cos\frac{2\pi x}{\ell}\right) - \left[(\ell - x) + \frac{\ell}{2\pi}\sin\frac{2\pi x}{\ell}\right] + \frac{2}{\pi}\left(\frac{\ell}{2} - x\right)\right\}$$

(2.5.5.3)

$$U = \frac{1}{2EI}\int_0^\ell M^2 dx = \frac{q^2a^2}{2EI}\int_0^\ell \left\{(\ell - x)\left(1 - \cos\frac{2\pi x}{\ell}\right) - \left[(\ell - x) + \frac{\ell}{2\pi}\sin\frac{2\pi x}{\ell}\right] + \frac{2}{\pi}\left(\frac{\ell}{2} - x\right)\right\}^2 dx$$

$$= (767/277200)\,\ell^{11}q^2a^2/2EI$$

$$V_q = -\frac{1}{2}\int_0^\ell \int_x^\ell q\left(\frac{dy}{dx}\right)^2 d\xi\, dx = -\frac{a^2q}{2}\int_0^\ell (\ell - x)(\ell^3 - 9\ell x^2 + 8x^3)^2 dx$$

$$= \frac{3\ell^8 qa^2}{28}$$

$$\frac{\partial U}{\partial a} + \frac{\partial V}{\partial a} = \frac{\ell^{11}q}{EI}\left(\frac{767}{277200}\right) - \frac{6\ell^8}{28} = 0 \Rightarrow q$$

$$= 77.4\frac{EI}{\ell^3} \text{ (3.8\% greater than the exact solution)}$$

2.6. ELASTICALLY SUPPORTED BEAM-COLUMNS

As an example of the stability of a bar on elastic supports, consider a prismatic continuous beam simply supported at the ends on rigid supports and having several intermediate elastic supports. A similar problem was considered in Section 1.9 in which the bar was considered to be rigid so that the strain energy stored was in the elastic supports only. Let q = force developed in the spring= ky. Then the work done by the spring is $(1/2)qy = (1/2)ky^2$. Rotational spring can also be considered at any support. Total potential energy function of the system becomes

$$\Pi = U + V = \frac{EI}{2}\int_0^\ell (y'')^2 dx - \frac{P}{2}\int_0^\ell (y')^2 dx + \frac{1}{2}\sum_{i=1}^{n} k_i y_i^2 \quad (2.6.1)$$

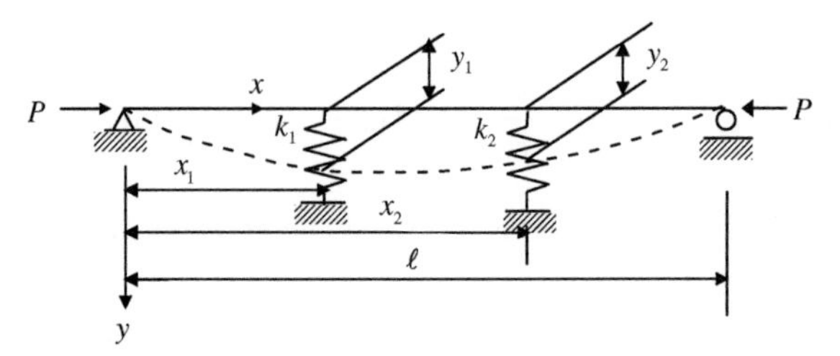

Figure 2-12 Column resting on elastic supports

Let $k_1 = k_2 = k$ and $x_1 = \ell/3,\ x_2 = 2\ell/3$ to simplify the computation effort. It appears that at least three sine functions need to be considered for the three-span configuration shown in Fig. 2-12. Assume

$$y = a_1 \sin\frac{\pi x}{\ell} + a_2 \sin\frac{2\pi x}{\ell} + a_3 \sin\frac{3\pi x}{\ell} \tag{2.6.2}$$

$$\Pi = \frac{EI}{2}\int_0^\ell \left[-a_1\left(\frac{\pi}{\ell}\right)^2 \sin\frac{\pi x}{\ell} - a_2\left(\frac{2\pi}{\ell}\right)^2 \sin\frac{2\pi x}{\ell} - a_3\left(\frac{3\pi}{\ell}\right)^2 \sin\frac{3\pi x}{\ell}\right]^2 dx$$

$$-\frac{P}{2}\int_0^\ell \left[a_1\left(\frac{\pi}{\ell}\right)\cos\frac{\pi x}{\ell} + a_2\left(\frac{2\pi}{\ell}\right)\cos\frac{2\pi x}{\ell} + a_3\left(\frac{3\pi}{\ell}\right)\cos\frac{3\pi x}{\ell}\right]^2 dx$$

$$+\frac{1}{2}k\left[\left(a_1 \sin\frac{\pi}{3} + a_2 \sin\frac{2\pi}{3}\right)^2 + \left(a_1 \sin\frac{2\pi}{3} + a_2 \sin\frac{4\pi}{3}\right)^2\right]$$

Noting that

$$\int_0^\ell \sin\frac{i\pi x}{\ell}\sin\frac{j\pi x}{\ell}\,dx = \begin{cases} 0 \text{ for } i \neq j \\ \dfrac{\ell}{2} \text{ for } i = j \end{cases} \quad \text{and}$$

$$\int_0^\ell \cos\frac{i\pi x}{\ell}\cos\frac{j\pi x}{\ell}\,dx = \begin{cases} 0 \text{ for } i \neq j \\ \dfrac{\ell}{2} \text{ for } i = j \end{cases}$$

$$\Pi = \frac{EI\pi^4}{4\ell^3}\left(a_1^2 + 16a_2^2 + 81a_3^2\right) - \frac{P\pi^2}{4\ell}\left(a_1^2 + 4a_2^2 + 9a_3^2\right)$$
$$+\frac{3}{4}k\left(a_1^2 + a_2^2\right)$$

$$\frac{\partial \Pi}{\partial a_1} = 0 = \frac{EI\pi^4}{4\ell^3}(2a_1) - \frac{P\pi^2}{4\ell}(2a_1) + \frac{3}{4}k(2a_1) = \left(\frac{EI\pi^4}{\ell^3} - \frac{P\pi^2}{\ell} + 3k\right)a_1$$

$$P_{cr} = \frac{4\ell}{\pi^2}\left(\frac{EI\pi^4}{4\ell^3} + \frac{3}{4}k\right) = \frac{\pi^2 EI}{\ell^2} + \frac{3k\ell}{\pi^2} = P_E\left(1 + \frac{3}{\pi^2}\frac{k\ell}{P_E}\right)$$

$$\frac{P_{cr}}{P_E} = 1 + \frac{3k\ell}{\pi^2 P_E} \tag{2.6.3}$$

$$\frac{\partial \Pi}{\partial a_2} = 0 = \frac{EI\pi^4}{4\ell^3}(32a_2) - \frac{P\pi^2}{4\ell}(8a_2) + \frac{3}{4}k(2a_2) = \left(\frac{4EI\pi^4}{\ell^3} - \frac{P\pi^2}{\ell} + \frac{3}{4}k\right)a_2$$

$$P_{cr} = \frac{\ell}{\pi^2}\left(\frac{4EI\pi^4}{\ell^3} + \frac{3}{4}k\right) = \frac{4\pi^2 EI}{\ell^2} + \frac{3k\ell}{4\pi^2} = P_E\left(4 + \frac{3}{4\pi^2}\frac{k\ell}{P_E}\right)$$

$$\frac{P_{cr}}{P_E} = 4 + \frac{3k\ell}{4\pi^2 P_E} \tag{2.6.4}$$

$$\frac{\partial \Pi}{\partial a_3} = 0 = \frac{EI\pi^4}{4\ell^3}(162a_3) - \frac{P\pi^2}{4\ell}(18a_3) = \left(\frac{9EI\pi^2}{\ell^3} - \frac{P}{\ell}\right)a_3$$

$$P_{cr} = \frac{\ell}{\pi^2}\left(\frac{9EI\pi^4}{\ell^3}\right) = \frac{9\pi^2 EI}{\ell^2}, \quad \frac{P_{cr}}{P_E} = 9 \tag{2.6.5}$$

The same results can be obtained by setting the coefficient determinant equal to zero.

$$\begin{vmatrix} \left(\frac{EI\pi^4}{\ell^3} - \frac{P\pi^2}{\ell} + 3k\right) & 0 & 0 \\ 0 & \left(\frac{4EI\pi^4}{\ell^3} - \frac{P\pi^2}{\ell} + \frac{3}{4}k\right) & 0 \\ 0 & 0 & \left(\frac{9EI\pi^2}{\ell^3} - \frac{P}{\ell}\right) \end{vmatrix} = 0$$

Equating (2.6.3) and (2.6.4) yields

$$1 + \frac{3}{\pi^2}\frac{k\ell}{P_E} = 4 + \frac{3}{4\pi^2}\frac{k\ell}{P_E} \Rightarrow \frac{\cancel{3}}{\pi^2}\left(1 - \frac{1}{4}\right)\frac{k\ell}{P_E} = \cancel{3} \Rightarrow \frac{k\ell}{P_E} = 13.16$$

Likewise, equating (2.6.4) and (2.6.5) gives

$$4 + \frac{3k\ell}{4\pi^2 P_E} = 9 \Rightarrow \frac{k\ell}{P_E} = 5\frac{4\pi^2}{3} = 65.8$$

Assuming $\ell = 3L$ and $P_{cr} = 9P_E$ for three equal spans,
$k = 65.8P_E/(3L) = 65.8P_{cr}/9/(3L) = 2.437P_{cr}/L = \beta P_{cr}/L$
This equivalent β value of 2.437 is slightly less than that ($\beta = 3$) obtained for three equal spans rigid body system, which is logical as the elastic strain energy stored in the deformed body shares a portion of the energy provided by the spring system. Note that practical bracing design is carried out based on the rigid body mechanics examined in Chapter 1 to be conservative. The P_{cr}/P_E versus $k\ell/P_E$ plot is shown in Fig. 2-13.

Next example is a propped (by a linear spring) column shown in Fig. 2-14. Considering only elastic deformation and neglecting the rigid body motion, the strain energy equation based on the assumed displacement function is to be derived.

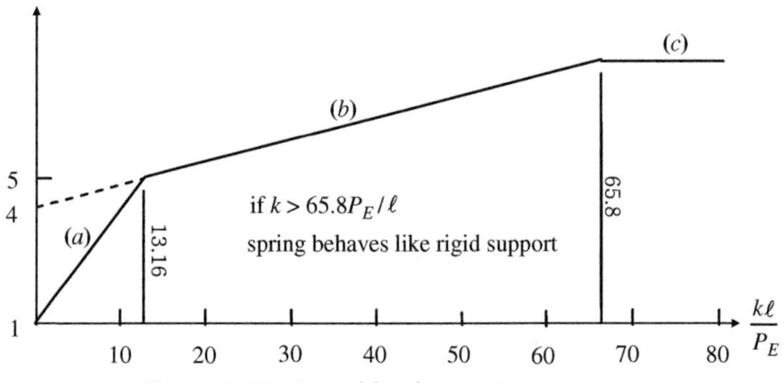

Figure 2-13 Critical load vs. spring constant

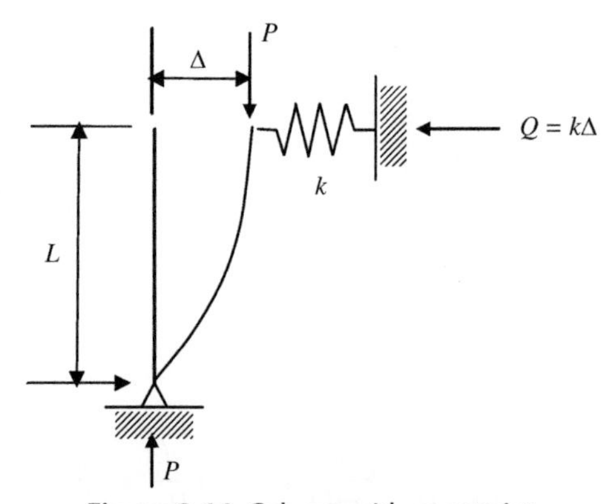

Figure 2-14 Column with constraint

Assume the deflection shape to be of the form

$$y = \delta \sin \frac{\pi x}{2L} + \lambda \sin \frac{\pi x}{L}$$

$$y' = \frac{\delta \pi}{2L} \cos \frac{\pi x}{2L} + \frac{\lambda \pi}{L} \cos \frac{\pi x}{L},$$

$$y'' = -\delta \left(\frac{\pi}{2L}\right)^2 \sin \frac{\pi x}{2L} - \lambda \left(\frac{\pi}{L}\right)^2 \sin \frac{\pi x}{L}$$

$$\begin{aligned}
\Pi &= U + V \\
&= \frac{EI}{2} \int_0^L \left[\delta^2 \left(\frac{\pi}{2L}\right)^4 \sin^2 \frac{\pi x}{2L} + \lambda^2 \left(\frac{\pi}{L}\right)^4 \sin^2 \frac{\pi x}{L} \right. \\
&\quad \left. + 2\delta\lambda \left(\frac{\pi}{2L}\right)^2 \left(\frac{\pi}{L}\right)^2 \sin \frac{\pi x}{2L} \sin \frac{\pi x}{L} \right] dx + \frac{k\delta^2}{2} \\
&\quad - \frac{P}{2} \int_0^L \left[\delta^2 \left(\frac{\pi}{2L}\right)^2 \cos^2 \frac{\pi x}{2L} + \lambda^2 \left(\frac{\pi}{L}\right)^2 \cos^2 \frac{\pi x}{L} \right. \\
&\quad \left. + 2\delta\lambda \left(\frac{\pi}{2L}\right) \left(\frac{\pi}{L}\right) \cos \frac{\pi x}{2L} + \cos \frac{\pi x}{L} \right] dx \\
&= \frac{EI}{2} \left[\delta^2 \left(\frac{\pi}{2L}\right)^4 \frac{L}{2} + \lambda^2 \left(\frac{\pi}{L}\right)^4 \frac{L}{2} \right] + \frac{k\delta^2}{2} - \frac{P}{2} \left[\delta^2 \left(\frac{\pi}{2L}\right)^2 \frac{L}{2} \right. \\
&\quad \left. + \lambda^2 \left(\frac{\pi}{L}\right)^2 \frac{L}{2} \right]
\end{aligned}$$

$$\frac{\partial \Pi}{\partial \delta} = 0 = EI \left[\delta \left(\frac{\pi}{2L}\right)^4 \frac{L}{2} \right] + k\delta - P \left[\delta \left(\frac{\pi}{2L}\right)^2 \frac{L}{2} \right] \Rightarrow k = \frac{P\pi^2}{8L} - \frac{EI\pi^4}{32L^3}$$

If $P_{cr} = \dfrac{\pi^2 EI}{L^2}$, then

$$k = \frac{P_{cr}}{L} \left(\frac{\pi^2}{8} - \frac{\pi^2}{32} \right) = \frac{0.9253 P_{cr}}{L} = \frac{\beta P_{cr}}{L} \Rightarrow \beta < 1.0$$

$$\frac{\partial \Pi}{\partial \lambda} = 0 = EI \left[\lambda \left(\frac{\pi}{2L}\right)^4 \frac{L}{2} \right] - P \left[\lambda \left(\frac{\pi}{2L}\right)^2 \frac{L}{2} \right] \Rightarrow$$

$$P = \frac{EI\pi^2}{4L^2} \quad \text{for } k = 0 \Leftarrow \text{not valid}$$

As expected, the value of β of 0.9253 is slightly less than 1.0 obtained in Chapter 1. It is customary in practice to neglect the contribution of strain energy from the elastic deformation of the column.

2.7. DIFFERENTIAL EQUATION METHOD

The critical load on columns of stepped (variable) cross section as used in telescopic power cylinders can be computed applying differential equations considering continuity at the junctures. In order to limit the computational complexity, only two-stepped columns shown in the sketch are considered. Multiple-stepped columns are best analyzed by a means of computerized structural analysis methods.

Consider the stepped cantilever column shown in Fig. 2-15(a). The bending moment of the column at any section along the member x-axis can be written for each segment as

$$EI_1 y_1'' = P(\delta - y_1) \quad \text{and} \quad EI_2 y_2'' = P(\delta - y_2)$$

Let $k_1^2 = \dfrac{P}{EI_1}$ and $k_2^2 = \dfrac{P}{EI_2}$, then the equations becomes

$$y_1'' + k_1^2 y_1 = k_1^2 \delta \tag{2.7.1}$$

$$y_2'' + k_2^2 y_2 = k_2^2 \delta \tag{2.7.2}$$

The total solutions of Eqs. (2.7.1) and (2.7.2) are

$$y_1 = \delta + C \cos k_1 x + D \sin k_1 x$$

$$y_2 = \delta + A \cos k_2 x + B \sin k_2 x$$

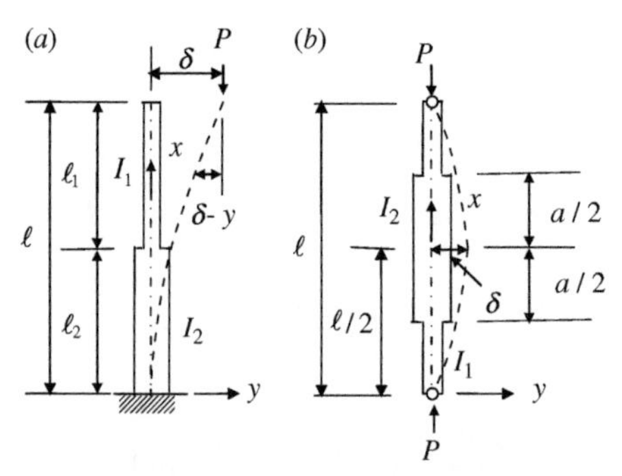

Figure 2-15 Stepped columns

In order to determine the integral constants A and B for segment 2, consider the following boundary conditions:

$$y_2 = 0 \quad \text{at } x = 0 \Rightarrow A = -\delta$$

$$y_2' = 0 \quad \text{at } x = 0 \Rightarrow B = 0 \Rightarrow y_2 = \delta(1 - \cos k_2 x)$$

At the top of the column for y_1, it requires that

$$\delta + C\cos k_1\ell + D\sin k_1\ell = \delta \Rightarrow C\cos k_1\ell + D\sin k_1\ell = 0 \Rightarrow$$

$$C = -D\tan k_1\ell$$

The continuity at the juncture requires that

$$\not{\delta} + C\cos k_1\ell_2 + D\sin k_1\ell_2 = \delta(1 - \cos k_2\ell_2) = \not{\delta} - \delta\cos k_2\ell_2$$

$$-\tan k_1\ell\cos k_1\ell_2 D + D\sin k_1\ell_2 = -\left(\frac{\sin k_1\ell}{\cos k_1\ell}\cos k_1\ell_2 - \sin k_1\ell_2\right)D$$

$$= -\delta\cos k_2\ell_2$$

$$D = \frac{\delta\cos k_2\ell_2\cos k_1\ell}{\sin k_1\ell\cos k_1\ell_2 - \sin k_1\ell_2\cos k_1\ell}$$

$$= \frac{\delta\cos k_2\ell_2\cos k_1\ell}{\sin k_1(\ell_1 + \ell_2)\cos k_1\ell_2 - \sin k_1\ell_2\cos k_1(\ell_1 + \ell_2)}$$

$$= \frac{\delta\cos k_2\ell_2\cos k_1\ell}{\sin k_1\ell_1}$$

$$C = -\tan k_1\ell\frac{\delta\cos k_2\ell_2\cos k_1\ell}{\sin k_1\ell_1} = \frac{\delta\cos k_2\ell_2\sin k_1\ell}{\sin k_1\ell_1}$$

The continuity condition that the two segments of the deflected curve have the same slope at the juncture ($@x = \ell_2$) gives

$$\delta k_2\sin k_2\ell_2 = -Ck_1\sin k_1\ell_1 + Dk_1\cos k_1\ell_2$$

$$= -\frac{\delta\cos k_2\ell_2\sin k_1\ell}{\sin k_1\ell_1}k_1\sin k_1\ell_2$$

$$+\frac{\delta\cos k_2\ell_2\cos k_1\ell}{\sin k_1\ell_1}k_1\cos k_1\ell_2$$

Rearranging gives

$$\begin{aligned} k_2 \sin k_2\ell_2 \sin k_1\ell_1 &= k_1 \cos k_2\ell_2(\sin k_1\ell_1 \cos k_1\ell_2 \\ &\quad + \cos k_1\ell_1 \sin k_1\ell_2) \sin k_1\ell_2 \\ &\quad + k_1 \cos k_2\ell_2(\cos k_1\ell_1 \cos k_1\ell_2 \\ &\quad - \sin k_1\ell_1 \sin k_1\ell_2) \cos k_1\ell_2 \\ &= k_1(\cos k_1\ell_1 \cos k_2\ell_2) \end{aligned}$$

which leads to $\tan k_1\ell_1 \tan k_2\ell_2 = \dfrac{k_1}{k_2} \Leftarrow$ stability condition equation.

The same stability condition equation can be obtained by setting the coefficient determinant equal to zero. There are a total of four integral constants to be determined. As the governing differential equation is in second order, only one boundary condition at each support is to be used. Hence, the other two conditions are to be extracted from the continuity condition as used above.

$$y_2' = 0 \quad \text{at } x = 0 \Rightarrow B = 0 \Rightarrow y_2 = \delta + A\cos k_2 x \tag{a}$$

$$y_1 = \delta \ (\text{or } y_1'' = 0) \quad \text{at } x = \ell \Rightarrow C\cos k_1\ell + D\sin k_1\ell = 0 \tag{b}$$

$$y_1 = y_2 \quad \text{at } x = \ell_2 \Rightarrow A\cos k_2\ell_2 - C\cos k_1\ell_2 - D\sin k_1\ell_2 = 0 \tag{c}$$

$$y_1' = y_2' \quad \text{at } x = \ell_2 \Rightarrow Ak_2\sin k_2\ell_2 - Ck_1\sin k_1\ell_2 + Dk_1\cos k_1\ell_2 = 0 \tag{d}$$

Setting the determinant for the coefficients, A, C, and D equal to zero yields the identical stability condition equation. This process can be expedited using a computer program capable of symbolic computations, such as **Maple®**.

Knowing I_1/I_2 and ℓ_1/ℓ_2, the solution of the transcendental equation can be found. By substituting $a/2$ for ℓ_2 and $\ell/2$ for ℓ, the result obtained can be directly applied to the column shown in sketch (b). Coefficient m for $P_{cr} = mEI_2/\ell^2$ is given in Table 2-2.

Upon executing the transcendental equation identified as the stability condition equation above, the table is somewhat confusing. The table should be used for the case shown in sketch (b). For the case of stepped columns shown in sketch (a), values for m should be taken from Table 2-3.

Table 2-2 Buckling coefficients for stepped columns, Fig. 2-15(b)

I_1/I_2	a/ℓ			
	0.2	0.4	0.6	0.8
0.01	0.15	0.27	0.60	2.26
0.1	1.47	2.40	4.50	8.59
0.2	2.80	4.22	6.69	9.33
0.4	5.09	6.68	8.51	9.67
0.6	6.98	8.19	9.24	9.78
0.8	8.55	9.18	9.63	9.84

Table 2-3 Buckling coefficients, Fig. 2-15(a)

I_1/I_2	ℓ_2/ℓ			
	0.2	0.4	0.6	0.8
0.01	0.038	0.068	0.150	0.563
0.1	0.367	0.600	1.124	2.147
0.2	0.699	1.056	1.674	2.332
0.4	1.272	1.669	2.127	2.419
0.6	1.745	2.046	2.311	2.446
0.8	2.138	2.294	2.408	2.459

Consider a stepped cantilever column similar to that shown in Fig. 2-15 (a). The length of each segment is 20 inches. The cross-sectional area of the bottom segment is 4 in^2 and the upper segment is 1 in^2 The modulus of elasticity of the material is assumed to be 29,000 ksi. The stability condition equation now becomes

$$\tan(80k_2)\tan(20k_2) = 4$$

Maple® gives $k_2 = 0.0184315$, which leads to $P_{cr} = 13.136$ kips. A computer program based on the differential equation such as STSTB (Yoo 1980) also gives the same critical load. However, a modern-day finite element program such as ABAQUS (2006) gives the critical load of 9.928 kips. The lower value (32%) is considered to be much more realistic. Since a large portion of the cross-sectional area (at least 75%) of the segment 2 around the juncture cannot participate in carrying the load due to discontinuity, the realistic critical load is expected to be less than that computed assuming all parts of segment 2 are effective as used in the case of modeling the differential equation method. This is a classical example of the class of the analytical techniques employed. A solution is as good as the assumptions employed.

There is another important lesson to learn. In the early days of telescopic power cylinder development, manufacturers reduced the elastic critical load substantially by a "knock down factor" of up to five. Yet, they witnessed a large number of field failures in that the piston was digging into the cylinder wall. Years later, they discovered that this was caused by the slack of the phenolic ring due to wear, thereby providing an initial imperfection.

2.8. METHODS OF SUCCESSIVE APPROXIMATION

2.8.1. Solution of Buckling Problems by Finite Differences

Because of the rapid development of finite element method in structural mechanics, the application of finite differences became only of historical interest. However, it was probably one of the main numerical techniques for solving complex structural mechanics problems in bygone years, and it is still frequently applied in other discipline areas such as hydraulics. The finite difference technique is merely replacing the derivatives in a differential equation and is solving the resulting linear simultaneous equation numerically. Hence, one must have the differential equation(s) and accompanying boundary conditions to apply finite differences (cf. in the finite element method, one does not need to have the differential equations and accompanying boundary conditions). The finite difference technique applied to a one-dimensional problem is illustrated. A one-dimensional field can be approximated by a Taylor series expansion as

$$f(x) = f(a) + f'(a)\frac{(x-a)}{1!} + f''(a)\frac{(x-a)^2}{2!} + f'''(a)\frac{(x-a)^3}{3!} + \cdots + f^{(n)}(a)\frac{(x-a)^n}{n!} \tag{2.8.1}$$

Accuracy of FD Method

At $x = a + \lambda$

$$y_r = y_o + \lambda y_0' + \frac{\lambda^2}{2!}y_o'' + \frac{\lambda^3}{3!}y_o''' + \frac{\lambda^4}{4!}y_o^{iv} + \frac{\lambda^5}{5!}y_o^{v} + \cdots \tag{2.8.2}$$

At $x = a + 2\lambda$

$$y_{rr} = y_o + 2\lambda y_o' + \frac{4\lambda^2}{2!}y_o'' + \frac{8\lambda^3}{3!}y_o''' + \frac{16\lambda^4}{4!}y_o^{iv} + \frac{32\lambda^5}{5!}y_o^{v} + \cdots \tag{2.8.3}$$

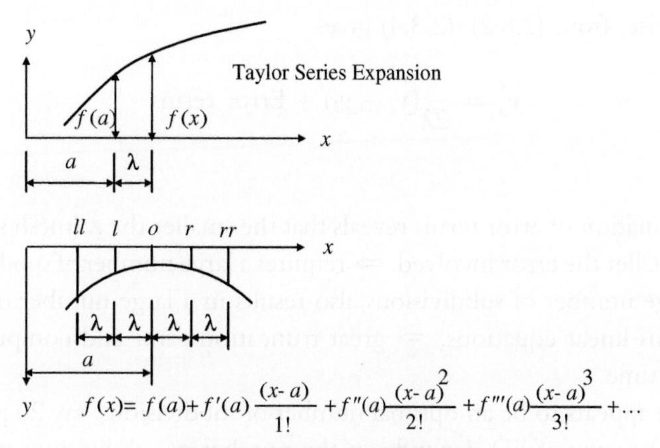

Figure 2-16 Finite differences

At $x = a - \lambda$

$$y_l' = y_o - \lambda y_0' + \frac{\lambda^2}{2!}y_o'' - \frac{\lambda^3}{3!}y_o''' + \frac{\lambda^4}{4!}y_o^{iv} - \frac{\lambda^5}{5!}y_o^{v} + \cdots \quad (2.8.4)$$

At $x = a - 2\lambda$

$$y_{ll} = y_o - 2\lambda y_o' + \frac{4\lambda^2}{2!}y_o'' - \frac{8\lambda^3}{3!}y_o''' + \frac{16\lambda^4}{4!}y_o^{iv} - \frac{32\lambda^5}{5!}y_o^{v} + \cdots \quad (2.8.5)$$

Adding Eqs. (2.8.2) and (2.8.4) gives

$$y_r + y_l = 2y_o + \frac{2\lambda^2}{2!}y_o'' + \frac{2\lambda^4}{4!}y_o^{iv} + \frac{2\lambda^6}{6!}y_o^{vi} +$$

$$y_o'' = \frac{1}{\lambda^2}(y_r - 2y_o + y_l) - \frac{1}{12}\lambda^2 y_o^{iv} - \frac{1}{360}\lambda^4 y_o^{vi} - \cdots$$

$$\text{Error terms} = -\frac{1}{12}\lambda^2 y_o^{iv} - \frac{1}{360}\lambda^4 y_o^{vi} - \cdots$$

Adding Eqs. (2.8.3) and (2.8.5) gives

$$y_{rr} + y_{ll} = 2y_o + \frac{8\lambda^2}{2!}y_o'' + \frac{32\lambda^4}{4!}y_o^{iv} + \frac{64\lambda^6}{6!}y_o^{vi} + \cdots$$

Substituting y_o'' expression into the above and rearranging yields

$$y_o^{iv} = \frac{1}{\lambda^4}(y_{rr} - 4y_r + 6y_o - 4y_l + y_{ll}) + \text{Error terms}$$

Similarly, subtracting Eq. (2.8.5) from Eq. (2.8.3) gives

$$y_o''' = \frac{1}{2\lambda^3}(y_{rr} - 2y_r + 2y_l - y_{ll}) + \text{Error terms}$$

Likewise, from (2.8.2)–(2.8.4) gives

$$y'_o = \frac{1}{2\lambda}(y_r - y_l) + \text{Error terms}$$

Note

1. Examination of error terms reveals that the smaller the λ (mesh spacing), the smaller the error involved. ⇒ requires a large number of subdivisions.
2. A large number of subdivisions also results in a large number of simultaneous linear equations. ⇒ great truncation error and computational CPU time.
3. There appears to be an optimal number of subdivisions, say 20 per span.
4. The accuracy of FD depends on the number of subdivisions per span, not the absolute numerical value of mesh spacing, λ. A span of 2" in length with three nodal points gives $\lambda = 1$" while a 100' span with 11 nodal points yields $\lambda = 12$," which gives a better solution.

Example 1 Consider a both-end pinned prismatic column. The column is subdivided into four equal segments, and a node point is assigned on each quarter point and the ends. Mesh equations will be generated at three interior (load) points based on the governing differential equation and proper boundary conditions.

$$y'' + k^2 y = 0, \quad \text{with } k^2 = \frac{P}{EI}, \quad y'' \doteq \frac{1}{\lambda^2}(y_r - 2y_o + y_l)$$

The finite difference mesh equation (or load point) at $x = \lambda$ is:

$$\frac{y_2 - 2y_1 + 0}{\lambda^2} + k^2 y_1 = 0 \Rightarrow \left(\lambda^2 k^2 - 2\right) y_1 + y_2 = 0 \tag{a}$$

The same at $x = 2\lambda$ is:

$$\frac{y_3 - 2y_2 + y_1}{\lambda^2} + k^2 y_2 = 0 \Rightarrow \left(\lambda^2 k^2 - 2\right) y_2 + y_1 + y_3 = 0 \tag{b}$$

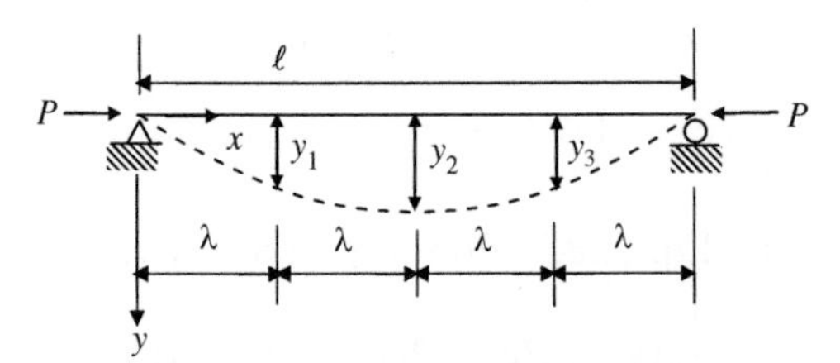

Figure 2-17 Five-node finite difference model

At $x = 3\lambda$

$$\frac{0 - 2y_3 + y_2}{\lambda^2} + k^2 y_3 = 0 \Rightarrow (\lambda^2 k^2 - 2)y_3 + y_2 = 0 \qquad \text{(c)}$$

In a matrix form

$$\begin{bmatrix} \lambda^2 k^2 & 1 & 0 \\ 1 & \lambda^2 k^2 & 1 \\ 0 & 1 & \lambda^2 k^2 \end{bmatrix} \begin{Bmatrix} y_1 \\ y_2 \\ y_3 \end{Bmatrix} = \begin{Bmatrix} 0 \\ 0 \\ 0 \end{Bmatrix}$$

Setting the coefficient determinant equal to zero yields

$$(\lambda^2 k^2 - 2)^3 - 2(\lambda^2 k^2 - 2) = 0$$

Let $(\lambda^2 k^2 - 2) = R \Rightarrow R^3 - 2R = 0 \Rightarrow R = 0 \text{ or } \pm\sqrt{2}$

$$(\lambda^2 k^2 - 2) = -\sqrt{2} \Rightarrow k^2 = \frac{2 - \sqrt{2}}{\lambda^2} = \frac{.58579}{(\ell/4)^2} \Rightarrow$$

$$P_{cr}^1 = \frac{9.3725EI}{\ell^2} \left(5\% \text{ less than } \frac{\pi^2 EI}{\ell^2} \right)$$

$$(\lambda^2 k^2 - 2) = 0 \Rightarrow k^2 = \frac{2}{\lambda^2} = \frac{2}{(\ell/4)^2} \Rightarrow$$

$$P_{cr}^2 = \frac{32EI}{\ell^2} \left(18.9\% \text{ less than} \frac{4\pi^2 EI}{\ell^2} \right)$$

$$(\lambda^2 k^2 - 2) = \sqrt{2} \Rightarrow k^2 = \frac{2 + \sqrt{2}}{\lambda^2} = \frac{3.4142}{(\ell/4)^2} \Rightarrow$$

$$P_{cr}^3 = \frac{54.627EI}{\ell^2} \left(38.5\% \text{ less than } \frac{9\pi^2 EI}{\ell^2} \right)$$

Note

1. The finite difference method gives lower-bound solutions.
2. More subdivisions ⇒ better convergence.
3. Higher modes deviate more from the exact solution.

The two-dimensional Taylor series expansion of a function, $f(x, y)$, near a point $P(x_p, y_p)$ is given by

$$\begin{aligned} f(x,y) &= f(x_p, y_p) + f_x(x_p, y_p)(x - x_p) + f_y(x_p, y_p)(y - y_p) \\ &+ \frac{1}{2!}\Big[f_{xx}(x_p, y_p)(x - x_p)^2 + f_{yy}(x_p, y_p)(y - y_p)^2 \\ &+ 2 f_{xy}(x_p, y_p)(x - x_p)(y - y_p) \Big] \end{aligned}$$

$$+\frac{1}{3!}\left[\begin{array}{l} f_{xxx}(x_p,y_p)(x-x_p)^3 + f_{yyy}(x_p,y_p)(y-y_p)^3 + 3f_{xxy}(x_p,y_p)(x-x_p)^2(y-y_p) \\ \quad + 3f_{xyy}(x_p,y_p)(x-x_p)(y-y_p)^2 \end{array}\right]$$

$$+\frac{1}{4!}\left[\begin{array}{l} f_{xxxx}(x_p,y_p)(x-x_p)^4 + f_{yyyy}(x_p,y_p)(y-y_p)^4 + 4f_{xxxy}(x_p,y_p)(x-x_p)^3(y-y_p) \\ \quad + 4f_{xyyy}(x_p,y_p)(x-x_p)(y-y_p)^3 + 6f_{xxyy}(x_p,y_p)(x-x_p)^2(y-y_p)^2 \end{array}\right]$$

$$+\frac{1}{5!}\left[\quad\right] + \ldots\ldots$$

The three-dimensional Taylor series expansion of a function, $f(x,y,z)$, near a point $P(x_p, y_p, z_p)$ is given by

$$\begin{aligned} f(x,y,z) &= f(x_p,y_p,z_p) + f_x(x_p,y_p,z_p)(x-x_p) + f_y\big(x_p,y_p,z_p\big)\big(y-y_p\big) \\ &\quad + f_z\big(x_p,y_p,z_p\big)(z-z_p) \end{aligned}$$

$$+\frac{1}{2!}\left[\begin{array}{l} f_{xx}(x_p,y_p,z_p)(x-x_p)^2 + f_{yy}(x_p,y_p,z_p)(y-y_p)^2 + f_{zz}(x_p,y_p,z_p)(z-z_p)^2 \\ \quad + 2f_{xy}(x_p,y_p,z_p)(x-x_p)(y-y_p) + 2f_{yz}(x_p,y_p,z_p)(y-y_p)(z-z_p) \\ \quad + 2f_{xz}(x_p,y_p,z_p)(x-x_p)(z-z_p) \end{array}\right]$$

$$+\frac{1}{3!}\left[\begin{array}{l} f_{xxx}(x_p,y_p,z_p)(x-x_p)^3 + f_{yyy}(x_p,y_p,z_p)(y-y_p)^3 + f_{zzz}(x_p,y_p,z_p)(z-z_p)^3 \\ \quad + 3f_{xxy}(x_p,y_p,z_p)(x-x_p)^2(y-y_p) + 3f_{xyy}(x_p,y_p,z_p)(x-x_p)(y-y_p)^2 \\ \quad + 3f_{yyz}(x_p,y_p,z_p)(y-y_p)^2(z-z_p) + 3f_{yzz}(x_p,y_p,z_p)(y-y_p)(z-z_p)^2 \\ \quad + 3f_{xxz}(x_p,y_p,z_p)(x-x_p)^2(z-z_p) + 3f_{xzz}(x_p,y_p,z_p)(x-x_p)(z-z_p)^2 \end{array}\right]$$

$$+\frac{1}{4!}\left[\begin{array}{l} f_{xxxx}(x_p,y_p,z_p)(x-x_p)^4 + f_{yyyy}(x_p,y_p,z_p)(y-y_p)^4 + f_{zzz}(x_p,y_p,z_p)(z-z_p)^4 \\ \quad + 4f_{xxxy}(x_p,y_p,z_p)(x-x_p)^3(y-y_p) + 4f_{xxxz}(x_p,y_p,z_p)(x-x_p)^3(z-z_p) \\ \quad + 4f_{xyyy}(x_p,y_p,z_p)(x-x_p)(y-y_p)^3 + 4f_{yyyz}(x_p,y_p,z_p)(y-y_p)^3(z-z_p) \\ \quad + 4f_{xzzz}(x_p,y_p,z_p)(x-x_p)(z-z_p)^3 + 4f_{yzzz}(x_p,y_p,z_p)(y-y_p)(z-z_p)^3 \\ \quad + 6f_{xxyy}(x_p,y_p,z_p)(x-x_p)^2(y-y_p)^2 + 6f_{yyzz}(x_p,y_p,z_p)(y-y_p)^2(z-z_p)^2 \\ \quad + 6f_{xxzz}(x_p,y_p,z_p)(x-x_p)^2(z-z_p)^2 \end{array}\right] \cdots$$

where $f_x = \dfrac{\partial f}{\partial x}$, $f_{xy} = \dfrac{\partial^2 f}{\partial x \partial y}$, $f_{xxx} = \dfrac{\partial^3 f}{\partial x^3}$, etc.

The transformation of a partial derivative into a finite difference equation can be accomplished in a manner similar to that used for the ordinary derivatives shown earlier.

2.8.2. Newmark's Procedure

Newmark (1943) published a procedure of computing deflection, moments, and buckling loads. Although this procedure is old, it is still an effective method, particularly for nonprismatic members subjected to complex loading, including the elastic buckling loads for multiple stage telescopic power cylinders. As his procedure is reasonably fast converging, it does not usually require iterations more than three times. Experience has shown that the simplified equations for the linearly varying loads are equally effective in all problems solved.

For an infinitesimal element shown in Fig. 2-18(a), an equilibrium consideration immediately yields the following relationships:

$$\frac{dv}{dx} = -q \tag{2.8.6}$$

$$\frac{dm}{dx} = -v \tag{2.8.7}$$

The moment-area theorems to be reviewed in Chapter 3 give the following relationships:

$$d\theta = \frac{mdx}{EI} \tag{2.8.8}$$

$$y = \theta dx \tag{2.8.9}$$

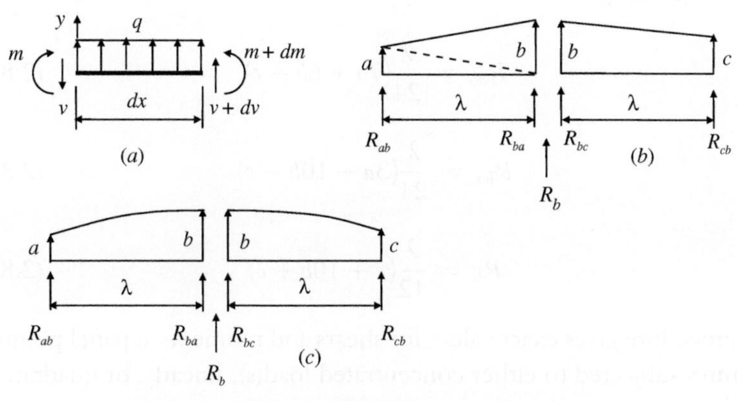

Figure 2-18 Equivalent concentrated reactions

Equations (2.8.6)–(2.8.9) can be converted as

$$\Delta v = \int q dx \tag{2.8.10}$$

$$\Delta m = \int v dx \tag{2.8.11}$$

$$\Delta \theta = \int \frac{m dx}{EI} \tag{2.8.12}$$

$$\Delta y = \int \theta dx \tag{2.8.13}$$

The equivalent panel point loads for a linearly varying load shown in Fig. 2-18(b) can be computed using Eqs. (2.8.14)–(2.8.16).

$$R_{ab} = \frac{\lambda}{6}(2a + b) \tag{2.8.14}$$

$$R_{ba} = \frac{\lambda}{6}(a + 2b) \tag{2.8.15}$$

$$R_b = \frac{\lambda}{6}(a + 4b + c) \tag{2.8.16}$$

Likewise, equivalent panel point loads for any distributed loads shown in Fig. 2-18(c) following higher order curves can be computed by Eqs. (2.8.17)–(2.8.19).

$$R_{ab} = \frac{\lambda}{24}(7a + 6b - c) \tag{2.8.17}$$

$$R_{ba} = \frac{\lambda}{24}(3a + 10b - c) \tag{2.8.18}$$

$$R_b = \frac{\lambda}{12}(a + 10b + c) \tag{2.8.19}$$

The procedure gives exact values for shears and moments at panel points for structures subjected to either concentrated load(s), linearly, or quadratically varying load(s).

Example 1

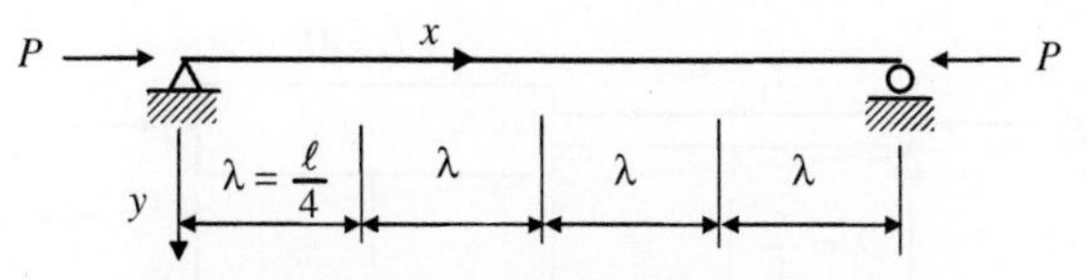

Figure 2-19 Four-node Newmark example

										Common Factor
Assumed y_1	0.00		0.70		1.00		0.70		0.00	1
$m = Py_1$	0.00		0.70		1.00		0.70		0.00	P
$y'' = -m/EI$	0.00		−0.70		−1.00		−0.70		0.00	P/EI
R[1] (α)			−0.80		−1.14		−0.80			$P\lambda/1.2EI$
Average slope		13.7		5.7		−5.7		−13.7		$P\lambda/12EI$
$\Delta y = \theta\lambda$		13.7		5.7		−5.7		−13.7		$P\lambda^2/12EI$
y_2	0.00		13.70		19.40		13.70		0.00	$P\lambda^2/12EI$
y_1/y_2	0.00		5.11		5.15		5.11		0.00	$0.12EI/P\lambda^2$
Average y_1/y_2					51.23					$0.012EI/P\lambda^2$
$m = Py_2$	0.00		13.70		19.40		13.70		0.00	P
$y'' = -m/EI$	0.00		−13.70		−19.40		−13.70		0.00	P/EI
R[1] (α)			−15.64		−22.14		−15.64			$P\lambda/1.2EI$
Average slope		267.1		110.7		−110.7		−267.1		$P\lambda/12EI$
$\Delta y = \theta\lambda$		267.1		110.7		−110.7		−267.1		$P\lambda^2/12EI$
y_3	0.00		26.71		37.78		26.71		0.00	$P\lambda^2/1.2EI$
y_2/y_3			51.29		51.38		51.29			$0.012EI/P\lambda^2$
Average y_2/y_3					51.32					$0.012EI/P\lambda^2$

[1] $\alpha_1 = (\lambda/12)(0 - 7 - 1) = -(8\lambda/12), \alpha_2 = (\lambda/12)(-7 - 10 - 7)$
$= -(11.4\lambda/12), \alpha_3 = \alpha_1 \Leftarrow$ parabolic equation

To find the average slope at the end panel, $\theta = (1/2)(8 + 11.4 + 8) = 13.7$

[1] $$\alpha_1 = \frac{\lambda}{12}(0 - 13.7 - 19.4) = -\frac{156.4\lambda}{12},$$

$$\alpha_2 = \frac{\lambda}{12}(-13.7 - 19.4 - 13.7) = -\frac{221.4\lambda}{12}, \alpha_3 = \alpha_1$$

To find the average slope at the end panel, $\theta = (1/2)(156.4 + 221.1 + 156.4) = 267.1$

At the end of the first cycle, $(y_1/y_2) = 51.23\,(0.012EI/P\lambda^2)$,

$$P_{cr} = ([51.23 \times 0.012EI]/\lambda^2) = (9.836EI/\ell^2)$$

At the end of the second cycle, $(y_2/y_3) = 51.32\,(0.012EI/P\lambda^2)$,

$$P_{cr} = ([51.32 \times 0.012EI]/\lambda^2) = 9.853EI/\ell^2$$

The critical value converges to the exact value of $P_E = (9.87EI/\ell^2)$.

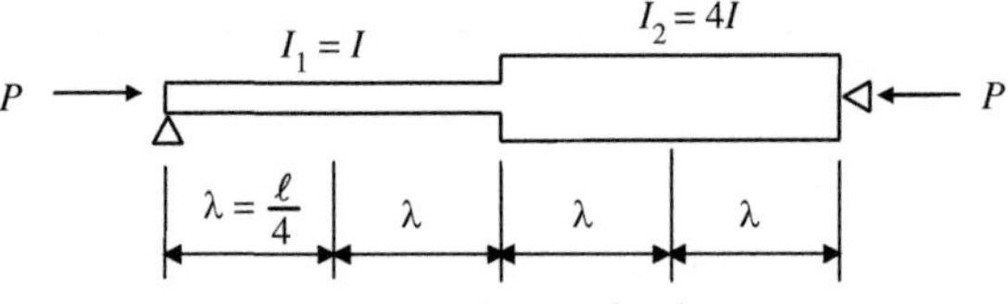

Figure 2-20 Stepped column

Example 2

										Common Factor
Assumed y_1	0		64		100		64		0	1
$m = Py_1$	0		64		100		64		0	P
$y'' = -m/EI$	0		−64		−100		−16		0	P/EI
					−25					
$\overset{1}{R}$ (α)			−356		−330		−89			$P\lambda/6EI$
Trial slope		400		44		−286		−375		$P\lambda/6EI$
Trial Δy		400		44		−286		−375		$P\lambda^2/6EI$
$\bar{y}_2$	0		400		444		158		−217	$P\lambda^2/6EI$
Lin Cor, $\Delta\bar{y}_2$	0		54		108		165		217	$P\lambda^2/6EI$
y_2	0		454		552		325		0	$P\lambda^2/6EI$
y_1/y_2			14		18		20			$6EI/100\lambda^2P$
Average					17					$6EI/100\lambda^2P$
End of 1st y_3	0		2934		3500		1955			$P\lambda^2/6EI$
y_2/y_3			16		16		17			$6EI/100\lambda^2P$
Average					16					$6EI/100\lambda^2P$
End 2nd y_4	0		1822		2120		1201			$P\lambda^2/6EI$
			1		5		7			
y_3/y_4			16		16		16			$6EI/100\lambda^2P$
Average					16					$6EI/100\lambda^2P$

$$R_b = \frac{\lambda}{6}(a + 4b + c) = \frac{\lambda}{6}[0 + 4(-64) + (-100)]$$

$$= \frac{\lambda}{6}(-356) \Leftarrow \text{linear equation}$$

$$R_{cb} = \frac{\lambda}{6}(b + 2c) = \frac{\lambda}{6}(-64 - 2 \times 100) = \frac{\lambda}{6}(-264)$$

$$R_{cd} = \frac{\lambda}{6}(2c + d) = \frac{\lambda}{6}(-25 \times 2 - 16) = \frac{\lambda}{6}(-66)$$

$$R_d = \frac{\lambda}{6}(c + 4d + e) = \frac{\lambda}{6}(-25 - 4 \times 16 + 0) = \frac{\lambda}{6}(-89)$$

$$\left(\frac{y_3}{y_4}\right)_{avg} = 16\frac{6EI}{100\lambda^2 P} \Rightarrow P_{cr} = \frac{16 \times 6EI}{100(\ell/4)^2} = \frac{15.36EI}{\ell^2}$$

The convergence trend may be monotonic or oscillatory.

2.9. MATRIX METHOD

2.9.1. Derivation of Element Geometric Stiffness Matrix

Consider a prismatic column shown in Fig. 1-12. The axial strain of a point at a distance y from the neutral axis is

$$\varepsilon_x = \frac{du}{dx} - y\frac{d^2v}{dx^2} + \frac{1}{2}\left(\frac{dv}{dx}\right)^2 \tag{2.9.1}$$

where u and v are displacement components in the x and y directions, respectively, and

du/dx = axial strain;

$-y(d^2v)/(dx^2)$ = strain produced by curvature; and

$1/2[(dv)/(dx)]^2$ = nonlinear part of the axial strain.

With $dV = dAdx$, the element strain energy is

$$U = \frac{1}{2}\int_V \varepsilon\sigma dV = \frac{1}{2}\int_\ell \int_A E\,\varepsilon_x^2 dAdx \tag{2.9.2}$$

where E = modulus of elasticity.

Substituting Eq. (2.9.1) into Eq. (2.9.2) and recalling that

$$\int_A dA = A, \int_A y\,dA = 0, \int_A y^2\,dA = I, \quad \text{and} \quad \int_A E\frac{du}{dx}dA = P$$

where P is the axial force, positive in tension, leads the strain energy to be written:

$$U = \frac{1}{2}\int_0^\ell EA\left(\frac{du}{dx}\right)^2 dx + \frac{1}{2}\int_0^\ell EI\left(\frac{d^2v}{dx^2}\right)^2 dx + \frac{1}{2}\int_0^\ell P\left(\frac{dv}{dx}\right)^2 dx \tag{2.9.3}$$

The first integral in Eq. (2.9.3) yields the stiffness matrix for a bar element associated with the kinematic degrees of freedom u_1 and u_2. The second integral yields the stiffness matrix for a beam element. The third integral sums the work done by the external load P when differential elements dx are stretched by an amount $[(dv/dx)^2 \times dx/2]$ (there exists another interpretation of the third integral: a change in the potential energy of the applied load during buckling). The third integral leads to the derivation of the element geometric stiffness matrix K_G.

The lateral displacement field v of the beam and its derivative dv/dx are

$$v = \lfloor N \rfloor \{\Delta\} \tag{2.9.4}$$

$$\frac{dv}{dx} = \frac{d\lfloor N \rfloor}{dx}\{\Delta\} = \lfloor G \rfloor\{\Delta\} \tag{2.9.5}$$

where

$$\lfloor \Delta \rfloor = \lfloor v_1 \quad \theta_1 \quad v_2 \quad \theta_2 \rfloor \tag{2.9.6}$$

$$\lfloor N \rfloor = \left\lfloor 1 - \frac{3x^2}{\ell^2} + \frac{2x^3}{\ell^3} \quad x - \frac{2x^2}{\ell} + \frac{x^3}{\ell^2} \quad \frac{3x^2}{\ell^2} - \frac{2x^3}{\ell^3} \quad -\frac{x^2}{\ell} + \frac{x^3}{\ell^2} \right\rfloor \tag{2.9.7}$$

$$\lfloor G \rfloor = \left\lfloor -\frac{6x}{\ell^2} + \frac{6x^2}{\ell^3} \quad 1 - \frac{4x}{\ell} + \frac{3x^2}{\ell^2} \quad \frac{6x}{\ell^2} - \frac{6x^2}{\ell^3} \quad -\frac{2x}{\ell} + \frac{3x^2}{\ell^2} \right\rfloor \tag{2.9.8}$$

The third integral is expanded as

$$\frac{1}{2}\lfloor \Delta \rfloor [K_G]\{\Delta\} = \frac{1}{2}\lfloor \Delta \rfloor \left[P \int_0^\ell \{G\}\lfloor G \rfloor dx \right] \{\Delta\} \tag{2.9.9}$$

Hence,

$$K_{G11} = P \int_0^\ell \left(-\frac{6x}{\ell^2} + \frac{6x^2}{\ell^3} \right)^2 dx = \frac{6P}{5}$$

$$K_{G12} = P \int_0^\ell \left(-\frac{6x}{\ell^2} + \frac{6x^2}{\ell^3} \right)\left(1 - \frac{4x}{\ell} + \frac{3x^2}{\ell^2} \right) dx = \frac{P}{10}$$

Other elements are evaluated likewise.

$$K_G = \frac{P}{30\ell}\begin{bmatrix} 36 & 3\ell & -36 & 3\ell \\ 3\ell & 4\ell^2 & -3\ell & -\ell^2 \\ -36 & -3\ell & 36 & -3\ell \\ 3\ell & -\ell^2 & -3\ell & 4\ell^2 \end{bmatrix} \tag{2.9.10}$$

It should be noted that P in Eq. (2.9.10) is positive when it is tension.

2.9.2. Application

Consider a propped (fixed-pinned) column shown in Fig. 2-21. The prismatic column length is L. Using the numbering scheme, one obtains the following stiffness relationship: As the global coordinate system and the local

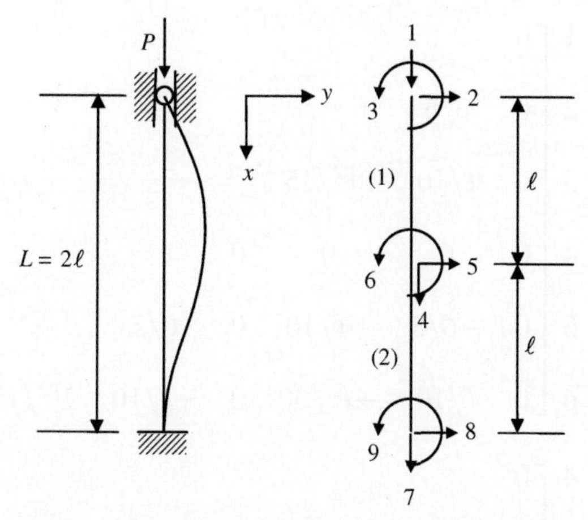

Figure 2-21 Column model, degrees-of-freedom

coordinate system are identical, there is no need for coordinate transformation.

Let $\phi = A\ell^2/I$.

Superimposing element stiffness matrices of bar element and beam element, one obtains an element stiffness matrix for a two-dimensional frame element.

$$K_E^{(1)} = \frac{EI}{\ell^3} \begin{matrix} 1 \\ 2 \\ 3 \\ 4 \\ 5 \\ 6 \end{matrix} \begin{bmatrix} \phi & & & & & \\ 0 & 12 & & & & \\ 0 & 6\ell & 4\ell^2 & & & \\ -\phi & 0 & 0 & \phi & & \\ 0 & -12 & -6\ell & 0 & 12 & \\ 0 & 6\ell & 2\ell^2 & 0 & -6\ell & 4\ell^2 \end{bmatrix} \tag{2.9.11}$$

$$K_E^{(2)} = \frac{EI}{\ell^3} \begin{matrix} 4 \\ 5 \\ 6 \\ 7 \\ 8 \\ 9 \end{matrix} \begin{bmatrix} \phi & & & & & \\ 0 & 12 & & & & \\ 0 & 6\ell & 4\ell^2 & & & \\ -\phi & 0 & 0 & \phi & & \\ 0 & -12 & -6\ell & 0 & 12 & \\ 0 & 6\ell & 2\ell^2 & 0 & -6\ell & 4\ell^2 \end{bmatrix} \tag{2.9.12}$$

$$K_G^{(1)} = -\frac{P}{\ell}\begin{array}{c} 1 \\ 2 \\ 3 \\ 4 \\ 5 \\ 6 \end{array}\begin{bmatrix} 0 & & & & & \\ 0 & 6/5 & & & & \\ 0 & \ell/10 & 2\ell^2/15 & & & \\ 0 & 0 & 0 & 0 & & \\ 0 & -6/5 & -\ell/10 & 0 & 6/5 & \\ 0 & \ell/10 & -\ell^2/30 & 0 & -\ell/10 & 2\ell^2/15 \end{bmatrix} \tag{2.9.13}$$

$$K_G^{(2)} = -\frac{P}{\ell}\begin{array}{c} 4 \\ 5 \\ 6 \\ 7 \\ 8 \\ 9 \end{array}\begin{bmatrix} 0 & & & & & \\ 0 & 6/5 & & & & \\ 0 & \ell/10 & 2\ell^2/15 & & & \\ 0 & 0 & 0 & 0 & & \\ 0 & -6/5 & -\ell/10 & 0 & 6/5 & \\ 0 & \ell/10 & -\ell^2/30 & 0 & -\ell/10 & 2\ell^2/15 \end{bmatrix} \tag{2.9.14}$$

The elastic stiffness matrices K_E and the stability matrices K_G can now be assembled, reduced, and rearranged, separating the degrees of freedom associated with the axial deformations and the flexural deformations, respectively. Assembling the element stiffness matrices to construct the structural stiffness matrix is of course to combine the element contribution to the global stiffness. Reducing the assembled stiffness matrix is necessary to eliminate the rigid body motion, thereby making the structural stiffness matrix nonsingular.

$$K_E = \frac{EI}{\ell^3}\begin{array}{c} 1 \\ 4 \\ 3 \\ 5 \\ 6 \end{array}\begin{bmatrix} \phi & -\phi & 0 & 0 & 0 \\ -\phi & 2\phi & 0 & 0 & 0 \\ 0 & 0 & 4\ell^2 & -6\ell & 2\ell^2 \\ 0 & 0 & -6\ell & 24 & 0 \\ 0 & 0 & 2\ell^2 & 0 & 8\ell^2 \end{bmatrix} \tag{2.9.15}$$

$$K_G = -\frac{P}{\ell}\begin{matrix}1\\4\\3\\5\\6\end{matrix}\begin{bmatrix}0 & 0 & 0 & 0 & 0\\ 0 & 0 & 0 & 0 & 0\\ 0 & 0 & 2\ell^2/15 & -\ell/10 & -\ell^2/30\\ 0 & 0 & -\ell/10 & 12/5 & 0\\ 0 & 0 & -\ell^2/30 & 0 & 4\ell^2/15\end{bmatrix} \tag{2.9.16}$$

Noting that K_G^* is equal to K_G for $P = 1$, one can set up the stability determinant $|K_E + \lambda K_G^*| = 0$. This leads to

$$\begin{vmatrix}\phi & -\phi & 0 & 0 & 0\\ -\phi & 2\phi & 0 & 0 & 0\\ 0 & 0 & 4\ell^2 - \frac{2}{15}\frac{\lambda\ell^2}{EI} & -6\ell + \frac{1}{10}\frac{\lambda\ell^3}{EI} & 2\ell^2 + \frac{1}{30}\frac{\lambda\ell^4}{EI}\\ 0 & 0 & -6\ell + \frac{1}{10}\frac{\lambda\ell^3}{EI} & 24 - \frac{12}{5}\frac{\lambda\ell^2}{EI} & 0\\ 0 & 0 & 2\ell^2 + \frac{1}{30}\frac{\lambda\ell^4}{EI} & 0 & 8\ell^2 - \frac{4}{15}\frac{\lambda\ell^4}{EI}\end{vmatrix} = 0 \tag{2.9.17}$$

Let $\mu = \lambda\ell^2/EI$.
Then, Eq. (2.9.7) simplifies to

$$\begin{vmatrix}\phi & -\phi & 0 & 0 & 0\\ -\phi & 2\phi & 0 & 0 & 0\\ 0 & 0 & 2\left(2 - \frac{\mu}{15}\right) & -6 + \frac{\mu}{10} & 2 + \frac{\mu}{30}\\ 0 & 0 & -6 + \frac{\mu}{10} & 12\left(2 - \frac{\mu}{5}\right) & 0\\ 0 & 0 & 2 + \frac{\mu}{30} & 0 & 4\left(2 - \frac{\mu}{15}\right)\end{vmatrix} = 0$$

Expanding this determinant, one obtains a cubic equation in μ

$$3\mu^3 - 220\mu^2 + 3,840\mu - 14,400 = 0$$

The lowest root of this equation is $\mu = 5.1772 \Rightarrow 5.1772 = \lambda\ell^2/EI$

Hence,

$$P_{cr} = \frac{5.1772EI}{\ell^2} = \frac{5.1772EI}{(0.5L)^2} = \frac{20.7088EI}{L^2} = \frac{2.098\pi^2 EI}{L^2}$$

$$= 1.026P_{exact} = 1.026\left(\frac{20.19EI}{L^2}\right)$$

Considering the fact that only two elements were used to model the column, this (2.6% difference) is a fairly good performance.

2.10. FREE VIBRATION OF COLUMNS UNDER COMPRESSIVE LOADS

In Chapter 1, deflection-amplification-type buckling and bifurcation-type buckling were discussed. In order to reach the solution of the critical load of the column problem, three different approaches were applied. In the deflection-amplification-type problem, the concern is: What is the value of the compressive load for which the static deflections of a slightly crooked column become excessive? In the bifurcation-type buckling problem, two general approaches were taken: eigenvalue method and energy method. In the eigenvalue method, the concern is: What is the value of the compressive load for which a perfect column bifurcates into a nontrivial equilibrium configuration? In the energy method, the concern is: What is the value of the compressive load for which the potential energy of the column ceases to be positive definite? As illustrated in Fig. 1-1, the body will return to its undeformed position upon release of the disturbing action if the potential energy is positive and the system is in stable equilibrium. On the other hand, if the potential energy of the system is not positive, the disturbed body will remain at the displaced position or be displaced further upon the release of the disturbing action.

All of these approaches are based on static concepts. The fourth approach is based on the dynamic concept. In this approach the concern is: What is the value of the compressive load for which the free vibration of the perfect column ceases to occur?

It will be demonstrated that the natural frequency of the column is altered depending on the presence of the axial compressive load on the column.

The governing differential equation of a prismatic column is given by

$$EI\frac{\partial^{iv} y}{\partial x^4} + P\frac{\partial^2 y}{\partial x^2} = -m\frac{\partial^2 y}{\partial t^2} \tag{2.10.1}$$

where m is the mass per unit length of the column and the right-hand side of Eq. (2.10.1) is the inertia force per unit length of the column. Note that the inertia force always develops in the opposite direction of the positive acceleration.

Invoking the method of separation of variables, the deflection as a function of the position coordinate x and time t is given by

$$y(x,t) = Y(x)T(t) \tag{2.10.2}$$

Substituting Eq. (2.10.2) into Eq. (2.10.1) gives

$$EIY^{iv}T + PY''T = -mYT'' \tag{2.10.3}$$

Dividing both sides of Eq. (2.10.3) by YT yields

$$EI\frac{Y^{iv}}{Y} + P\frac{Y''}{Y} = -m\frac{T''}{T} \tag{2.10.4}$$

The left-hand side of Eq. (2.10.4) is independent of t, and the right-hand side of Eq. (2.10.4) is independent of x and is equal to the expression on the left. Being independent of both x and t, and yet identically equal to each other, each side of Eq. (2.10.4) must be a constant. Let this constant be α so that

$$EI\frac{Y^{iv}}{Y} + P\frac{Y''}{Y} = -m\frac{T''}{T} = \alpha \tag{2.10.5}$$

Equation (2.10.5) will be separated into two homogeneous ordinary differential equations as

$$Y^{iv} + k^2Y'' - \alpha Y = 0 \tag{2.10.6}$$

$$T'' + \omega^2 T = 0 \tag{2.10.7}$$

where

$$k^2 = \frac{P}{EI} \tag{2.10.8}$$

$$\omega^2 = \frac{\alpha EI}{m} \tag{2.10.9}$$

By way of Eq. (2.10.9), it is seen that α is a nonzero, positive constant. Following the procedure of the characteristic equation, the general solutions for the two ordinary linear differential equations with constant coefficients, Eqs. (2.10.6) and Eq. (2.10.7), are obtained. The general solution for Eq. (2.10.6) is

$$Y(x) = A_1 \cos \alpha_1 x + A_2 \sin \alpha_1 x + A_3 \cosh \alpha_2 x + A_4 \sinh \alpha_2 x \tag{2.10.10}$$

where

$$\alpha_1^2, \alpha_2^2 = \frac{k^2 + \sqrt{k^4 + 4\alpha}}{2}, \frac{-k^2 + \sqrt{k^4 + 4\alpha}}{2} \tag{2.10.11}$$

The general solution for Eq. (2.10.7) is

$$T(t) = B_1 \cos \omega t + B_2 \sin \omega t \tag{2.10.12}$$

For a simply supported column, the boundary conditions to determine the integral constants are

$$\begin{aligned} Y(0) &= 0 \quad Y''(0) = 0 \\ Y(\ell) &= 0 \quad Y''(\ell) = 0 \end{aligned} \tag{2.10.13}$$

The first and second conditions yield

$$\begin{aligned} A_1 + A_3 &= 0 \\ -\alpha_1^2 A_1 + \alpha_2^2 A_3 &= 0 \end{aligned} \tag{2.10.14}$$

By virtue of Eq. (2.10.11), Eq. (2.10.14) can only be satisfied when

$$A_1 = A_3 = 0 \tag{2.10.15}$$

unless $\alpha_1 = \alpha_2 = 0$, which corresponds to the case of $P = 0$, which is a trivial case. The third and fourth conditions give

$$\begin{aligned} A_2 \sin \alpha_1 \ell + A_4 \sinh \alpha_2 \ell &= 0 \\ -\alpha_1^2 A_2 \sin \alpha_1 \ell + \alpha_2^2 A_4 \sinh \alpha_2 \ell &= 0 \end{aligned} \tag{2.10.16}$$

For a nontrivial solution for A_2 and A_4, the coefficient determinant must vanish.

$$\begin{vmatrix} \sin \alpha_1 \ell & \sinh \alpha_2 \ell \\ -\alpha_1^2 \sin \alpha_1 \ell & \alpha_2^2 \sinh \alpha_2 \ell \end{vmatrix} = 0 \tag{2.10.17}$$

Expanding the determinant gives

$$\left(\alpha_1^2 + \alpha_2^2\right) \sin \alpha_1 \ell \sinh \alpha_2 \ell = 0 \tag{2.10.18}$$

Except for the case $\alpha = 0$ $(\alpha_2 = 0)$, which is a trivial case, Eq. (2.10.18) is satisfied only when

$$\sin \alpha_1 \ell = 0 \tag{2.10.19}$$

or

$$\alpha_1 \ell = n\pi \tag{2.10.20}$$

Substituting Eq. (2.10.11) into Eq. (2.10.20) for α_1, the constant α is computed

$$\alpha = \left(\frac{n\pi}{\ell}\right)^4 \left(1 - \frac{k^2\ell^2}{n^2\pi^2}\right) \tag{2.10.21}$$

Substituting Eq. (2.10.21) into Eq. (2.10.9) gives the natural frequencies of the column

$$\omega_n = \sqrt{\frac{EI}{m}}\left(\frac{n\pi}{\ell}\right)^2 \sqrt{\left(1 - \frac{k^2\ell^2}{n^2\pi^2}\right)} \tag{2.10.22}$$

Rearranging Eq. (2.10.22) gives

$$m\omega_n^2 = \frac{n^2\pi^2}{\ell^2}\left(\frac{n^2\pi^2}{\ell^2}EI - P\right)(n = 1, 2, \ldots) \tag{2.10.23}$$

Substituting Eq. (2.10.19) into the second Eq. (2.10.16) yields

$$A_4 = 0 \tag{2.10.24}$$

and the vibration mode of the column is determined from Eq. (2.10.10) as

$$Y_n(x) = A_2 \sin\frac{n\pi x}{\ell} \tag{2.10.25}$$

Two initial conditions determine the other integral constants, B_1 and B_2 in Eq. (2.10.12). Assume the vibration is initiated by an initial displacement such that

$$y(x,0) = w(x) \quad \text{and} \quad \frac{\partial y(x,0)}{\partial t} = 0 \tag{2.10.26}$$

Then

$$\begin{aligned} Y(x)(B_1 \cos\omega t + B_2 \sin\omega t)|_{t=0} &= w(x) \\ Y(x)(-B_1 \sin\omega t + B_2 \cos\omega t)|_{t=0} &= 0 \end{aligned} \tag{2.10.27}$$

from which one obtains the following:

$$B_1 Y(x) = w(x) \quad \text{and} \quad B_2 = 0 \tag{2.10.28}$$

Hence, the general solution of Eq. (2.10.2) for the simply supported column is given by an infinite sum of natural vibration modes

$$y(x,t) = \sum_{n=1}^{\infty} C_n \sin\frac{n\pi x}{\ell} \cos\omega_n t \tag{2.10.29}$$

where $C_n = A_2B_2$. The coefficient C_n can be determined from the first condition of Eq. (2.10.28)

$$\sum_{n=1}^{\infty} C_n \sin\frac{n\pi x}{\ell} = w(x) \tag{2.10.30}$$

Since Eq. (2.10.30) is a Fourier series expansion for the given initial deflection, the coefficient can be readily determined by use of the orthogonality condition

$$C_n = \frac{2}{\ell}\int_0^{\ell} w(x)\sin\frac{n\pi x}{\ell}\,dx \quad (n = 1, 2, ...) \tag{2.10.31}$$

As the initial deflection $w(x)$is assumed to be known, Eq. (2.10.31) can be evaluated. Note that $\int_0^{\ell}\sin^2 nx\ dx = \ell/2$. The general solution of the free vibration of a simply supported column is

$$y(x,t) = \frac{2}{\ell}\sum_{n=1}^{\infty}\left[\int_0^{\ell} w(\xi)\sin\frac{n\pi\xi}{\ell}d\xi\right]\sin\frac{n\pi x}{\ell}\cos\omega_n t \tag{2.10.32}$$

It is of interest to note in Eq. (2.10.23) that the frequency of the vibration of the compressed column is reduced due to the presence of the compressive load. Once the load P reaches P_E, the frequency becomes equal to zero and the column vibrates with an infinitely long period.

2.11. BUCKLING BY A NONCONSERVATIVE LOAD

Consider a free-standing prismatic cantilever column that is loaded by a follower force, P, a force that turns its direction so as to always remain tangential to the deflection curve at the column top as shown in Fig. 2-22. Such a load is called a tangential load or nonconservative load. As was done in all three static approaches, assume a nontrivial (neighboring) equilibrium position and establish a static equilibrium equation. The moment at any point along the slightly (small deflection) deflected column is

$$EIy'' = P(y_\ell - y) - Py'_\ell(\ell - x) \tag{2.11.1}$$

Differentiating twice gives

$$EIy^{iv} + Py'' = 0 \tag{2.11.2}$$

and the general solution of Eq. (2.11.2) is

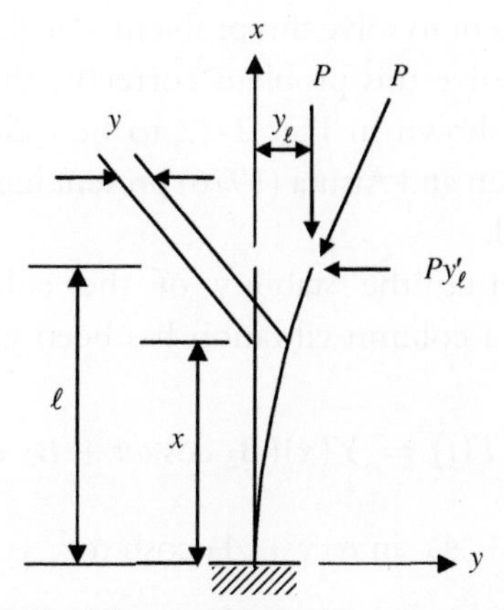

Figure 2-22 Column subjected to a tangential load

$$y = A\cos kx + B\sin kx + Cx + D \tag{2.11.3}$$

where

$$k^2 = \frac{P}{EI} \tag{2.11.4}$$

The boundary conditions to determine the integral constants are

$$\begin{aligned} y(0) &= y'(0) = 0 \\ y''(\ell) &= y'''(\ell) = 0 \end{aligned} \tag{2.11.5}$$

The resulting coefficient determinant must vanish for nontrivial equilibrium configuration

$$\begin{vmatrix} 1 & 0 & 0 & 1 \\ 0 & k & 1 & 0 \\ \cos k\ell & \sin k\ell & 0 & 0 \\ \sin k\ell & -\cos k\ell & 0 & 0 \end{vmatrix} = 0 \tag{2.11.6}$$

However, the expansion of determinant yields $\sin^2 k\ell + \cos^2 k\ell = 1$. It follows then that the only solutions for the integral constants are $A = B = C = D = 0$ or $y(x) = 0$. This means that there is no nontrivial (neighboring) equilibrium configuration for $P > 0$; that is, it cannot buckle in a static manner. However, this does not mean that the column cannot buckle at all; this is a striking conclusion that was incorrectly drawn in one of the early studies of this problem. In other words, static approaches are insufficient to solve this problem. Ziegler (1968) credits Beck (1952) as the first to solve this problem correctly, thereby prompting the free-standing column shown in Fig. 2-22 to be called "Beck's column." Bolotin (1963) and Chen and Atsuta (1976) present highlights of the present problem in some detail.

Consider investigating the stability of the column by a dynamic approach. Solution of a column vibration has been given by Eq. (2.10.2), which is rewritten as

$$y(x,t) = Y(x)T(t) = Y(x)(B_2 \cos \omega t + B_2 \sin \omega t) \tag{2.11.7}$$

$$Y(x) = A_1 \cos \alpha_1 x + A_2 \sin \alpha_1 x + A_3 \cosh \alpha_2 x + A_4 \sinh \alpha_2 x \tag{2.11.8}$$

where

$$\alpha_1^2, \alpha_2^2 = \frac{k^2 + \sqrt{k^4 + 4\alpha}}{2}, \frac{-k^2 + \sqrt{k^4 + 4\alpha}}{2} \tag{2.11.9}$$

and

$$\alpha = \frac{m\,\omega^2}{EI} \tag{2.11.10}$$

From the boundary conditions at the fixed end given in Eq. (2.11.5), one obtains

$$\begin{aligned} A_1 + A_3 &= 0 \\ \alpha_1 A_2 + \alpha_2 A_4 &= 0 \end{aligned} \tag{2.11.11}$$

Substituting Eq. (2.11.11) into Eq. (2.11.8) yields

$$Y(x) = A_1(\cos \alpha_1 x - \cosh \alpha_2 x) + A_2\left(\sin \alpha_1 x - \frac{\alpha_1}{\alpha_2} \sinh \alpha_2 x\right) \tag{2.11.15}$$

The other two boundary conditions at the free end, Eq. (2.11.5), applied to Eq. (2.11.15) give two additional relationships.

$$\begin{aligned} A_1\left(\alpha_1^2 \cos \alpha_1 x + \alpha_2^2 \cosh \alpha_2 x\right) + A_2\left(\alpha_2^2 \sin \alpha_2 x + \alpha_1 \alpha_2 \sinh \alpha_2 x\right) &= 0 \\ A_1\left(\alpha_1^3 \sin \alpha_1 x - \alpha_2^3 \cosh \alpha_2 x\right) - A_2\left(\alpha_2^3 \cos \alpha_2 x + \alpha_1 \alpha_2^2 \cosh \alpha_2 x\right) &= 0 \end{aligned} \tag{2.11.16}$$

For a nontrivial solution, the determinant of Eq. (2.11.16) for coefficients, A_1 and A_2, must vanish.

$$\alpha_1^4 + \alpha_2^4 + 2\,\alpha_1^2\alpha_2^2 \cos \alpha_1 \ell \cosh \alpha_2 \ell + \alpha_1 \alpha_2\left(\alpha_1^2 - \alpha_2^2\right) \sin \alpha_1 \ell \sinh \alpha_2 \ell = 0 \tag{2.11.17}$$

Substituting Eq. (2.11.9) for α_1 and α_2 into Eq. (2.11.17) gives

$$\left(k^4 + 2\alpha\right) + 2\alpha \cos \alpha_1 \ell \cosh \alpha_2 \ell + \sqrt{\alpha} k^2 \sin \alpha_1 \ell \sinh \alpha_2 \ell = 0 \tag{2.11.18a}$$

$$k^4 + 2\alpha(1 + \cos \alpha_1 \ell \cosh \alpha_2 \ell) + \sqrt{\alpha} k^2 \sin \alpha_1 \ell \sinh \alpha_2 \ell = 0 \tag{2.11.18b}$$

For any given column EI, m and ℓ are known along with α_1 and $\alpha_1$which are functions of P and ω, according to Eq. (2.11.9). Thus Eq. (2.11.18b) gives the relationship between axial compressive force, $P = k^2 EI$, and the frequency, $\omega = \sqrt{\alpha EI/m}$. The plot of Eq. (2.11.18b) is given in Fig. 2-23.

Equation (2.11.18b) consists of the infinity of branches, every one of them originating in the fourth quadrant and reaching a maximum in the first one. Figure 2-23 shows the first and second branches; the second has a higher maximum than the first, and the others have higher maxima than the second. The points of intersection of the various branches with the axis, $\omega^2 m\ell^4/EI$, supply the circular frequencies of the flexural vibration of the unloaded cantilever column ($P = 0$), and the first four frequencies are found to be 12.36, 485.52, 3806.55, and 14617.27 as shown in Fig. 2-23. A given load P corresponds to a vertical line in Fig. 2-23. Its points of intersection with the curves yield the circular frequencies of the loaded column. For small values of P, the corresponding oscillations are harmonic. However, if P is sufficiently increased, the vertical line ceases to intersect the first branches, and the roots of Eq. (2.11.18b) corresponding to P become complex. It is seen from Eq. (2.11.7) that this implies unbounded amplitudes, that is, self-excited oscillations. The column thus becomes unstable

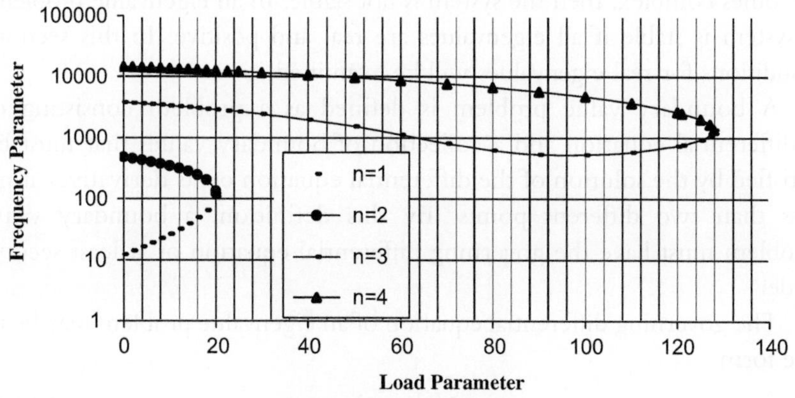

Figure 2-23 Critical loads

when the load parameter, $P\ell^2/EI$, reaches the maximum of the first branch in Fig. 2-23, which is found to be 20.05095 by **Maple®**.

Therefore the buckling load is

$$P_1 = \frac{20.05EI}{\ell^2} = \frac{2.031\pi^2 EI}{\ell^2} \tag{2.11.19}$$

Despite the early researchers' misguided conclusions, the column really buckles by a follower load, and the smallest critical load is approximately eight times the Euler load for the case of cantilever column.

As shown in Fig. 2-23, the second critical load is found to be

$$P_2 = \frac{127.811EI}{\ell^2} = \frac{12.95\pi^2 EI}{\ell^2} \tag{2.11.20}$$

As stated by Bolotin (1963) and Ziegler (1968), the vertical line corresponding to a level of the tangential load intersects two distinct points on each branch until the load reaches the maximum where the two points meet. This is confirmed for the second branch in Fig. 2-23.

No definite conclusion can be drawn regarding the practical value of the critical tangential load since no method has been devised for applying a tangential force to a cantilever column undergoing flexural oscillations, although an idea of attaching a rocket engine of thrust P at the end of the pulsating column has been proposed, which is highly impractical.

2.12. SELF-ADJOINT BOUNDARY VALUE PROBLEMS

As it was shown in the previous section, when the frequency of the system becomes complex, then the system is not stable. In an eigenvalue problem, a system is stable if all eigenvalues are real and positive. In this section, conditions for real eigenvalues will be examined.

A boundary value problem is defined as a problem consisting of a differential equation and a collection of boundary values that must be satisfied by the solution of the differential equation or its derivatives at no less than two different points. By this definition, a boundary value problem must have the governing differential equation of at least second order.

The governing differential equation of an eigenvalue problem may be in the form

$$L[y] = \lambda y \tag{2.12.1}$$

where $L[\,]$ is a linear differential operator. In the case of a prismatic column vibration problem, for example, the differential equation for the deflection has the form

$$L[y] = EIy^{iv} + Py'' \quad \text{and} \quad \lambda = m\omega^2 \tag{2.12.2}$$

Only when λ takes specific value, λ_n, Eq. (2.12.1) has solutions, $y_n(x)$. λ_n are eigenvalues, and $y_n(x)$ are the corresponding eigenfunctions of the system. Assume the eigenvalue λ_i is complex, then the corresponding eigenfunction y_i is also complex such that

$$L[y_i] = \lambda_i y_i \tag{2.12.3}$$

The pair of complex conjugates must also satisfy the equation

$$L[\overline{y}_i] = \overline{\lambda}_i \overline{y}_i \tag{2.12.4}$$

where the bar denotes the complex conjugate, so that

$$\begin{aligned} \lambda_i &= a_i + ib_i, \quad \overline{\lambda}_i = a_i - ib_i \\ y_i &= u + iv_i, \quad \overline{y}_i = u - iv_i \end{aligned} \tag{2.12.5}$$

where $i = \sqrt{-1}$. Executing inner (scalar) products of two functions y_i and $\overline{y}_i$ from Eqs. (2.12.3) and (2.12.4) and integrating over the domain of the column yield

$$\begin{aligned} \int_0^\ell \overline{y}_i L[y_i]dx &= \lambda_i \int_0^\ell \overline{y}_i y_i dx \\ \int_0^\ell y_i L[\overline{y}_i]dx &= \overline{\lambda}_i \int_0^\ell y_i \overline{y}_i dx \end{aligned} \tag{2.12.6}$$

Subtracting the second equation from the first gives

$$\int_0^\ell \overline{y}_i L[y_i]dx - \int_0^\ell y_i L[\overline{y}_i]dx = (\lambda_i - \overline{\lambda}_i) \int_0^\ell \overline{y}_i y_i dx \tag{2.12.7}$$

If the eigenvalues are real, it follows

$$(\lambda_i - \overline{\lambda}_i) = 0 \tag{2.12.8}$$

Hence,

$$\int_0^\ell \overline{y}_i L[y_i]dx = \int_0^\ell y_i L[\overline{y}_i]dx \tag{2.12.9}$$

Equation (2.12.9) is known to be the condition for the linear operator L to be symmetric, and if it is bounded, it is also called self-adjoint. In a system dealing with only bounded (continuous) operators, these two terms become synonymous. Operator L is defined to be positive definite in Section 2.4 if $(Ly, y) > 0$ for any admissible function y (except $y = 0$). If the system is a discrete one and the eigenvalue equation is written in matrix form.

$$[L]\{y\} = \lambda\{y\} \tag{2.12.10}$$

Equation (2.12.9) can be rewritten as

$$\begin{aligned} 0 &= \{\bar{y}\}^T[L]\{y\} - \{y\}^T[L]\{\bar{y}\} = \left(\{\bar{y}\}^T[L]\{y\}\right)^T - \{y\}^T[L]\{\bar{y}\} \\ &= \{y\}^T[L]^T\{\bar{y}\} - \{y\}^T[L]\{\bar{y}\} = \{y\}^T\left([L]^T - [L]\right)\{\bar{y}\} \end{aligned} \tag{2.12.11}$$

or

$$[L]^T = [L] \tag{2.12.12}$$

which implies that L is a symmetric matrix.

In the case of the column vibration problem given by Eq. (2.12.2), an inner product of two functions, $\bar{y}$ and $L[y]$, integrated over the domain is

$$\int_0^\ell \bar{y}L[y]dx = \int_0^\ell \bar{y}\left[EIy^{iv} + Py''\right] dx \tag{2.12.13}$$

Integrating by parts the right-hand side of Eq. (2.12.13) (four times the first term and twice the second term) gives

$$\begin{aligned} \int_0^\ell \bar{y}L[y]\, dx &= \int_0^\ell \bar{y}\left(EIy^{iv} + Py''\right)dx \\ &= -y(EI\bar{y}''' + P\bar{y}')_0^\ell + \bar{y}(EIy''' + Py')_0^\ell - \bar{y}'(EIy'')_0^\ell \\ &\quad + y'(EI\bar{y}'')_0^\ell + \int_0^\ell yL[\bar{y}]dx \end{aligned} \tag{2.12.14}$$

Invoking the self-adjoint condition of Eq. (2.12.9), the sum of integrated terms in Eq. (2.12.14), which is referred to as conjunct or concomitant, vanishes.

$$-y(EI\bar{y}''' + P\bar{y}')_0^\ell + \bar{y}(EIy''' + Py')_0^\ell - \bar{y}'(EIy'')_0^\ell + y'(EI\bar{y}'')_0^\ell = 0 \tag{2.12.15}$$

In fact, each term of Eq. (2.12.15) vanishes for any combination of column end support conditions: free, pinned, and fixed. This is called the self-adjoint boundary conditions for a column.

Consider now the case of a column loaded tangentially. The boundary conditions given in Eq. (2.11.5) do not satisfy the self-adjoint boundary condition for a column given by Eq. (2.12.15). Thus a cantilever column loaded by a tangential force does not provide a self-adjoint boundary condition, and hence, all the eigenvalues are not necessarily real. This does not render the problem to be a self-adjoint boundary value problem. Hence, the problem is not a properly posed one, and a unique solution is not guaranteed by any one of the classical solution methods.

Another way of discerning whether or not a system of boundary conditions is conservative is to evaluate the potential energy of the boundary forces at elastically constrained supports. If the potential energy thus computed is path-independent, then it is said to be a system of conservative boundary conditions.

REFERENCES

ABAQUS. (2006). *ABAQUS Analysis User's Manual*. Pawtucket, RI: ABAQUS.

Beck, M. (1952). Die Knicklast des einseitig eingespannten, tangential gedrückten Stabes. *Z. Angew. Math. Phys., 3*, 225.

Bolotin, V.V. (1963). *Nonconservative Problems of the Theory of Elastic Stability*. Translated from the Russian by T.K. Lusher, English Translation edited by G. Herrmann, Pergamon Press, London, UK.

Chen, W. F., & Atsuta, T. (1976). *Theory of Beam-columns, Vol. 1*. New York: Inplane Behavior and Design, McGraw-Hill.

Forsyth, A. R. (1960). *Calculus of Variations*. New York: Dover.

Fung, Y. C., & Tong, P. (2001). *Classical and Computational Solid Mechanics*. Singapore: World Scientific.

Hjelmstad, K. D. (2005). *Fundamentals of Structural Mechanics* (2nd ed.). New York: Springer.

Hoff, N. J. (1956). *The Analysis of Structures*. New York: John Wiley and Sons.

Langhaar, H. L. (1962). *Energy Methods in Applied Mechanics*. New York: John Wiley and Sons.

Love, A. E. H. (1944). *A Treatise of the Mathematical Theory of Elasticity*. New York, NY: Dover.

Mikhlin, S. G. (1964). *Variational Methods in Mathematical Physics*. Macmillan Company.

Newmark, N. (1943). Numerical Procedure for Computing Deflection, Moments, and Buckling Loads. *Transaction, ASCE, Vol. 108*, 1161.

Saada, A. S. (1974). *Elasticity Theory and Applications* (2nd ed.). Malabar, FL: Krieger Publishing Company.

Shames, I. H., & Dym, C. L. (1985). *Energy and Finite Element Methods in Structural Mechanics*. New York: Hemisphere.

Sokolnikoff, I. S. (1956). *Mathematical Theory of Elasticity* (2nd ed.). New York: McGraw-Hill.

Tauchert, T. R. (1974). *Energy Principles in Structural Mechanics*. New York: McGraw-Hill.

Timoshenko, S. P., & Gere, J. M. (1961). *Theory of Elastic Stability*. New York: McGraw-Hill.

Washizu, K. (1974). *Variational Methods in Elasticity and Plasticity* (2nd ed.). Oxford, UK: Pergamon Press, Ltd.

Wikipedia. (2009). The Free Encyclopedia, http://en.wikipedia.org/wiki/Einstein_notation, February 6, 2009.
Yoo, C. H. (1980). Bimoment Contribution to Stability of Thin-Walled Assemblages. *Computers and Structures, Vol. 11*(No.5), 465–471.
Ziegler, H. (1968). *Principles of Structural Stability*. Waltham, MA: Blaisdell Publishing Co.

PROBLEMS

2.1 Determine an approximate value for the critical load of a propped column. The column is hinged at the top loaded end and fixed at its base. Use the energy method. Assume the deflected shape of the column by the deflection curve of a uniformly loaded propped beam whose boundary conditions are the same as those of the column.

2.2 Find the critical load for a rigid bar system loaded as shown in Fig. P2-2. Assume the two rigid bars of length $\ell/2$ are connected by a hinge and displacements remain small.

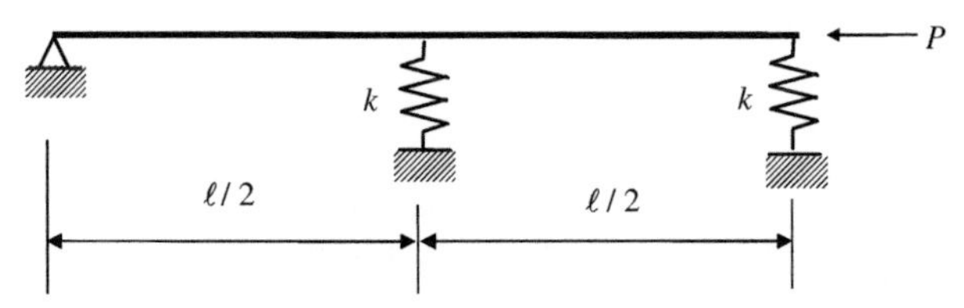

Figure P2-2 Spring-supported rigid bar

2.3 Use the principle of minimum potential energy to derive the governing differential equation of equilibrium and the natural boundary conditions for a prismatic column resting on an elastic foundation with a foundation modulus k_f. Then, compute the critical load for the pinned column shown in Fig. P2-3.

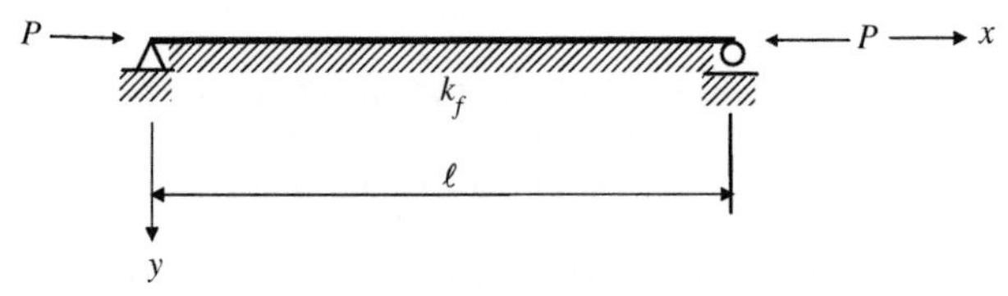

Figure P2-3 Pinned column resting on elastic foundation

2.4 Determine the critical loads of columns (a) and (b) shown in Fig. P2-4 by the Rayleigh method, using both C1 and C2 methods.
Hint: For (a), assume ϕ is the elastic curve caused by a 1 lb load applied laterally at a point $\ell/3$ from the origin of the coordinate system. The

corresponding deflection curve can be found in any structural analysis textbook such as *AISC, Manual of Steel Construction.*

For (b), assume $y = \delta(1 - \cos(\pi x/2\ell))$. Compute for $I_2/I_1 = 5$ and $\ell_1/\ell_2 = 4$. Check the result by that obtained from DE method.

$$\left(U = \int_0^{\ell_2} \frac{m^2 dx}{2EI_2} + \int_{\ell_2}^{\ell} \frac{m^2 dx}{2EI_1}\right)$$

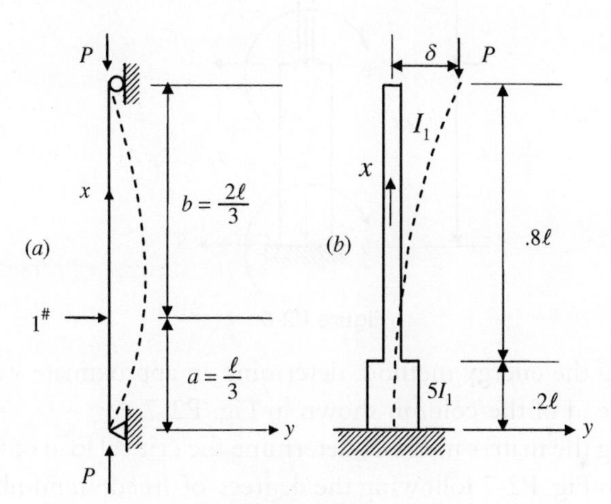

Figure P2-4

2.5 (a) Find the critical load of the stepped column shown in Fig. P2-5(a) using four and eight segments (Newmark method) and compare the results using the DE method. $(P_{cr} = 6.5\ (EI_o/\ell^2))$

(b) Find the critical load of the tapered column using four and eight segments (Newmark method) and compare the result with the solution by the C2 method. $(P_{cr} = 2.5\ (\pi^2 EI_o/\ell^2))$

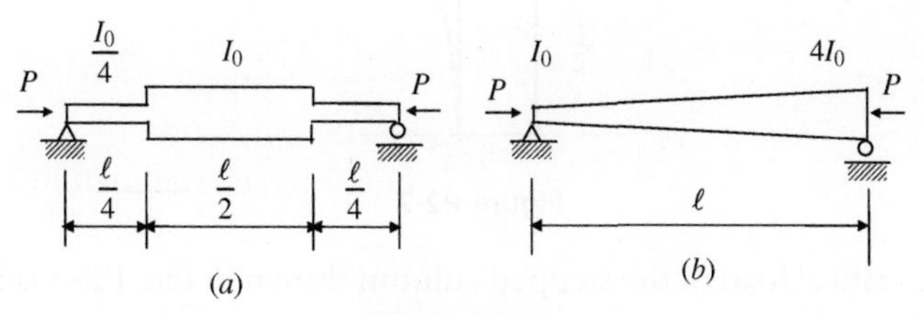

Figure P2-5

2.6 Compute the buckling load of a stepped column shown by the matrix method illustrated in the class. Use the numbering scheme shown in Fig. P2-6.

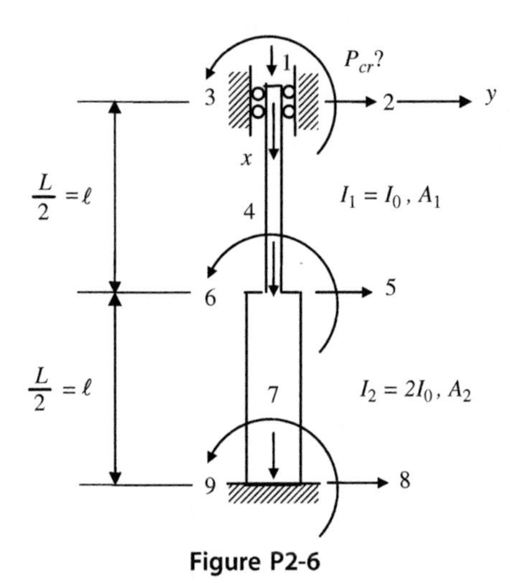

Figure P2-6

2.7 (a) Using the energy method, determine an approximate value for the critical load of the column shown in Fig. P2-7.
(b) Using the matrix method, determine the critical load of the column shown in Fig. P2-7 following the degrees-of-freedom numbering used in the previous problem.

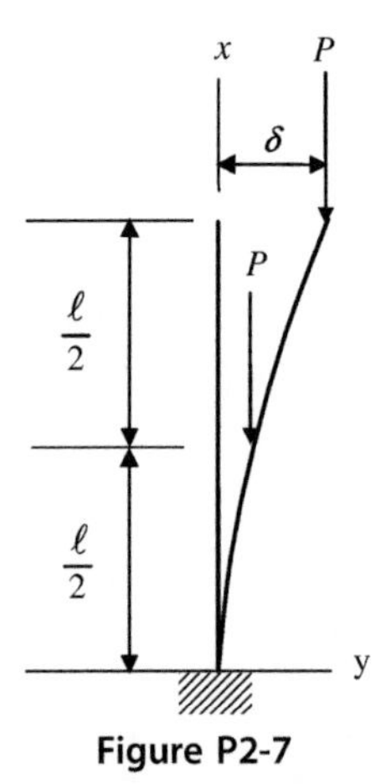

Figure P2-7

2.8 If the critical load of the stepped column shown in Fig. P2-8 is $50EI/\ell^2$ and the coefficient, α, of the linear spring constant, $k = \alpha\, EI/\ell$, is 51, determine the coefficient, β, of the rotational spring constant, $\theta = \beta\, EI/\ell$. Use the energy method. (*Hint:* The Ritz method appears to be the best.)

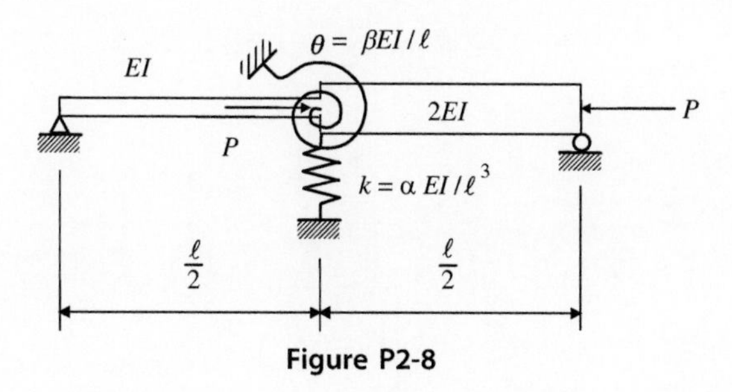

Figure P2-8

2.9 A linear system of eigenvalue problem is given by $My - \lambda Ny = 0$ with self-adjoint boundary conditions. If M and N are linear differential operators and λ is the eigeTRUN nvalue to be found, prove that all eigenvalues are positive.

CHAPTER 3

Beam-Columns

Contents

3.1. TRANSVERSELY LOADED BEAM SUBJECTED TO AXIAL COMPRESSION

3.1.1. The Concept of Amplification

This chapter seeks to familiarize the student with buckling of some simple structural members and frames, and it presents a few methods that can be successfully used to arrive at the critical condition. A more comprehensive treatment of the buckling analysis of structures may be found in the books by Bleich (1952), Britvec (1973), and Bazant and Credolin (1991). Since one of the methods employed for analysis of the structural stability is based on the theory of beam-columns, a brief review will be in order (see also Timoshenko and Gere 1961).

A slender member meeting the Euler-Bernoulli-Navier hypotheses under transverse loads and inplane compressive load (see Fig. 3-1) is called a beam-column. An exact analysis of a beam-column can only be accomplished by solving the governing differential equation or its derivatives (for example, slope-deflection equations).

Consider a very simple case of a beam-column shown in Fig. 3-1. The beam-column is subjected simultaneously to a transverse load Q at its mid-span and a concentric compressive force P. Since the response of

Stability of Structures
ISBN 978-0-12-385122-2, doi:10.1016/B978-0-12-385122-2.10003-X

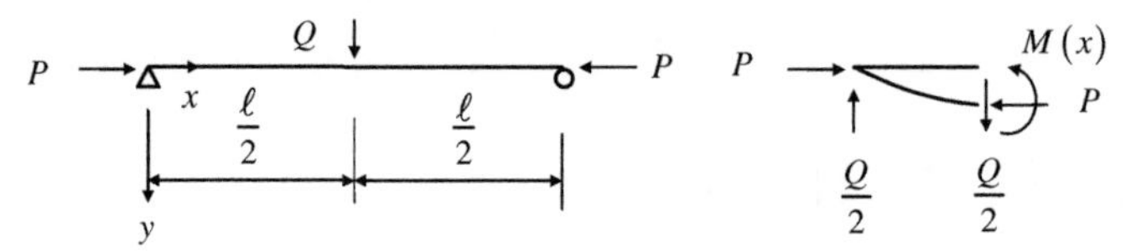

Figure 3-1 Simple beam-column

a beam-column under these loads is no longer linear, the method of superposition does not apply even if the final results are within the elastic limit.

Summing moments at a point x from the origin gives

$$M(x) - Py - \frac{Q}{2}x = 0 \quad \text{for } 0 \le x \le \ell/2 \quad \text{with } M(x) = -EIy'' \tag{3.1.1}$$

$$\text{or} \quad y'' + k^2 y = -\frac{Q}{2}\frac{x}{EI} = -\frac{Qx}{2P}k^2 \quad \text{with } k^2 = \frac{P}{EI}$$

The general solution to this differential equation is $y = y_h + y_P$ The homogeneous solution has been given earlier. The particular solution can be obtained by the method of undetermined coefficients. Assume the particular solution to be of the form

$$y_P = C + Dx \quad \text{with } y_P' = D,\ y_P'' = 0$$

Substituting these derivatives into the differential equation yields

$$0 + k^2(C + Dx) = -\frac{Qx}{2P}k^2$$

Hence,

$$C = 0 \quad \text{and} \quad D = -\frac{Q}{2P} \Rightarrow y_P = -\frac{Q}{2P}x$$

The total solution is

$$y = A\cos kx + B\sin kx - \frac{Qx}{2P}$$

The two constants of integration can be determined from the following boundary conditions:

$$y = 0 \quad \text{at } x = 0 \Rightarrow A = 0$$

$$y' = 0 \quad \text{at } x = \ell/2$$

(Note : the boundary condition, $y = 0$ at $x = \ell$, cannot be used here as $0 \le x \le \ell/2$)

$$y' = Bk\cos kx - \frac{Q}{2P},\ 0 = Bk\cos\frac{k\ell}{2} - \frac{Q}{2P} \Rightarrow B = \frac{Q}{2Pk\cos\frac{k\ell}{2}}$$

$$y = \frac{Q\sin kx}{2Pk\cos\frac{k\ell}{2}} - \frac{Qx}{2P} \quad \text{for } 0 \le x \le \frac{\ell}{2} \quad \text{with } P_{cr} = P_E = \frac{\pi^2 EI}{\ell^2}$$

By observation, the maximum lateral deflection occurs at the midspan.

$$y_{\max}\Big|_{x=\frac{\ell}{2}} = \frac{Q}{2Pk}\left(\tan\frac{k\ell}{2} - \frac{k\ell}{2}\right) \quad \text{with } u = \frac{k\ell}{2} = \frac{\ell}{2}\sqrt{\frac{P}{EI}}$$

$$y_{\max}\Big|_{x=\frac{\ell}{2}} = \frac{Qk^3\ell^3}{16Pku^3}\left(\tan\frac{k\ell}{2} - \frac{k\ell}{2}\right) = \frac{Q\ell^3}{48EI}\left[\frac{3(\tan u - u)}{u^3}\right] = \frac{Q\ell^3}{48EI}X(u) \tag{3.1.2}$$

$$y_{\max}\Big|_{x=\frac{\ell}{2}} = \delta_{\max} = \frac{Q\ell^3}{48EI} \quad \text{when } P = 0$$

$$u^2 = \frac{\ell^2}{4}\frac{P}{EI} \Rightarrow P = \frac{4EIu^2}{\ell^2} \quad \text{and} \quad P_E = \frac{\pi^2 EI}{\ell^2}$$

$$\frac{P}{P_E} = \frac{4EIu^2}{\ell^2}\frac{\ell^2}{\pi^2 EI} = \frac{4u^2}{\pi^2}, \quad X(u) = \frac{3(\tan u - u)}{u^3}$$

3.1.2. Stress Amplification in Columns

The behavior of a compression member under increasing load can be seen most clearly by calculating the bending stresses and lateral deflections that occur as the axial load is gradually applied. Consider a perfectly straight, slender member supporting a nominal axial load P. The ends of the member are assumed free to rotate in this case. If the member were perfectly straight and homogeneous and the load were perfectly centered, the stress in the column at any section would be simply $\sigma_a = P/A$, where A is the cross-sectional area of the column. No actual member ever would be perfectly straight and homogeneous, nor would the load be perfectly centered. Even when great efforts are made to achieve such perfection in laboratory tests, it is not completely

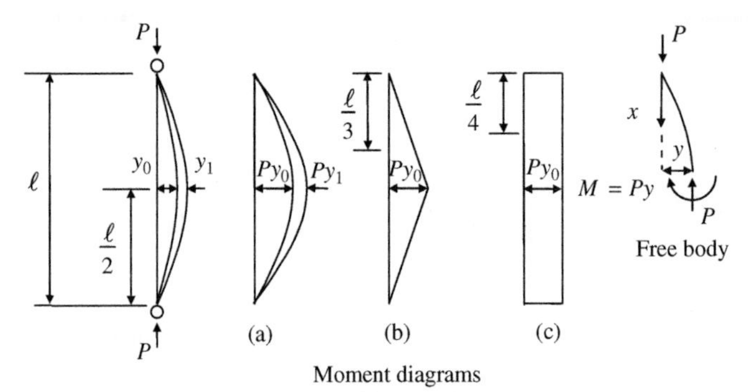

Figure 3-2 Load and moment diagrams of imperfect column.

attained. Therefore, the actual case is best represented by assuming a slight initial imperfection of loading or an initial crookedness represented by a deflection Y_0 at midheight of the member as shown in Fig. 3-2.

When a load P is acting on the column, the stresses in the extreme fibers at the midheight section are

$$\sigma = \frac{P}{A} \pm \frac{Mc}{I} \tag{3.1.3}$$

where c is the distance measured from the centroidal axis. At any section of the column, the bending moment is the load times the eccentricity, and the bending moment diagram has the same shape as the curve of the deflected member (see Fig. 3-2). This bending moment produces a further deflection at the midheight.

$$y_1 = \frac{1}{10}\frac{Py_0\ell^2}{EI} \tag{3.1.4}$$

The constant 1/10 is taken as a mean value for a deflected curve of more or less uniform curvature as shown. From the moment-area theorem part two, the midheight deflection can be computed from two extreme cases of moment diagrams, namely, triangular moment diagram and rectangular moment diagram, as shown in Fig. 3-2.

$$y_1 = \frac{Py_0}{EI} \times \frac{\ell}{2} \times \frac{1}{2} \times \frac{2}{3} \times \frac{\ell}{2} = \frac{1}{12}\frac{Py_0\ell^2}{EI} \quad \text{for triangle} \tag{3.1.5}$$

$$y_1 = \frac{Py_0}{EI} \times \frac{\ell}{2} \times \frac{\ell}{4} = \frac{1}{8}\frac{Py_0\ell^2}{EI} \quad \text{for rectangle} \tag{3.1.6}$$

For other cases of initial curvature or eccentricity, the constant may vary between limits of 1/8 and 1/12 (If it were known to be a sine function, then the factor would be $1/\pi^2 = 1/9.8696$ and the correct Euler load would result.) Because of the added deflection y_1, there will be an increased bending moment Py_1, and additional deflection y_2.

$$y_2 = \frac{1}{10}\frac{Py_1\ell^2}{EI} \tag{3.1.7}$$

Continuing this process, the total deflection becomes

$$\begin{aligned} y &= y_0 + y_1 + y_2 + \ldots \\ &= y_0 + \frac{P\ell^2}{10EI}y_0 + \frac{P\ell^2}{10EI}y_1 + \ldots \\ &= y_0 + \left(\frac{P\ell^2}{10EI}\right)y_0 + \left(\frac{P\ell^2}{10EI}\right)^2 y_0 + \ldots \\ &= y_0\left[1 + \left(\frac{P\ell^2}{10EI}\right) + \left(\frac{P\ell^2}{10EI}\right)^2 + \left(\frac{P\ell^2}{10EI}\right)^3 + \ldots\right] \end{aligned} \tag{3.1.8}$$

The series in the bracket is the multiplier by which the initial deflection y_0 is increased under load P to give the final deflection y at that load. For values of P less than $10EI/\ell^2$ (Euler buckling load), the terms in the series are less than unity and the series is convergent, having the limit

$$\frac{1}{1 - \dfrac{P\ell^2}{10EI}} \tag{3.1.9}$$

The final deflection is thus

$$y = y_0 \frac{1}{1 - \dfrac{P\ell^2}{10EI}} \tag{3.1.10}$$

This requires a slight modification that will be discussed later if the curve of initial deflection differs greatly from the uniform curvature assumed.

For any value of P such that $(P\ell^2/10EI) \geq 1$, the series is divergent. This indicates that any small initial deflection will be indefinitely magnified at the load $P = P_E = (10EI/\ell^2)$ or greater.

Let denote $P_E = \pi^2 EI/\ell^2 \doteq 10EI/\ell^2$. Then the total stress of a column at mid height is

$$\sigma = \frac{P}{A} \pm \frac{Py_0c}{I} \frac{1}{1 - \dfrac{P}{P_E}} \tag{3.1.11}$$

Let $\sigma_a = \dfrac{P}{A}$ and $\sigma_{cr} = \dfrac{P_E}{A} = \dfrac{\pi^2 EAr^2}{A\ell^2} = \dfrac{\pi^2 E}{\left(\dfrac{\ell}{r}\right)^2}$

Further, recall that

$$\frac{1}{1 - \dfrac{P}{P_E}} = \frac{1}{1 - \dfrac{\sigma_a}{\sigma_{cr}}}$$

and

$$\frac{Py_0c}{I} = \frac{Py_0c}{Ar^2} = \sigma_a \frac{y_0c}{r^2} = \sigma_a \left(\frac{c}{r}\right)^2 \frac{y_0}{c}$$

The total stress is then

$$\sigma = \sigma_a \pm \sigma_a \left(\frac{c}{r}\right)^2 \frac{y_0}{c} \frac{1}{1 - \dfrac{\sigma_a}{\sigma_{cr}}} \tag{3.1.12}$$

Thus, the magnitude of the bending stress, the second term in Eq. (3.1.12), depends on P represented in σ_a; the shape of the cross section (c/r); and the initial curvature (y_0/c).

As the critical stress, $\sigma_{cr} = \pi^2 E/(\ell/r)^2$, is a function of the stiffness of the material of the column and the slenderness ratio, it is convenient to make the expression for stress dimensionless by dividing Eq. (3.1.12) by σ_{cr}.

Thus

$$\frac{\sigma}{\sigma_{cr}} = \frac{\sigma_a}{\sigma_{cr}} \pm \frac{\sigma_a}{\sigma_{cr}} \left(\frac{c}{r}\right)^2 \frac{y_0}{c} \frac{1}{1 - \dfrac{\sigma_a}{\sigma_{cr}}} \tag{3.1.13}$$

The value of shape factor (c/r) ranges from 1.0 for a section in which all (most) of the area is assumed concentrated in the flanges to $\sqrt{3}$ for rectangular section and 2.0 for a solid circular section. Rolled shapes generally used for columns have $(c/r)^2$ in the vicinity of 1.4 about the strong axis and

3.8 about the weak axis. S shapes (wide flange shapes with sloped flanges) run the values of 5.0 and over about the weak axis.

Reasonable values of (y_0/c) are more difficult to estimate since the initial crookedness may be the result of either lack of straightness of the member itself or imperfection of the alignment of loading through the connections. Pending better establishment of the values, the combined constant $[(c/r)^2 (y_0/c)]$ has been assumed to range from 0.01 to 1.0.

Since it is usually more convenient to express the initial crookedness y_0 in terms of the length of the member, $[(c/r)^2(y_0/c)]$ may be written as $[(c/r)^2(y_0/c)] = (y_0/\ell)(\ell/r)(c/r)$, where $y_0/\ell =$ lack of straightness, $\ell/r =$ slenderness ratio, and $c/r =$ shape factor. The acceptable tolerances for straightness of rolled shapes are listed in some specifications (AISC 2005).

3.2. BEAM-COLUMNS WITH CONCENTRATED LATERAL LOADS

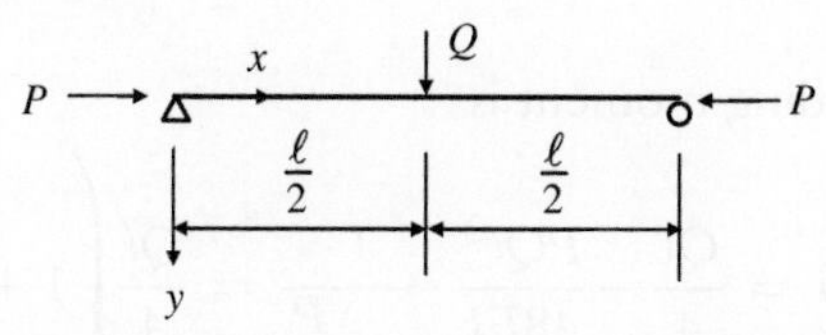

Figure 3-3 Beam-column with concentrated lateral load

The previous section showed that the deflection at the midspan of a simple beam-column subjected to a lateral load shown in Fig. 3-3 is

$$\delta = y_{\max} = \delta_0 \frac{3(\tan u - u)}{u^3}$$

where

$$\delta_0 = \frac{Q\ell^3}{48EI}, u = \frac{k\ell}{2}, \quad \text{and} \quad k = \sqrt{\frac{P}{EI}}$$

Recall the power series expansion of tan u given by

$$\tan u = u + \frac{u^3}{3} + \frac{2u^5}{15} + \frac{17u^7}{315} + \ldots$$

Hence,

$$\delta = \delta_0 \left(1 + \frac{2u^2}{5} + \frac{17u^4}{105} + \ldots \right)$$

Noting

$$u^2 = \frac{k^2\ell^2}{4} = \frac{P\ell^2}{4EI}\frac{\pi^2}{\pi^2} = 2.46\,\frac{P}{P_E}$$

$$\delta = \delta_0\left[1 + 0.984\,\frac{P}{P_e} + 0.998\left(\frac{P}{P_e}\right)^2 + ...\right]$$

$$\doteq \delta_0\left[1 + \frac{P}{P_E} + \left(\frac{P}{P_E}\right)^2 + ...\right]$$

$$= \delta_0\,\frac{1}{1 - \dfrac{P}{P_E}} \Leftarrow \text{from power series sum for } \frac{P}{P_E} < 1$$

where

$\dfrac{1}{1 - \dfrac{P}{P_E}}$ is called amplification factor or magnification factor.

The maximum bending moment is

$$M_{\max} = \frac{Q\ell}{4} + P\delta = \frac{Q\ell}{4} + \frac{PQ\ell^3}{48EI}\,\frac{1}{1 - \dfrac{P}{P_E}} = \frac{Q\ell}{4}\left(1 + \frac{P\ell^2}{12EI}\,\frac{1}{1 - \dfrac{P}{P_E}}\right)$$

$$= \frac{Q\ell}{4}\left(1 + 0.82\,\frac{P}{P_E}\,\frac{1}{1 - \dfrac{P}{P_E}}\right)$$

or

$$M_{\max} = \frac{Q\ell}{4}\left(\frac{1 - 0.18\,\dfrac{P}{P_E}}{1 - \dfrac{P}{P_E}}\right) \tag{3.2.1}$$

where

$$\left(\frac{1 - 0.18\dfrac{P}{P_E}}{1 - \dfrac{P}{P_E}}\right) \tag{3.2.2}$$

is amplification factor for bending moment due to a concentrated load.

The variation of δ with Q as given by the amplification factor is plotted on the left side of Figure 3-4 for $P = 0$, $P = 0.4\,P_{cr}$, and $P = 0.7\,P_{cr}$. The curves show that the relation between Q and δ is linear even when $P \neq 0$, provided P is a constant. However, if P is allowed to vary, as is the case on the right side of Figure 3-4, the load-deflection relation is not linear. This is true regardless of whether Q remains constant (dashed curve) or increases as P increases (solid curve). The deflection of a beam-column is thus a linear function of Q but a nonlinear function of P.

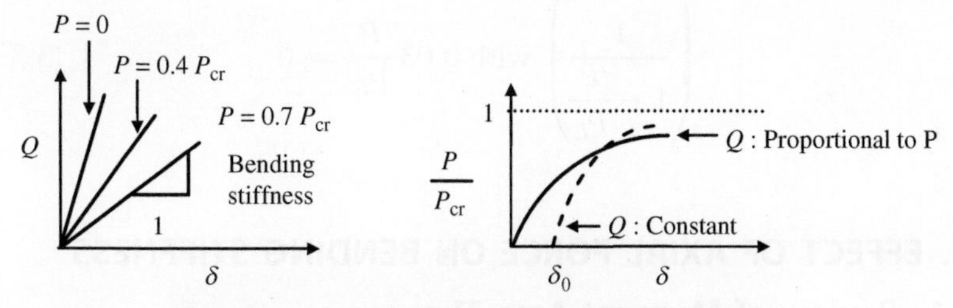

Figure 3-4 Lateral displacements of beam-column

3.3. BEAM-COLUMNS WITH DISTRIBUTED LATERAL LOADS

In the case of a simple beam-column subjected to a uniform lateral load, the midspan deflection is amplified in a similar manner as in the case of a concentrated load. That is

$$\delta = \delta_0 \left(\frac{1}{1 - \dfrac{P}{P_E}} \right) \tag{3.3.1}$$

$$M_{\max} = M_0 \left(\frac{1 + 0.03\dfrac{P}{P_E}}{1 - \dfrac{P}{P_E}} \right) \tag{3.3.2}$$

where

$$\left. \begin{aligned} \delta_0 &= \frac{5w\ell^4}{384EI} \\ M_0 &= \frac{w\ell^2}{8} \end{aligned} \right\} \text{for } P = 0$$

Conservatively, the AISC ASD (1989) part suggests these moment-amplification factors to be used as

$$\left(\frac{1-0.2\dfrac{P}{P_E}}{1-\dfrac{P}{P_E}}\right) \tag{3.3.3}$$

and

$$\left(\frac{1}{1-\dfrac{P}{P_E}}\right) \text{ with } 0.03\,\frac{P}{P_E}=0 \tag{3.3.4}$$

3.4. EFFECT OF AXIAL FORCE ON BENDING STIFFNESS

3.4.1. Review of Moment-Area Theorems

Kinney (1957) gives credit to Professor Charles E. Greene of the University of Michigan who invented the moment-area method in 1873, although the concept of the conjugate beam method, which is a more commonly known terminology of the elastic weights that is the basis of the moment-area method, was presented by Otto Mohr[1] in 1868.

Theorem 1: The change in slope between any two points on the elastic curve equals the area of the M/EI (sometimes called the elastic weight) diagram between these two points.

Note the right-hand rule adopted in Fig. 3-5. Hence, the counterclockwise rotation is taken to be positive. The counterclockwise angle measured from the tangent drawn to the elastic curve at the point A to the tangent at B is denoted as θ_{AB} and is given by

$$\theta_{AB}=\int_A^B \frac{M}{EI}dx \tag{3.4.1}$$

$$\theta_{BA}=\int_B^A \frac{M}{EI}\,dx=-\int_A^B \frac{M}{EI}\,dx=-\theta_{AB} \tag{3.4.2}$$

Theorem 2: The vertical translation $t_{A/B}$ of the tangent drawn to the elastic curve at A from B is equal to the sum of the M/EI diagram between A and B

[1] Mohr (1835–1918) was a great German structural engineer.

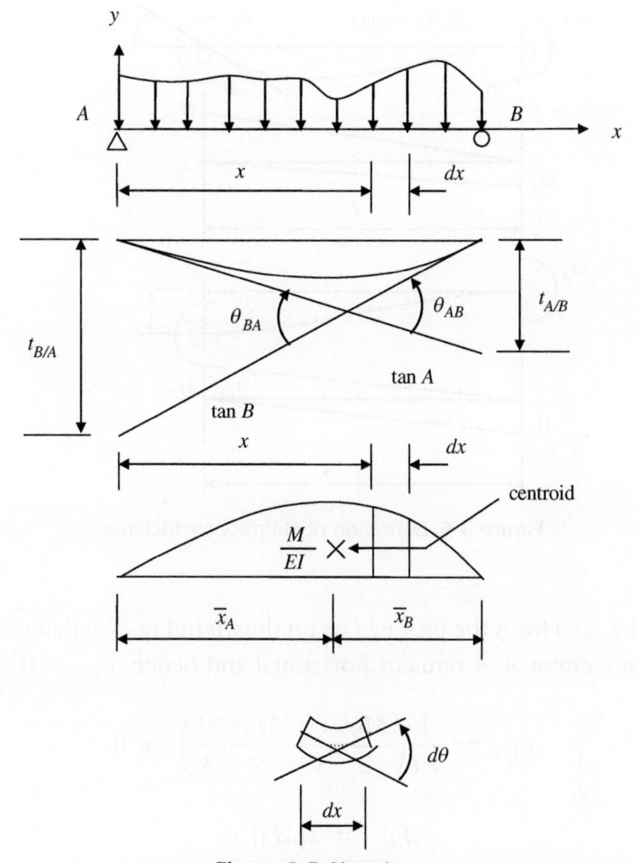

Figure 3-5 Notations

multiplied by the horizontal distance from the centroid of the M/EI diagram to B.

Hence, $t_{A/B}$ is given by

$$t_{A/B} = \bar{x}_B \int_A^B \frac{M}{EI}\, dx \tag{3.4.3}$$

where $\bar{x}_B$ is the horizontal distance from the centroid of the M/EI diagram between A and B to the point B. Likewise, the vertical translation of the tangent drawn to the elastic curve at B from A is defined by $t_{A/B}$ and is given by

$$t_{B/A} = \bar{x}_A \int_A^B \frac{M}{EI}\, dx \tag{3.4.4}$$

Example 1 Determine the stiffness coefficients shown in Fig. 3-6 using the moment-area theorems.

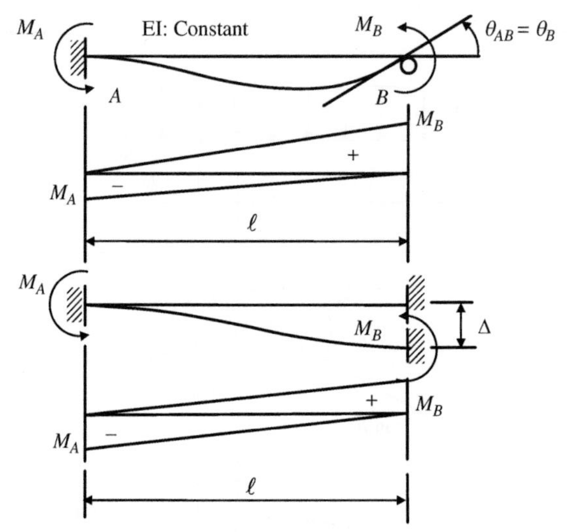

Figure 3-6 Definition of stiffness coefficients

It can be seen from the propped beam shown in Fig. 3-6 that the tangent to the elastic curve at A remains horizontal and hence, $t_{A/B} = 0$

$$t_{A/B} = \frac{1}{EI}\left(\frac{M_B\ell}{2}\frac{\ell}{3} - \frac{M_A\ell}{2}\frac{2\ell}{3}\right) = 0$$

$$|M_B| = 2|M_A| \text{ or}$$

$$-M_A = \frac{1}{2}M_B$$

$$\theta_B = \theta_{AB} = \frac{1}{EI}\left(\frac{M_B\ell}{2} - \frac{M_A\ell}{2}\right) = \frac{1}{EI}\left(\frac{M_B\ell}{2} - \frac{M_B}{2}\frac{\ell}{2}\right)$$

$$= \frac{M_B\ell}{4EI} \text{ or}$$

$$M_B = \frac{4EI}{\ell}\theta_B \tag{3.4.5}$$

As is evident from fixed beam at both ends subjected to a vertical translation at B shown in Fig. 3-6, both of the tangents remain horizontal and hence $\theta_{A/B} = 0$

$$\theta_{AB} = 0 = \frac{1}{EI}\left(\frac{M_B\ell}{2} - \frac{M_A\ell}{2}\right)$$

$$|M_B| = |M_A| \text{ or } M_A = -M_B$$

$$t_{A/B} = -\Delta = \frac{1}{EI}\left(\frac{M_B\ell}{2}\frac{\ell}{3} - \frac{M_B\ell}{2}\frac{2\ell}{3}\right) = -\frac{M_B\ell^2}{6EI}$$

$$M_B = \frac{6EI}{\ell^2}\Delta \tag{3.4.6}$$

3.4.2. Slope-Deflection Equation without Axial Force

Maney (1915) is credited as the first to publish the modern slope-deflection equations where deformations are treated as unknowns instead of stresses and reactions. A typical derivation process will be traced here as it will be used again in the development of the slope-deflection equations that include the effect of axial compression on the bending stiffness.

From the deformations of a beam shown in Fig. 3-7, the moment at a distance x from the origin is expressed as:

$$M_x = M_{ab} - (M_{ab} + M_{ba})\frac{x}{\ell}$$

Know $y'' = -\dfrac{M_x}{EI}$

Taking successive derivatives of the above equation gives

$$EIy^{iv} = 0$$

The general solution of the differential equation is

$$y = A + Bx + Cx^2 + Dx^3 \tag{3.4.2.1}$$

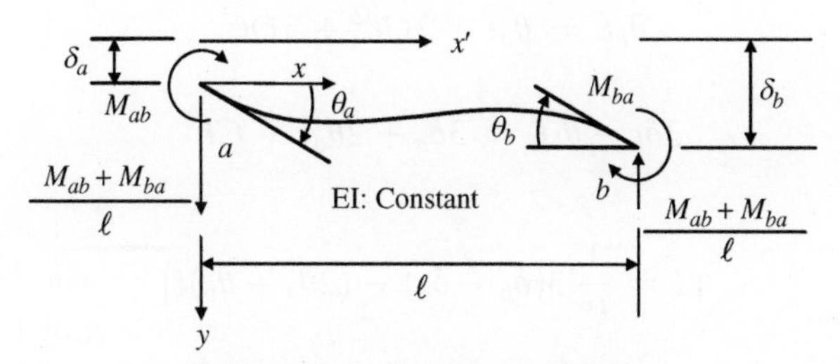

Figure 3-7 Deformations of beam

$$y' = B + 2Cx + 3Dx^2 \tag{3.4.2.2}$$

$$y'' = 2C + 6Dx \tag{3.4.2.3}$$

The four kinematic boundary conditions available are

$$y = \delta_a \quad \text{at } x = 0 \quad \text{and} \quad y = \delta_b \quad \text{at } x = \ell$$

$$y' = \theta_a \quad \text{at } x = 0 \quad \text{and} \quad y' = \theta_b \quad \text{at } x = \ell$$

Substituting these boundary conditions into Eqs. (3.4.2.1) and (3.4.2.2) gives

$$\delta_a = A, \ \ \theta_a = B \tag{3.4.2.4}$$

$$\delta_b = \delta_a + \theta_a \ell + C\ell^2 + D\ell^3 \tag{3.4.2.5}$$

$$\theta_b \ell = \theta_a \ell + 2C\ell^2 + 3D\ell^3 \tag{3.4.2.6}$$

(3.4.2.5) × 2 − (3.4.2.6) gives

$$2\delta_b = 2\theta_a \ell + 2C\ell^2 + 2D\ell^3 + 2\delta_a$$

$$\theta_b \ell = \theta_a \ell + 2C\ell^2 + 3D\ell^3$$

$$2\delta_b - \theta_b \ell = 2\delta_a + \theta_a \ell - D\ell^3$$

from which

$$D = \frac{1}{\ell^3}[-2(\delta_b - \delta_a) + (\theta_a + \theta_b)\ell] \tag{3.4.2.7}$$

(3.4.2.5) × 3 - (3.4.2.6) gives

$$3\delta_b = 3\theta_a \ell + 3C\ell^2 + 3D\ell^3 + 3\delta_a$$

$$\theta_b \ell = \theta_a \ell + 2C\ell^2 + 3D\ell^3$$

$$3\delta_b - \theta_b \ell = 3\delta_a + 2\theta_a \ell + C\ell^2$$

from which

$$C = \frac{1}{\ell^2}[3(\delta_b - \delta_a) - (2\theta_a + \theta_b)\ell] \tag{3.4.2.8}$$

Substituting Eqs. (3.4.2.7) and (3.4.2.8) into Eq. (3.4.2.3) yields

$$y'' = \frac{2}{\ell^2}[3(\delta_b - \delta_a) - (2\theta_a + \theta_b)\ell] + \frac{6}{\ell^3}[-2(\delta_b - \delta_a) + (\theta_a + \theta_b)\ell]x$$

$$y''(0) = \frac{2}{\ell^2}[3(\delta_b - \delta_a) - (2\theta_a + \theta_b)\ell] = -\frac{M_{ab}}{EI} \tag{3.4.2.9}$$

$$y''(\ell) = \frac{2}{\ell^3}[3(\delta_b - \delta_a) - (2\theta_a + \theta_b)\ell]\ell + \frac{6}{\ell^2}[-2(\delta_b - \delta_a) + (\theta_a + \theta_b)\ell]\ell$$

$$= -\frac{M_{ab}}{EI} + \frac{6}{\ell^2}[-2(\delta_b - \delta_a) + (\theta_a + \theta_b)\ell] = \frac{M_{ba}}{EI} \tag{3.4.2.10}$$

From Eq. (3.4.2.9), one obtains

$$M_{ab} = \frac{2EI}{\ell}\left[2\theta_a + \theta_b - \frac{3}{\ell}(\delta_b - \delta_a)\right] \tag{3.4.2.11}$$

From Eq. (3.4.2.10), one obtains

$$M_{ba} = \frac{2EI}{\ell}\left[2\theta_b + \theta_a - \frac{3}{\ell}(\delta_b - \delta_a)\right] \tag{3.4.2.12}$$

If any fixed end moments exist prior to releasing the joint constraints such as $M_{ab\ fixed}$ and $M_{ba\ fixed}$, then final member end moments become

$$M_{AB} = \frac{2EI}{\ell}\left[2\theta_a + \theta_b - \frac{3}{\ell}(\delta_b - \delta_a)\right] + M_{ab\ fixed} \tag{3.4.2.13}$$

$$M_{ba} = \frac{2EI}{\ell}\left[2\theta_b + \theta_a - \frac{3}{\ell}(\delta_b - \delta_a)\right] + M_{ba\ fixed} \tag{3.4.2.14}$$

Example 1 Consider a frame shown in Fig. 3-8. If each member is inextensible, then the frame is a 6-degree indeterminate structure. (*Note:* This elementary slope-deflection equation cannot handle member extensibility.) Assume just for simplicity that all four members are identical. Dimensions of the frame, cross-sectional properties, and material constant are = 100 in., $I = 1$ in.^{4}, $A = 10^6$ in.^{2}, $E = 29{,}000$ ksi, respectively. Member *ab* is subjected to a uniform load, $w = 0.1$ kip/in.

There is only one unknown, θ_b (kinematic degree-of-freedom).

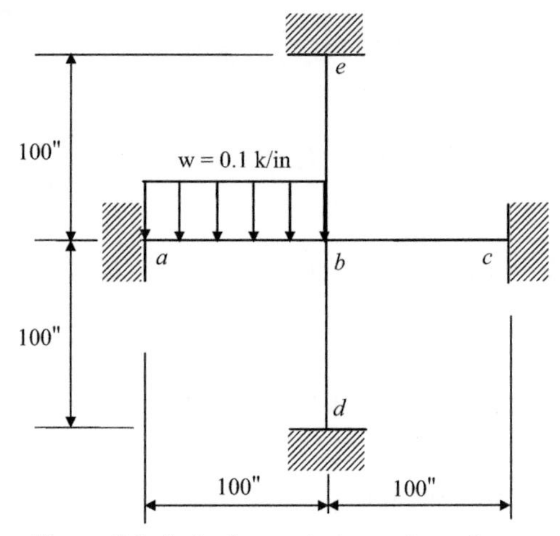

Figure 3-8 A six-degree indeterminate frame

The fixed end moments prior to releasing the constraint are

$$M_{ab\ fixed} = +\frac{w\ell^2}{12} \quad \text{and} \quad M_{ba\ fixed} = -\frac{w\ell^2}{12}$$

By virtue of Eq. (3.4.2.13), the moment of the member ab at the b end becomes $M_{ba} = (4EI/\ell)\,\theta_b - (w\ell^2/12)$

$$M_{bc} = \frac{4EI}{\ell}\theta_b$$

$$M_{bd} = \frac{4EI}{\ell}\theta_b$$

$$M_{be} = \frac{4EI}{\ell}\theta_b$$

The sum of moments at b must be equal to zero for equilibrium. Hence,

$$\frac{16EI}{\ell}\theta_b - \frac{w\ell^2}{12} = 0 \Rightarrow \theta_b = \frac{w\ell^3}{192EI}$$

Substituting θ_b into Eqs. (3.4.2.11), (3.4.2.12), (3.4.2.13), and (3.4.2.14) yields

$$M_{ba} = -\frac{3w\ell^2}{48} = -62.5 \ k - in.$$

$$M_{bc} = \frac{w\ell^2}{48} = 20.83 \ k - in.$$

$$M_{bd} = \frac{w\ell^2}{48} = 20.83 \ \ k - in.$$

$$M_{be} = \frac{w\ell^2}{48} = 20.83 \ \ k - in.$$

$$M_{ab} = \frac{3w\ell^2}{32} = 93.75 \ \ k - in.$$

The slope-deflection equations are very effective when applied to problems with a small number of kinematic degrees of freedom.

3.4.3. Effects of Axial Loads on Bending Stiffness

The classical slope-deflections equations that are introduced in any standard text on indeterminate structures (Parcel and Moorman 1955; Kinney 1957) give the moments, M_{ab} and M_{ba}, induced at the ends of member AB as a function of end rotations θ_a and θ_b and by a displacement Δ of one end to the other. In conventional linear structural analysis (first-order analysis), it is customary to ignore the effect of axial forces on the bending stiffness of flexural members. It can be shown that the effect of amplification is negligibly small as long as the axial load remains small in comparison with the critical load of the member. When the ratio of the axial load to the critical load becomes sizable, however, the bending stiffness is reduced markedly due to the axial compression, and it is no longer acceptable to neglect this reduction. As the first-order analysis results may become dangerously unconservative, modern design specifications call for a mandatory second-order analysis (AISC 2005).

It is expedient to introduce $\Delta = \delta_b - \delta_a$ with $\delta_a = 0$ to avoid the rigid body translation. The moment of the beam-column shown in Figure 3-9 at a distance x from the origin is

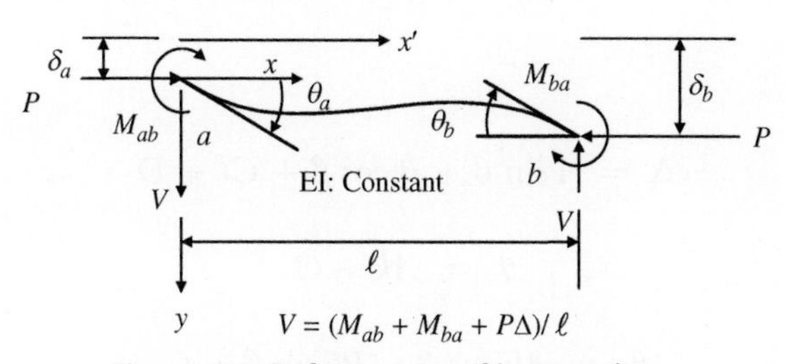

Figure 3-9 Deformations of beam-column

$$M_x = M_{ab} + Py - (M_b + M_{ba} + P\Delta)\frac{x}{\ell}$$

$$y'' = -\frac{M_x}{EI}$$

$$EIy'' + Py = -M_{ab} + (M_{ab} + M_{ba} + P\Delta)\frac{x}{\ell}$$

Taking successive derivatives on both sides yields

$$EIy^{iv} + Py'' = 0$$

Let $k^2 = \dfrac{P}{EI}$

The simplified differential equation is

$$y^{iv} + k^2 y'' = 0$$

for which the general solution is

$$y = A \sin kx + B \cos kx + Cx + D$$

The proper geometric boundary conditions are

$$y(0) = 0, \quad y(\ell) = \Delta, \quad y'(0) = \theta_a, \quad \text{and} \quad y'(\ell) = \theta_b$$

The proper natural boundary conditions are

$$y''(0) = -\frac{M_{ab}}{EI}, \quad \text{and} \quad y''(\ell) = \frac{M_{ba}}{EI}$$

Applying the geometric boundary conditions to eliminate the integral constants, A, B, C, D, and solving for M_{ab} and M_{ba} gives

$$0 = B + D$$

Let $\beta = k\ell$

$$\Delta = A \sin \beta + B \cos \beta + C\ell + D$$

$$\theta_a = Ak + C$$

$$\theta_b = Ak \cos \beta - Bk \sin \beta + C$$

The matrix equation for the integral constants becomes

$$\begin{bmatrix} 0 & 1 & 0 & 1 \\ \sin\beta & \cos\beta & \ell & 1 \\ k & 0 & 1 & 0 \\ k\cos\beta & -k\sin\beta & 1 & 0 \end{bmatrix} \begin{Bmatrix} A \\ B \\ C \\ D \end{Bmatrix} = \begin{Bmatrix} 0 \\ \Delta \\ \theta_a \\ \theta_b \end{Bmatrix}$$

Applying Cramer's rule yields

$$A = \frac{\begin{vmatrix} 0 & 1 & 0 & 1 \\ \Delta & \cos\beta & \ell & 1 \\ \theta_a & 0 & 1 & 0 \\ \theta_b & -k\sin\beta & 1 & 0 \end{vmatrix}}{\begin{vmatrix} 0 & 1 & 0 & 1 \\ \sin\beta & \cos\beta & \ell & 1 \\ k & 0 & 1 & 0 \\ k\cos\beta & -k\sin\beta & 1 & 0 \end{vmatrix}} = \frac{D_a}{D_d}$$

$$D_a = \theta_a \begin{vmatrix} 1 & 0 & 1 \\ \cos\beta & \ell & 1 \\ -k\sin\beta & 1 & 0 \end{vmatrix} + \begin{vmatrix} 0 & 1 & 1 \\ \Delta & \cos\beta & 1 \\ \theta_b & -k\sin\beta & 0 \end{vmatrix}$$

$$= \theta_a(\cos\beta + \beta\sin\beta - 1) + \theta_b - k\sin\beta\,\Delta - \theta_b\cos\beta$$

$$= \theta_a(\cos\beta + \beta\sin\beta - 1) + \theta_b(1 - \cos\beta) - k\sin\beta\;\Delta$$

$$D_d = -\begin{vmatrix} \sin\beta & \ell & 1 \\ k & 1 & 0 \\ k\cos\beta & 1 & 0 \end{vmatrix} - \begin{vmatrix} \sin\beta & \cos\beta & \ell \\ k & 0 & 1 \\ k\cos\beta & -k\sin\beta & 1 \end{vmatrix}$$

$$= -k + k\cos\beta - k(\cos^2\beta + \sin^2\beta) + k\,\beta\sin\beta + k\cos\beta$$

$$= -2k + 2k\cos\beta + k\;\beta\sin\beta = k(2\cos\beta + \beta\sin\beta - 2)$$

$$B = \frac{\begin{vmatrix} 0 & 0 & 0 & 1 \\ \sin\beta & \Delta & \ell & 1 \\ k & \theta_a & 1 & 0 \\ k\cos\beta & \theta_b & 1 & 0 \end{vmatrix}}{D_d} = \frac{D_b}{D_d}$$

$$\begin{aligned} D_b &= -\begin{vmatrix} \sin\beta & \Delta & \ell \\ k & \theta_a & 1 \\ k\cos\beta & \theta_b & 1 \end{vmatrix} \\ &= -\theta_a \sin\beta - \theta_b\beta - k\cos\beta\ \Delta + \theta_a\beta\cos\beta + k\ \Delta + \theta_b\sin\beta \\ &= \theta_a(\beta\cos\beta - \sin\beta) + \theta_b(\sin\beta - \beta) + \Delta(k - k\cos\beta) \end{aligned}$$

$$y' = Ak\cos kx - Bk\sin kx + C$$

$$y'' = -Ak^2\sin kx - Bk^2\cos kx$$

$$\begin{aligned} M_{ab} &= -EIy''(0) = EIBk^2 \\ &= \left[\frac{EIk^2}{k(2\cos\beta + \beta\sin\beta - 2)}\right][(\beta\cos\beta - \sin\beta)\theta_a + (\sin\beta - \beta)\theta_b \\ &\quad + (k - k\cos\beta)\Delta] \\ &= \left[\frac{EI\beta}{\ell(2\cos\beta + \beta\sin\beta - 2)}\right]\left[(\beta\cos\beta - \sin\beta)\theta_a + (\sin\beta - \beta)\theta_b\right. \\ &\quad \left. + (\beta - \beta\cos\beta)\frac{\Delta}{\ell}\right] \end{aligned} \tag{3.4.11}$$

Let

$$S_1 = S = \frac{\beta(\beta\cos\beta - \sin\beta)}{2\cos\beta + \beta\sin\beta - 2} \tag{3.4.12}$$

$$S_2 = \frac{(\sin\beta - \beta)}{2\cos\beta + \beta\sin\beta - 2} \tag{3.4.13}$$

Recall identities

$$\sin\beta = 2\sin(\beta/2)\cos(\beta/2)$$

$$\cos\beta = \cos^2(\beta/2) - \sin^2(\beta/2) = 1 - 2\sin^2(\beta/2)$$

Dividing the numerator and denominator of S_1 by $\sin\beta$ gives

$$S_1 = S = \frac{\beta(\beta\cot\beta - 1)}{2\cot\beta - \dfrac{2}{\sin\beta} + \beta} = \frac{\beta(\beta\cot\beta - 1)}{den1 + \beta}$$

where

$$\begin{aligned} den1 &= 2\cot\beta - \frac{2}{\sin\beta} = \frac{2\cos\beta - 2}{\sin\beta} = \frac{2[1 - 2\sin^2(\beta/2) - 1]}{2\sin(\beta/2)\cos(\beta/2)} \\ &= -2\tan(\beta/2) \end{aligned}$$

$$S_1 = S = \frac{\beta(\beta\cot\beta - 1)}{-2\tan(\beta/2) + \beta}$$

$$S_1 = S = \frac{1 - \beta\cot\beta}{\dfrac{2\tan(\beta/2)}{\beta} - 1} \tag{3.4.14}$$

$$\text{Let } S_2 = C = \frac{\beta(\sin\beta - \beta)}{2\cos\beta + \beta\sin\beta - 2}$$

Taking the same procedure used above gives

$$S_2 = C = \frac{\beta(1 - \beta\,\mathrm{cosec}\,\beta)}{2\cot\beta - \dfrac{2}{\sin\beta} + \beta} = \frac{\beta(1 - \beta\,\mathrm{cosec}\,\beta)}{-2\tan(\beta/2) + \beta}$$

$$S_2 = C = \frac{\beta\,\mathrm{cosec}\,\beta - 1}{\dfrac{2\tan(\beta/2)}{\beta} - 1} \tag{3.4.15}$$

$$\text{Let } S_3 = SC = \frac{\beta(\beta - \beta\cos\beta)/\ell}{2\cos\beta + \beta\sin\beta - 2}$$

Again dividing the numerator and denominator of S_3 by $\sin\beta$ gives

$$S_3 = SC = \frac{\beta(\beta \operatorname{cosec} \beta - \beta \cot \beta)/\ell}{2 \cot \beta - \dfrac{2}{\sin \beta} + \beta} = \frac{\beta(\beta \operatorname{cosec} \beta - \beta \cot \beta)/\ell}{-2 \tan(\beta/2) + \beta}$$

$$= \frac{(\beta \cot \beta - \beta \operatorname{cosec} \beta)/\ell}{\dfrac{2 \tan(\beta/2)}{\beta} - 1} = \frac{[-(1 - \beta \cot \beta) - (\beta \operatorname{cosec} \beta - 1)]/\ell}{\dfrac{2 \tan(\beta/2)}{\beta} - 1} \tag{3.4.16}$$

$$S_3 = SC = -\frac{S_1 + S_2}{\ell} = -\frac{S + C}{\ell}$$

Recall $M_{ab} = M(0) = -EIy''(0)$.

But $M_{ba} = -M(\ell) = EIy''(\ell)$ (note the negative sign!)

$$y'' = -Ak^2 \sin kx - Bk^2 \cos kx$$

$$M_{ba} = +EIy''(\ell)$$

$$= \left[\frac{-EIk^2}{k(2 \cos \beta + \beta \sin \beta - 2)}\right] \left\{ \begin{aligned} &\sin \beta[\theta_a(\cos \beta + \beta \sin \beta - 1) + \theta_b(1 - \cos \beta) - \Delta \ k \sin \beta] \\ &\quad + \cos \beta[\theta_a(\beta \cos \beta - \sin \beta) + \theta_b(\sin \beta - \beta) \\ &\quad + \Delta(k - k \cos \beta)] \end{aligned} \right\}$$

$$= \left(\frac{-EIk}{2 \cos \beta + \beta \sin \beta - 2}\right) \left[\begin{aligned} &\theta_a(\cos \beta \sin \beta + \beta \sin^2 \beta - \sin \beta + \beta \cos^2 - \cos \beta \sin \beta) \\ &\quad + \theta_b(\sin \beta - \cos \beta \sin \beta + \cos \beta \sin \beta - \beta \cos \beta) \\ &\quad + \Delta(k \cos \beta - k \cos^2 \beta - k \sin^2 \beta) \end{aligned} \right]$$

$$= \left(-\frac{EI\beta}{\ell}\right) \frac{[\theta_a(\beta - \sin \beta) + \theta_b(\sin \beta - \beta \cos \beta) + \Delta(k \cos \beta - k)]}{(2 \cos \beta + \beta \sin \beta - 2)}$$

$$= \left(\frac{EI}{\ell}\right) \frac{[\theta_a\beta(\sin \beta - \beta) + \theta_b\beta(\beta \cos \beta - \sin \beta) + \Delta\beta(\beta - \beta \cos \beta)/\ell]}{(2 \cos \beta + \beta \sin \beta - 2)} \tag{3.4.17}$$

Examination of Eqs. (3.4.11) and (3.4.17) reveals that they can be rewritten as follows:

$$M_{ab} = \frac{EI}{\ell}\left[S_1\theta_a + S_2\theta_b - (S_1 + S_2)\frac{\Delta}{\ell}\right] \tag{3.4.18}$$

$$M_{ba} = \frac{EI}{\ell}\left[S_2\theta_a + S_1\theta_b - (S_1 + S_2)\frac{\Delta}{\ell}\right] \tag{3.4.19}$$

If $M_{ab} = 0$ (when the support A is either pinned or roller), then

$$M_{ab} = \frac{EI}{\ell}\left[S_1\theta_a + S_2\theta_b - \left(S_1 + S_2\right)\frac{\Delta}{\ell}\right] = 0$$

$$\theta_a = \frac{1}{S_1}\left[-S_2\theta_b + (S_1 + S_2)\frac{\Delta}{\ell}\right]$$

Substituting θ_a into M_{ba} yields

$$M_{ba} = \frac{EI}{\ell}\left[\left(S_1 - \frac{S_2^2}{S_1}\right)\theta_b - \left(S_1 + S_2\right)\left(1 - \frac{S_2}{S_1}\right)\frac{\Delta}{\ell}\right]$$

Let $\overline{S} = \frac{1}{S_1}(S_1^2 - S_2^2)$, then

$$\overline{M}_{ba} = \frac{EI}{\ell}\left[\overline{S}\theta_b - \overline{S}\frac{\Delta}{\ell}\right]$$

$$\begin{aligned}
\overline{S} &= \frac{1}{S_1}\left(S_1^2 - S_2^2\right) \\
&= \left[\frac{-2\tan(\beta/2) + \beta}{\beta(\beta\cot\beta - 1)}\right]\left[\frac{\beta^2(\beta\cot\beta - 1)^2}{(-2\tan(\beta/2) + \beta)^2} - \frac{\beta^2(1 - \operatorname{cosec}\beta)^2}{(-2\tan(\beta/2) + \beta)^2}\right] \\
&= \frac{\beta}{(\beta\cot\beta - 1)[-2\tan(\beta/2) + \beta]}[(\beta\cot\beta - 1)^2 - (1 - \operatorname{cosec}\beta)^2] \\
&= \frac{\beta^2}{(\beta\cot\beta - 1)[-2\tan(\beta/2) + \beta]}[-\beta + 2\tan(\beta/2)] = \frac{\beta^2}{1 - \beta\cot\beta}
\end{aligned} \tag{3.4.20}$$

When $P \Rightarrow 0$, then $\beta \Rightarrow 0$. Limiting values (for $P = 0$): $S_1 = 4$, $S_2 = 2$. Values for various β can be readily evaluated using **Maple**®.

Example 1 Stiffness coefficients of a beam-column shown in Figure 3-10.

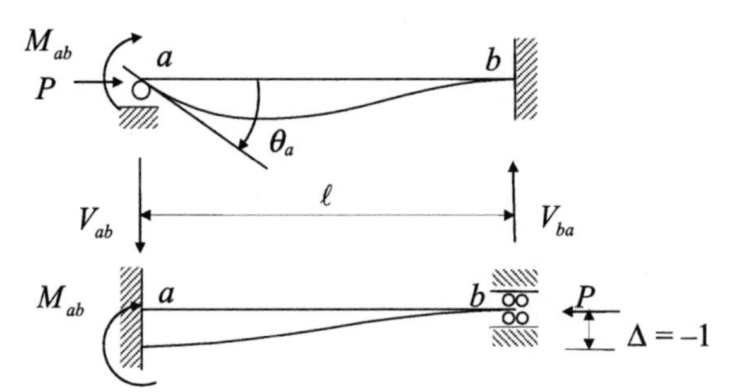

Figure 3-10 Stiffness coefficients of beam-column

For the first case

$$\theta_a = 1 \text{ radian}, \theta_b = 0, \delta_a = \delta_b = 0$$

then

$$M_{ab} = \frac{EI}{\ell} S_1 \quad \text{and} \quad M_{ba} = \frac{EI}{\ell} S_2$$

$$V_{ab} = -\frac{M_{ab} + M_{ba}}{\ell} \quad \text{and} \quad V_{ba} = -V_{ab}$$

$$S_1 = M_{ab} \frac{\ell}{EI}$$

The numerical value of S_1 shown in Figure 3-11 is a measure of bending moment depending on the magnitude of the axial force

$$k\ell = \phi = \beta = \ell\sqrt{\frac{P}{EI}} = \pi\sqrt{\frac{P\ell^2}{\pi^2 EI}} = \pi\sqrt{\frac{P}{P_E}}$$

The critical load of a propped column is

$$P_{cr} = \frac{2.04\pi^2 EI}{\ell^2}$$

Hence,

$$k\ell = \sqrt{2.04}\pi = 4.49$$

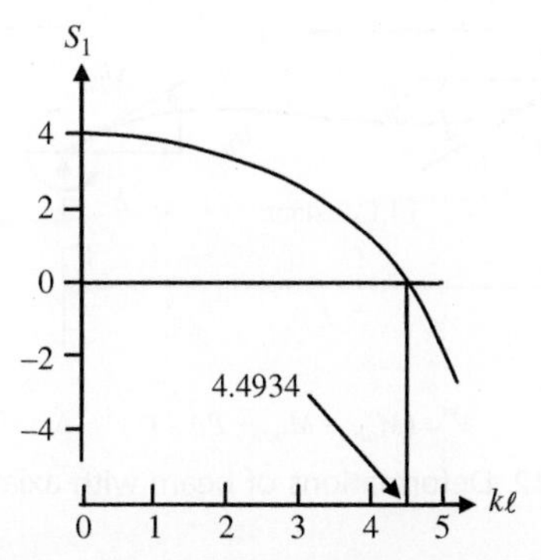

Figure 3-11 S_1 vs $k\ell$ in a propped beam-column

If $k\ell > 4.49$, S_1 becomes negative. θ_a is resisted by the adjacent member(s). Or the propped beam-column will undergo buckling failure unless the adjacent member(s) provide stability against failure.

For the second case

$$\Delta = -1, \theta_a = \theta_b = 0$$

$$M_{ab} = M_{ba} = \frac{EI}{\ell^2} S_3 \quad \text{and} \quad V_{ab} = -\frac{2EI}{\ell^3} S_3 = -V_{ba}$$

Again, it would be difficult to say who should be given the credit for first developing the slope-deflection equations, including the effect of axial compression. Bazant and Credolin (1991) introduce James (1935), who presented the stiffness matrix relating the end moments and the member rotations in a work dealing with the moment-distribution method. The stiffness coefficients S_1, S_2, and S_3 take slightly different forms depending on the extent of manipulations. Because these coefficients serve as the basis for stability analysis of frames, they are also called stability functions. Horne and Merchant (1965) give credit to Berry (1916) for being the first who suggested various types of stability functions and James (1935) for being the first who calculated S_1 and S_2. Before the advancement of modern digital computers, evaluating these functions would have been a formidable task. Winter et al. (1948), Niles and Newell (1948), Goldberg (1954), Livesley and Chandler (1956), and Timoshenko and Gere (1961) published tables and charts of these stability functions.

3.4.4. Slope-Deflection Equations with Axial Tension

As was done earlier, it is expedient to let $\Delta = \delta_b - \delta_a$ with $\delta_a = 0$ in order to avoid the rigid body translation. The moment of the bean-column shown in Figure 3-12 at a distance x from the origin is

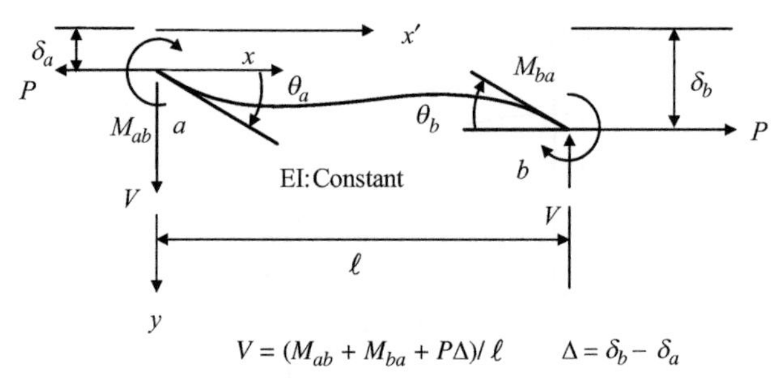

Figure 3-12 Deformations of beam with axial tension

$$M_x = M_{ab} - Py - (M_{ab} + M_{ba} - P\Delta)\frac{x}{\ell}$$

$$y'' = -\frac{M_x}{EI}$$

$$EIy'' - Py = -M_b + (M_{ab} + M_{ba} - P)\frac{x}{\ell}$$

Let $k^2 = \frac{P}{EI}$. Then,

$$y'' - k^2 y = f(x)$$

where $f(x)$ is a linear function of x.

Upon taking successive derivatives of the differential equation, the differential equation becomes

$$y^{iv} - k^2 y'' = 0$$

The general solution of this differential equation is

$$y = A \sinh kx + B \cosh kx + Cx + D$$

The proper geometric boundary conditions are

$$y(0) = \delta_a = 0, \quad y(\ell) = \delta_b = \Delta, \quad y'(0) = \theta_a, \quad y'(\ell) = \theta_b$$

The appropriate natural boundary conditions are

$$y''(0) = -\frac{M_{ab}}{EI}, \quad y''(\ell) = \frac{M_{ba}}{EI}$$

Eliminating the integral constants, A, B, C, D and solving for M_{ab} and M_{ba} gives

$$0 = B + D$$

$$\Delta = A \sinh k\ell + B \cosh k\ell + C\ell + D$$

$$y' = Ak \cosh kx + Bk \sinh kx + C$$

$$\theta_a = Ak + C$$

$$\theta_b = Ak \cosh k\ell + Bk \sinh k\ell + C$$

Let $\beta = k\ell$

$$\begin{bmatrix} 0 & 1 & 0 & 1 \\ \sinh\beta & \cosh\beta & \ell & 1 \\ k & 0 & 1 & 0 \\ k\cosh\beta & k\sinh\beta & 1 & 0 \end{bmatrix} \begin{Bmatrix} A \\ B \\ C \\ D \end{Bmatrix} = \begin{Bmatrix} 0 \\ \Delta \\ \theta_a \\ \theta_b \end{Bmatrix}$$

Applying Cramer's rule yields

$$A = \frac{\begin{vmatrix} 0 & 1 & 0 & 1 \\ \Delta & \cosh\beta & \ell & 1 \\ \theta_a & 0 & 1 & 0 \\ \theta_b & k\sinh\beta & 1 & 0 \end{vmatrix}}{\begin{vmatrix} 0 & 1 & 0 & 1 \\ \sinh\beta & \cosh\beta & \ell & 1 \\ k & 0 & 1 & 0 \\ k\cosh\beta & k\sinh\beta & 1 & 0 \end{vmatrix}} = \frac{D_a}{D_d}$$

$$D_a = \theta_a \begin{vmatrix} 1 & 0 & 1 \\ \cosh\beta & \ell & 1 \\ k\sinh\beta & 1 & 0 \end{vmatrix} + \begin{vmatrix} 0 & 1 & 1 \\ \Delta & \cos\beta & 1 \\ \theta_b & k\sinh\beta & 0 \end{vmatrix}$$

$$= \theta_a\left(k\sinh\beta \begin{vmatrix} 0 & 1 \\ \ell & 1 \end{vmatrix} - \begin{vmatrix} 1 & 1 \\ \cosh\beta & 1 \end{vmatrix} \right) + \theta_b \begin{vmatrix} 1 & 1 \\ \cosh\beta & 1 \end{vmatrix} - k\sinh\beta \begin{vmatrix} 0 & 1 \\ \Delta & 1 \end{vmatrix}$$

$$= \theta_a(\cosh\beta - \beta\sinh\beta - 1) + \theta_b(1 - \cosh\beta) + \Delta k \sinh\beta$$

$$D_d = -\begin{vmatrix} \sinh\beta & \ell & 1 \\ k & 1 & 0 \\ k\cosh\beta & 1 & 0 \end{vmatrix} - \begin{vmatrix} \sinh\beta & \cosh\beta & \ell \\ k & 0 & 1 \\ k\cosh\beta & k\sinh\beta & 1 \end{vmatrix}$$

$$= -\begin{vmatrix} k & 1 \\ k\cosh\beta & 1 \end{vmatrix} + k\begin{vmatrix} \cosh\beta & \ell \\ k\sinh\beta & 1 \end{vmatrix} + \begin{vmatrix} \sinh\beta & \cosh\beta \\ k\cosh\beta & k\sinh\beta \end{vmatrix}$$

$$= -k + k\cosh\beta + k\cosh\beta - k\beta\sinh\beta - k(\cosh^2\beta - \sinh^2\beta)$$

$$= -2k + 2k\cosh\beta - k\beta\sinh\beta$$

$$A = \frac{D_a}{D_d}$$

$$B = \frac{\begin{vmatrix} 0 & 0 & 0 & 1 \\ \sinh\beta & \Delta & \ell & 1 \\ k & \theta_a & 1 & 0 \\ k\cosh\beta & \theta_b & 1 & 0 \end{vmatrix}}{D_d} = \frac{D_b}{D_d}$$

$$D_b = -\begin{vmatrix} \sinh\beta & \Delta & \ell \\ k & \theta_a & 1 \\ k\cosh\beta & \theta_b & 1 \end{vmatrix}$$

$$= -(\theta_a\sinh\beta + \beta\theta_b + k\Delta\cosh\beta - \theta_a\beta\cosh\beta - \theta_b\sinh\beta - k\Delta)$$

$$= \theta_a(\beta\cosh\beta - \sinh\beta) + \theta_b(\sinh\beta - \beta) + \Delta(k - k\cosh\beta)$$

$$y' = Ak\cosh kx + Bk\sinh kx + C$$

$$y'' = Ak^2\sinh kx + Bk^2\cosh kx$$

$$
\begin{aligned}
M_{ab} &= -EIy''(0) \\
&= -EI\frac{k^2[(\beta\cosh\beta - \sinh\beta)\theta_a + (\sinh\beta - \beta)\theta_b + (k - k\cosh\beta)\Delta]}{k(-2 + 2\cosh\beta - \beta\sinh\beta)} \\
&= \frac{EI}{\ell}\frac{\beta[(\beta\cosh\beta - \sinh\beta)\theta_a + (\sinh\beta - \beta)\theta_b + (\beta - \beta\cosh\beta)\frac{\Delta}{\ell}]}{(2 - 2\cosh\beta + \beta\sinh\beta)}
\end{aligned}
\tag{3.4.21}
$$

Let $S_1' = S' = \dfrac{\beta(\beta\cosh\beta - \sinh\beta)}{2 - 2\cosh\beta + \beta\sinh\beta}$

Recall identities

$$\sinh\beta = 2(\sinh\beta/2)(\cosh\beta/2)$$

$$\cosh\beta = \cosh^2\beta/2 + \sinh^2\beta/2 = 1 + 2\sinh^2\beta/2$$

Dividing the numerator and denominator of $S_1' = S'$ by $\sinh\beta$ gives

$$S_1' = S'\frac{\beta(\beta\coth\beta - 1)}{\dfrac{2}{\sinh\beta} - 2\coth\beta + \beta} = \frac{\beta(\beta\coth\beta - 1)}{I + \beta}$$

Let $I = \dfrac{2(1 - \cosh\beta)}{\sinh\beta} = \dfrac{2[1 - (1 + 2\sinh^2\beta/2)]}{2\sinh\beta/2\cosh\beta/2} = -2\tanh\beta/2$

Substituting I into $S_1' = S'$ gives

$$S_1' = S' = \frac{\beta(\beta\coth\beta - 1)}{\beta - 2\tanh\beta/2}$$

$$S_1' = S' = \frac{\beta\coth\beta - 1}{1 - \dfrac{2\tanh\beta/2}{\beta}} \tag{3.4.22}$$

Let $S_2' = C' = \dfrac{\beta(\sinh\beta - \beta)}{2 - 2\cosh\beta + \beta\sinh\beta}$

Dividing the numerator and denominator of $S_2' = C'$ by $\sinh\beta$ gives

$$S_2' = C' = \frac{\beta(1-\beta\,\text{csch}\,\beta)}{\dfrac{2}{\sinh\beta} - 2\coth\beta + \beta} = \frac{\beta(1-\beta\,\text{csch}\,\beta)}{1+\beta} = \frac{\beta(1-\beta\,\text{csch}\,\beta)}{\beta - 2\tanh\beta/2}$$

$$S_2' = C' = \frac{1-\beta\,\text{csch}\,\beta}{1-\dfrac{2\tanh\beta/2}{\beta}} \tag{3.4.23}$$

Let $S_3' = SC' = \dfrac{\beta(\beta-\beta\cosh\beta)/\ell}{2-2\cosh\beta+\beta\sinh\beta}$

Dividing the numerator and denominator of $S_3' = SC'$ by $\sinh\beta$ gives

$$S_3' = SC' = \frac{\beta(\beta\,\text{csch}\,\beta - \beta\coth\beta)/\ell}{\dfrac{2}{\sinh\beta} - 2\coth\beta + \beta} = \frac{\beta(\beta\,\text{csch}\,\beta - \beta\coth\beta)/\ell}{I+\beta}$$

$$= \frac{\beta(\beta\,\text{csch}\,\beta - \beta\coth\beta)/\ell}{\beta - 2\tanh\beta/2}$$

$$S_3' = SC' = \frac{(\beta\,\text{csch}\,\beta - \beta\coth\beta)/\ell}{1-\dfrac{2\tanh\beta/2}{\beta}}$$

$$= \frac{[-(\beta\coth\beta-1)-(1-\beta\,\text{csch}\,\beta)]/\ell}{1-\dfrac{2\tanh\beta/2}{\beta}} \tag{3.4.24}$$

$$S_3' = SC' = -\frac{S_1'+S_2'}{\ell} = -\frac{S'+C'}{\ell}$$

Recall

$$M_{ab} = M(0) = -EIy''(0)$$

But

$$M_{ba} = -M(\ell) = EIy''(\ell)(\text{note the negative sign!})$$

$$y'' = Ak^2\sinh kx + Bk^2\cosh kx$$

$$
\begin{aligned}
M_{ba} &= EIy''(\ell) \\
&= \frac{EIk}{(-2 + 2\cosh\beta - \beta\sinh\beta)} \\
&\times \left\{ \begin{aligned} &\sinh\beta[\theta_a(\cosh\beta - \beta\sinh\beta - 1) + \theta_b(1 - \cosh\beta) + \Delta k\sinh\beta] \\ &+ \cosh\beta\left[\begin{aligned}&\theta_a(\beta\cosh\beta - \sinh\beta) + \theta_b(\sinh\beta - \beta) \\ &+ \Delta(k - k\cosh\beta)\end{aligned}\right] \end{aligned} \right\} \\
&= \frac{EI\beta}{\ell}\frac{[\theta_a(\beta - \sinh\beta) + \theta_b(\sinh\beta - \beta\cosh\beta) + \Delta(k\cosh\beta - k)]}{(-2 + 2\cosh\beta - \beta\sinh\beta)} \\
&= \frac{EI\beta}{\ell}\frac{[\theta_a(\sinh\beta - \beta) + \theta_b(\beta\cosh\beta - \sinh\beta) + \frac{\Delta}{\ell}(\beta - \beta\cosh\beta)]}{(2 - 2\cosh\beta + \beta\sinh\beta)}
\end{aligned}
\tag{3.4.25}
$$

Examination of Eqs. (3.4.21) and (3.4.25) reveals that they can be rewritten as follows:

$$
M_{ab} = \frac{EI}{\ell}\left[S_1'\theta_a + S_2'\theta_b - (S_1' + S_2')\frac{\Delta}{\ell}\right] \tag{3.4.26}
$$

$$
M_{ba} = \frac{EI}{\ell}\left[S_2'\theta_a + S_1'\theta_b - (S_1' + S_2')\frac{\Delta}{\ell}\right] \tag{3.4.27}
$$

3.5. ULTIMATE STRENGTH OF BEAM-COLUMNS

Up to this point in the study of beam-columns, the subject of failure was not considered; hence, it was possible to limit the discussion to elastic behavior. It is the modern trend that the design specifications are developed using the probability-based load and resistance factor design concepts: The load-carrying capacity of each structural member all the way up to its ultimate strength needs to be evaluated. Since the ultimate strength of a structural member frequently involves yielding, it becomes necessary to work with the complexities of inelastic behavior in the analysis. It was pointed out in Chapter 1 that problems involving inelastic behavior do not possess

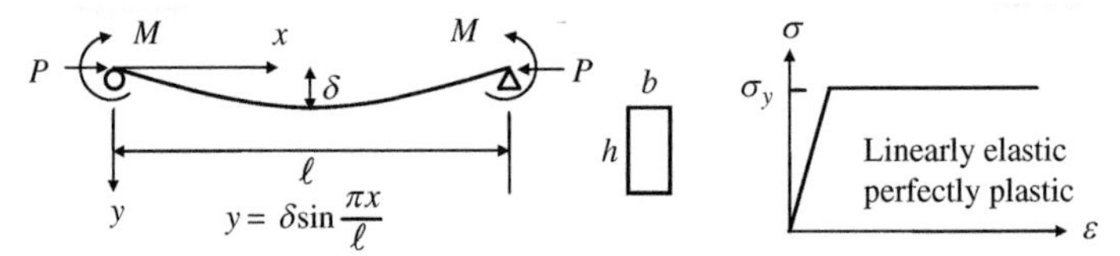

Figure 3-13 Idealized beam-column of Jezek

closed-form solutions. They must either be solved numerically, or approximate answers must be sought by making simplifying assumptions. In this section, the latter approach is entailed.

Consider the simply supported, symmetrically loaded beam-column shown in Fig. 3-13. Jezek (1934, 1935, 1936) demonstrated that a closed-form solution for the load-deflection behavior beyond the proportional limit can be obtained when the following assumptions are made:

1. The cross section of the member is rectangular as shown in Fig. 3-13.
2. The material obeys linearly elastic and perfectly plastic stress-strain relationships.
3. The bending deflection of the member takes the form of a half-sine wave.

Inelastic bending is difficult to analyze because the stress-strain relation varies in a complex manner both along the member and across the section once the proportional limit has been exceeded. In addition to these major idealizations, which simplify the analysis greatly, the following assumptions are also made:

4. Deformations are finite but still small enough so that the curvature can be approximated by the second derivative of the deflected curve.
5. The member is initially straight.
6. Bending takes place about the major principal axis.

The residual stress that cannot be avoided in rolled and/or fabricated metal sections is ignored in the analysis.

Based on the coordinate system shown in Fig. 3-13, the external bending moment at a distance x from the origin is

$$M_{ext} = M + Py \tag{3.5.1}$$

Since Eq. (3.5.1) is an external equilibrium equation, it is valid regardless of whether or not the elastic limit of the material is exceeded.

The relationship between the load and deflection up to the proportional limit is

$$M + Py = -EIy'' = -EI\left(-\delta\frac{\pi^2}{\ell^2}\right)\sin\frac{\pi x}{\ell}$$

or

$$M + Py = EI\frac{\delta\pi^2}{\ell^2}\sin\frac{\pi x}{\ell} \tag{3.5.2}$$

This relationship at the midspan becomes

$$M + P\delta = \frac{EI\delta\pi^2}{\ell^2} \tag{3.5.3}$$

Assuming that M is proportional to P, one introduces the notation $e = M/P$; then the above moment equation becomes

$$P(e + \delta) = \frac{\delta EI\pi^2}{\ell^2} = \delta P_E \tag{3.5.4}$$

Dividing both sides of Eq. (3.5.4) by h yields

$$P\left(\frac{e}{h} + \frac{\delta}{h}\right) = \frac{\delta}{h}P_E$$

or

$$\frac{\delta}{h} = \frac{e}{h}\frac{1}{\frac{\sigma_E}{\sigma_0} - 1} \tag{3.5.5}$$

where

$\sigma_E = P_E/bh$ is the Euler stress and $\sigma_0 = P/bh$ is the average axial stress.

Equation (3.5.5) gives the load versus deflection relationship in the elastic range. In order to determine the load at which Eq. (3.5.5) becomes invalid, one must evaluate the maximum stress in the member.

$$\sigma_{max} = \frac{P}{bh} + \frac{M + P\delta}{\frac{bh^2}{6}} = \sigma_0 + \sigma_0\frac{6(e+\delta)}{h} \tag{3.5.6}$$

or

$$\sigma_{max} = \sigma_0\left[1 + \frac{6(e+\delta)}{h}\right] \tag{3.5.7}$$

If the stress given by Eq. (3.5.7) equals the yield stress, the elastic load versus deflection relationship given by Eq. (3.5.5) becomes invalid.

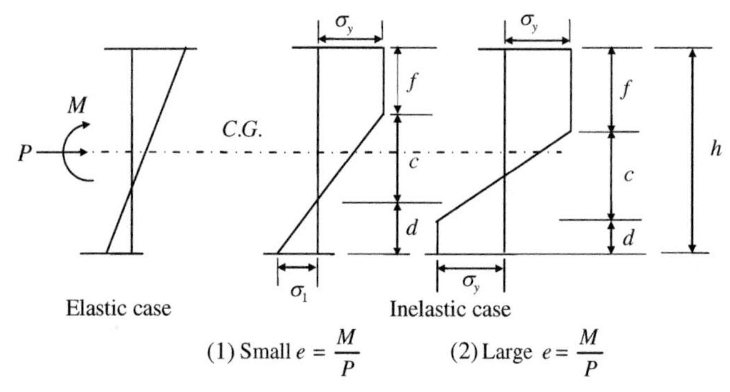

Figure 3-14 Stress distributions for beam-column

As yielding propagates inward, the inner elastic core as indicated in Fig. 3-14 for the inelastic case, the moment curvature relationship expressed in Eq. (3.5.2) becomes invalid and a new moment-curvature expression needs to be developed. Depending on the eccentricity, two different stress distributions are possible. If the ratio $e = M/P$ is relatively small, yielding occurs only on the concave side of the member prior to reaching its ultimate strength range. On the other hand, if the eccentricity is relatively large, both the convex and concave sides of the member will have started to yield before the maximum load is reached, as shown in Fig. 3-14. To simplify the analysis, the discussion is limited herein to small values of e only. Bleich (1952) discusses the case of large values of e.

Summing the horizontal forces in case (1) of Fig. 3-14 yields

$$P = b\left(\sigma_y f + \frac{\sigma_y c}{2} - \frac{\sigma_1 d}{2}\right)$$

Dividing both sides by bh yields

$$\sigma_0 = \frac{1}{h}\left(\sigma_y f + \frac{\sigma_y c}{2} - \frac{\sigma_1 d}{2}\right) \tag{3.5.8}$$

Summing the moment about the centroidal axis gives

$$M_{int} = \left[\sigma_y f\left(\frac{h}{2} - \frac{f}{2}\right) + \frac{\sigma_y c}{2}\left(\frac{h}{2} - f - \frac{c}{3}\right) + \frac{\sigma_1 d}{2}\left(\frac{h}{2} - \frac{d}{3}\right)\right] \tag{3.5.9}$$

Noting that $f + c + d = h$, c value can be determined from Eqs. (3.5.8) and (3.5.9). After some lengthy algebraic manipulations, one obtains

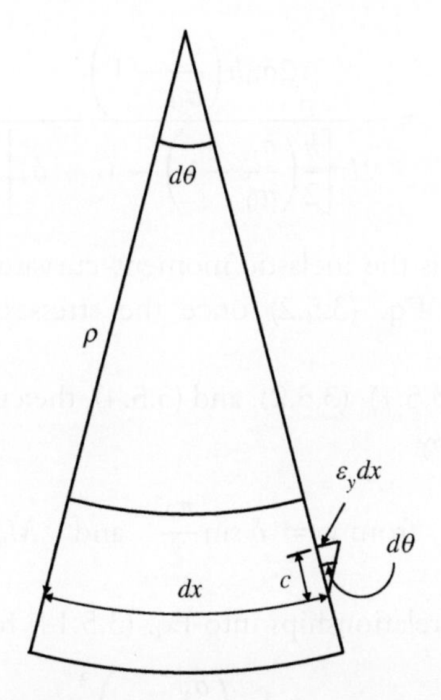

Figure 3-15 Similar triangle relationship

$$c = \frac{9\left[\frac{h}{2}\left(\frac{\sigma_y}{\sigma_0} - 1\right) - \frac{M_{int}}{P}\right]^2}{2\sigma_0 h\left(\frac{\sigma_y}{\sigma_0} - 1\right)^3} \tag{3.5.10}$$

From the similar triangle relationship shown in Fig. 3-15, the following relationship can be readily established:

$$\frac{\rho}{dx} = \frac{c}{\varepsilon_y dx} \tag{3.5.11}$$

where ρ is the radius of curvature.
Thus

$$\frac{1}{\rho} = \frac{\varepsilon_y}{c} \doteq \frac{d^2y}{dx^2} \tag{3.5.12}$$

or

$$y'' = \frac{\sigma_y}{cE} \tag{3.5.13}$$

Substituting c given by Eq. (3.5.10) into Eq. (3.5.13) gives the moment-curvature relationship in the inelastic range

$$y'' = \frac{2\sigma_0 h\left(\dfrac{\sigma_y}{\sigma_0} - 1\right)^3}{9E\left[\dfrac{h}{2}\left(\dfrac{\sigma_y}{\sigma_0} - 1\right) - (e + \delta)\right]^2} \tag{3.5.14}$$

Equation (3.5.14) is the inelastic moment-curvature relation that must be used in place of Eq. (3.5.2) once the stresses have exceeded the proportional limit.

By virtue of Eqs. (3.5.1), (3.5.2), and (3.5.4), the curvature and moment at midspan are given by

$$y''\Big|_{\ell/2} = \delta\frac{\pi^2}{\ell^2} \quad \text{from } y = \delta \sin\frac{\pi x}{\ell} \quad \text{and} \quad M_{int} = P(e + \delta)$$

Substituting these relationships into Eq. (3.5.14) for y'' above gives

$$\delta\frac{\pi^2}{\ell^2} = \frac{2\sigma_0 h\left(\dfrac{\sigma_y}{\sigma_0} - 1\right)^3}{9E\left[\dfrac{h}{2}\left(\dfrac{\sigma_y}{\sigma_0} - 1\right) - (e + \delta)\right]^2} \text{ or}$$

$$\delta\left[\frac{h}{2}\left(\frac{\sigma_y}{\sigma_0} - 1\right) - e - \delta\right]^2 = \frac{2h\ell^2\sigma_0}{9E\pi^2}\left(\frac{\sigma_y}{\sigma_0} - 1\right)^3 \text{ or}$$

$$\frac{\delta}{h}\left[\frac{1}{2}\left(\frac{\sigma_y}{\sigma_0} - 1\right) - \frac{e}{h} - \frac{\delta}{h}\right]^2 = \frac{2\ell^2\sigma_0}{9E\pi^2 h^2}\left(\frac{\sigma_y}{\sigma_0} - 1\right)^3 \tag{3.5.15}$$

Since $\sigma_E = \pi^2 EI/(A\ell^2) = \pi^2 Eh^2/(12\ell^2)$, Eq. (3.5.15) can be rewritten in the form

$$\frac{\delta}{h}\left[\frac{1}{2}\left(\frac{\sigma_y}{\sigma_0} - 1\right) - \frac{e}{h} - \frac{\delta}{h}\right]^2 = \frac{1}{54}\frac{\sigma_0}{\sigma_E}\left(\frac{\sigma_y}{\sigma_0} - 1\right)^3 \tag{3.5.16}$$

Equation (3.5.16) gives the load versus deflection relationship in the inelastic range.

Example 1 Consider a simply supported rectangular steel beam-column with the following dimensions and properties:

$$\ell = 120 \text{ in.}, r = 1.0 \text{ in.}, e = 1.15 \text{ in.}, \sigma_y = 34\ ksi, E = 30 \times 10^3\ ksi$$

Determine the ultimate load-carrying capacity of the member.

Table 3-1 Load-deflection data for beam-column

Elastic Range, Eq. (3.5.5)			Inelastic Range, Eq. (3.5.16)	
σ_0 (ksi)	δ/h	σ_{max}	σ_0 (ksi)	δ/h
2	0.036	6.4	8.0	0.21
4	0.080	14.0	8.5	0.24
6	0.137	23.0	9.0	0.30
8	0.212	34.0	9.1	0.35
10	0.314	invalid	9.0	0.40
12	0.463		–	–
14	0.710		–	–
16	1.150		–	–

$$r = \sqrt{\frac{I}{A}} = \sqrt{\frac{bh^3}{12bh}} = \sqrt{\frac{h^2}{12}} = 1 \Rightarrow h = 2\sqrt{3}\text{ in.} \quad \text{and}$$

$$\sigma_E = \frac{\pi^2 E}{\left(\frac{\ell}{r}\right)^2} = \frac{\pi^2 \times 30 \times 10^3}{120^2} = 20.6\ ksi$$

The load-deflection data for the elastic range evaluated using Eq. (3.5.5) are given in Table 3-1. Corresponding to each set of σ_0 and δ/h listed in the table; the maximum stress is also evaluated using Eq. (3.5.7). It is evident that the maximum stress in the member reaches 34 ksi, the yield stress, at approximately $\sigma_0 = 8$ ksi. Hence, Eq. (3.5.16) must be used for deflections for axial stresses in excess of 8 ksi. The load-deflection data for the inelastic range computed using Eq. (3.5.16) are also listed in Table 3-1. The entire load-deflection curve is plotted in Fig. 3-16. It is of interest to observe the load-deflection behavior of this beam-column. For the load to produce

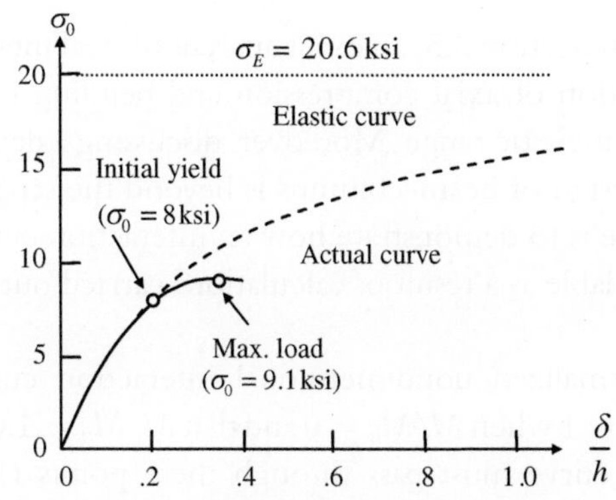

Figure 3-16 Load-deflection curve for beam-column

the average axial stress of 8 ksi, the material obeys Hooke's law and the deflections are relatively small. However, once the stress exceeds 8 ksi, yielding starts spreading rapidly, and there occurs a noticeable decrease in the stiffness of the member. At the average axial stress of 9.1 ksi, the member is no longer able to resist any increase in load. The average axial stress, σ_0, of 9.1 ksi represents the ultimate strength of the member.

It has been demonstrated here that the load-deflection characteristic of a simple rectangular section under a simplified assumption of linearly elastic and perfectly plastic stress-strain relation is fairly complex. Although Chwalla (1934, 1935) improved the stress-strain curve of Jezek (1934) by adopting a curved stress-strain diagram (but ignoring residual stresses), the limitation of the deflection shape of a sinusoidal form could be a liability. Therefore, today accurate determinations of the ultimate strength of beam-columns are best obtained by finite element nonlinear incremental analyses without ignoring important parameters such as initial imperfections, residual stresses, and strain-hardening effects that are known to affect the ultimate strength considerably. In view of the fact that determining the maximum load of a beam-column is extremely complex and time consuming, the load at which yielding begins has often been used in place of the ultimate load as the limit of structural usefulness. The load corresponding to initial yielding is an attractive design criterion because it is relatively simple to obtain and is conservative. However, it is sometimes too conservative. There are a few semi-empirical design interactive equations. For rolled shapes, a similar procedure can be programmed to execute.

3.6. DESIGN OF BEAM-COLUMNS

As demonstrated in Section 3.5, an exact analysis of steel members subjected to a combined action of axial compression and bending is very complex, particularly in the inelastic range. Moreover, discussing a detailed procedure involved in the design of beam-columns is beyond the scope of this book. The intention here is to demonstrate how an interaction curve is generated using the data available as a result of calculations carried out in the previous section.

In creating a normalized nondimensional interaction curve, it is quite obvious that $P/P_u = 1$ when $M/M_u = 0$ and that $M/M_u = 1$ when $P/P_u = 0$. Thus the desired curve must pass through these points (1,0), (0,1). The simplest curve that satisfies this condition is a straight line

$$\frac{P}{P_u} + \frac{M}{M_u} = 1 \tag{3.6.1}$$

P = axial load acting on the member at failure when both axial compression and bending are present.

P_u = ultimate load of the member when only axial compression is present.

M = maximum primary bending moment acting on the member at failure when both bending and axial compression exist; this excludes the amplified moment.

M_u = ultimate bending moment when only bending exists.

Although Eq. (3.6.1) may represent an interaction reasonably well where instability cannot occur (i.e., $K\ell/r = 0$), all theoretically and experimentally obtained failure loads fall below the curve. Hence, it is an unconservative upper-bound interaction curve. Obviously, the moment included in Eq. (3.6.1) is only the primary moment. As shown in Section 3.2, the presence of an axial compressive force amplifies the primary bending moment by an amplification factor. If this factor is reflected in Eq. (3.6.1), one obtains

$$\frac{P}{P_u} + \frac{M}{M_u\left(1 - \dfrac{P}{P_E}\right)} = 1 \tag{3.6.2}$$

Example 1 Revisit the rectangular beam-column examined in Section 3.5. For the member the average axial stress at the ultimate strength was found to be $\sigma_0 = 9.1$ ksi, and the corresponding Euler stress is $\sigma_E = 20.6$ ksi. Thus

$$\frac{P}{P_E} = \frac{\sigma_0}{\sigma_E} = 0.44$$

$$M = \frac{\sigma_0(bh)e}{P}, M_u = \frac{\sigma_y bh^2}{4}, \frac{e}{h} = 0.33 \Rightarrow \sigma_{max} = 34\ ksi$$

$$\frac{M}{M_u} = \frac{\sigma_0(bh)e}{\sigma_{max}\dfrac{bh^2}{4}} = \frac{4\sigma_0}{\sigma_{max}}\frac{e}{h} = \frac{4(9.1)}{34} \times 0.33 = 0.35$$

The coordinates of the interaction point are shown in Fig. 3-17. Chajes (1974) has demonstrated that Eq. (3.6.2) agrees fairly well with calculations similar to those leading to the interaction point shown by Jezek (1935, 1936) for rectangular columns of various values of e/h.

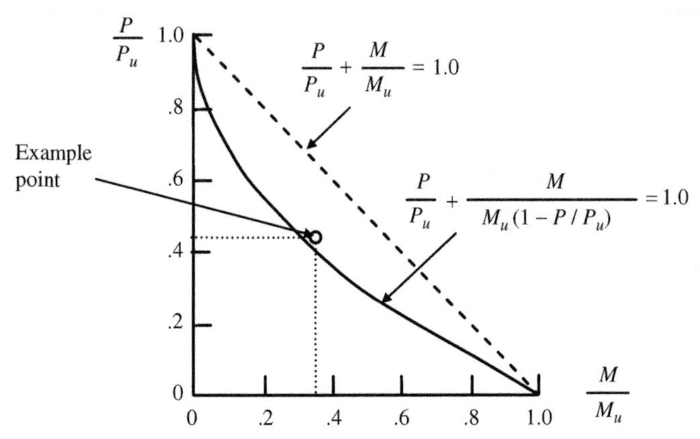

Figure 3-17 Ultimate strength interaction equation for beam-column

In studying Eq. (3.6.2), the student is reminded that the strength P_u when $M_u = 0$ is based on the slenderness ratio with respect to the major axis ($K\ell/r_x$), which implies that the member was assumed to fail by instability in the plane of bending. This can present a serious limitation in general applications as the plane of bending and the plane of instability frequently do not coincide in most beam-columns. Although Eq. (3.6.2) could have served adequately in the 1930s, an attempt to use only one interaction equation as the guide for the design of beam-columns in modern-day applications is grossly inadequate. Research published by Ketter (1961) has affected specification-writing bodies for many years, particularly the AISC as demonstrated by Salmon and Johnson (1996). Chen and Atsuta (1976, 1977) published a comprehensive treatise of beam-columns in two volumes. In the current AISC (2005) specification, the design of beam-columns is addressed in Chapter H, where members subjected to axial force and flexure about one or both axes with or without torsion are classified into:

H1. Doubly and Singly Symmetric Members Subject to Flexure and Axial Force.

H2. Unsymmetric and Other Members Subject to Flexure and Axial Force

H3. Members Under Torsion and Combined Torsion, Flexure, Shear and/or Axial Force. In the current AISC specification, and it appears to be the case for the upcoming edition, a second-order analysis is mandated for all beam-columns.

AISC (1989) interaction formulas include C_{mx}, C_{my} factors to account for loading, sway condition, amplification, and single or reverse curvature.

REFERENCES

AISC. (1989). *Manual of Steel Construction* (9th ed.). Chicago, IL: American Institute of Steel Construction.

AISC. (2005). *Specification for Structural Steel Buildings, ANSI/AISC 360–05*. Chicago, IL: American Institute of Steel Construction.

Bazant, Z. P., & Cedolin, L. (1991). *Stability of Structures, Elastic, Inelastic, Fracture, and Damage Theories*. Oxford: Oxford University Press.

Berry, A. (1916). *The Calculation of Stresses in Aeroplane Spars*, Trans. Roy. Aer. Soc., No.1., London

Bleich, F. (1952). *Buckling Strength of Metal Structures*. New York: McGraw-Hill.

Britvec, S. J. (1973). *The Stability of Elastic Systems*. Elmsford, NY: Pergamon Press.

Chajes, A. (1974). *Principles of Structural Stability Theory*. Englewood Cliffs, NJ: Prentice-Hall.

Chen, W. F., & Atsuta, T. (1976). *Theory of Beam-columns, Vol. 1 Inplane Behavior and Design*. New York: McGraw-Hill.

Chen, W. F., & Atsuta, T. (1977). *Theory of Beam-columns, Vol. 2 Space Behavior and Design*. New York: McGraw-Hill.

Chwalla, E. (1934). *Ügber die experimentelle Untersuchung des Tragverhaltensgedrhckter Stäbe aus Baustahl, Vol. 7*. Berlin, Germany: Der Stahlbau.

Chwalla, E. (1935). *Der Einfluss der Querschnittsform auf das Tragvermögen aussermittig gedrückter Baustahlstäbe, Vol. 8*. Berlin, Germany: Der Stahlbau.

Goldberg, J. E. (1954). Stiffness Charts for Gusseted Members Under Axial Load, Transaction. *ASCE, 119* (Paper No. 2657), 43–54.

Horne, M. R., & Merchant, W. (1965). *The Stability of Frames* (1st ed.). Oxford: Pergamon Press.

James, B.W. (1935). Principal Effects of Axial Load by Moment Distribution Analysis of Rigid Structures, NACA, Technical Note No. 534 (Sec. 2.1).

Jezek, K. (1934). "Die Tragfähigkeit des exzentrisch beanspruchten und des querbelasteten Druckstabes aus einen ideal plastischen Material," Sitzunggsberichte der Akademie der Wissenschaften in Wien, Abt. IIa, Vol. 143.

Jezek, K. (1935). *Näherungsberechnung der Tragkraft exzentrisch gedrhckter Stahlstäbe, Vol. 8*. Berlin, Germany: Der Stahlbau.

Jezek, K. (1936). *Die Tragfähigkeit axial gedrückter und auf Biegung beanspruchter Stahlstäbe, Vol. 9*. Berlin, Germany: Der Stahlbau.

Ketter, R. L. (1961). Further Studies of the Strength of Beam-Columns. *Journal of the Structural Div., ASCE, Vol. 87* (No. ST6), 135–152.

Kinney, J. S. (1957). *Indeterminate Structural Analysis*. Reading, MA: Addison-Wesley.

Livesley, R. K., & Chandler, D. B. (1956). *Stability Functions for Structural Frameworks*. Manchester, UK: Manchester University Press.

Maney, G. A. (1915). *Studies in Engineering—No. 1*. Minneapolis, MN: University of Minnesota.

Niles, A. S., & Newell, J. S. (1948). *Airplane Structures, Vol. II* (3rd ed.). New York: John Wiley and Sons.

Parcel, J. I., & Moorman, R. B. B. (1955). *Analysis of Statically Indeterminate Structures*. New York: John Wiley and Sons.

Salmon, C. G., & Johnson, J. E. (1996). *Steel Structures, Design and Behavior* (4th ed.). New York: Harper Collins.

Timoshenko, S. P., & Gere, J. M. (1961). *Theory of Elastic Stability* (2nd ed.). New York: McGraw-Hill.

Winter, G., Hsu, P. T., Koo, B., & Loh, M. H. (1948). *Buckling of Trusses and Rigid Frames*. Ithaca, NY: Cornel Engineering Experiment Station Bulletin, No. 36.

PROBLEMS

3.1 Obtain expressions for the maximum deflection and the maximum moment of a prismatic beam-column subjected to a uniformly distributed load as shown in Figure P3-1.

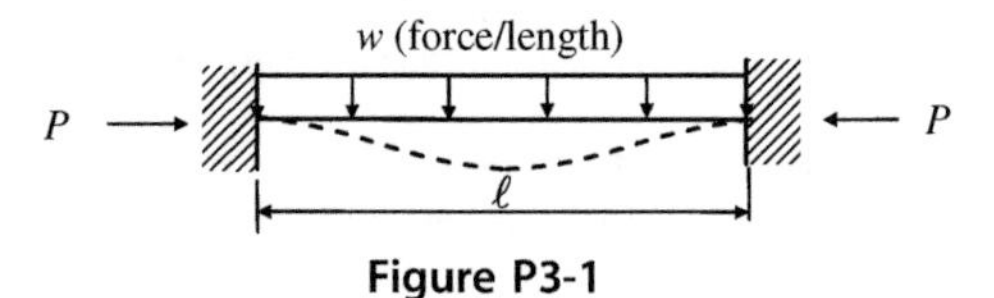

Figure P3-1

3.2 Determine the expression for the maximum deflection and maximum moment of a both ends clamped that is subjected to a concentrated load at midspan as shown in Figure P3-2.

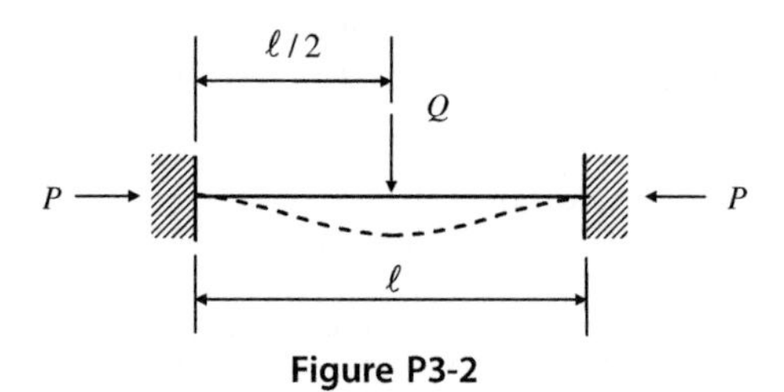

Figure P3-2

3.3 Determine the maximum moment for a beam-column shown in Figure P3-3 that is bent in (a) single curvature and (b) reverse curvature when $P/P_E = 0.2$ with $P_E = \pi^2\ EI/\ell^2$.

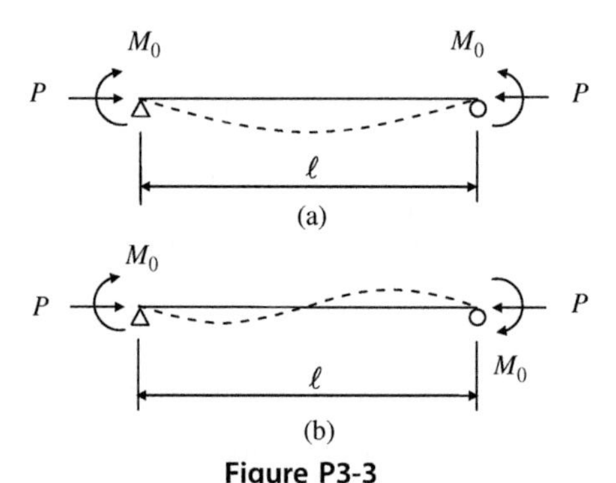

Figure P3-3

Discuss the problem.

3.4 A simply supported beam-column is subjected to an axial load, *P*, and a linearly varying load, W_0, as shown in Figure P3-4.

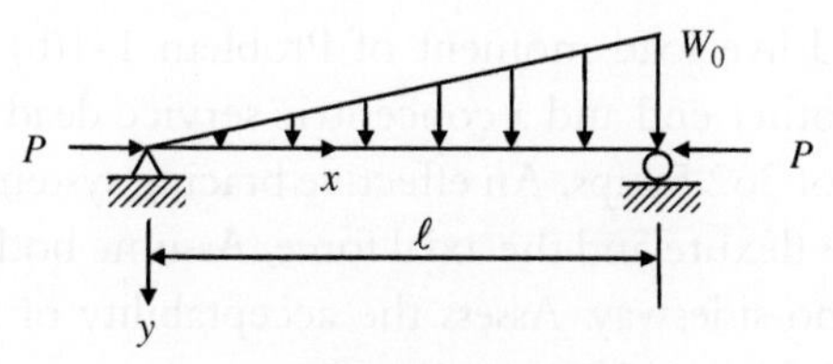

Figure P3-4

(a) Determine the equations for the deflection and moment at any point along the length by solving the governing differential equation.

(b) Develop an elastic interaction curve (P/P_y vs M/M_y or $W_0\ell^2/M_y$) for $\ell/r = 120, \sigma_y = 33$ ksi, $E = 30 \times 10^3$ ksi using the results obtained in (a).

(c) Using an approximation that the deflection computed on the basis for no axial force is amplified by the factor, $[1/(1 - P/P_y]$, determine an approximate interaction curve for the data given in (b).

3.5 Show by repeated applications of L'Hôpital's rule that
(a) $\lim_{P\to 0}(S_1) = 4$ **(b)** $\lim_{P\to 0}(S_2) = 2$ **(c)** $\lim_{P\to 0}(\overline{S}) = 3$

3.6 Solve Problem 1-1(b) using slope-deflection equations derived in this chapter.

3.7 A W16X67 steel beam-column (Grade 50) shown in Figure P3-7 is subjected to a linearly varying primary service dead load moment of

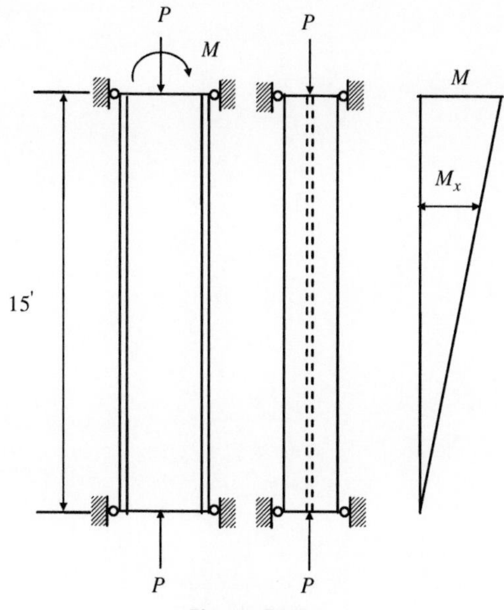

Figure P3-7

15 ft-kips and live load moment of Problem 1-1(b) at one end with none on the other end and a concentric service dead load of 87.5 kips and live load of 262.5 kips. An effective bracing system is available at 15 ft for both the flexure and the axial force. Assume both ends are pinned and there is no sidesway. Assess the acceptability of the design (a) by Eq. 3.6.2 and (b) by AISC specification.

3.8 Would the final results, internal forces, and deflections be different or the same if the axial force is applied first followed by the transverse load or vice versa in a beam-column? State the reason for your answer.

Continuous Beams and Rigid Frames

Contents

4.1. INTRODUCTION

In the study of isolated column stability in Chapter 1, the member ends are idealized to be pinned, fixed, or free. However, members in a real framed structure are usually part of a larger framework, and their ends are elastically restrained by the adjacent members to which they are framed. In this chapter, the investigation is extended to consider the behavior of framed members.

In a framework, the members are usually rigidly connected at joints. Therefore, no single compression member can buckle independently from the adjacent members. Hence, it is often necessary to investigate the stability of the entire structure just to obtain the critical load of one or two members that are part of a larger framework.

Using an example problem, it will be demonstrated that the effect of a second-order analysis becomes significant over that of the first-order analysis when the compressive axial load is, say, greater than 10% of the critical load of the member.

4.2. CONTINUOUS BEAMS

Slope-deflection equations with axial forces have been derived in the previous chapter. These equations will be used to solve elastic stability

Stability of Structures
ISBN 978-0-12-385122-2, doi:10.1016/B978-0-12-385122-2.10004-1

problems of continuous beams and rigid frames in this chapter. Recall the slope-deflection equations.

$$M_{ab} = \frac{EI}{\ell}\left[S_1\theta_a + S_2\theta_b + (S_1 + S_2)\frac{\delta_a - \delta_b}{\ell}\right] \tag{4.2.1}$$

$$M_{ba} = \frac{EI}{\ell}\left[S_2\theta_a + S_1\theta_b + (S_1 + S_2)\frac{\delta_a - \delta_b}{\ell}\right] \tag{4.2.2}$$

In the case when the end "b" is hinged, $M_{ba} = 0$ and M_{ab} is modified as a result of eliminating θ_b as

$$M'_{ab} = \frac{EI}{\ell}S_3\left(\theta_a + \frac{\delta_a - \delta_b}{\ell}\right) \tag{4.2.3}$$

where

$$S_1 = \frac{1 - \beta\cot\beta}{\dfrac{2\tan\beta/2}{\beta} - 1}$$

$$S_2 = \frac{\beta\,\mathrm{cosec}\,\beta - 1}{\dfrac{2\tan\beta/2}{\beta} - 1}$$

$$S_3 = \frac{\beta^2}{1 - \beta\cot\beta}$$

$$\beta = k\ell = \sqrt{\frac{P\ell^2}{EI}} \Leftarrow \text{referred to as buckling parameter.}$$

It can be shown by applying the L'Hôpital's rule that the limit values for S_1, S_2, and S_3 are 4, 2, and 3, respectively, when $\beta = 0$ (or $P = 0$).

Example 1 Determine P_{cr} for the structure shown in Figure 4-1.
For AB:

$$M'_{ba} = \left(\frac{EI}{\ell}S_3\right)_1\theta_b$$

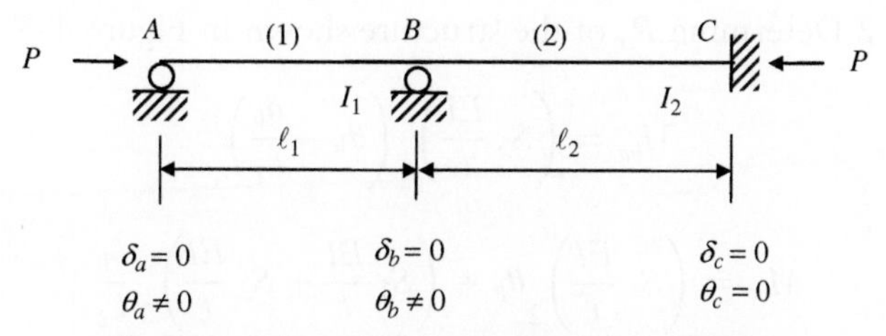

Figure 4-1 Two-span continuous beam-column

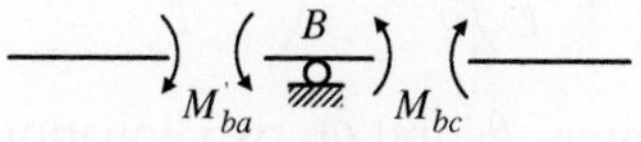

B

M_{ba} M_{bc}

*Recall sign convention in the derivation

Figure 4-2 Moment equilibrium at joint B

For BC:

$$M_{bc} = \left(\frac{EI}{\ell} S_1\right)_2 \theta_b$$

$$\sum M_b = 0 \Rightarrow M'_{ba} + M_{bc} = 0 \Rightarrow \left[\left(S_3 \frac{EI}{\ell}\right)_1 + \left(S_1 \frac{EI}{\ell}\right)_2\right] \theta_b = 0$$

Since $\theta_b \neq 0$, the stability condition equation is

$$\left(S_3 \frac{EI}{\ell}\right)_1 + \left(S_1 \frac{EI}{\ell}\right)_2 = 0.$$

Hence,

$$\frac{\beta_1^2}{1 - \beta_1 \cot \beta_1} \frac{EI_1}{\ell_1} + \frac{1 - \beta_2 \cot \beta_2}{\dfrac{2 \tan \beta_2/2}{\beta_2} - 1} \frac{EI_2}{\ell_2} = 0$$

As β_1 and β_2 are functions of P, the smallest value of P which satisfies the above equation is P_{cr}.

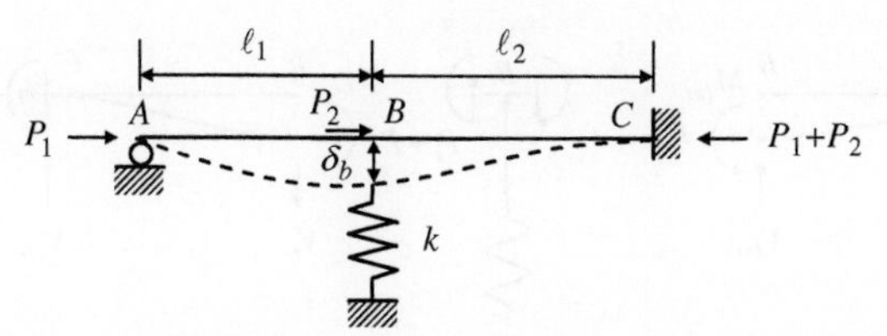

Figure 4-3 Elastically constrained two-span beam-column

Example 2 Determine P_{cr} of the structure shown in Figure 4-3.

$$M_{ba} = \left(S_3 \frac{EI}{\ell}\right)_1 \left(\theta_b - \frac{\delta_b}{\ell_1}\right)$$

$$M_{bc} = \left(S_1 \frac{EI}{\ell}\right)_2 \theta_b + \left(S_1 \frac{EI}{\ell} + S_2 \frac{EI}{\ell}\right)_2 \frac{\delta_b}{\ell_2}$$

$$M_{cb} = \left(S_2 \frac{EI}{\ell}\right)_2 \theta_b + \left(S_1 \frac{EI}{\ell} + S_2 \frac{EI}{\ell}\right)_2 \frac{\delta_b}{\ell_2}$$

As there are two unknowns, θ_b and δ_b, two equations are needed. The first equation is provided by moment equilibrium. The moment equilibrium condition at joint B gives

$$M'_{ba} + M_{bc} = 0 \tag{4.2.4}$$

One additional equation can be derived considering equilibrium of the vertical forces at joint B. Consider the free-body diagram of each span and joint equilibrium at B shown in Fig. 4-4.

$$\sum M_a = 0 = V_{ba}\ell_1 - M'_{ba} - P_1\delta_b$$

$$V_{ba} = \frac{M'_{ba} + P_1\delta_b}{\ell_1} \tag{4.2.5}$$

$$\sum M_c = 0 = -V_{bc}\ell_2 + M_{bc} + M_{cb} - (P_1 + P_2)\delta_b$$

$$V_{bc} = \frac{M_{bc} + M_{cb} - (P_1 + P_2)\delta_b}{\ell_2} \tag{4.2.6}$$

$$\sum F_{vertical} = 0 \quad \text{at joint B}$$

$$V_{ba} - k\delta_b - V_{bc} = 0 \tag{4.2.7}$$

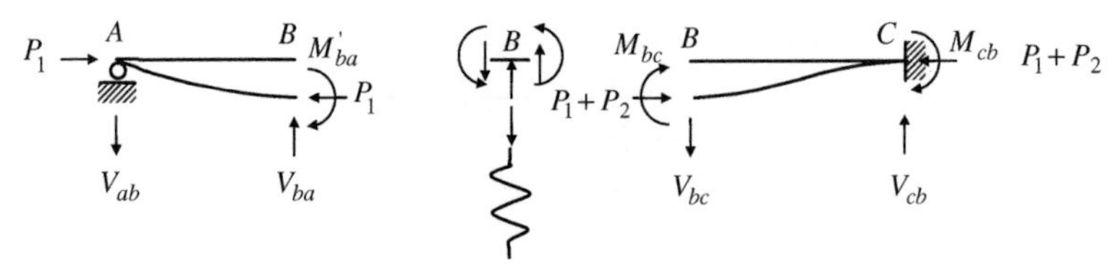

Figure 4-4 Free-body diagram

Equations (4.2.4) and (4.2.7), along with Eqs. (4.2.5) and (4.2.6), yield the following general form of equations:

$$a_{11}\theta_b + a_{12}\delta_b = 0$$

$$a_{21}\theta_b + a_{22}\delta_b = 0$$

Set the coefficient determinant equal to zero for a nontrivial solution (or stability condition equation). The resulting equation is a transcendental equation in β $(k\ell)$. The roots in β lead to the critical loads.

4.3. BUCKLING MODES OF FRAMES

Consider first the frame in which sidesway is prevented by bracing either internally or externally. It is obvious that the upper end of each column is elastically restrained by the beam to which the column is rigidly framed, and that the critical load of the column depends not only on the column stiffness, but also on the stiffness of the beam. It would be very informative to assume the beam stiffness to be either infinitely stiff or infinitely flexible as these two conditions constitute the upper and lower bounds of the connection rigidities. When the beam is assumed to be infinitely stiff, the beam must then remain straight while the frame deforms as shown in part (a), (1) Sidesway prevented, Fig. 4-5. Under this condition, the columns behave as if they were fixed at both ends, and the critical load of the column is equal to four times the Euler load of the same column pinned at its both ends. As the other extreme case of the opposite side, the beam can be assumed to be infinitely flexible. The frame then deforms as shown in part (b), (1) Sidesway prevented, Fig. 4-5, and the columns behave as if they were pinned at the top, and the critical load is the same as that of the propped column: approximately twice that of the Euler load of the same column pinned at both ends.

For an actual frame, the stiffness of the beam must be somewhere between the two extreme cases examined above. The critical load on the column in such a frame can be bounded as follows:

$$4P_E > P_{cr} > 2P_E \tag{4.3.1}$$

where P_{cr} is the critical load of the column and P_E is the Euler load of the same column pinned at both ends.

It is just as informative to apply the same logic to frames in which sidesway is permitted. If the beam is assumed to be infinitely stiff, the frame buckles in the manner shown in part (a), (2) Sidesway permitted, Fig. 4-5. The upper ends of the columns are permitted to translate, but they cannot rotate by

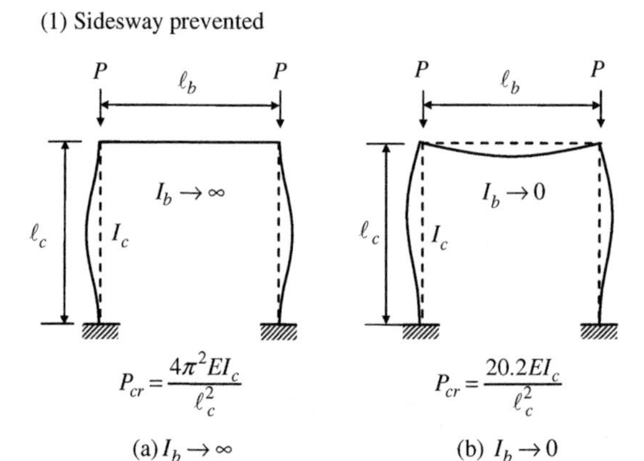

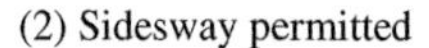

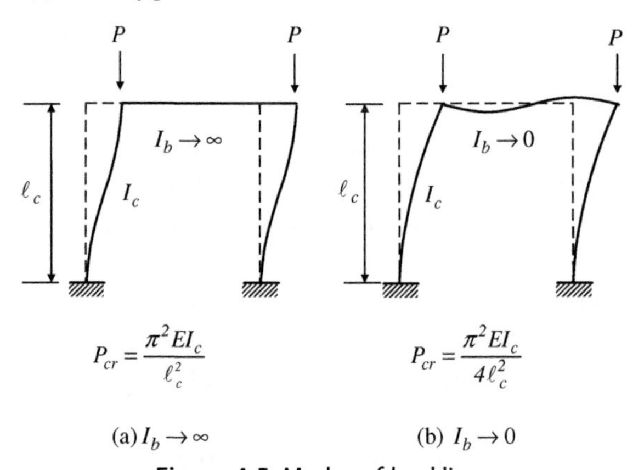

Figure 4-5 Modes of buckling

definition. Hence, the critical load on each column in the frame is equal to the Euler load of the same column pinned at both ends. On the other extreme, if the beam is assumed infinitely flexible, the upper ends of the columns are both permitted to rotate and translate as shown in part (b), (2) Sidesway permitted, Fig. 4-5. In this extreme case, each column acts as if it were a cantilever column, and the critical load on each column is equal to one-fourth the Euler load of the same column pinned at both ends. The critical load on each column of the frame in which sidesway is permitted can be bounded as follows:

$$P_E > P_{cr} > \frac{1}{4} P_E \tag{4.3.2}$$

Hence,

$$P_{cr}\big|_{\text{braced frame}} > P_{cr}\big|_{\text{unbraced frame}} \tag{4.3.3}$$

A portal frame will always buckle in the sidesway permitted mode unless it is braced. Unlike the braced frame where sidesway is inherently prohibited, both the sidesway permitted and prevented modes are theoretically possible in the unbraced frame under the loading condition shown in Fig. 4-5. The unbraced frame, however, will buckle first at the smallest critical load, which is the one corresponding to the sidesway permitted mode. This conclusion is valid for multistory frames as well as for single-story frames as shown by Bleich (1952). The reason appears to be obvious as the effective length of the compression member in an unbraced frame is always increased due to the frame action, while that in the braced frame is always reduced unless the beams in the frame are infinitely flexible. The same conclusion can be extended to the case of buckling of an equilateral triangle, which will be detailed later.

4.4. CRITICAL LOADS OF FRAMES

4.4.1. Review of the Differential Equation Method

In the previous section, the qualitative aspects of the buckling characteristics of a single-story single-bay portal frame are illustrated. It is now desired to determine the critical load of such a frame by means of neighboring equilibrium (neutral equilibrium). Depending on whether or not the frame is braced, buckling will take place in the symmetric or the antisymmetric mode. An antisymmetric buckling is considered first here.

It is assumed that a set of usual assumptions normally employed in the classical analysis of linear elastic structures under the small displacement theory is valid. The sidesway buckling mode shape assumed and the forces acting on each member are identified in Fig. 4-6(a) and (b), respectively. The moment of the left vertical member at a point x from the origin based on the coordinate shown in Fig. 4-6(c) is (moment produced by the continuity shear developed in BC is neglected)

$$M(x) = M_{ab} - Py = M_{int} = EI_1 y'' \tag{4.4.1}$$

or

$$y'' + k_1^2 y = \frac{M_{ab}}{EI_1} \tag{4.4.2}$$

where $k_1^2 = \dfrac{P}{EI_1}$

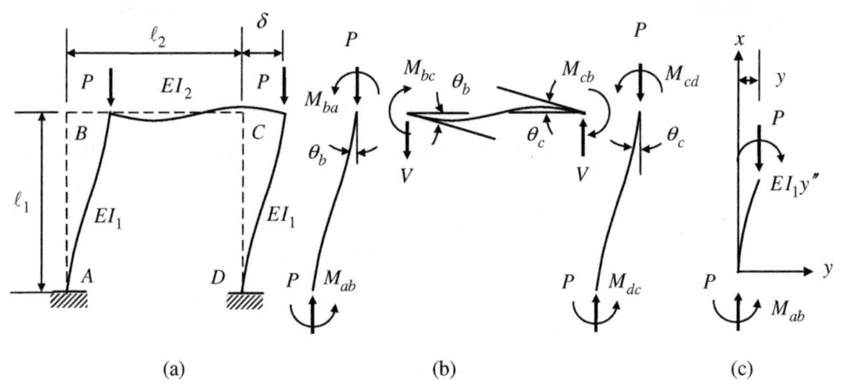

Figure 4-6 Buckling of unbraced frame

The general solution of Eq. (4.4.2) is given by

$$y = A \sin k_1 x + B \cos k_1 x + \frac{M_{ab}}{P} \tag{4.4.3}$$

Two independent boundary conditions are needed to determine the integral constants, A and B. They are

$$y = 0 \quad \text{at } x = 0$$

from which

$$B = -\frac{M_{ab}}{P}$$

and

$$y' = 0 \quad \text{at } x = 0$$

which leads to

$$A = 0$$

Hence,

$$y = \frac{M_{ab}}{P}(1 - \cos k_1 x) \tag{4.4.4}$$

Denoting the horizontal displacement at the top of the column ($x = \ell_1$) by δ, then

$$\delta = \frac{M_{ab}}{P}(1 - \cos k_1 \ell_1) \tag{4.4.5}$$

Summing the moment of member AB at A gives

$$P\delta - M_{ab} - M_{ba} = 0 \tag{4.4.6}$$

It is tacitly assumed that the same lateral displacement occurs at points B and C, as the horizontal force, if any, in member BC is small enough to be ignored. Hence, there is no horizontal force at B which leads to zero shear in member AB. Substituting Eq. (4.4.6) into Eq. (4.4.5) gives

$$M_{ab} \cos k_1 \ell_1 + M_{ba} = 0 \tag{4.4.7}$$

Since it is assumed that there is no axial compression presented in member BC, the slope-deflection equations without axial force apply.
Hence,

$$M_{bc} = \frac{2EI_2}{\ell_2}(2\theta_b + \theta_c) \tag{4.4.8}$$

Since θ_b ia equal to θ_c and they are positive based on the coordinate system employed in Fig. 4-6(c), Eq. (4.4.8) reduces to

$$M_{bc} = \frac{6EI_2}{\ell_2}\theta_b \tag{4.4.9}$$

The compatibility condition at joint B requires that θ_b in Eq. (4.4.9) be equal to the slope of Eq. (4.4.4) at $x = \ell_1$.

Hence,

$$\frac{M_{bc}\ell_2}{6EI_2} = \frac{M_{ab}}{k_1 EI_1} \sin k_1 \ell_1 \tag{4.4.10}$$

or

$$\frac{6I_2}{k_1 I_1 \ell_2} M_{ab} \sin k_1 \ell_1 - M_{bc} = 0 \tag{4.4.11}$$

Equations (4.4.7) and (4.4.11) are the required equations to solve the frame. Ordinarily, a frame with n unknowns would require n equations. However, in this case, as the two vertical members are identical, which leads to only two unknowns, namely, δ and θ_b instead of three unknowns (δ, θ_b, θ_c), two equations suffice. Setting the coefficient determinant equal to zero gives

$$\frac{\tan k_1 \ell_1}{k_1 \ell_1} = -\frac{I_1 \ell_1}{6 I_2 \ell_2} \tag{4.4.12}$$

The critical load of the frame is the smallest root of this transcendental equation.
For

$$I_2 = I_1 = I, \quad \ell_2 = \ell_1 = \ell$$

Equation (4.4.12) reduces to

$$\frac{\tan kl}{kl} = -\frac{1}{6}$$

From **Maple**® or BISECT or any other transcendental equation solver,

$$kl = 2.71646$$

$$\text{and } P_{cr} = \frac{7.38EI}{\ell^2}$$

which is $9.87EI/\ell^2 > 7.38EI/\ell^2 > 9.87EI/(4\ell^2)$ as expected from Eq. (4.4.2).

The next case to be examined is a portal frame in which sidesway is prevented either by internal bracing or external supports shown in Fig. 4-7.

Consider the symmetric buckling shown in Fig. 4-7(a). Based on the assumed deformation mode shown in Fig. 4-7(a), a continuity shear is developed in member AB. That is

$$V_{ab} = \frac{(M_{ab} - M_{ba})}{\ell_1} \tag{4.4.13}$$

Hence, the moment at a distance x from the origin (joint A) is

$$-EI_1y'' - Py + M_{ab} - \frac{M_{ab} - M_{ba}}{\ell_1}x = 0 \tag{4.4.14}$$

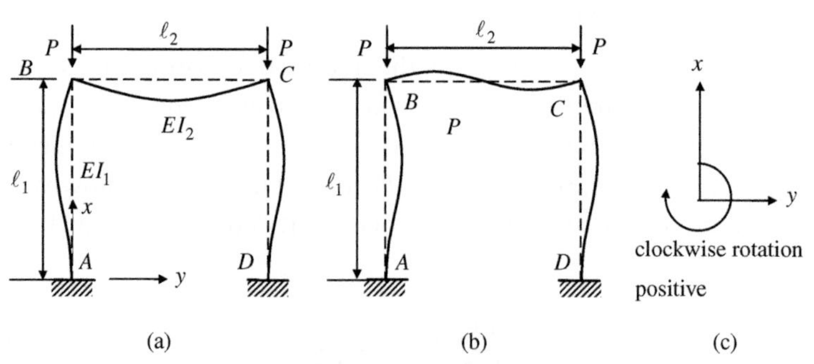

Figure 4-7 Braced frame

or

$$y'' + k_1^2 y = \frac{M_{ab}}{EI_1}\left(1 - \frac{x}{\ell_1}\right) + \frac{M_{ba}}{EI_1}\left(\frac{x}{\ell_1}\right) \tag{4.4.15}$$

where $k_1^2 = P/EI_1$. The general solution of Eq. (4.4.15) is

$$y = A \sin k_1 x + B \cos k_1 x + \frac{M_{ab}}{P}\left(1 - \frac{x}{\ell_1}\right) + \frac{M_{ba}}{P}\left(\frac{x}{\ell_1}\right) \tag{4.4.16}$$

Two boundary conditions are needed to determine the integral constant, A and B. They are

$$y = 0 \quad \text{at } x = 0$$

which leads

$$B = -\frac{M_{ab}}{P}$$

and

$$y' = 0 \quad \text{at } x = 0$$

from which

$$A = \frac{M_{ab} - M_{ba}}{k_1 \ell_1 P}$$

Hence,

$$y = \frac{M_{ab}}{P}\left(\frac{1}{k_1\ell_1}\sin k_1 x - \cos k_1 x + 1 - \frac{x}{\ell_1}\right) + \frac{M_{ba}}{P}\left(\frac{x}{\ell_1} - \frac{1}{k_1\ell_1}\sin k_1 x\right) \tag{4.4.17}$$

As the top end of member AB is assumed not to be able to move laterally, that is, $y = 0$ at $x = \ell_1$, Eq. (4.4.17) becomes

$$M_{ab}(\sin k_1\ell_1 - k_1\ell_1 \cos k_1\ell_1) + M_{ba}(k_1\ell_1 - \sin k_1\ell_1) = 0 \tag{4.4.18}$$

Applying the slope-deflection equation assuming no axial forces resulting from the continuity shear generated from the vertical members are transmitted to the horizontal member due to either internal bracing or external supports, it reads

$$M_{bc} = \frac{2EI_2}{\ell_2}(2\theta_b + \theta_c) \tag{4.4.19}$$

Since $\theta_c = -\theta_b$, Eq. (4.4.19) reduces to

$$M_{bc} = \frac{2EI_2}{\ell_2}\theta_b \tag{4.4.20}$$

Compatibility of slope at joint B requires that θ_b of the horizontal member be equal to $-y'$ at $x = \ell_1$ of the vertical member for the consistent sign convention adopted in Fig. 4-7(c). It is noted here that the condition of $M_{ba} = M_{bc}$ has been used in the above derivation starting from Eq. (4.4.13). Thus

$$\frac{M_b\ell_2}{2EI_2} = -\frac{M_a}{P}\left(\frac{1}{\ell_1}\cos k_1\ell_1 + k_1\sin k_1\ell_1 - \frac{1}{\ell_1}\right) - \frac{M_b}{P}\left(\frac{1}{\ell_1} - \frac{1}{\ell_1}\cos k_1\ell_1\right)$$

which is rearranged to

$$M_a(\cos k_1\ell_1 + k_1\ell_1 \sin k_1\ell_1 - 1) + M_b\left(1 - \cos k_1\ell_1 + \frac{I_1\ell_1k_1^2\ell_2}{2I_2}\right) = 0 \tag{4.4.21}$$

For a nontrivial solution, set the determinant for the coefficient matrix equal to zero. The resulting transcendental equation is

$$2 - 2\cos k_1\ell_1 - k_1\ell_1\sin k_1\ell_1 + \frac{\ell_2I_1k_1}{2I_2}(\sin k_1\ell_1 - k_1\ell_1\cos k_1\ell_1) = 0 \tag{4.4.22}$$

By setting $I_1 = I_2 = I$ and $\ell_1 = \ell_2 = \ell$ in Eq. (4.4.22), the smallest root is $k\ell = 5.018$ and

$$P_{cr} = \frac{25.18EI}{\ell^2}$$

This load is considerably larger than that of the same frame $(7.34EI/\ell^2)$ where sidesway is permitted. The critical load also satisfies Eq. (4.4.1), as expected.

4.4.2. Application of Slope-Deflection Equations to Frame Stability

Although the differential equation method examined in the previous section is theoretically applicable to any frame, it becomes prohibitively complex in actuality, particularly in a frame of many kinematic degrees of freedom. In order to show the versatility of the slope-deflection equations, the same example examined above will be revisited.

It is assumed again that the axial compression in member BC would be negligibly small.
Since $\theta_a \equiv 0$, the moment at the top joint of member AB is

$$M_{ba} = (S_1\overline{k})_1\theta_b \tag{4.4.23}$$

where $\overline{k}_i = [(EI)/\ell]_i$
The moment in the horizontal member is

$$M_{bc} = (S_1\overline{k})_2\theta_b + (S_2\overline{k})_2\theta_c \tag{4.4.24}$$

As $\theta_c = -\theta_b$ for the buckling mode shown in Fig. 4-7(a), Eq. (4.4.24) reduces to

$$M_{bc} = [(S_1\overline{k})_2 - (S_2\overline{k})_2]\theta_b \tag{4.4.25}$$

Since there is no axial force in member BC, $(S_1)_2 = 4$ and $(S_2)_2 = 2$.
For joint equilibrium M_{ba} and M_{bc} are the same in magnitude and opposite in sign. Thus

$$\sum M_b = 0 \Rightarrow M_{ba} + M_{bc} = 0 \tag{4.4.26}$$

For $I_2 = I_1 = I$ and $\ell_2 = \ell_1 = \ell$, Eq. (4.4.26) reduces to

$$S_1\frac{EI}{\ell}\theta_b = \left[(4-2)\frac{EI}{\ell}\right]\theta_b \tag{4.4.27}$$

from which

$$S_1 = 2 \tag{4.4.28}$$

Equation (4.4.28) will lead to the critical load of $P_{cr} = (25.18EI)/\ell^2$.
For the buckling mode shown in Fig. 4-7(b), $\theta_b = \theta_c$. By keeping θ_b and θ_c as unknown independent variables, the analysis can be generalized. The unknown moments at joint C are

$$M_{cd} = (S_1\overline{k})_1\theta_c \tag{4.4.29}$$

and

$$M_{cb} = (S_1\overline{k})_2\,\theta_c + (S_2\overline{k})_2\,\theta_b \tag{4.4.30}$$

The equilibrium condition at B requires

$$\sum M_b = 0 \Rightarrow [(S_1\bar{k})_1 + (S_1\bar{k})_2]\theta_b + (S_2\bar{k})_2\theta_c = 0$$

from which

$$(S_1 + 4)\theta_b + 2\theta_c = 0 \tag{4.4.31}$$

Likewise, moment equilibrium at C demands

$$\sum M_c = 0 \Rightarrow [(S_1\bar{k})_1 + (S_1\bar{k})_2]\theta_c + (S_2\bar{k})_2\theta_b = 0 \tag{4.4.32}$$

from which

$$2\theta_b + (S_1 + 4)\theta_c = 0 \tag{4.4.33}$$

Setting the determinant of the coefficient matrix of the unknowns θ_b and θ_c for the stability condition gives

$$\begin{vmatrix} S_1 + 4 & 2 \\ 2 & S_1 + 4 \end{vmatrix} = 0 \Rightarrow (S_1 + 4)^2 - 4 = 0 \Rightarrow S_1 = -2, -6$$

By **Maple**®, BISECT or any other transcendental equation solver, one can obtain $k\ell = 5.01818$ and 5.52718 for $S_1 = -2$ and -6.

The smallest root for $k\ell = 5.01818$ gives the critical load of $25.18EI/\ell^2$ for the buckling mode shown in Fig. 4-7(a), and $k\ell = 5.52718$ yields the critical load of $30.55EI/\ell^2$ for the buckling mode shown in Fig. 4-7(b). It is of interest to note that the critical load is larger for the antisymmetric buckling mode than that for the symmetric buckling mode within the same braced frame. This difference can be explained by examining the buckling mode shapes shown in Fig. 4-7. In the antisymmetric buckling mode, the beam deformed in such a manner as to create an inflection point at the middle of the member (reducing the effective length by half), thereby increasing its stiffness. The increased stiffness of the beam, in turn, provides a little bit more constraint at the top of the column, which would shorten the effective length of the column.

The next example is buckling of a rigidly connected equilateral triangle shown in Fig. 4-8.

Take the counterclockwise moment and rotation as positive quantities as adopted in the derivation of the slope-deflection equations in Chapter 3. As the joints are assumed rigid, the original subtended angle of 60 degrees will be maintained throughout the history of deformations. Hence,

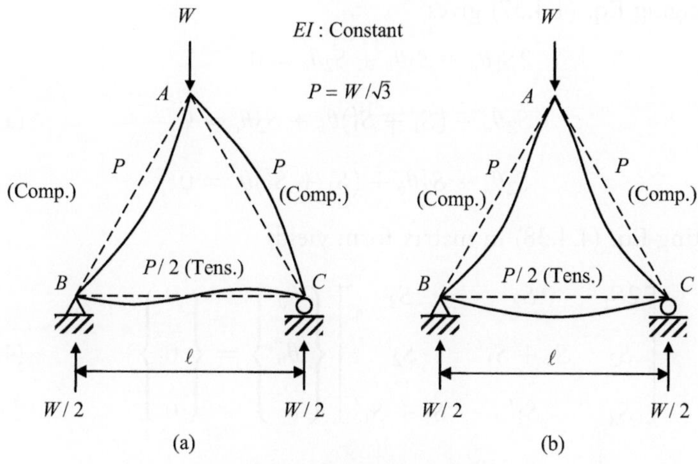

Figure 4-8 Equilateral triangle

$$\begin{aligned} \theta_{ab} &= \theta_{ac} = \theta_a \\ \theta_{ba} &= \theta_{bc} = \theta_b \\ \theta_{cb} &= \theta_{ca} = \theta_c \end{aligned} \tag{4.4.34}$$

The moment at each end of each member is then given by

$$\begin{aligned} M_{ab} &= \bar{k}(S_1\theta_a + S_2\theta_b), \quad M_{ac} = \bar{k}(S_1\theta_a + S_2\theta_c) \\ M_{ba} &= \bar{k}(S_1\theta_b + S_2\theta_a), \quad M_{bc} = \bar{k}(S_1'\theta_b + S_2'\theta_c) \\ M_{ca} &= \bar{k}(S_1\theta_c + S_2\theta_a), \quad M_{cb} = \bar{k}(S_1'\theta_c + S_2'\theta_b) \end{aligned} \tag{4.4.35}$$

where $\bar{k} = EI/\ell$, and S_1' and S_2' reflect the tensile force in member BC.

The compatibility of the rigid joint requires the following moment-equilibrium condition at each joint:

$$\begin{aligned} M_{ab} + M_{ac} &= 0 \\ M_{ba} + M_{bc} &= 0 \\ M_{ca} + M_{cb} &= 0 \end{aligned} \tag{4.4.36}$$

Substituting Eq. (4.4.35) into Eq. (4.4.36) yields

$$\begin{aligned} (S_1\theta_a + S_2\theta_b) + (S_1\theta_a + S_2\theta_c) &= 0 \\ (S_1\theta_b + S_2\theta_a) + (S_1\theta_b + S_2\theta_c)' &= 0 \\ (S_1\theta_c + S_2\theta_a) + (S_1\theta_c + S_2\theta_b)' &= 0 \end{aligned} \tag{4.4.37}$$

Rearranging Eq. (4.4.37) gives

$$\begin{aligned}2S_1\theta_a + S_2\theta_b + S_2\theta_c = 0\\ S_2\theta_a + (S_1 + S_1')\theta_b + S_2'\theta_c = 0\\ S_2\theta_a + S_2'\theta_b + (S_1 + S_1')\theta_c = 0\end{aligned} \tag{4.4.38}$$

Rewriting Eq. (4.4.38) in matrix form yields

$$\begin{bmatrix} 2S_1 & S_2 & S_2 \\ S_2 & S_1 + S_1' & S_2' \\ S_2 & S_2' & S_1 + S_1' \end{bmatrix} \begin{Bmatrix} \theta_a \\ \theta_b \\ \theta_c \end{Bmatrix} = \begin{Bmatrix} 0 \\ 0 \\ 0 \end{Bmatrix} \tag{4.4.39}$$

Setting the determinant of the augmented matrix equal to zero for the stability condition (a nontrivial solution) gives

$$\det = 0 = (S_1 + S_1' - S_2')\left[S_1(S_1 + S_1' + S_2') - S_2^2\right] = 0 \tag{4.4.40}$$

Two buckling modes are indicated by Eq. (4.4.40).

$$S_1 + S_1' - S_2' = 0 \quad \text{or} \quad S_1(S_1 + S_1' + S_2') - S_2^2 = 0$$

From **Maple**® or BISECT, $S_1\,(S_1 + S_1' + S_2') - S_2^2 = 0$ gives $k\ell = 4.0122 \Rightarrow P_{cr} = 16.1EI/\ell^2$

From **Maple**® or BISECT, $S_1 + S_1' - S_2' = 0$ gives $k\ell = 5.3217 \Rightarrow P_{cr} = 28.32EI/\ell^2$

For $k\ell = 4.0122$, $S_1 = 1.1490$, $S_2 = 3.0150$, $S_1' = 4.9763$, $S_2' = 1.7861$. Substituting these values into the matrix equation, Eq. (4.4.39) gives

$$\begin{bmatrix} 2.298 & 3.015 & 3.015 \\ 3.015 & 6.1253 & 1.7861 \\ 3.015 & 1.7861 & 6.1253 \end{bmatrix} \begin{Bmatrix} \theta_a \\ \theta_b \\ \theta_c \end{Bmatrix} = \begin{Bmatrix} 0 \\ 0 \\ 0 \end{Bmatrix} \tag{4.4.41}$$

Recall that the determinant was equal to zero. Hence, the augmented matrix in Eq. (4.4.41) is a singular matrix and therefore, cannot be inverted.

One can only obtain the normalized eigenvector or mode shape. An eigenvector can just show the deformation shape of the structure in a neighboring equilibrium position. Hence, the exact magnitude of the mode shape in eigenvalue problems is immaterial.

Let $\theta_a = 1$ and expand the first and second rows of the matrix equation, Eq. (4.4.41), to yield,

$$2.298 + 3.015\,\theta_b + 3.015\,\theta_c = 0$$

from which

$$\theta_b = \frac{1}{3.015}(-2.298 - 3.015\,\theta_c) \tag{4.4.42}$$

and

$$3.015 + 6.1253\,\theta_b + 1.7861\,\theta_c = 0 \tag{4.4.43}$$

Substituting Eq. (4.4.42) into Eq. (4.4.43) yields

$$3.015 + \frac{6.1253}{3.015}(-2.298 - 3.015\,\theta_c) + 1.7861\,\theta_c = 0$$

from which

$$\theta_c = -0.381 \tag{4.4.44}$$

Substituting Eq. (4.4.44) into Eq. (4.4.43) gives

$$\theta_b = -0.381 \tag{4.4.45}$$

The buckling mode shape is given in Fig. 4-9.

For $k\ell = 5.3217$, $S_1 = -3.9419$, $S_2 = 6.2624$, $S_1{}' = 5.6170$, $S_2{}' = 1.6751$

Substituting these values into the matrix equation, Eq. (4.4.39) gives

$$\begin{bmatrix} -7.8838 & 6.2624 & 6.2624 \\ 6.2624 & 1.6751 & 1.6751 \\ 6.2624 & 1.6751 & 1.6751 \end{bmatrix} \begin{Bmatrix} \theta_a \\ \theta_b \\ \theta_c \end{Bmatrix} = \begin{Bmatrix} 0 \\ 0 \\ 0 \end{Bmatrix} \tag{4.4.46}$$

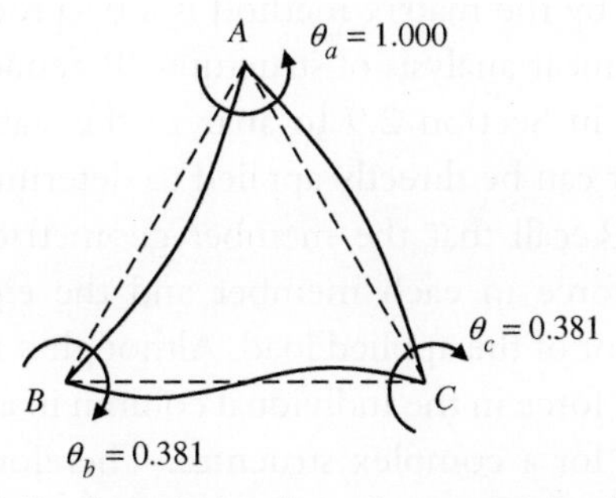

Figure 4-9 Equilateral triangle antisymmetric buckling mode

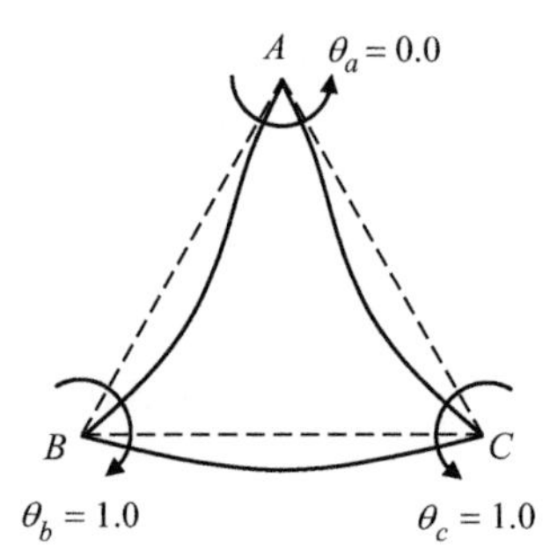

Figure 4-10 Equilateral triangle symmetric buckling mode

Again the augmented matrix in Eq. (4.4.46) is a singular matrix. One can only obtain the relative deformation shape of the structure in the neighboring equilibrium (buckled) position. By virtue of Fig. 4-9, a symmetrical mode shape is expected in this case. Let $\theta_a = 0$ and expand the second and third rows of the matrix equation, Eq. (4.4.46), to yield,

$$\theta_b = -\theta_c = 1.0 \tag{4.4.47}$$

The buckling mode shape is given graphically in Fig. 4-10.

Although there is no joint translation at the loaded vertex of the triangle, the critical load corresponding to the antisymmetric buckling mode is less than that corresponding to the symmetric buckling mode. Examining the symmetric buckling mode shape shown in Fig. 4-10 reveals that an inflection point exists in the compression member, thereby making the effective column length considerably smaller than that in the antisymmetric buckling mode. This makes the compression members in the symmetric buckling mode carry a greater load.

4.5. STABILITY OF FRAMES BY MATRIX ANALYSIS

The stability analysis by the matrix method is a by-product of research on the incremental nonlinear analysis of structures (Przemieniecki 1968). The matrix method used in Section 2.9 to analyze the stability of an isolated compression member can be directly applied to determine the critical load of an entire frame. Recall that the member geometric stiffness matrix is a function of axial force in each member and the eigenvalue is merely a proportionality factor of the applied load. Although it is intuitively simple to recognize the axial force in the individual column in a simple structure, it may not be the case for a complex structure. Therefore, it is required to conduct a static analysis of the structure under a given set of loading for which the critical value is sought to determine the axial force in each member.

As an illustration, consider the stability of the simple portal frame shown in Fig. 4-11(a). The portal frame is unbraced. Each member has a length of ℓ and bending rigidity EI, and the frame is clamped at its base and is loaded as shown.

Positive member (local) and structure (global) kinematic degrees of freedom and corresponding force are defined in Figs. 4-11(b) and 4-11(c). According to Eqs. (2.9.11) and (2.9.13), the member stiffness matrices for the column are

$$[k_1] = [k_3]$$

$$= \frac{EI}{\ell^3}\begin{bmatrix} 12 & 6\ell & -12 & 6\ell \\ 6\ell & 4\ell^2 & -6\ell & 2\ell^2 \\ -12 & -6\ell & 12 & -6\ell \\ 6\ell & 2\ell^2 & -6\ell & 4\ell^2 \end{bmatrix} - \frac{P}{\ell}\begin{bmatrix} 6/5 & \ell/10 & -6/5 & \ell/30 \\ \ell/10 & 2\ell^2/15 & -\ell/10 & -\ell^2/30 \\ -6/5 & -\ell/10 & 6/5 & -\ell/10 \\ \ell/10 & -\ell^2/30 & -\ell/10 & 2\ell^2/15 \end{bmatrix} \tag{4.5.1}$$

and the member stiffness matrix for the beam is

$$[k_2] = \frac{EI}{\ell^3}\begin{bmatrix} 12 & 6\ell & -12 & 6\ell \\ 6\ell & 4\ell^2 & -6\ell & 2\ell^2 \\ -12 & -6\ell & 12 & -6\ell \\ 6\ell & 2\ell^2 & -6\ell & 4\ell^2 \end{bmatrix} \tag{4.5.2}$$

Note that there is no member geometric stiffness matrix in Eq. (4.5.2), as the axial force is assumed equal to zero in the beam.

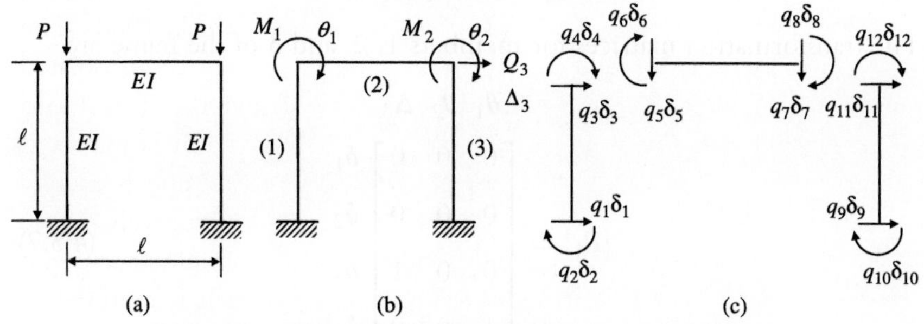

Figure 4-11 Global and local coordinates of portal frame

In order to obtain the structure (global) stiffness matrix, the member stiffness matrices are first transformed to structure coordinates and then combined and reduced (to eliminate the rigid body motion) for an automatic programming scheme. However, for a manual operation as is being carried out here, the reduction process can be eliminated by arranging the transformation matrices to reflect the unsuppressed global degrees of freedom. Since each member has four degrees of freedom and there are three global degrees of freedom, the size of each member transformation matrix must be 4 × 3. The member degrees of freedom and the structure degrees of freedom are related as

$$[\delta] = [B_n][\Delta] \tag{4.5.3}$$

where the subscript n indicates the member number shown in Fig. 4-11(b) and $[\Delta]$ and $[\delta]$ are given by

$$[\Delta] = \begin{Bmatrix} \theta_1 \\ \theta_2 \\ \Delta_3 \end{Bmatrix} \tag{4.5.4}$$

and

$$[\delta] = \begin{Bmatrix} \delta_1 \\ \delta_2 \\ \delta_3 \\ \delta_4 \end{Bmatrix} \tag{4.5.5}$$

The member stiffness matrix is related to the structure stiffness matrix by a triple matrix product as

$$[K_n] = [B_n]^T [k_n][B_n] \tag{4.5.6}$$

The transformation matrices for members 1, 2, and 3 of the frame are

$$[B_1] = \begin{bmatrix} 0 & 0 & 0 \\ 0 & 0 & 0 \\ 0 & 0 & 1 \\ 1 & 0 & 0 \end{bmatrix} \begin{matrix} \delta_1 \\ \delta_2 \\ \delta_3 \\ \delta_4 \end{matrix} \quad (\text{columns: } \theta_1 \;\; \theta_2 \;\; \Delta_3) \tag{4.5.7}$$

$$[B_2] = \begin{array}{c} \begin{matrix} \theta_1 & \theta_2 & \Delta_3 \end{matrix} \\ \begin{bmatrix} 0 & 0 & 0 \\ 1 & 0 & 0 \\ 0 & 0 & 0 \\ 0 & 1 & 0 \end{bmatrix} \end{array} \begin{matrix} \delta_5 \\ \delta_6 \\ \delta_7 \\ \delta_8 \end{matrix} \tag{4.5.8}$$

$$[B_3] = \begin{array}{c} \begin{matrix} \theta_1 & \theta_2 & \Delta_3 \end{matrix} \\ \begin{bmatrix} 0 & 0 & 0 \\ 0 & 0 & 0 \\ 0 & 0 & 1 \\ 1 & 0 & 0 \end{bmatrix} \end{array} \begin{matrix} \delta_9 \\ \delta_{10} \\ \delta_{11} \\ \delta_{12} \end{matrix} \tag{4.5.9}$$

Executing the matrix triple products indicated in Eq. (4.5.6) using these transformation matrices, the member stiffness matrices in Eqs. (4.5.1) and (4.5.2) transform

$$[K_1] = \frac{EI}{\ell^3}\begin{bmatrix} 4\ell^2 & 0 & -6\ell \\ 0 & 0 & 0 \\ -6\ell & 0 & 12 \end{bmatrix} - \frac{P}{\ell}\begin{bmatrix} 2\ell^2/15 & 0 & -\ell/10 \\ 0 & 0 & 0 \\ -\ell/10 & 0 & 6/5 \end{bmatrix} \tag{4.5.10}$$

$$[K_2] = \frac{EI}{\ell^3}\begin{bmatrix} 4\ell^2 & 2\ell^2 & 0 \\ 2\ell^2 & 4\ell^2 & 0 \\ 0 & 0 & 0 \end{bmatrix} \tag{4.5.11}$$

$$[K_3] = \frac{EI}{\ell^3}\begin{bmatrix} 0 & 0 & 0 \\ 0 & 4\ell^2 & -6\ell \\ 0 & -6\ell & 2\ell^2 \end{bmatrix} - \frac{P}{\ell}\begin{bmatrix} 0 & 0 & 0 \\ 0 & 2\ell^2/15 & -\ell/10 \\ 0 & -\ell/10 & 6/5 \end{bmatrix} \tag{4.5.12}$$

The structure stiffness matrix by assembling the transformed member stiffness matrices is

$$[K] = \frac{EI}{\ell^3}\begin{bmatrix} 8\ell^2 & 2\ell^2 & -6\ell \\ 2\ell^2 & 8\ell^2 & -6\ell \\ -6\ell & -6\ell & 24 \end{bmatrix} - \frac{P}{\ell}\begin{bmatrix} 2\ell^2/15 & 0 & -\ell/10 \\ 0 & 2\ell^2/15 & -\ell/10 \\ -\ell/10 & -\ell/10 & 12/5 \end{bmatrix} \tag{4.5.13}$$

Let

$$\lambda = \frac{P\ell^2}{30EI} \tag{4.5.14}$$

Then, the structure stiffness matrix reduces to

$$[K] = \frac{EI}{\ell^3}\begin{bmatrix} (8-4\lambda)\ell^2 & 2\ell^2 & (-6+3\lambda)\ell \\ 2\ell^2 & (8-4\lambda)\ell^2 & (-6+3\lambda)\ell \\ (-6+3\lambda)\ell & (-6+3\lambda)\ell & 24-72\lambda \end{bmatrix} \tag{4.5.15}$$

At the critical load, the determinant of the stiffness matrix must vanish. The resulting equation in terms of λ is

$$90\lambda^3 - 383\lambda^2 + 428\lambda - 84 = 0 \tag{4.5.16}$$

The smallest root of this equation by **Maple®** is $\lambda_1 = 0.24815$, from which

$$P_{cr} = \frac{7.44EI}{\ell^2} \tag{4.5.17}$$

This result is only 0.87% higher than the exact value of $7.38EI/\ell^2$ obtained in Section 4.3. A monotonic convergence to the exact value is guaranteed in a computer analysis by taking a refined grid of the structure.

4.6. SECOND-ORDER ANALYSIS OF A FRAME BY SLOPE-DEFLECTION EQUATIONS

The current AISC (2005) specification stipulates that "any second-order elastic analysis method that considers both $P-\Delta$ and $P-\delta$ effects may be used." Since both the joint rotation ($P-\delta$ effect) and joint translation ($P-\Delta$ effect) are reflected by the slope-deflection equations with axial force

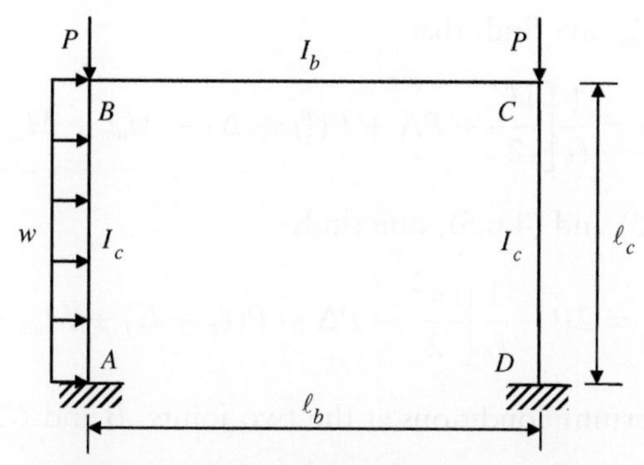

Figure 4-12 Portal frame with horizontal load

by a means of stability functions, S_1 and S_2, an elastic analysis using the slope-deflection equations is considered to be acceptable second-order analysis.

As an illustration, consider the portal frame shown in Fig. 4-12. The frame is subjected to a concentrated load of 275 kips each at the top of the column and a uniformly distributed load of 1 kip/ft. These are factored loads. The length of the column (W 8 × 31) is 13 feet, and the beam (W 10 × 33) is 20 feet long. Use $E = 30{,}000$ ksi, $\sigma_y = 60$ ksi.

To be consistent with the assumptions normally adopted in the longhand analysis of the slope-deflection equations, the axial force (less than 1% of the axial force in the column) in the beam is ignored and the shortening of the column is also neglected. As a result of the simplifying assumptions, the chord rotation of each member becomes

$$\rho_{ab} = \rho_{cd} = \rho = \Delta/\ell_c \Rightarrow \rho_{bc} = 0 \tag{4.6.1}$$

where Δ is the horizontal translation of the beam. Horizontal equilibrium for the entire frame gives

$$\sum H = 0 \Rightarrow H_a + H_d = w\ell_c \tag{4.6.2}$$

where H_a and H_d are the horizontal reactions at joints A and D, respectively. Vertical equilibrium for the entire frame yields

$$\sum V = 0 \Rightarrow R_a + R_d = 2P \tag{4.6.3}$$

where R_a and R_d are the vertical reactions at joints A and D, respectively.

The moment equilibrium condition for the entire frame about the point A gives

$$\frac{w\ell_c^2}{2} + P(\rho\ell_c) + P(\ell_b + \rho\ell_c) - R_d\ell_b + M_{ab} + M_{dc} = 0 \tag{4.6.4}$$

From Eq. (4.6.4), one finds that

$$R_d = \frac{1}{\ell_b}\left[\frac{w\ell_c^2}{2} + P\Delta + P(\ell_b + \Delta) + M_{ab} + M_{dc}\right] \quad (4.6.5)$$

From Eqs. (4.6.3) and (4.6.5), one finds

$$R_a = 2P - R_d = 2P - \frac{1}{\ell_b}\left[\frac{w\ell_c^2}{2} + P\Delta + P(\ell_b + \Delta) + M_{ab} + M_{dc}\right] \quad (4.6.6)$$

Moment equilibrium conditions at the two joints, B and C, are

$$M_{ba} + M_{bc} = 0 \quad (4.6.7)$$

$$M_{cb} + M_{cd} = 0 \quad (4.6.8)$$

Equilibrium $\left(\sum M_b = 0\right)$ of the isolated left column gives

$$R_a\Delta + H_a\ell_c + M_{ab} + M_{ba} - \frac{w\ell_c^2}{2} = 0 \quad (4.6.9)$$

Likewise, equilibrium of the isolated right column gives

$$R_d\Delta + H_d\ell_c + M_{dc} + M_{cd} = 0 \quad (4.6.10)$$

Summing Eqs. (4.6.9) and (4.6.10) yields

$$(R_a + R_d)\Delta + (H_a + H_b)\ell_c + M_{ab} + M_{ba} + M_{cd} + M_{dc} = \frac{w\ell_c^2}{2}$$

Substituting Eqs. (4.6.2) and (4.6.3) into the above equation gives

$$2P\Delta + \frac{w\ell_c^2}{2} + M_{ab} + M_{ba} + M_{cd} + M_{dc} = 0 \quad (4.6.11)$$

From slope-deflection equations with and without the effect of axial forces, one finds

$$k^2 = \frac{P}{EI} = \frac{275}{30000 \times 110} = 83.33 \times 10^{-6}, \quad k = 9.13 \times 10^{-3}$$

$$u = \frac{k\ell_c}{2} = \frac{9.13 \times 10^{-3} \times 156}{2} = 0.712, \quad \frac{3(\tan u - u)}{u^2 \tan u} = 1.035515$$

$$\frac{w\ell_c^2}{12}\left[\frac{3(\tan u - u)}{u^2 \tan u}\right] = \frac{1 \times 156^2}{12 \times 12} \times 1.035515 = 175 \text{ k-in.}$$

It is noted that the fixed-end moments in a member with a compressive force must be evaluated reflecting the effect of the amplification as suggested by Horne and Merchant (1965).

$$M_{ab} = \left(\frac{EI}{\ell}\right)_c [S_1\theta_a + S_2\theta_b - (S_1 + S_2)\rho] - 175 \tag{4.6.12}$$

$$M_{ba} = \left(\frac{EI}{\ell}\right)_c [S_1\theta_b + S_2\theta_a - (S_1 + S_2)\rho] + 175 \tag{4.6.13}$$

$$M_{bc} = \left(\frac{EI}{\ell}\right)_b (4\theta_b + 2\theta_c) \tag{4.6.14}$$

$$M_{cb} = \left(\frac{EI}{\ell}\right)_b (4\theta_c + 2\theta_b) \tag{4.6.15}$$

$$M_{cd} = \left(\frac{EI}{\ell}\right)_c [S_1\theta_c + S_2\theta_d - (S_1 + S_2)\rho] \tag{4.6.16}$$

$$M_{dc} = \left(\frac{EI}{\ell}\right)_c [S_1\theta_d + S_2\theta_c - (S_1 + S_2)\rho] \tag{4.6.17}$$

For $W\,8 \times 31 \Rightarrow A = 9.12\ \text{in}^2 \Rightarrow P_y = A \times \sigma_y = 9.12 \times 60 = 547$ kips

$$\beta = k\ell_c = 0.00913 \times 156 = 1.424$$

From **Maple**®

$$S_1 = 3.7221 \quad \text{and} \quad S_2 = 2.0721$$

Substituting these numerical values into moment equations yields

$$\begin{aligned} M_{ab} &= \frac{30000 \times 110}{156}(2.0721\theta_b - 5.7942\rho) - 175 \\ &= 43833\theta_b - 122570\rho - 175 \end{aligned} \tag{4.6.12a}$$

$$\begin{aligned} M_{ba} &= \frac{30000 \times 110}{156}(3.7221\theta_b - 5.7942\rho) + 175 \\ &= 78737\theta_b - 122570\rho + 175 \end{aligned} \tag{4.6.13a}$$

$$M_{bc} = \frac{30000 \times 171}{240}(4\theta_b + 2\theta_c) = 855000\theta_b + 427500\theta_c \quad (4.6.14a)$$

$$M_{cb} = \frac{30000 \times 171}{240}(4\theta_c + 2\theta_b) = 427500\theta_b + 855000\theta_c \quad (4.6.15a)$$

$$M_{cd} = \frac{30000 \times 110}{156}(3.7221\theta_c - 5.7942\rho) = 787370\theta_c - 122570\rho \quad (4.6.16a)$$

$$M_{dc} = \frac{30000 \times 110}{156}(2.0721\theta_c - 5.7942\rho) = 438330\theta_c - 122570\rho \quad (4.6.17a)$$

Substituting Eqs. (4.6.12a) through (4.6.17a) into Eqs. (4.6.7), (4.6.8), and (4.6.11) yields

$$1642370\theta_b + 427500\theta_c - 122570\rho = -175 \quad (4.6.18)$$

$$427500\theta_b + 1642370\theta_c - 122570\rho = 0 \quad (4.6.19)$$

$$-122570\theta_b - 122570\theta_c + 404480\rho = 1014 \quad (4.6.20)$$

Solving Eqs. (4.5.18), (4.5.19), and (4.5.20) simultaneously by **Maple®** gives

$$\theta_b = 0.0009359 \text{ rad.}, \ \theta_c = 0.002376 \text{ rad.}, \ \rho = 0.00351\text{rad}$$

$$\Delta = 0.00351 \times 156 = 0.5476 \text{ in}$$

Substituting these values into the moment equation gives

$$M_{ab} = 43833 \times 0.0009359 - 122570 \times 0.00351 - 175 = -564.2 \text{ k-in}$$

$$M_{ba} = 78737 \times .0009359 - 122570 \times .00351 + 175 = -181.53 \text{ k-in}$$

$$M_{bc} = 85500 \times .0009359 + 42750 \times .002376 = 181.59 \text{ k-in}$$

$$M_{cb} = 42750 \times .0009359 + 85500 \times .002376 = 243.16 \text{ k-in}$$

$$M_{cd} = 78737 \times .002376 - 122570 \times .00351 = -243.14 \text{ k-in}$$

$$M_{dc} = 43833 \times .002376 - 122570 \times .00351 = -326.07 \text{ k-in}$$

Bifurcation buckling load. As the bifurcation buckling load is independent from any primary bending, the modified coefficient determinant can be set equal to zero to determine the P_{cr}. Substituting these numerical values into moment equations yields

$$\begin{aligned} M_{ab} &= \frac{30000 \times 110}{156}[S_2\theta_b - (S_1 + S_2)\rho] \\ &= 21153.8S_2\theta_b - 21153.8(S_1 + S_2)\rho \end{aligned} \tag{4.6.21}$$

$$\begin{aligned} M_{ba} &= \frac{30000 \times 110}{156}[S_1\theta_b - (S_1 + S_2)\rho] \\ &= 21153.8S_1\theta_b - 21153.8(S_1 + S_2)\rho \end{aligned} \tag{4.6.22}$$

$$M_{bc} = \frac{30000 \times 171}{240}(4\theta_b + 2\theta_c) = 855000\theta_b + 427500\theta_c \tag{4.6.23}$$

$$M_{cb} = \frac{30000 \times 171}{240}(4\theta_c + 2\theta_b) = 427500\theta_b + 855000\theta_c \tag{4.6.24}$$

$$\begin{aligned} M_{cd} &= \frac{30000 \times 110}{156}[S_1\theta_c - (S_1 + S_2)\rho] \\ &= 21153.8S_1\theta_c - 21153.8(S_1 + S_2)\rho \end{aligned} \tag{4.6.25}$$

$$\begin{aligned} M_{dc} &= \frac{30000 \times 110}{156}[S_2\theta_c - (S_1 + S_2)\rho] \\ &= 21153.8S_2\theta_c - 21153.8(S_1 + S_2)\rho \end{aligned} \tag{4.6.26}$$

Substituting Eqs. (4.6.24) through (4.6.26) into Eqs. (4.6.7), (4.6.8), and (4.6.11) yields

$$(21153.8S_1 + 85500)\theta_b + 427500\theta_c - 21153.8(S_1 + S_2)\rho = 0$$

$$427500\theta_b + (21153.8S_1 + 85500)\theta_c - 21153.8(S_1 + S_2)\rho = 0$$

$$\begin{aligned} &-21153.8(S_1 + S_2)\theta_b - 21153.8(S_1 + S_2)\theta_c \\ &-[312P - 84615.2(S_1 + S_2)]\rho = 0 \end{aligned}$$

The stability condition requires that the determinant of the augmented matrix vanish

$$\begin{vmatrix} (21153.8S_1 + 85500) & 42750 & -21153.8(S_1 + S_2) \\ 42750 & (21153.8S_1 + 85500) & -21153.8(S_1 + S_2) \\ -21153.8(S_1 + S_2) & -21153.8(S_1 + S_2) & -[312P - 84615.2(S_1 + S_2)] \end{vmatrix} = 0$$

Solving the expanded polynomial by **Maple®** gives

$$P_{cr} = 1{,}003.15 \text{ kips}$$

The maximum combined stress assuming the given loads are factored loads is

$$\sigma = \frac{P}{A} \pm \frac{M_{\max}c}{I} = \frac{275}{9.12} + \frac{564.2 \times 4}{110} = 30.15 + 20.52$$

$$= 50.67\ ksi\ <\ 60ksi = \sigma_y\ OK$$

From Eqs. (4.5.5) and (4.5.6), the vertical reactions are

$$R_d = \frac{1}{\ell_b}\left[\frac{w\ell_c^2}{2} + P\Delta + P(\ell_b + \Delta) + M_{ab} + M_{dc}\right]$$
$$= [1014 + 275 \times .5476 + 275(240 + .5476) - 564.2 - 326.07]/240$$
$$= 276.77 \text{ kips}$$

$$R_a = 2P - R_d = 550 - 276.77 = 273.23 \text{ kips}$$

Consider the free body of member AB. The shear at joint A is computed as

$$V_a = \frac{1}{156}(564.2 + 181.59 - 273.23 \times 0.5476 + 13 \times 6.5 \times 12)$$
$$= 10.32 \text{ kips},\ V_b = 2.68 \text{ kips}$$

$$M(x)_{ab} = -564.2 + 273.23 \times \frac{0.5476}{156}x + 10.32x - \frac{1}{2}\frac{1}{12}x^2$$

$$M(x = 66.2'')_{ab} = 0,\ M(x = 135.35'')_{\max} = 199.11 \text{ k-in}$$

$$V_{cd} = V_{dc} = (243.14 + 326.07 - 276.77 \times 0.5476)/156 = 2.68 \text{ kips}$$

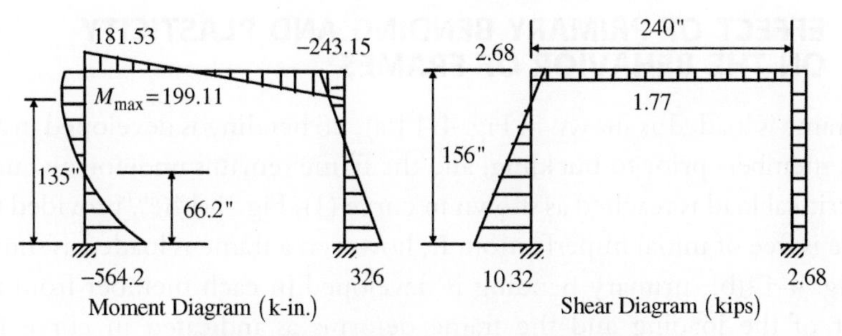

Figure 4-13 Moment and shear diagrams

Table 4-1 Comparison of analysis

	Slope-Deflection Equations	Matrix Method
M_a	−564.200000	475.00000
M_b	181.530000	117.00000
M_c	−243.150000	−173.00000
M_d	326.000000	249.00000
P_{ab}	273.230000	274.00000
P_{cd}	276.770000	276.00000
Δ	0.547600	0.40100
θ_b	0.000936	0.00048
θ_c	0.002376	0.00179
P_{cr}	1003.150000	1003.00000
H_a	10.320000	10.30000
H_d	2.680000	2.70000

Note: Units are k-in. and radian.

Table 4-1 shows the results of comparative analyses of the frame. The matrix method is considered to be the first-order analysis method.

Consider the amplification effect.

$$AF = \frac{1}{1 - P/P_{cr}} = \frac{1}{1 - \dfrac{275}{1003.15}} = 1.378$$

It would be interesting to note how closely the results of the first-order analysis can be amplified to simulate the second-order analysis results. The current AISC (2005) specification introduces an indirect second-order analysis incorporating B_1 and B_2 factors to the results of the first-order analysis results.

4.7. EFFECT OF PRIMARY BENDING AND PLASTICITY ON THE BEHAVIOR OF FRAMES

If a frame is loaded as shown in Fig. 4-14(a), no bending is developed in any of its members prior to buckling, and the frame remains undeformed until the critical load is reached as shown in curve (1), Fig. 4-14(c), provided the frame is free of initial imperfection. If, however, a frame is loaded as shown in Fig. 4-14(b), primary bending is developed in each member from the onset of the loading and the frame deforms as indicated in curve (2), Fig. 4-14(c). Frames with primary bending have been investigated experimentally as well as theoretically (Masur et al. 1961; Lu 1963). The somewhat consistent conclusion drawn from past studies is that primary bending does not significantly reduce the critical load of a frame as long as stresses remain elastic. An exception to this observation occurs when the beam is very long. In that case, the presence of primary bending reduces the symmetric buckling load of the frame possibly due to the excessive deflection of the beam, thereby further decreasing the elastic constraint at the top of the columns. As frames with such a long beam that can adversely affect the symmetric buckling load are rarely encountered in practice, it appears to be safe to conclude that the effect of primary bending can be ignored in computing the critical load of a frame. Primary bending is, therefore, only negligible in determination of the critical (ultimate) load and not in design; that is, it should be treated as beam-column in design. If P/P_E for the individual member exceeds 0.15, amplification effect must be considered.

It appears to be customary in steel design that most columns are designed with slenderness ratios between 40 and 80. Hence, inelastic buckling covers most column design, and the elastic buckling load does not control the

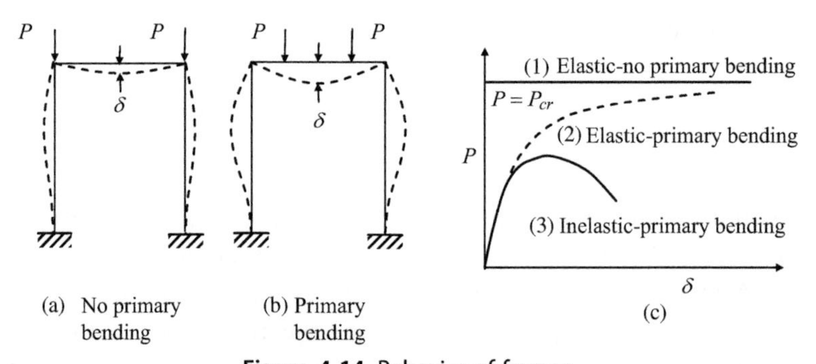

Figure 4-14 Behavior of frames

design. Frames having columns in this range will fail at a load that is smaller than the elastic critical load as shown in curve (2), Fig. 4-14(c). Frequently, an elastic second-order analysis as shown in curve (2), Fig. 4-14, is very deceptive unless the stress level is checked at every step.

If instability were the only factor leading to collapse, failure would occur at the critical load. If collapse were solely due to the plasticity effect, the frame would fail when it becomes a mechanism due to formation of plastic hinge(s). In the actual case, both instability and plasticity are present, and collapse occurs due to an interaction of these two at a load that is lower than either the critical load or the mechanism load. To predict this kind of failure load, Horne and Merchant (1965) proposed the following empirical interaction equation, known as the Rankine equation:

$$\frac{P_f}{P_e} + \frac{P_f}{P_p} = 1.0 \tag{4.7.1}$$

Equation (4.7.1) can be rearranged into a convenient form as

$$P_f = \frac{P_p P_e}{P_e + P_p} \tag{4.7.2}$$

where

P_f = failure load
P_e = elastic buckling load
P_p = plastic mechanism load

Although Horne and Merchant demonstrated the reasonableness of the proposed Rankine equation by a scattering chart of experimental data, the data do not appear to be representative of a wide spectrum of plausible cases. It appears that if P_e is greater than 3 times P_p, Eq. (4.7.1) overestimates the failure load. It was noted that if P_e is less than 3 times P_p, the scatter of points away from Eq. (4.7.2) becomes considerable. The derivation of Eq. (4.7.1) is conservative. Hence, Eq. (4.7.2) might be used to give rapid, but safe, estimates of P_f. Since an access to a general-purpose finite element code such as ABAQUS (2006) is readily available to most academics and practitioners and a much better estimate can be obtained with the computer, the attractiveness of Merchant's use of the Rankine equation is greatly diminished. Examples of refined analyses include Alvarez and Birnstiel (1969) and Ojalvo and Lu (1961). Further treatises of this important topic are presented by Galambos (1968).

4.8. STABILITY DESIGN OF FRAMES

A framed compression member is likely to be subjected to both bending and axial loading and must be designed as a beam-column using an interaction equation. Hence, the critical load of the member is required to be correctly determined. One way of determining the critical load is to carry out a three-dimensional stability analysis of the entire frame. However, an analysis of the entire frame is frequently too involved for routine design. Moreover, even where the best of analysis models are available, the designer still must account for uncertainties introduced by the variability in the magnitude and distribution of loads and in the strength and stiffness of members, connections, foundations, and so on. One very crude method of obtaining the critical load of a framed column is to estimate the degree of restraint at the ends of the member as shown in Fig. 4-15. When idealized boundary conditions are approximated, AISC (2005) recommends somewhat conservative K values for design. For braced frames, it is always conservative to take the K factor as unity. For unbraced frames, except perhaps for the flagpole-type column, case (e), Fig. 4-15, an arbitrary selection of K is not satisfactory for design. In the old days, a simple design methodology that would give a reasonable result for columns in a multistory building frame subjected to lateral load(s) was to assume an inflection point at the mid-height of each column. Treating the entire building frame as a flagpole-type cantilever column generally yields a poor result.

Today (2009), all major design specifications include the use of second-order analysis, although a unified approach to frame stability design has yet

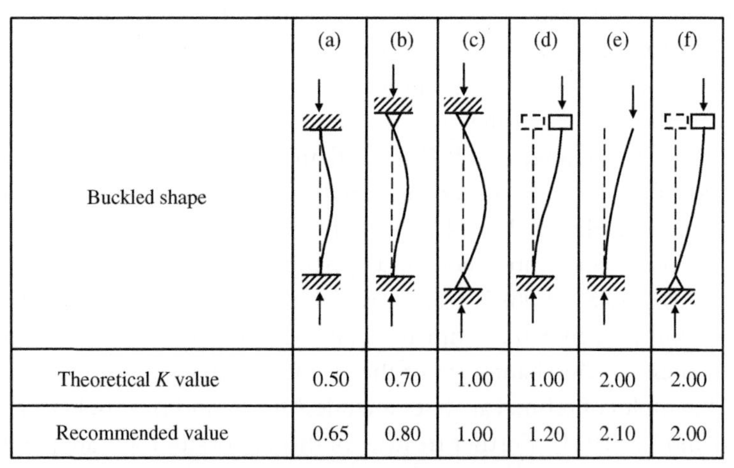

	(a)	(b)	(c)	(d)	(e)	(f)
Buckled shape						
Theoretical K value	0.50	0.70	1.00	1.00	2.00	2.00
Recommended value	0.65	0.80	1.00	1.20	2.10	2.00

Figure 4-15 Idealized boundary conditions

to emerge. The Canadian standards for steel structures, CAN3-S16.1 (CSA 1994), have eliminated the use of the effective-length concept (K-factors), and frame stability is solely to be checked through second-order analysis procedures incorporating notional lateral forces. The current (2005) AISC specification recognizes both the notional load analysis and the effective-length concept, along with a direct or indirect second-order analysis. The purpose of the notional loads is to account for the destabilizing effects of initial imperfections, nonideal conditions (incidental pattern gravity load effects, temperature gradients, foundation settlement, uneven column shortening, or any other effects that could induce sway that is not explicitly considered in the analysis), inelasticity in structural members, or combinations thereof. The magnitude of the notional lateral load, 0.002 times the story vertical loads, can be thought of as the continuity shear representing $P\Delta/\ell$ in which Δ is an initial out-of-plumbness in each story of 1/500 times the story height. Although the notional load procedure is considered to be an improved method of analysis, it still requires a stability analysis of the entire frame. In this regard, the effective-length concept still has a role to play in the design of framed columns.

The most common procedure for determining effective lengths is to use the Jackson and Moreland alignment charts originally developed by Julian and Lawrence (1959) and presented in detail by Kavanagh (1960). An improved approximate method of analyses of columns in frames was introduced by Kavanagh (1960). In the derivation, a number of simplifying assumptions were introduced. One of the major weaknesses was that the frame was assumed to behave in a purely elastic fashion. In light of the common practice of designing columns with a slenderness ratio between 40 and 80, this must be a serious shortcoming.

AISC (2005) Specification Commentary endorses the suggested adjustment of the G-factor by Yura (1971) and ASCE Task Committee on Effective Length (1997) when the column is inelastic. The derivation was based on the slope-deflection equation with axial forces.

The following assumptions are used in the development of the elastic stability equation:

1. Behavior is purely elastic.
2. All members are prismatic.
3. All columns reach their buckling loads simultaneously.
4. The structure consists of symmetrical rectangular frames.
5. At a joint, the restraining moment provided by the girder is distributed to the column in proportion to their stiffnesses.

6. The girders are elastically restrained at their ends by the columns, and at the onset of buckling, the rotations of the girder at its ends are equal in magnitude and opposite in direction with sidesway inhibited. If sidesway is uninhibited, rotations at opposite ends of the restraining girders are equal in magnitude and direction.
7. The girders carry no axial forces.

Assumption 6 leads to $-\theta_C = -\theta_D = -\theta_A$ and $\theta_E = \theta_F = -\theta_B$. From the slope-deflection equations with or without axial forces, one obtains

$$
\begin{aligned}
M_{ac} &= \frac{2EI_{blt}}{\ell_{bl}}\theta_A \\
M_{ad} &= \frac{2EI_{brt}}{\ell_{br}}\theta_A \\
M_{be} &= \frac{2EI_{blb}}{\ell_{bl}}\theta_B \\
M_{bf} &= \frac{2EI_{brb}}{\ell_{br}}\theta_B
\end{aligned}
\tag{4.8.1}
$$

and

$$
\begin{aligned}
M_{ba} &= \frac{EI_c(S_2\theta_A + S_1\theta_B)}{l_c} \\
M_{ab} &= \frac{EI_c(S_1\theta_A + S_2\theta_B)}{l_c}
\end{aligned}
\tag{4.8.2}
$$

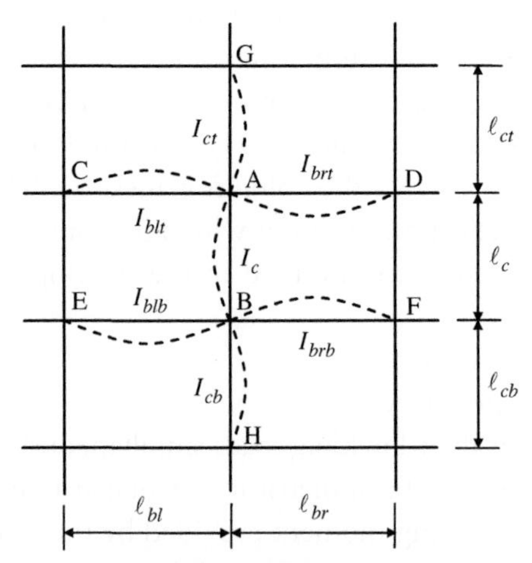

Figure 4-16 Sidesway inhibited

Although the stability relationship may be developed using either stiffness coefficients or flexibility coefficients, flexibility coefficients will be utilized here because they are easier to work with, as will be shown later. The stiffness coefficients are given by Eqs. (3.4.12) and (3.4.13).

$$S_1 = \frac{\beta(\beta \cos \beta - \sin \beta)}{2 \cos \beta + \beta \sin \beta - 2} \tag{3.4.12}$$

$$S_2 = \frac{(\sin \beta - \beta)}{2 \cos \beta + \beta \sin \beta - 2} \tag{3.4.13}$$

The flexibility relationships are

$$\theta_a = \frac{\ell}{EI}(f_1 M_a + f_2 M_b)$$
$$\theta_b = \frac{\ell}{EI}(f_1 M_b + f_2 M_a)$$

Inverting the stiffness relationship of Eq. (4.8.2) gives

$$f_1 = \frac{\sin \beta - \beta \cos \beta}{\beta^2 \sin \beta} \tag{4.8.3}$$

$$f_2 = \frac{\sin \beta - \beta}{\beta^2 \sin \beta} \tag{4.8.4}$$

The beam-column AB is elastically restrained. If the elastic restraints are u and v, then

$$\theta_a = -\frac{M_a}{u} \quad \text{and} \quad \theta_b = -\frac{M_b}{v} \tag{4.8.5}$$

The negative sign is required as the restraint moments are opposite to the positive direction of M_a and M_b.
Substituting Eqs. (4.8.3), (4.8.4), and (4.8.5) into the flexibility relationship yields

$$\begin{aligned} 0 &= \frac{M_a \ell_c}{EI_c}\left(\frac{\sin \beta - \beta \cos \beta}{\beta^2 \sin \beta} + \frac{EI_c}{u\ell_c}\right) + \frac{M_b \ell_c}{EI_c}\left(\frac{\sin \beta - \beta}{\beta^2 \sin \beta}\right) \\ 0 &= \frac{M_a \ell_c}{EI_c}\left(\frac{\sin \beta - \beta}{\beta^2 \sin \beta}\right) + \frac{M_b \ell_c}{EI_c}\left(\frac{\sin \beta - \beta \cos \beta}{\beta^2 \sin \beta} + \frac{EI_c}{v\ell_c}\right) \end{aligned} \tag{4.8.6}$$

The stability condition equation (or for nontrivial solution) requires that the coefficient determinant must be equal to zero.

$$\begin{aligned}&\frac{1}{uv}\left(\frac{EI_c}{\ell_c}\right)^2+\left(\frac{EI_c}{\ell_c}\right)\left(\frac{1}{u}+\frac{1}{v}\right)\frac{\sin\beta-\beta\cos\beta}{\beta^2\sin\beta}\\&+\left(\frac{\sin\beta-\beta\cos\beta}{\beta^2\sin\beta}\right)^2-\left(\frac{\sin\beta-\beta}{\beta^2\sin\beta}\right)^2=0\end{aligned}\tag{4.8.7}$$

which can be further simplified to

$$\frac{\beta^2}{uv}\left(\frac{EI_c}{\ell_c}\right)^2+\left(\frac{1}{u}+\frac{1}{v}\right)\left(\frac{EI_c}{\ell_c}\right)\left(1-\frac{\beta}{\tan\beta}\right)+\frac{2}{\beta}\tan\frac{\beta}{2}=1\tag{4.8.8}$$

The elastic restraint factors u and v must be determined. Consider the girders in Fig. 4-16. By virtue of assumptions 6 and 7, M_{ac} and M_{ad} are determined as follows:

$$M_{ac}=\theta_a\left(\frac{4EI_{blt}}{\ell_{bl}}\right)-\theta_a\left(\frac{2EI_{blt}}{\ell_{bl}}\right)=\theta_a\left(\frac{2EI_{blt}}{\ell_{bl}}\right)\tag{4.8.9}$$

$$M_{ad}=\theta_a\left(\frac{4EI_{brt}}{\ell_{br}}\right)-\theta_a\left(\frac{2EI_{brt}}{\ell_{br}}\right)=\theta_a\left(\frac{2EI_{brt}}{\ell_{rl}}\right)\tag{4.8.10}$$

or the sum of the reactive moments M_{abeam} developed due to beam stiffness is

$$M_{abeam}=\sum\frac{2EI_{abeam}}{\ell_{abeam}}\tag{4.8.11}$$

The moments at A on the column are

$$M_{ab}=\theta_a\left(\frac{EI_c}{\ell_c}\right)S_1-\theta_a\left(\frac{EI_c}{\ell_c}\right)S_2\tag{4.8.12}$$

$$M_{at}=\theta_a\left(\frac{EI_{ct}}{\ell_{ct}}\right)S_1-\theta_a\left(\frac{EI_{ct}}{\ell_{ct}}\right)S_2\tag{4.8.13}$$

The sum of the moments at a on the column AB is

$$M_{acol}=\sum\frac{EI_{acol}}{\ell_{acol}}(S_1-S_2)\theta_a\tag{4.8.14}$$

where $(S_1 - S_2)$ is assumed the same for all columns framing at joint A.

Solving for θ_a from Eq. (4.8.14) and substituting into Eq. (4.8.12) yields

$$M_{ab} = \left(\frac{EI_c}{\ell_c}\right)\frac{(S_1 - S_2)}{(S_1 - S_2)}\frac{M_{acol}}{\sum \frac{EI_{acol}}{\ell_{acol}}} \tag{4.8.15}$$

The term $(S_1 - S_2)$ cancels since it is assumed identical for all columns framing at joint A. As it is assumed that no external joint moment is acting at joint a, $M_{acol} = -M_{abeam}$. Substituting the negative of Eq. (4.8.11) for M_{acol} in Eq. (4.8.15) gives

$$M_{ab} = -\frac{2EI_c}{\ell_c}\frac{\sum \frac{EI_{abeam}}{\ell_{abeam}}}{\sum \frac{EI_{acol}}{\ell_{acol}}}\theta_a \tag{4.8.16}$$

Juxtaposing Eqs. (4.8.5) and Eq. (4.8.16) yields

$$u = \frac{2EI_c}{\ell_c}\frac{\sum \frac{EI_{abeam}}{\ell_{abeam}}}{\sum \frac{EI_{acol}}{\ell_{acol}}} \text{ (for joint } A) \tag{4.8.17}$$

and likewise

$$v = \frac{2EI_c}{\ell_c}\frac{\sum \frac{EI_{bbeam}}{\ell_{bbeam}}}{\sum \frac{EI_{bcol}}{\ell_{bcol}}} \text{ (for joint } B) \tag{4.8.18}$$

Defining, as in the AISC (2005) Commentary C2,

$$G_A(\text{or } G_{top}) = \frac{\sum \frac{EI_{abeam}}{\ell_{abeam}}}{\sum \frac{EI_{acol}}{\ell_{acol}}} \quad \text{and} \quad G_B(\text{or } G_{bottom}) = \frac{\sum \frac{EI_{bbeam}}{\ell_{bbeam}}}{\sum \frac{EI_{bcol}}{\ell_{bcol}}} \tag{4.8.19}$$

Hence, the elastic restraint factors become

$$u = \frac{2EI_c}{\ell_c}\left(\frac{1}{G_A}\right) \quad \text{and} \quad v = \frac{2EI_c}{\ell_c}\left(\frac{1}{G_B}\right) \tag{4.8.20}$$

It is noted that the stability parameter $\beta = k\ell$ in the stability functions is the critical load factor of a column in the frame having a length of ℓ. Comparing with the isolated pinned column,

$$\frac{\beta^2 EI}{\ell^2} = \frac{\pi^2 EI}{(K\ell)^2} \tag{4.8.21}$$

it may be realized that the effective length factor K may be expressed as

$$K = \frac{\pi}{\beta} \quad \text{or} \quad \beta = \frac{\pi}{K} \tag{4.8.22}$$

Substituting Eqs. (4.8.20) into Eq. (4.8.8) and replacing β with π/K gives

$$\frac{\pi^2 G_A G_B}{4K^2} + \left(\frac{G_A + G_B}{2}\right)\left[1 - \frac{\pi/K}{\tan(\pi/K)}\right] + \frac{2K}{\pi}\tan\left(\frac{\pi}{2K}\right) = 1 \tag{4.8.23}$$

where $K = \pi/\sqrt{(P\ell_c^2)/(EI_c)}$ is defined as the effective column length factor corresponding to $P_{cr} = \pi^2 EI_c/(K\ell_c)^2$.

Equation (4.7.23) is used to plot the nomograph shown in Fig. 4-17. AISC (2005) Commentary C2 recommends that for columns not rigidly connected to footing or foundation, G may be taken as 10, and for columns rigidly connected to properly designed footing, G may be taken as 1. When the far end of one of the girders framing into the column joint is fixed or pinned, adjustments on G may be necessary.

For girder far ends fixed, $\theta_b = 0$, and Eq. (4.8.11) becomes

$$M_{abeam} = \sum \frac{4EI_{abeam}}{\ell_{abeam}} \tag{4.8.24}$$

For girder far ends hinged, $\theta_b = -\theta_a / 2$, and Eq. (4.8.11) becomes

$$M_{abeam} = \sum \frac{3EI_{abeam}}{\ell_{abeam}} \tag{4.8.25}$$

Hence, it may be reflected in the evaluation of

$$\sum \frac{I_{girder}}{\ell_{girder}}$$

Consider next the case of an unbraced column AB shown in Fig. 4-18. The assumptions for the unbraced frame are the same as for the braced frame, except for assumption No. 6. For the unbraced frame, the girder (or beam) is assumed to be in reverse curvature, with the rotation at both ends equal in magnitude and direction. The definitions of elastic restraints u and v are the same as for the braced frame.

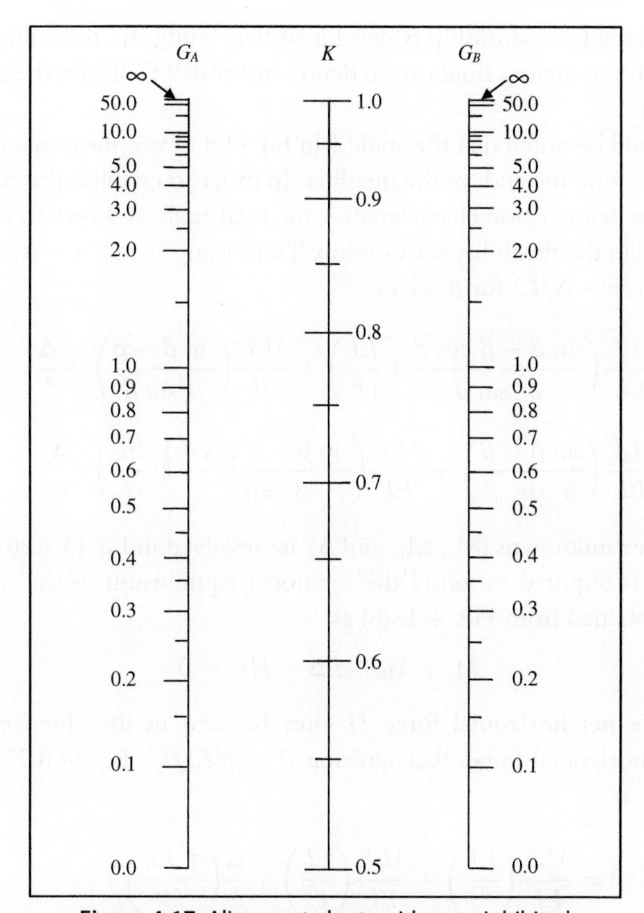

Figure 4-17 Alignment chart—sidesway inhibited

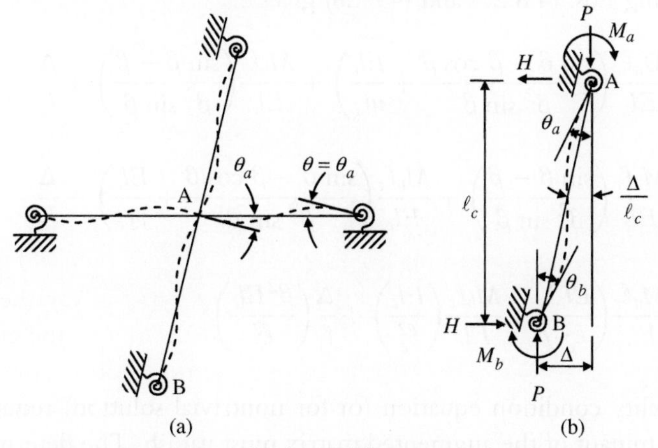

Figure 4-18 Portion of unbraced frame with elastic restraints

The stability relationship is developed here using the flexibility coefficients as they contain a single-term denominator that facilitates the algebraic operation.

It should be noted that the angle θ in Eq. (4.8.6) was measured from the axis connecting the ends of the member. In order to establish the consistent rigid-joint deformation characteristics, the total angle as shown in Fig. 4-18 (b) is used in the flexibility relationship. Thus, using $(-M_a/u - \Delta/\ell_c)$ for θ_a and $(-M_b/v - \Delta/\ell_c)$ for θ_b gives

$$0 = \frac{M_a \ell_c}{EI_c}\left(\frac{\sin\beta - \beta\cos\beta}{\beta^2\sin\beta} + \frac{EI_c}{u\ell_c}\right) + \frac{M_b \ell_c}{EI_c}\left(\frac{\sin\beta - \beta}{\beta^2\sin\beta}\right) + \frac{\Delta}{\ell_c}$$
$$0 = \frac{M_a \ell_c}{EI_c}\left(\frac{\sin\beta - \beta}{\beta^2\sin\beta}\right) + \frac{M_b \ell_c}{EI_c}\left(\frac{\sin\beta - \beta\cos\beta}{\beta^2\sin\beta} + \frac{EI_c}{v\ell_c}\right) + \frac{\Delta}{\ell_c} \tag{4.8.26}$$

Since three unknowns (M_a, M_b, and Δ) are involved in Eq. (4.7.26), a third equation is required to satisfy the rotational equilibrium of the structure. This is obtained from Fig. 4-18(b) as

$$M_a + M_b + P\Delta - H\ell_c = 0 \tag{4.8.27}$$

where the net horizontal force H must be zero in the absence of any external horizontal force. Recognizing $P = \beta^2 EI/\ell_c^2$, Eq. (4.8.27) can be rewritten as

$$0 = \frac{M_a \ell_c}{EI_c}\left(\frac{EI_c}{\ell_c^2}\right) + \frac{M_b \ell_c}{EI_c}\left(\frac{EI_c}{\ell_c^2}\right) + \frac{\Delta}{\ell_c}\left(\frac{\beta^2 EI_c}{\ell_c^2}\right) \tag{4.8.28}$$

Combining Eqs. (4.8.27) and (4.8.28) gives

$$0 = \frac{M_a \ell_c}{EI_c}\left(\frac{\sin\beta - \beta\cos\beta}{\beta^2\sin\beta} + \frac{EI_c}{u\ell_c}\right) + \frac{M_b \ell_c}{EI_c}\left(\frac{\sin\beta - \beta}{\beta^2\sin\beta}\right) + \frac{\Delta}{\ell_c}$$
$$0 = \frac{M_a \ell_c}{EI_c}\left(\frac{\sin\beta - \beta}{\beta^2\sin\beta}\right) + \frac{M_b \ell_c}{EI_c}\left(\frac{\sin\beta - \beta\cos\beta}{\beta^2\sin\beta} + \frac{EI_c}{v\ell_c}\right) + \frac{\Delta}{\ell_c} \tag{4.8.29}$$
$$0 = \frac{M_a \ell_c}{EI_c}\left(\frac{EI_c}{\ell_c^2}\right) + \frac{M_b \ell_c}{EI_c}\left(\frac{EI_c}{\ell_c^2}\right) + \frac{\Delta}{\ell_c}\left(\frac{\beta^2 EI_c}{\ell_c^2}\right)$$

The stability condition equation (or for nontrivial solution) requires that the determinant of the augmented matrix must vanish. The determinant is

$$\frac{u\ell_c\beta\cos\beta EI_c + v\ell_c\beta\cos\beta EI_c - \beta^2\sin\beta(EI_c)^2 + uv\ell_c^2\sin\beta}{uv\ell_c^2\sin\beta} = 0 \quad (4.8.30)$$

Combining first and second terms and third and fourth terms, respectively, and multiplying by tan β gives the stability equation as

$$\left[\frac{\beta^2}{uv}\left(\frac{EI_c}{\ell_c}\right)^2 - 1\right]\tan\beta - \left(\frac{1}{u}+\frac{1}{v}\right)\left(\frac{EI_c}{\ell_c}\right)\beta = 0 \quad (4.8.31)$$

It should be recalled that the girders framing into joint A are assumed to deform, making a reverse curvature as shown in Fig. 4-18(a). Hence, the moment of one girder at joint A is

$$M_a = \theta_a\left(\frac{4EI_b}{\ell_b}\right) + \theta_a\left(\frac{2EI_b}{\ell_b}\right) = \theta_a\left(\frac{6EI_b}{\ell_b}\right) \quad (4.8.32)$$

or the sum of restraining moments developed at joint A due to beam stiffness is

$$M_{abeam} = \sum\frac{6EI_{abeam}}{\ell_{abeam}}\theta_a \quad (4.8.33)$$

For the column in the unbraced frame, the assumptions behind Eq. (4.8.15) are still valid; and in the absence of any external moment, $M_{acol} = -M_{abeam}$. Substituting Eq. (4.8.33) into Eq. (4.8.15) yields

$$M_{acol} = -\frac{6EI_c}{\ell_c}\frac{\sum\frac{EI_{abeam}}{\ell_{abeam}}}{\sum\frac{EI_{acol}}{\ell_{acol}}} \quad (4.8.34)$$

Juxtaposing Eqs. (4.8.5) and Eq. (4.8.34) yields

$$u = \frac{6EI_c}{\ell_c}\left(\frac{1}{G_A}\right) \quad \text{and} \quad v = \frac{6EI_c}{\ell_c}\left(\frac{1}{G_B}\right) \quad (4.8.35)$$

Substituting Eq. (4.8.35) into Eq. (4.8.31), and realizing $\beta = \pi/K$ gives

$$\frac{G_A G_B(\pi/K)^2 - 36}{6(G_A + G_B)} = \frac{\pi/K}{\tan(\pi/K)} \quad (4.8.36)$$

Equation (4.8.36) is used to plot the nomograph shown in Fig. 4-19. Adjustments for inelasticity according to AISC Commentary C2 may be necessary. Although column design using the K-factors can be tedious and

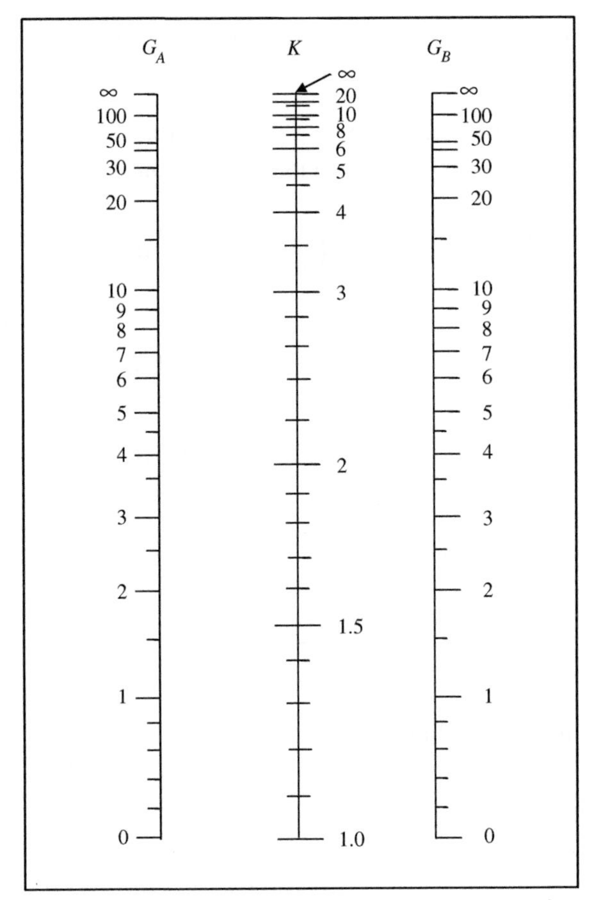

Figure 4-19 Alignment chart—sidesway permitted

confusing for complex building structures containing many leaning columns, particularly where column inelasticity is considered, the Jackson and Moreland nomographs are shown to give results very close to the theoretical values.

REFERENCES

ABAQUS. (2006). *ABAQUS Analysis User's Manual*. Pawtucket, RI: ABAQUS.

AISC. (2005). *Specification for Structural Steel Buildings* (13th ed.). Chicago, IL: American Institute of Steel Construction.

Alvarez, R. J., & Birnstiel, C. (1969). Inelastic Analysis of Multistory Multibay Frames. *Journal of the Structural Div., ASCE, Vol. 95*(No. ST11), 2477–2503.

ASCE Task Committee on Effective Length. (1997). *Effective Length and Notional Load Approaches for Assessing Frame Stability: Implications for American Steel Design*. New York: American Society of Civil Engineers.

Bleich, F. (1952). *Buckling Strength of Metal Structures*. New York: McGraw-Hill.

CSA. (1994). Limit State Design of Steel Structures, CAN/CSA-S16.1-M94, Canadian Standards Association, Rexdale, Ontario.

Galambos, T. V. (1968). *Structural Members and Frames*. Englewood Cliffs, NJ: Prentice-Hall.

Horne, M. R., & Merchant, W. (1965). *The Stability of Frames* (1st ed.). London: Pergamon Press.

Julian, O. G., & Lawrence, L. S. (1959). Notes on J and L Nomograms for Determination of Effective Lengths, unpublished.

Kavanagh, T. C. (1960). Effective Length of Framed Columns. *Journal of the Structural Div., ASCE, Vol. 86* (No. ST2), 1–21. (Also *in Transactions, ASCE*, 127(1962), Part II, 81–101.)

Lu, L. W. (1963). Stability of Frames Under Primary Bending Moment. *Journal of the Structural Div., ASCE, Vol. 89*(No. ST3), 35–62.

Masur, E. F., Chang, I. C., & Dennell, L. H. (1961). Stability of frames in the Presence of Primary Bending Moments. *Journal of the Engineering Mechanics Div, ASCE, Vol. 87*(No. EM4), 19–34.

Ojalvo, M., & Lu, L. W. (1961). Analysis of Frames Loaded into the Plastic Range. *Journal of the Engineering Mechanics Div., ASCE, Vol. 87*(No. EM4), 35–48.

Przemieniecki, J. S. (1968). *Theory of Matrix Structural Analysis*. New York: McGraw-Hill.

Yura, J. A. (1971). The Effective Length of Columns in Unbraced Frames. *Engineering Journal, AISC, Vol. 8*(No. 2), 37–42, (2nd Quarter).

PROBLEMS

4.1 Determine P_{cr} of the structure shown in Figure P4-1 for the given parameters:

$$\ell_2 = 1.5\ell_1$$
$$\ell_1 = 0.4\ell$$
$$\ell_2 = 0.6\ell$$
$$\ell = 156\ in$$

$$I_2 = I_1 = I = 109.7\ in^4$$

$$E = 30,000\ \text{ksi}$$

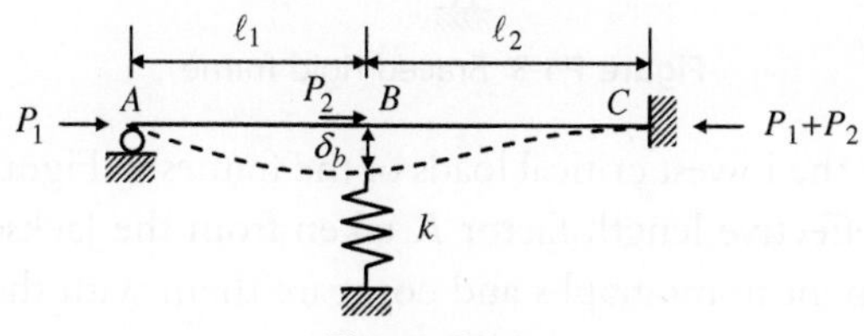

Figure P4-1

$$\text{Case } 1,\ k = 0,\ P_2 = 0,\ P = P_1 + P_2$$

$$\text{Case } 2, k = 0,\ P_1 = P_2,\ P = P_1 + P_2$$

$$\text{Case } 3, k = 1\ k/in,\ P_1 = P_2,\ P = P_1 + P_2$$

Compare the solutions by the Energy method and the slope-deflection equations and provide comments.

4.2 Using any method, including computer programs, determine the lowest three critical loads of the frame shown in Figure P4-2.

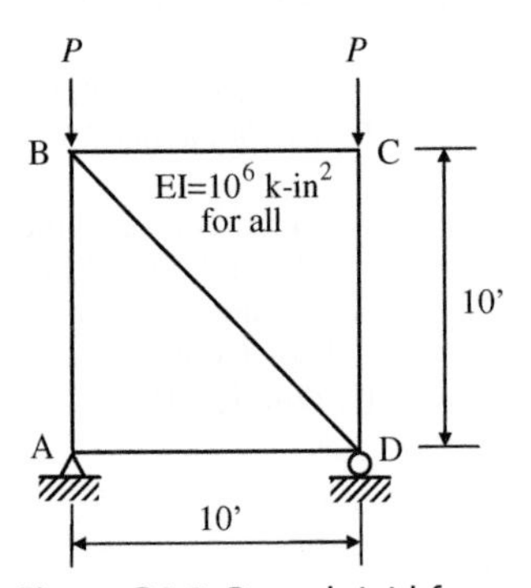

Figure P4-2 Braced rigid frame

4.3 Using any method, determine P_{cr} of the frame in Fig. P4-3 in terms of EI. E is constant for all members.

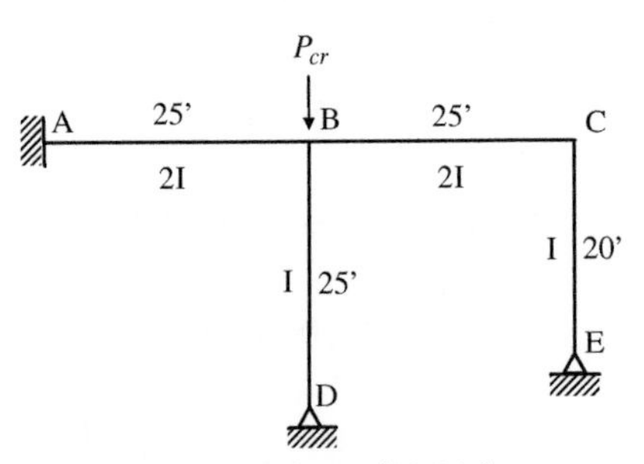

Figure P4-3 Braced rigid frame

4.4 Determine the lowest critical loads of the frames in Figures 4-6 and 4-7 using the effective length factor K taken from the Jackson and Moreland alignment nomographs and compare them with those theoretical values.

4-5 Using the matrix method, determine the critical load of the frame in Figure P4-5. Let each member consist of a single element.

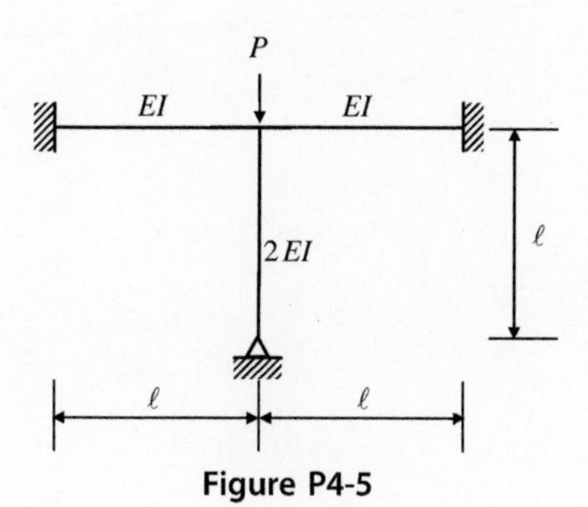

Figure P4-5

4-5 Using the matrix method, determine the critical load of the frame in Figure P4-5. Let each member consist of a single element.

Figure P4-5

Torsion in Structures

Contents

Stability of Structures
ISBN 978-0-12-385122-2, doi:10.1016/B978-0-12-385122-2.10005-3

5.1. INTRODUCTION

Torsion in structures is perhaps one of the least-well-understood subjects in structural mechanics. Purely torsional loading rarely occurs in structures except in the power-transmitting shafts of automobiles or generators. Frequently, torsion develops in structures along with bending from unintended eccentricities of transverse loading due to the limitation of workmanship or from unavoidable eccentricities as can be found in spandrel beams.

Generally, thin-walled sections do not behave according to the law of the plane sections employed by Euler-Bernoulli-Navier. A thin-walled section is referred to as a rolled shape in which the thickness of an element is less than one-tenth of the width. Many stocky rolled shapes do not meet this definition; however, the general theory of thin-walled section developed by Vlasov (1940, 1961) in the 1930s appears to be applicable without significant consequences.

A thin-walled section becomes "warped" when it is subjected to end couples (torsional moment). Hence, the cross section does not remain plane after deformation. Exceptions to this rule are tubular sections and thin-walled open sections in which all elements meet at a point, such as the cruciform, angle, and tee section.

Another distinct feature of the response of structural members to torsion is that the externally applied twisting moment is resisted internally by some combination of uniform (or pure, or St. Venant) torsion and nonuniform (or warping) torsion depending on the boundary conditions, that is, whether a member is free to warp or whether warping is restrained.

Thin-walled open sections are very weak against torsion and are susceptible to lateral-torsional buckling (or flexural-torsional buckling), which is affected by the torsional strength of the member, even though no intentional torsional loading is applied.

If warping does not occur or if warping is not restrained, the applied twisting moment is entirely carried by uniform torsion. When a member is free to warp, no internal normal stresses develop despite the warping deformation. This is tantamount to the fact that a heated rod will not develop any internal stresses if it is free to expand at one or both ends, despite the temperature-induced elongation of the rod.

If warping is restrained, the member develops additional shearing stresses, as well as normal stresses. Frequently, warping stresses are fairly high in magnitude, and they are not to be ignored.

5.2. UNIFORM TORSION AND ST. VENANT THEORY

The internal resisting torque due to shear stresses shown in Fig. 5-1 is computed by Eq. (5.2.1). The external twisting moment follows the right-hand screw rule, which is directing counterclockwise when observed from the positive end of the z-axis.

$$M_z = T = \int_A (-\tau_{zx} \cdot y + \tau_{zy} \cdot x) dA \tag{5.2.1}$$

From the free body of the infinitesimal element in Fig. 5-1, equilibrium equations can be established as:

$$\sum F_z = -\tau_{xz} \cdot dy \cdot dz + \left(\tau_{xz} + \frac{\partial \tau_{xz}}{\partial x} dx \right) dy \cdot dz - \tau_{yz} \cdot dx \cdot dz$$
$$+ \left(\tau_{yz} + \frac{\partial \tau_{yz}}{\partial y} dy \right) dx \cdot dz = 0$$

From which, it follows

$$\frac{\partial \tau_{xz}}{\partial x} + \frac{\partial \tau_{yz}}{\partial y} = 0 \tag{5.2.2}$$

Similarly,

$$\frac{\partial \tau_{zx}}{\partial x} + \frac{\partial \tau_{zy}}{\partial y} = 0 \; (\tau_{zx} = \tau_{xz} \text{ etc.}) \tag{5.2.3}$$

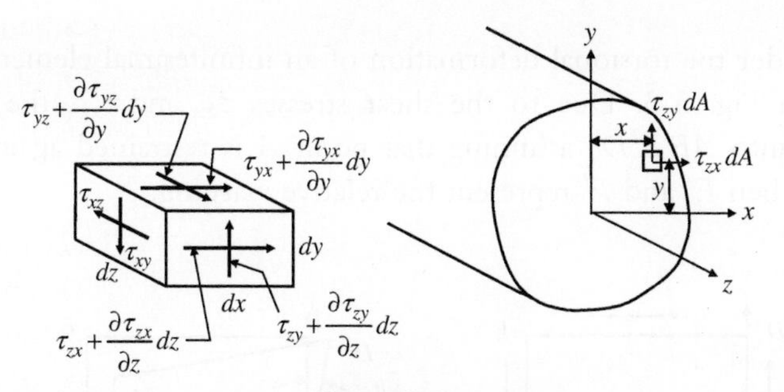

Figure 5-1

5.2.1. Geometry

Point A in Fig. 5-2 moves to point B under torsion such that OA and OB are the same. However, under the assumption of small displacement theory

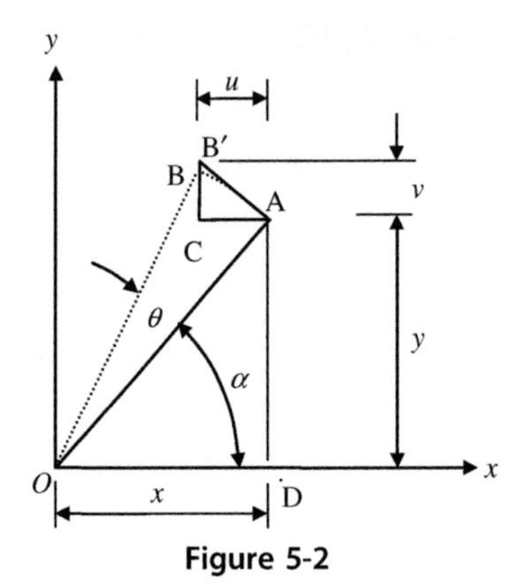

Figure 5-2

B and *B'* are considered the same where *AB'* is perpendicular to *OA*. The displacement components of point *A* along the *x*- and *y*-axes are represented by *u* and *v*, respectively.

From the similar triangle relation between ΔAOD and ΔABC

$$\frac{OA}{OD} = \frac{OA}{x} = \frac{AB'}{B'C} = \frac{AB'}{v} \Rightarrow v = \frac{AB'}{OA} x = \tan\theta x \simeq x \cdot \theta$$

Hence,

$$v = x\theta \tag{5.2.4}$$

Similarly,

$$u = -y\theta \tag{5.2.5}$$

Consider the torsional deformation of an infinitesimal element *AFED* shown in Fig. 5-3. Due to the shear stresses τ_{zx} and τ_{xz}, the element deforms into $AF'E'D'$, assuming that point *A* is restrained against translations. Then E' and F' represent the relative warping.

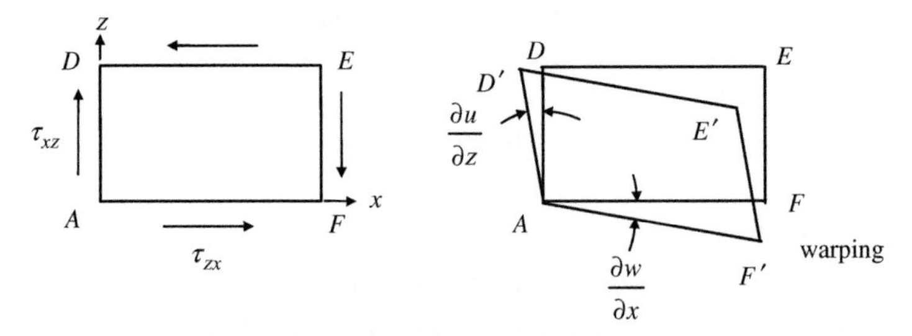

Figure 5-3

$$\gamma_{xz} = \gamma_{zx} = \angle\, FAF' + \angle\, DAD' = \frac{\partial w}{\partial x} + \frac{\partial u}{\partial z} \tag{5.2.6}$$

Similarly,

$$\gamma_{yz} = \gamma_{zy} = \frac{\partial v}{\partial z} + \frac{\partial w}{\partial y} \tag{5.2.7}$$

Differentiating Eq. (5.2.5) with respect to z gives

$$\frac{\partial u}{\partial z} = -y\frac{\partial \theta}{\partial z}$$

If the angular change is linear with respect to the member length, then

$$\frac{\partial u}{\partial z} = -y\frac{\theta}{\ell}$$

Substituting this into Eq. (5.2.6) yields

$$\gamma_{xz} = \gamma_{zx} = -y\frac{\theta}{\ell} + \frac{\partial w}{\partial x} \tag{5.2.8}$$

Equation (5.2.8) is the angular displacement per unit length.
For an elastic material, one has

$$\gamma_{xz} = \gamma_{zx} = \frac{\tau_{xz}}{G} = -\,y\frac{\theta}{\ell} + \frac{\partial w}{\partial x} \tag{5.2.9}$$

Likewise,

$$\gamma_{yz} = \gamma_{zy} = \frac{\tau_{yz}}{G} = \frac{x\theta}{\ell} + \frac{\partial w}{\partial y} \tag{5.2.10}$$

Differentiating Eq. (5.2.9) with respect to y gives

$$\frac{\partial \tau_{xz}}{\partial y G} = \frac{\partial \tau_{zx}}{\partial y G} = -\,\frac{\theta}{\ell} + \frac{\partial^2 w}{\partial x \partial y} \tag{5.2.11}$$

Differentiating Eq. (5.2.10) with respect to x yields

$$\frac{\partial \tau_{yz}}{\partial x G} = \frac{\theta}{\ell} + \frac{\partial^2 w}{\partial x \partial y} \tag{5.2.12}$$

From Eqs. (5.2.11) and (5.2.12), it follows immediately

$$\frac{\partial \tau_{xz}}{\partial y} - \frac{\partial \tau_{yz}}{\partial x} = -\,\frac{2G\theta}{\ell} \tag{5.2.13}$$

Taking partial derivatives of Eqs. (5.2.13) and (5.2.3) and adding yields

$$\frac{\partial^2 \tau_{xz}}{\partial x \partial y} - \frac{\partial^2 \tau_{yz}}{\partial x^2} = 0 \;\ldots\ldots(a) \quad \text{and} \quad \frac{\partial^2 \tau_{xz}}{\partial y^2} - \frac{\partial^2 \tau_{yz}}{\partial x \partial y} = 0\ldots\ldots(b)$$

$$\frac{\partial^2 \tau_{xz}}{\partial x^2} + \frac{\partial^2 \tau_{yz}}{\partial x \partial y} = 0 \;\ldots\ldots(c) \quad \text{and} \quad \frac{\partial^2 \tau_{xz}}{\partial x \partial y} + \frac{\partial^2 \tau_{yz}}{\partial y^2} = 0\ldots\ldots(d)$$

$$\left.\begin{aligned} (b)+(c) \quad & \frac{\partial^2 \tau_{xz}}{\partial x^2} + \frac{\partial^2 \tau_{xz}}{\partial y^2} = 0 \\ (d)-(a) \quad & \frac{\partial^2 \tau_{yz}}{\partial x^2} + \frac{\partial^2 \tau_{yz}}{\partial y^2} = 0 \end{aligned}\right\} \tag{5.2.14}$$

5.2.2. Stress Function

The analysis of uniform torsion is greatly simplified by the fortuitous fact that certain relationships exist between the torsion problem and the deformations of a soap film stretched across an opening equal in size and shape to the cross section for which torsional behavior is sought. The membrane analogy introduced by Prandtl (1903) is applicable not only to solid sections but also to open and closed thin-walled cross sections. Also, the membrane analogy can be extended to inelastic and fully plastic ranges if the concept of the soap-film is replaced by constant-slope surfaces. This was indicated by Prandtl according to Nadai (1923) who coined the term *sand-heap analogy*. Nadai (1950) also carried out many interesting experiments illustrating the sand-heap analogy to plastic torsion.
Let

$$\left.\begin{aligned} \tau_{xz} = \tau_{zx} &= \frac{\partial \phi}{\partial y} \\ \tau_{yz} = \tau_{zy} &= -\frac{\partial \phi}{\partial x} \end{aligned}\right\} \tag{5.2.15}$$

It is noted that $\phi = f(x, y)$ in Eq. (5.2.15) is a stress function introduced by Prandtl (1903).

Substituting Eq. (5.2.15) into Eq. (5.2.13) gives

$$\frac{\partial^2 \phi}{\partial x^2} + \frac{\partial^2 \phi}{\partial y^2} = -\frac{2G\theta}{\ell} \tag{5.2.16}$$

It can be readily shown that the stress function ϕ is constant along the boundary of the cross section of the twisted bar (Timoshenko and Goodier 1951) by considering the stress-free state. Further, as the constant can be chosen arbitrarily without affecting the stress, it is expedient to take it equal to zero. Hence, it follows that

$$\int dx \int \frac{\partial \phi}{\partial y}\, dy = \int dy \int \frac{\partial \phi}{\partial x}\, dx = 0$$

Substituting Eq. (5.2.15) into Eq. (5.2.1), one obtains

$$M_z = T = \int_A (-\tau_{zx} y + \tau_{zy} x)\, dA$$

$$= -\iint \left(\frac{\partial \phi}{\partial y} y + \frac{\partial \phi}{\partial x} x \right) dxdy = -\iint \frac{\partial \phi}{\partial y} ydxdy - \iint \frac{\partial \phi}{\partial x} xdxdy$$

Integrating by parts the above equation and observing ϕ is equal to zero along the cross-sectional boundary, one obtains

$$-\iint y \frac{\partial \phi}{\partial y}\, dxdy = -\int dx \int y \frac{\partial \phi}{\partial y}\, dy = -\int dx \left(\phi y - \int \phi dy \right)$$
$$= -\int \phi y dx + \iint \phi dxdy = \iint \phi dxdy$$

$$-\iint x \frac{\partial \phi}{\partial x}\, dxdy = -\int dy \int x \frac{\partial \phi}{\partial x}\, dx = -\int dy \left(\phi x - \int \phi dx \right)$$
$$= -\int \phi x dy + \iint \phi dxdy = \iint \phi dxdy$$

Hence,

$$M_z = T = 2 \iint \phi dxdy \tag{5.2.17}$$

Equation (5.2.17) indicates that one-half of the torque is due to the stress component τ_{zx} and the other half to τ_{zy} and the torque is equal to twice the volume under the stress function ϕ.

5.3. MEMBRANE ANALOGY

In the solution of the torsional problems, the membrane analogy introduced by Prandtl (1903) proved to be very useful. Consider a homogeneous

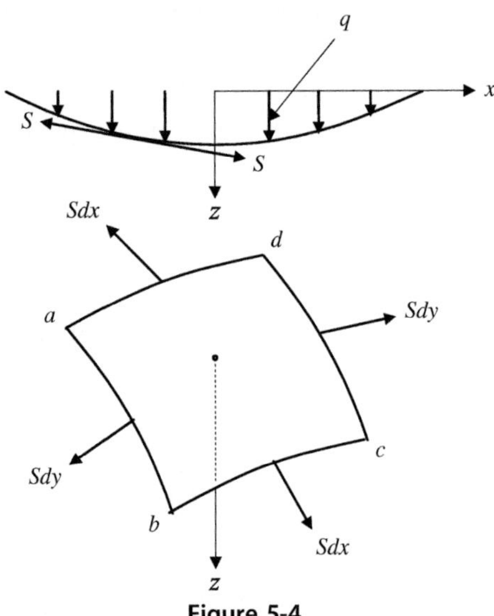

Figure 5-4

membrane in Fig. 5-4 supported at the edges with the same outline as that of the cross section of the twisted member, subjected to a uniform membrane stretching at the edges and a uniform pressure. Let q be the lateral pressure per unit area and S be the uniform tension per unit width of the membrane shown in Fig. 5-4. The vertical component of the tensile force acting on the side ab is $-Sdy(\partial z/\partial x)$. Likewise, the one on the side cd is $Sdy[\partial z/\partial x + (\partial^2 z/\partial x^2)dx]$. In a similar manner the vertical components of the tensile forces acting on the sides ad and bc can be determined (Rees 2000) as $-Sdx(\partial z/\partial y)$ and $Sdx[\partial z/\partial y + (\partial^2 z/\partial y^2)dy]$, respectively.

The equation of equilibrium of the element is

$$q \cdot dx \cdot dy - Sdy(\partial z/\partial x) + Sdy[\partial z/\partial x + (\partial^2 z/\partial x^2)dx]$$

$$- Sdx(\partial z/\partial y) + Sdx[\partial z/\partial y + (\partial^2 z/\partial y^2)dy] = 0$$

From which

$$\frac{\partial^2 z}{\partial x^2} + \frac{\partial^2 z}{\partial y^2} = -\frac{q}{S} \tag{5.3.1}$$

Comparing Eqs. (5.2.16) and (5.3.1) reveals that there is a remarkable similarity between the shape of the membrane and the stress distribution in torsion. Analogies between membrane and torsion are summarized in Table 5-1.

Table 5-1 Analogies

Membrane *Deflection z*	**Torsion** *Stress function ϕ*
$\frac{\partial^2 z}{\partial x^2} + \frac{\partial^2 z}{\partial y^2} = -\frac{q}{S}$	$\frac{\partial^2 \phi}{\partial x^2} + \frac{\partial^2 \phi}{\partial y^2} = -\frac{2G\theta}{\ell} = -2G\theta'$
Slopes $\frac{\partial z}{\partial x}, \frac{\partial z}{\partial y}$	Stresses τ_{xz}, τ_{yz}
volume $V = \iint z dx dy$	twisting moment $T = 2\iint \phi dx dy$

Note: θ' = rotation/unit length

5.4. TWISTING OF THIN RECTANGULAR BARS

As the shear stresses due to the uniform torsion of thin-walled open sections vary linearly through the thinner dimension, the shape of the membrane shown in Fig. 5-5 must be a parabola symmetric with respect to the z-axis.

Let the equation of the parabola be $z = Ay^2$. Since $z = z_0$ at $y = t/2$

$$A = \frac{4z_0}{t^2}$$

Hence,

$$z = \frac{4z_0}{t^2} y^2 \tag{a}$$

and the shear stress is

$$\frac{dz}{dy} = \frac{8}{t^2} z_0 y$$

$$\tau_{max} = \frac{4z_0}{t} \tag{b}$$

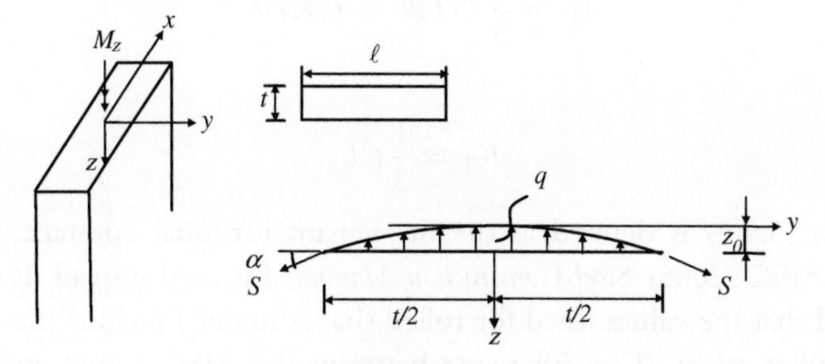

Figure 5-5

Neglecting the corner effect, the equilibrium of the forces in Fig. 5-5 in the vertical direction gives

$$qt\ell - 2S\sin\alpha = 0$$

Since $\sin\alpha \doteq \alpha \doteq \tan\alpha$ for a small angle

$$\alpha = \left.\frac{\partial z}{\partial y}\right|_{z=\frac{t}{2}} = \frac{4z_0}{t} \tag{c}$$

Hence,

$$qt\ell - 2S\ell\frac{4z_0}{t} = 0 \Rightarrow \frac{q}{S} = \frac{8z_0}{t^2} \tag{d}$$

The volume of the membrane is

$$\begin{aligned} V &= z_0 t\ell - \iint z\,dxdy = z_0 t\ell - \frac{4z_0}{t^2}\int_0^\ell dx \int_{-t/2}^{t/2} y^2 dy \\ &= z_0 t\ell - \frac{4z_0}{3t^2}(\ell)\left(y^3\right)\Big|_{-t/2}^{t/2} = \frac{2}{3}z_0 t\ell \end{aligned} \tag{e}$$

Then the torsional moment is

$$M_z = 2\,V = \frac{4}{3}t\,z_0\ell \tag{f}$$

$$z_0 = \frac{3M_z}{4t\ell}$$

Substituting into Eq. (d) gives

$$\frac{q}{S} = \frac{8z_0}{t^2} = \frac{8}{t^2}\frac{3M_z}{4t\ell} = \frac{6M_z}{t^3\ell} = 2G\theta'$$

From which

$$M_z = \frac{1}{3}t^3\ell G\theta' = GK_T\theta' \tag{5.4.1}$$

where

$$K_T = \frac{1}{3}t^3\ell \tag{5.4.2a}$$

Equation (5.4.2) is defined as the St. Venant torsional constant. In the current AISC (2005) *Steel Construction Manual*, J is used instead. It should be noted that the values listed for rolled shapes under J include the corner and/or fillet effect. The difference between the AISC values and those computed by simplified formula neglecting the corner effect is of no practical importance.

For an open cross section consisting of a series of rectangular elements, the St. Venant torsional constant is evaluated by

$$K_T = \frac{1}{3}\sum_{i=1}^{n} b_i t_i^3 \tag{5.4.2b}$$

where n is the number of elements, b is the length, and t is the thickness of each element, respectively. The thickness t is always smaller than the length b of each element.

The maximum shearing stress is given by equation (b) above as

$$\tau_{\max} = \frac{4z_0}{t}$$

Substituting the expression for z_0 gives

$$\tau_{\max} = \frac{4}{t}\frac{3}{4}\frac{M_z}{t\ell} = \frac{3M_z}{t^2\ell} = \frac{M_z t}{K_T} \tag{5.4.3}$$

$$\tau_{\max} = \frac{3}{t^2\ell}M_z = \frac{3}{t^2\ell}\frac{t^3\ell}{3}G\theta' = G\theta' t \tag{5.4.4}$$

5.5. TORSION IN THE INELASTIC RANGE

A solid circular shaft is considered here to illustrate the application of the membrane analogy for torsion in the elastic and inelastic range.

5.5.1. Elastic Torque

Based on the cylindrical coordinate system shown in Fig. 5-6, the equation for the membrane is given by

$$z = z_0\frac{r^2}{R^2} \tag{a}$$

Hence, the equation of the dome under the membrane is

$$z = z_0 - z_0\frac{r^2}{R^2} = z_0\left(1 - \frac{r^2}{R^2}\right) \tag{b}$$

$$dV = r d\theta dz dr$$

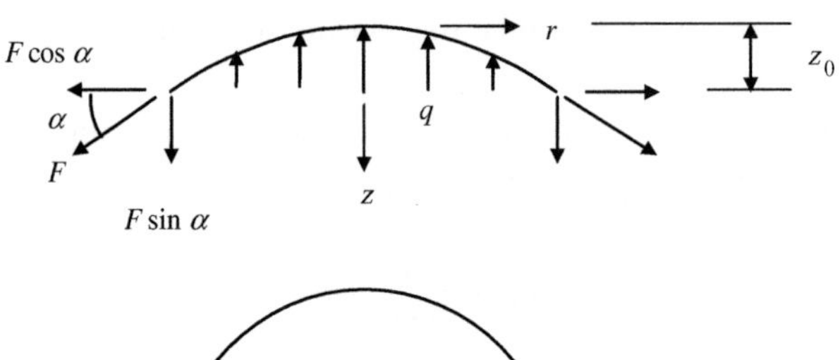

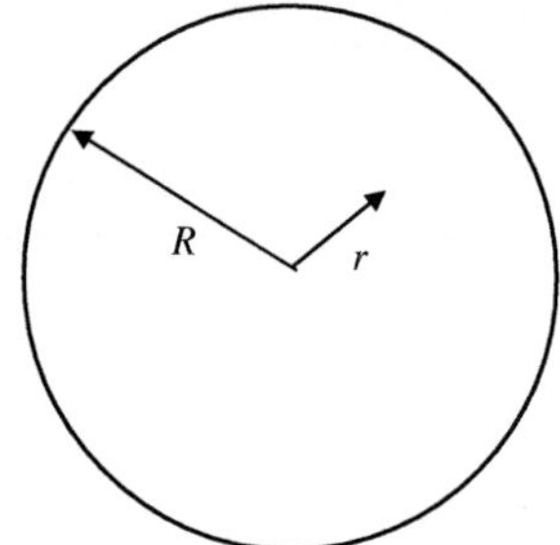

Figure 5-6

$$V = \int_0^{2\pi}\int_0^{R}\int_0^{z_0\left(1-\frac{r^2}{R^2}\right)} dz\,dr\,rd\theta = z_0\int_0^{2\pi} d\theta \int_0^{R}\left(r-\frac{r^3}{R^2}\right)dr$$
$$= \frac{z_0R^2}{4}\int_0^{2\pi} d\theta = \frac{z_0R^2}{4}2\pi = \frac{\pi R^2 z_0}{2}$$

$$M_{ze} = T_e = 2V = \pi R^2 z_0 \tag{c}$$

$$\frac{dz}{dr}\Big|_{r=R} = \tau_{\max} = \frac{2z_0}{R} = \tan\alpha \tag{d}$$

Equilibrium

$$\sum F_z = 0 \Rightarrow q\pi R^2 = F(2\pi R)\sin\alpha = 2\pi RF\frac{2z_0}{R} = 4\pi z_0 F$$

$$\frac{q}{F} = \frac{4z_0}{R^2}\left(\text{cf.}\,\frac{q}{F} = 2G\theta' = \frac{4z_0}{R^2}\right)$$

$$z_0 = \frac{G\theta' R^2}{2} \tag{e}$$

Substituting Eq. (e) into Eq. (c), one gets

$$M_{ze} = \pi R^2\frac{G\theta' R^2}{2} = \frac{\pi R^4 G\theta'}{2} \tag{f}$$

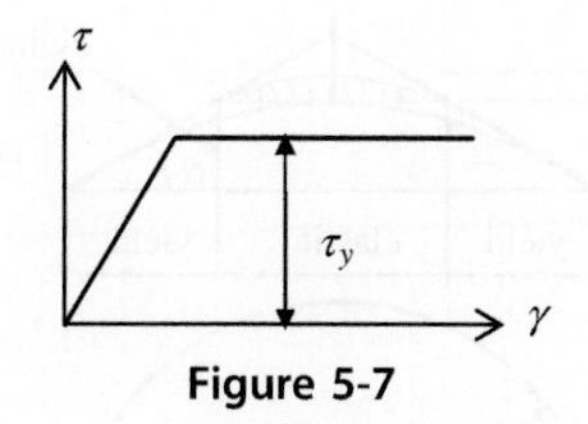

Figure 5-7

Recalling the polar moment of inertia (J) or the St. Venant torsional constant (K_T) of a solid circle is $\pi R^4/2$, the elastic twisting moment of a circular shaft is given by

$$M_{ze} = GJ\theta' = GK_T\theta' \tag{5.5.1}$$

5.5.2. Elastic Limit

If the stress–strain relationship is linearly elastic and perfectly plastic as shown in Fig. 5-7, the maximum elastic torque is limited by the first yield shear stress at the circumference of the cross section

$$\tau_{\max} = \tau_y = \frac{2z_0}{R} \quad \text{and} \quad z_0 = \frac{\tau_y R}{2} = \frac{R^2 G\theta'}{2} \tag{g}$$

where θ' is the rotation per unit length.

From Eq. (g), it follows that

$$\theta'_y = \frac{\tau_y}{GR} \tag{h}$$

Substituting Eq. (h) into Eq. (f) gives

$$M_{zy} = \frac{\pi R^4 G\theta'}{2} = \frac{\pi R^4 G}{2}\frac{\tau_y}{GR} = \frac{\pi R^3}{2}\tau_y \tag{5.5.2}$$

5.5.3. Plastic Torque

The membrane analogy is applicable to the case of fully plastic torque. The membrane is replaced by a surface of constant slope, a cone, which resembles the sand-heap on a circle. The volume of the cone with the base radius of R and height of z_1 as shown in Fig. 5-8 is $V = (z_1/3)\pi R^2$. The fully plastic torque is twice the volume of the cone. Hence,

$$M_{zp} = 2V = \frac{2}{3}\pi R^2 z_1 = \frac{2}{3}\pi R^2 \tau_y R = \frac{2}{3}\pi R^3 \tau_y \tag{5.5.3}$$

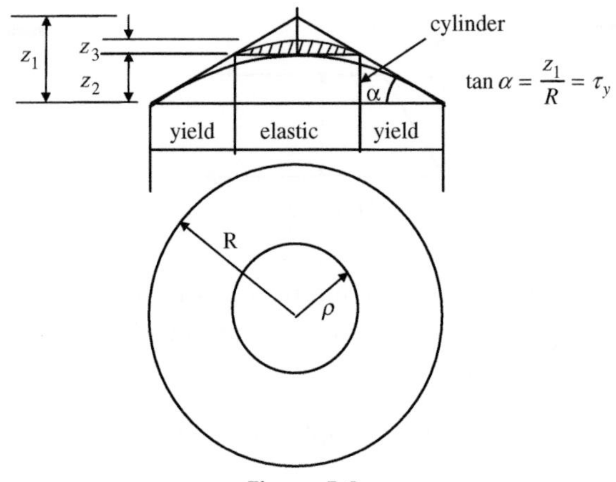

Figure 5-8

Hence, the shape factor for torsion for a solid circular section is

$$S.F. = \frac{M_{zp}}{M_{zy}} = \frac{2\pi R^3 \tau_y}{3} \frac{2}{\pi R^3 \tau_y} = \frac{4}{3} = 1.33 \tag{5.5.4}$$

5.5.4. Elasto-Plastic Torque

The elasto-plastic torque here refers to the case when the progression of the yielding is terminated leaving an elastic core of radius ρ as shown in Fig. 5-8. The volume of the shape can be computed by Eq. (i) considering the three shapes shown in Fig. 5-9.

$$V = \frac{\pi R^2}{3} z_1 - \frac{\pi \rho^2}{3}(z_1 - z_2) + \frac{\pi \rho^2}{2} z_3 \tag{i}$$

From the geometry of the shape shown in Fig. 5-8, z_1, z_2, and z_3 are determined as follows:

$$\begin{aligned} z_1 &= R\tau_y \\ z_2 &= (R - \rho)\tau_y \\ z_3 &= \frac{\rho}{2}\tau_y = \frac{\rho^2 G\theta'}{2} \end{aligned} \tag{j}$$

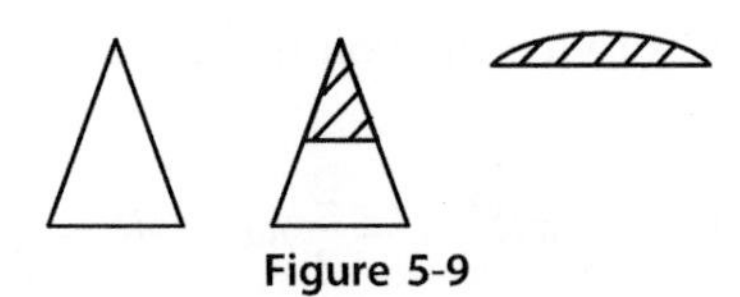

Figure 5-9

Substituting Eq. (j) into Eq. (i) gives

$$V = \frac{\pi R^2}{3} R\tau_y - \frac{\pi \rho^2}{3} \rho\tau_y + \frac{\pi \rho^2}{2} \frac{\rho}{2} \tau_y = \frac{\pi R^3}{3} \tau_y \left[1 - \left(\frac{\rho}{R}\right)^3 + \frac{3}{4}\left(\frac{\rho}{R}\right)^3\right]$$

$$= \frac{\pi R^3}{3} \tau_y \left[1 - \frac{1}{4}\left(\frac{\rho}{R}\right)^3\right] = \frac{\pi R^3}{3} \tau_y \left[1 - \frac{1}{4R^3}\left(\frac{\tau_y}{G\theta'}\right)^3\right]$$

The elasto-plastic torque is

$$M_{zep} = 2V = \frac{2\pi R^3}{3} \tau_y \left[1 - \frac{1}{4R^3}\left(\frac{\tau_y}{G\theta'}\right)^3\right] \tag{5.5.5}$$

Dividing Eq. (5.5.5) by Eq. (5.5.2) yields

$$\frac{M_{zep}}{M_{zy}} = \frac{\dfrac{2\pi R^3}{3} \tau_y \left[1 - \dfrac{1}{4R^3}\left(\dfrac{\tau_y}{G\theta'}\right)^3\right]}{\dfrac{\pi R^3}{2} \tau_y} = \frac{4}{3}\left[1 - \frac{1}{4R^3}\left(\frac{\tau_y}{G\theta'}\right)^3\right]$$

Substituting Eq. (h) into the above gives

$$\frac{M_{zep}}{M_{zy}} = \frac{4}{3}\left[1 - \frac{1}{4}\frac{1}{R^3}\left(\frac{GR\theta'_y}{G\theta'}\right)^3\right] = \frac{4}{3}\left[1 - \frac{1}{4\left(\dfrac{\theta'}{\theta'_y}\right)^3}\right] \quad \text{for } \theta'_y \le \theta' \le \alpha \tag{5.5.6}$$

and

$$\frac{M_z}{M_{zy}} = \frac{\theta'}{\theta'_y} \quad \text{for } 0 \le \theta' \le \theta'_y$$

The plot of these equations is given in Fig. 5-10.

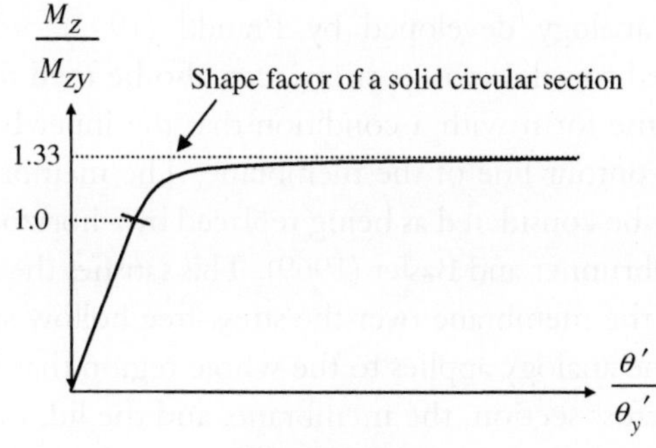

Figure 5-10

5.5.5. Uniform Torsion of Other Solid Sections

Torsion in other solid cross sections such as triangle, square, rectangle, and prestressed bridge girders (bulb tees and AASHTO girders) is primarily resisted by St. Venant torsion. Using the Prandtl stress function, the St. Venant torsional constant (K_T) can be computed for cross sections with relatively simple boundaries such as equilateral triangle, rectangle, or even an ellipse as demonstrated in textbooks on the theory of elasticity (for example, Saada 1974; Sokolnokoff 1956; Timoshenko and Goodier 1951), it would be impractical at best to apply the same procedure to bulb tees and AASHTO girders.

The shape of the soap bubble (membrane) is controlled by the second-order partial differential equation, as shown in Table 5-1

$$\frac{\partial^2 z}{\partial x^2} + \frac{\partial^2 z}{\partial y^2} = -\frac{q}{S} \tag{5.5.7}$$

where z = ordinate of membrane, x,y = planar coordinates, q = lateral pressure under membrane, and S = membrane tension. The St. Venant torsional constant (K_T) is related to the volume, V, of the membrane by

$$K_T = \frac{4SV}{q} \tag{5.5.8}$$

Equation (5.5.7) has been transformed into central differences by a Taylor series expansion and program (Yoo 2000). Fortran source code can be downloaded from the senior author's Web pages. Access codes are available from the back flap of the book. Illustrations of input and output schemes are inserted into the source code by a liberal use of comment statements.

5.6. TORSION IN CLOSED THIN-WALLED CROSS SECTIONS

The membrane analogy developed by Prandtl (1903), which has been successfully applied to solid cross sections, can also be used for hollow cross sections in the same form with a condition that the inner boundary has to correspond to a contour line of the membrane. The membrane across the-hollow space may be considered as being replaced by a horizontal plane lid as illustrated by Kollbrunner and Basler (1969). This satisfies the requirement of the zero slope of the membrane over the stress-free hollow space.

The membrane analogy applies to the whole region that is contained by the plane of the cross section, the membrane, and the lid, even though the true membrane is only stretched across the effective area of the cross section.

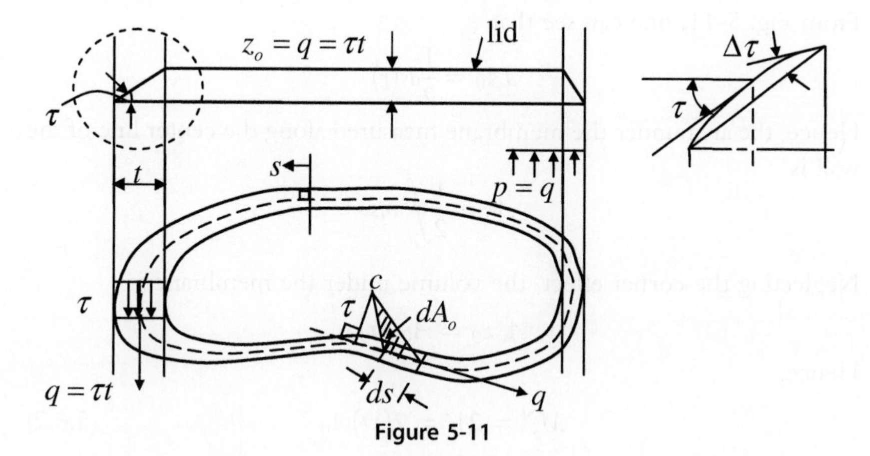

Figure 5-11

The gradient (slope) of the torsional stress functions, $\phi(x, y)$, is no longer a continuous vector function.

For a thin-walled cross section, the analogy may be considerably simplified due to the following two reasons:

1. It is admissible to work on an average slope of the membrane at the centerline of the wall, which implies a constant shear stress distribution across the wall. Then, the height of the lid from the plane of the cross section can be expressed by $z_0 = \tau t = \text{constant} = q$ (shear flow).

2. The average direction of the contour lines, which are identical to the shear stress trajectories, is assumed to be equal to the direction of the centerline of the wall, which implies that the shear force per unit length, q, is tangential to the centerline of the wall. The constant shear flow, q, obeys the conservation law of the hydrodynamic analogy, that is, the sum of the entering shear flows at a node (joint) must be exactly the same as the sum of the discharging shear flows.

Reviewing Fig. 5-11, one immediately notices that

$$\tau = \frac{z_0}{t} \Rightarrow \tau t = q = \text{shear flow} = z_0 \tag{5.6.1}$$

Again, from Fig. 5-11, it becomes obvious that

$$dM_z^{St} = (\tau t) ds(r)$$

Hence,

$$M_z^{St} = \tau t \oint r ds$$

From Fig. 5-11, one can see that

$$dA_0 = \frac{1}{2}ds(r)$$

Hence, the area under the membrane measured along the center line of the wall is

$$A_0 = \frac{1}{2}\oint rds$$

Neglecting the corner effect, the volume under the membrane is

$$A_0 z_0 = A_0(\tau t)$$

Hence,

$$M_z^{St} = 2V = 2(\tau t)A_0 \tag{5.6.2}$$

and the shear stress for the closed cross section is

$$\tau_c = \frac{M_z^{St}}{2A_0\, t} \tag{5.6.3}$$

From the small displacement theory (microgeometry holds), the following geometric relationship is obvious from Fig. 5-12:

$$\sin\phi = \frac{S_v}{S} = \frac{z_0}{t}$$

The equilibrium of forces in the vertical direction and recalling the analogy derived in the membrane analogy requires that

$$\sum F_y = 0 = pA_0 - \oint S\frac{z_0}{t}\,ds \Rightarrow \frac{p}{S} = \frac{z_0}{A_0}\oint \frac{ds}{t} = 2G\theta'$$

$$2G\theta' = \frac{z_0}{A_0}\oint \frac{ds}{t} \tag{5.6.4}$$

Recall the shear flow is given by $\tau t = z_0$ from Fig. 5-12.

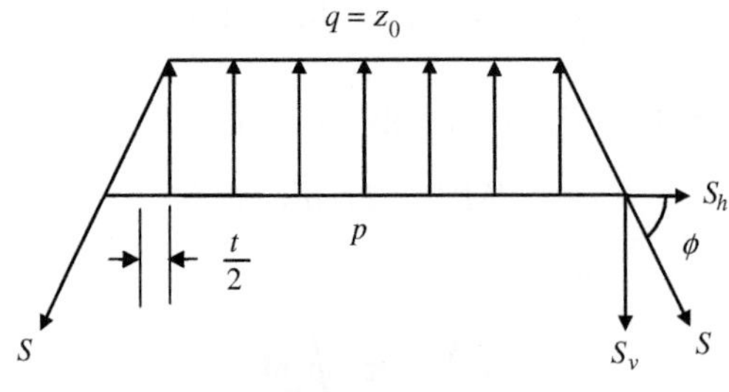

Figure 5-12

The shear flow is also defined in Eq. (5.6.2) as

$$\tau t = \frac{M_z^{St}}{2A_0}$$

Hence, the height of the membrane is

$$z_0 = \frac{M_z^{St}}{2A_0}$$

It should be noted that the shear flows become indeterminate for a multicellular section. As a consequence, the torsional properties of a multicellular section become indeterminate, too.

For a single-cell section

$$2G\theta' = \frac{M_z^{St}}{2A_0^2}\oint \frac{ds}{t} \Rightarrow M_z^{St} = 4A_0^2 G\theta' / \oint \frac{ds}{t} \tag{5.6.5}$$

Generally,

$$M_z^{St} = GK_T\theta' \text{ and } \theta' = \frac{d\theta}{dz}$$

Therefore the St. Venant torsional constant for closed cross sections is

$$K_{Tc} = \frac{4A_0^2}{\oint \frac{ds}{t}} \tag{5.6.6}$$

The torsional shear stress in a closed cross section is computed from

$$\tau_c = \frac{M_z^{St}}{2A_0\, t}$$

and the general differential relationship for the St. Venant torsion is

$$M_z^{St} = GK_T\theta'$$

Therefore, the shear stress of the closed cross section under the St. Venant (uniform) torsion is also computed by

$$\tau_c = \frac{GK_{Tc}\theta'}{2A_0\, t} \tag{5.6.7}$$

Notice that the thickness of the wall in a closed cross section is constant at a location along the length of the member (prismatic, not a variable). It varies only along the perimeter of the cross section. The corresponding shear stress, torsional moment, and the St. Venant torsional constant to the open and closed cross section shown in Fig. 5-13 are tabulated in Tables 5-2 and 5-3.

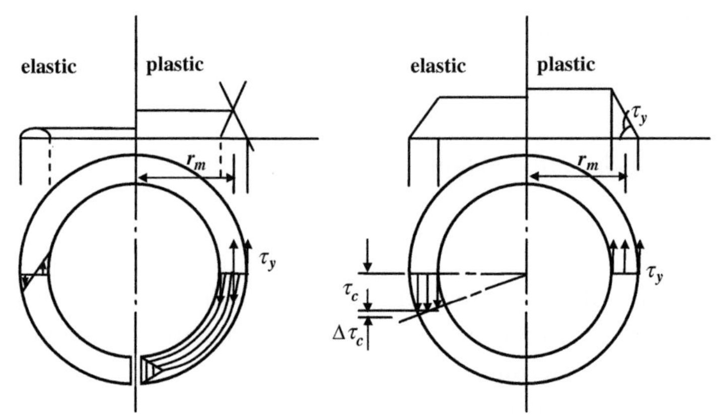

Figure 5-13 Membrane analogy applied to open and closed cross sections

Table 5-2 Torsional values of open and closed cross sections

	Open	Closed
Torsional constant	$K_{T0} = \frac{1}{3}(2\pi) r_m t^3$	$K_{Tc} = \frac{4(\pi r_m)^2}{\frac{2\pi r_m}{t}} = 2\pi r_m^3 t$
Shear stress	$\tau_0 = \frac{M_z^{St} t}{K_{T0}} = \frac{3M_z^{St}}{2\pi r_m t^2}$	$\tau_c = \frac{Mz^{St}}{2A_0 t} = \frac{Mz^{St}}{2\pi r_2^m t}$
Elastic torque	$M_{zyo}^{St} = \frac{\tau_y K_{T0}}{t}$	$M_{zyc}^{St} = \tau_y (2\pi r_m^2) t$
Plastic torque	$M_{zpo}^{St} = \tau_y \pi r_m t^2$	$M_{zpc}^{St} = M_{zyc}^{St}$
Shape factor $(M_{zp}^{St}/M_{zy}^{St})$	1.5	1.0

Table 5-3 Ratios of torsional values

	In the case of equality between		
	$\tau_0 = \tau_c$	$M_{zo}^{St} = M_{zc}^{St}$	$\theta_0' = \theta_c'$
Shear stresses τ_0/τ_c	1	$\frac{3r_m}{t}$	$\frac{t}{r_m}$
Torsional moment M_{zo}^{St}/M_{zc}^{St}	$\frac{t}{3r_m}$	1	$\frac{1}{3}\left(\frac{t}{r_m}\right)^2$
Specific rotation θ_0'/θ_c'	$\frac{r_m}{t}$	$3\left(\frac{r_m}{t}\right)^2$	1

5.7. NONUNIFORM TORSION OF W SHAPES

In Section 5.1, it is stated that an externally applied twisting moment is resisted internally by some combination of uniform (or pure, or St. Venant) torsion and nonuniform (or warping) torsion depending on the boundary conditions, that is, whether a member is free to warp or whether warping is restrained

$$M_z = M_z^{St} + M_z^w \tag{5.7.1}$$

5.7.1. St. Venant Torsion

$$M_z^{St} = GK_T\theta' = CJ\frac{d\theta}{dz} \tag{5.7.2}$$

where J is the symbol for the pure torsional constant used in current AISC (2005).

5.7.2. Warping Torsion

As a consequence of the assumptions used by Vlasov (1961) regarding nonuniform torsion, the following two distinctions are noticed for a doubly symmetric W shape or even a singly symmetric I-shaped section:

1. Web remains undeformed ⇒ torsion is resisted by flanges only.
2. Shear deformation in flanges is neglected.

Figures 5-14 depicts lateral deformation of flanges known as flange bending.

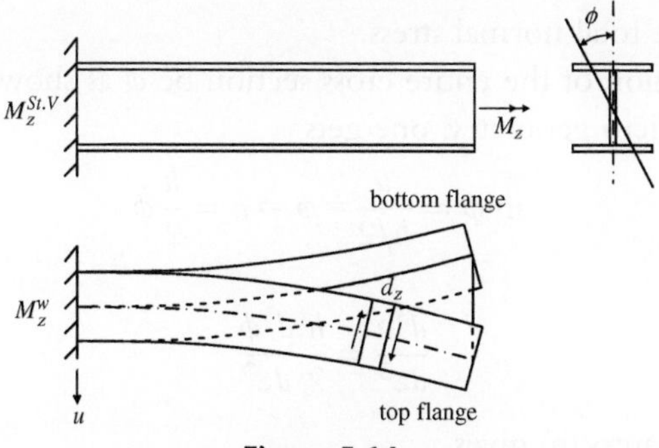

Figure 5-14

The flange bending moment for one flange, M_f, is given by

$$\frac{M_f}{EI_f} = +\frac{d^2u}{dz^2}$$

or

$$M_f = +EI_f\left(\frac{d^2u}{dz^2}\right) \tag{a}$$

where $I_f \doteq I_y/2$

The flange bending stress in the flanges which is called the warping normal stress is given by

$$\sigma_w = \frac{M_f x}{I_f}$$

The flange shear, V_f, is

$$V_f = -\frac{dM_f}{dz} = -EI_f\left(\frac{d^3u}{dz^3}\right) \tag{b}$$

The vertical bending stress and the warping normal stress are combined as shown in Figure 5-15 where the vertical bending stress along the web is not shown.

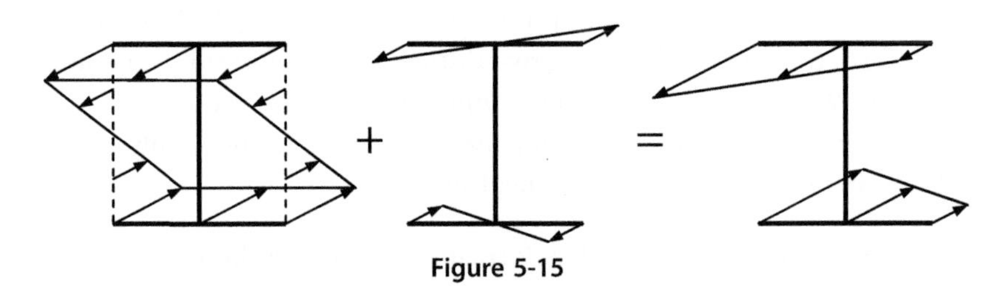

Figure 5-15

It is not unusual that the normal stress due to flange bending exceeds more than 50% of the total normal stress.

Let the rotation of the entire cross section be ϕ as shown in Fig. 5-16. Invoking the micro geometry, one gets

$$\sin\phi = \frac{u}{h/2} \doteq \phi \Rightarrow u = \frac{h}{2}\phi$$

$$\frac{d^3u}{dz^3} = \frac{h}{2}\frac{d^3\phi}{dz^3} \tag{c}$$

Substituting (c) into (b) gives

$$V_f = -\frac{EI_y}{2}\left(\frac{d^3\phi}{dz^3}\right)\frac{h}{2} = -\frac{EI_y h}{4}\frac{d^3\phi}{dz^3} \tag{d}$$

From Fig. 5-17, it follows immediately that the warping moment can be written as

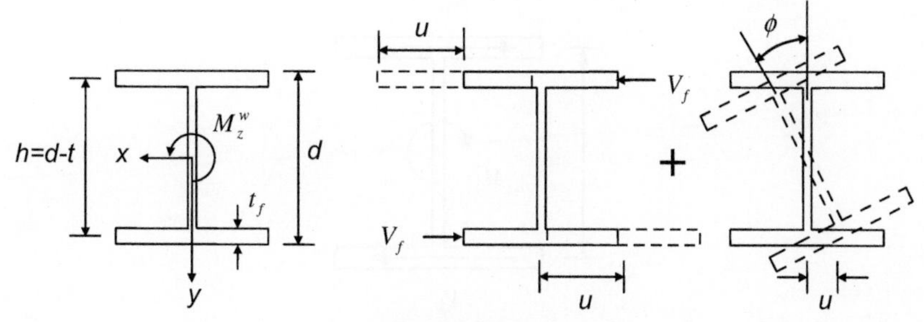

Figure 5-16

$$M_z^w = V_f \times h = -\frac{EI_y}{4}h\phi'''h \tag{5.7.3}$$

Defining the warping constant as

$$I_w = \frac{I_y}{4}h^2 \tag{5.7.4}$$

then

$$M_z^w = -EI_w\phi'''$$

Hence, the total moment is

$$\begin{aligned} M_z &= M_z^{St} + M_z^w \\ &= GK_T\phi' - EI_w\phi''' \\ &= C\phi' - C_1\phi''' \end{aligned} \tag{5.7.5}$$

where $C = GK_T = GJ$; $C_1 = EI_w = EC_w$; C and C_1 were introduced by Timoshenko for the St. Venant torsional rigidity and warping rigidity, respectively. *J and* C_w appear for the first time in the *AISC Manual* (7th ed., 1970). When a structure is subjected to an eccentrically applied load (combined bending and torsion), it can be resolved as shown in Fig. 5-18 and analyzed separately. It should be noted that Eq. (5.7.5) is good only for concentrated torques shown in Fig. 5-18.

5.7.3. General Equations

Consider a general case where torque varies along the z axis as shown in Figure 5-19, in which m_z is the rate of change of torque.
The equilibrium gives

$$\sum M_z = 0 = -M_z + m_z d_z + M_z + dM_z = 0$$

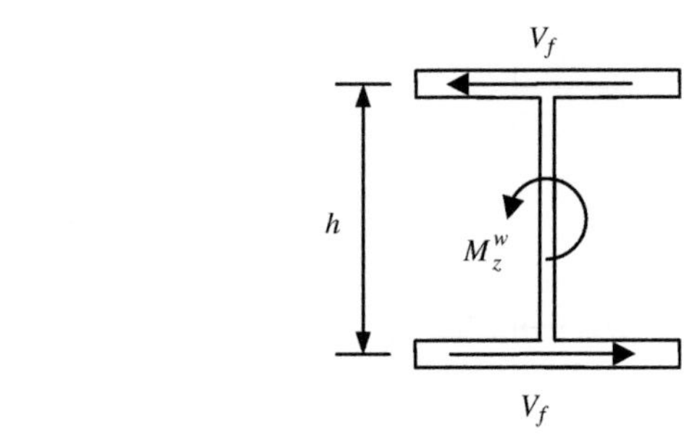

Figure 5-17

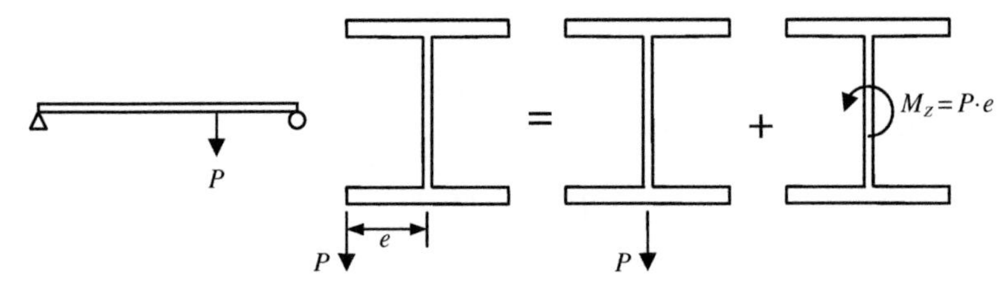

Figure 5-18

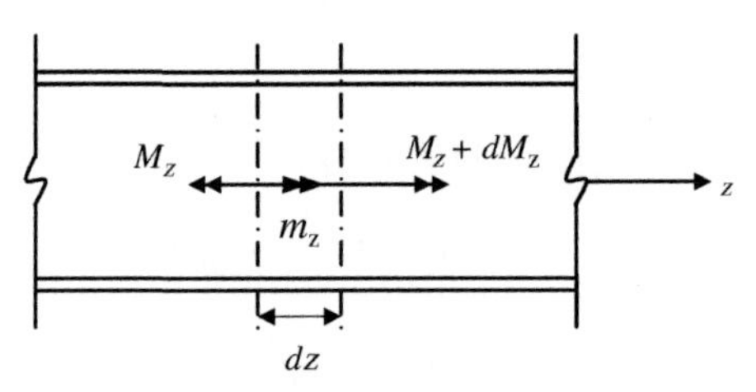

Figure 5-19

$$m_z = -\frac{dM_z}{dz}$$

Differentiating Eq. (5.7.5) with respect to z, one obtains

$$\frac{dM_z}{dz} = -m_z = GK_T\phi'' - EI_w\phi^{iv}$$

$$m_z = EI_w\phi^{iv} - GK_T\phi'' \quad = C_1\phi^{iv} - C\phi'' \tag{5.7.6}$$

or

$$\phi^{iv} - \frac{GK_T}{EI_w}\phi'' = \frac{m_z}{EI_w}$$

which is similar to Eq. (1.4.8)

$\phi^{iv} - (GK_T/EI_w)\phi'' = (m_z/EI_w)$ (Similar form $y^{iv} + Py''/EI) = w(z)/EI$

5.7.4. Solution of Differential Equations

Concentrated Torque

The governing differential equation is given by Eq. (5.7.5)

$$\phi''' - \frac{GK_T}{EI_w}\phi' = \frac{-M_z}{EI_w}$$

Let

$$\lambda^2 = \frac{GK_T}{EI_w}$$

Homogeneous solution:

$$\phi''' - \lambda^2\phi' = 0$$

Assume $\phi_h = e^{mz}$, then $\phi_h' = me^{mz}$, $\phi_h'' = m^2e^{mz}$, $\phi_h''' = m^3e^{mz}$
Substituting these equations gives

$$m(m^2 - \lambda^2) = 0, \; m = 0, \; m = \pm\lambda$$

Hence,

$$\phi_h = c_1e^{oz} + c_2e^{-\lambda z} + c_3e^{\lambda z}$$

Particular solution:

Assume $\phi_p = Az$, then $\phi_p' = A$, $\phi_p''' = 0$

Substituting these gives

$$-\lambda^2A = -\frac{M_z}{EI_w} \Rightarrow A = \frac{M_z}{\lambda^2EI_w}$$

Hence

$$\phi_p = \frac{M_z}{\lambda^2EI_w}z$$

Total solution $\phi = \phi_h + \phi_p$

$$\phi = c_1e^{oz} + c_2e^{-\lambda z} + c_3e^{\lambda z} + \frac{M_z}{\lambda^2EI_w}z$$

Recall identities:

$$e^{\lambda z} = \cosh\lambda z + \sinh\lambda z$$

$$e^{-\lambda z} = \cosh\lambda z - \sinh\lambda z$$

Then

$$\phi = c_1 + c_2(\cosh \lambda z - \sinh \lambda z) + c_3(\cosh \lambda z + \sinh \lambda z) + \frac{M_z}{\lambda^2 EI_w} z$$

or

$$\phi = A + B \cosh \lambda z + C \sinh \lambda z + \frac{M_z}{\lambda^2 EI_w} z \tag{5.7.7}$$

Distributed Torque

Equation (5.7.6) can be rearranged as follows:

$$\phi^{iv} - \lambda^2 \phi'' = \frac{m_z}{EI_w}$$

Homogeneous solution:
Assume the homogeneous solution to be of the form $\phi = ce^{mz}$

$$\phi' = cme^{z}, \phi'' = cm^2 e^{mz}, \phi''' = cm^3 e^{mz}, \phi^{iv} = cm^4 e^{mz}$$

Then, one obtains

$$m^2(m^2 - \lambda^2) = 0$$

The solutions are $m_1 = 0,\ m_2 = 0,\ m_3 = \lambda,\ m_4 = -\lambda$
Hence

$$\phi_h = c_1 e^{oz} + c_2 z e^{oz} + c_3 e^{\lambda z} + c_4 e^{-\lambda z}$$

Particular solution:
Assume the particular solution to be of the form, $\phi_p = c_5 + c_6 z + c_7 z^2$.
Then

$$\phi_p' = c_6 + 2c_7 z,\ \phi_p'' = 2c_7,\ \phi_p''' = \phi_p^{iv} = 0$$

Substituting these equations gives

$$0 - 2c_7\lambda^2 = \frac{m_z}{EI_w} \Rightarrow c_7 = -\frac{m_z}{2\lambda^2 EI_w}$$

Hence

$$\phi_p = c_5 + c_6 z - \frac{m_z}{2\lambda^2 EI_w} z^2$$

Total solution $\phi_T = \phi_h + \phi_p$

$$\phi = c_1 + c_2 z + c_3 e^{\lambda z} + c_4 e^{-\lambda z} + c_5 + c_6 z - \frac{m_z}{2\lambda^2 EI_w} z^2$$

or

$$\phi = A + Bz + C \cosh \lambda z + D \sinh \lambda z - \frac{m_z}{2\lambda^2 EI_w} z^2$$

or

$$\phi = A + Bz + C \cosh \lambda z + D \sinh \lambda z - \frac{m_z}{2GK_T} z^2 \quad (5.7.8)$$

5.7.5. Boundary Conditions

The integral constants in Eqs. (5.7.7) and (5.7.8) are to be determined by boundary conditions given in Table 5-4.

At fixed supports:

$\phi = \phi' = 0$, which implies warping is restrained and hence warping stresses may develop.

At pinned supports:

$\phi = \phi'' = 0$, which implies warping is not restrained.

At free ends:

$\phi'' = \phi''' = 0$, which implies warping is not restrained.

At interior supports of continuous beam:

$$\phi_\ell = \phi_r,\ \phi'_\ell = \phi'_r,\ \phi''_\ell = \phi''_r,\ \text{but}\ \phi'''_\ell \neq \phi'''_r$$

Table 5-4 Torsional Boundary Conditions

Function ϕ	Physical Condition	Torsional Condition
$\phi = 0$	No twist	Pinned or fixed
$\phi' = 0$	Warping restraint	Fixed end, warping exists
$\phi'' = 0$	Free warping	Pinned or free end, no warping
$\phi''' = 0$	-	Flange shear = 0

5.7.6. Stresses Due to Torsion

A classical analysis of stresses due to torsion is illustrated by Heins and Seaburg (1963) and Seaburg and Carter (1997).

- ***St. Venant's Stress***

  St. Venant's Stress is shown in Fig. 5-20. The maximum shear stress due to St. Venant torsion is

$$\tau_{\max} = \frac{tM_z^{St}}{K_T} = tG\phi'$$

where

$$M_z^{St} = GK_T\phi'$$

Then, the maximum stresses in the flange and web are

$$\tau_{\max}(web) = t_w G\phi', \; \tau_{\max}(flg) = t_f G\phi' \tag{5.7.9}$$

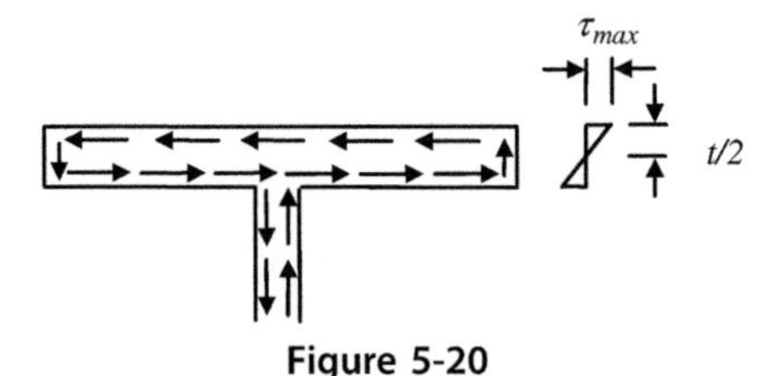

Figure 5-20

- ***Warping Stresses***

The maximum flange bending stress (warping normal stress)developed due to warping torsion is given by

$$\sigma_w = -(m_f/I_f)(b/2)$$

where

$$m_f = +\frac{EI_y}{2}\left(\frac{d^2u}{dz^2}\right) = +\frac{EI_y}{4}h\phi''$$

Hence,

$$\sigma_w = -\frac{Eh}{4}b\phi'' = -\frac{E(d-t)b}{4}\phi'' \tag{5.7.10}$$

From the elementary mechanics of materials, the maximum shear stress developed in the flanges due to the warping shear force (V_f) shown in Figure 5-21 is given by

$$\tau_{w\max} = \frac{3}{2}\frac{V_f}{A_f}$$

where

$$V_f = -\frac{EI_y}{4}h\phi''', A_f = b_f t_f, \; I_y \doteq \frac{b_f^3 t_f}{6}, \; h = d - t_f$$

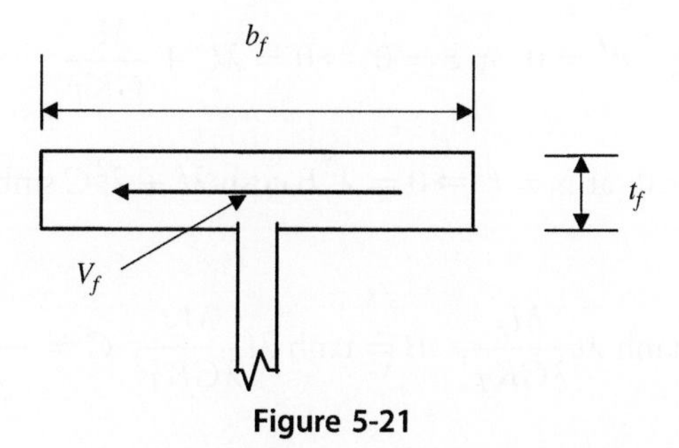

Figure 5-21

Hence

$$\tau_{w\max} = -\frac{Eb_f^2(d - t_f)}{16}\phi''' \qquad (5.7.11)$$

Example 1

Consider a cantilever subjected to a concentrated torque at the free end as shown in Figure 5-22.

The general solution is

$$\phi = A + B\cosh\lambda z + C\sinh\lambda z + \frac{M_z}{GK_T}z \qquad (5.7.12)$$

$$\phi' = \lambda B\sinh\lambda z + \lambda C\cosh\lambda z + \frac{M_z}{GK_T}$$

$$\phi'' = \lambda^2 B\cosh\lambda z + \lambda^2 C\sinh\lambda z$$

The boundary conditions are:

$$\phi = 0 \quad \text{at } z = 0, \; \phi' = 0 \quad \text{at } z = 0, \; \phi'' = 0 \quad \text{at } z = \ell$$

Then

$$\phi = 0 \text{ at } z = 0 \Rightarrow 0 = A + B \Rightarrow A = -B$$

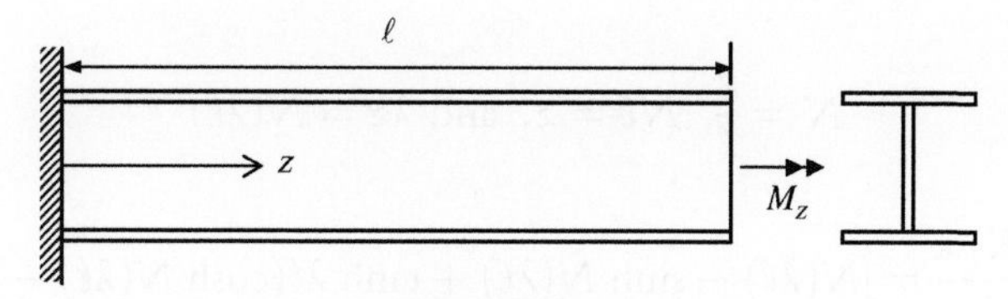

Figure 5-22

$$\phi' = 0 \;\text{ at } z = 0 \Rightarrow 0 = \lambda C + \frac{M_z}{GK_T}$$

$$\phi'' = 0 \;\text{ at } z = \ell \Rightarrow 0 = \lambda^2 B \cosh \lambda\ell + \lambda^2 C \sinh \lambda\ell$$

Hence

$$A = -\tanh \lambda\ell \frac{M_z}{\lambda GK_T}, \quad B = \tanh \lambda\ell \frac{M_z}{\lambda GK_T}, \quad C = -\frac{M_z}{\lambda GK_T}$$

The solution is

$$\phi = \frac{M_z}{GK_T \lambda}[\lambda z - \sinh \lambda z + \tanh \lambda\ell(\cosh\lambda z - 1)]$$

or

$$\phi \frac{GK_T \lambda}{M_z} = \lambda z - \sinh \lambda z + \tanh \lambda\ell(\cosh \lambda z - 1)$$

Differentiating gives

$$\phi' \frac{GK_T}{M_z} = 1 - \cosh \lambda z + \tanh \lambda\ell \sinh \lambda z$$

$$\phi'' \frac{GK_T}{M_z \lambda} = -\sinh \lambda z + \tanh \lambda\ell \cosh \lambda z$$

$$\phi''' \frac{GK_T}{M_z \lambda^2} = -\cosh \lambda z + \tanh \lambda\ell \sinh \lambda z$$

Let $\lambda = \sqrt{\dfrac{GK_T}{EI_w}}$ and $a = \sqrt{\dfrac{EI_w}{GK_T}}$

where a has a length dimension. Then $\lambda\ell = \ell/a$

Let N be the fraction varying from 0.1 to 1.0. Then,

$$N = \frac{z}{\ell}, \quad N\ell = z, \;\text{ and } \lambda z = N(\lambda\ell)$$

$$\phi \frac{GK_T \lambda}{M_z} = [N(\lambda\ell) - \sinh N(\lambda\ell) + \tanh \lambda\ell(\cosh N(\lambda\ell) - 1)]$$

Assume $\lambda\ell = 2.0$ (if the section properties and span length are given, the exact value can be computed). Then,

$$\lambda z = N(2.0), \quad N = 0.1 \sim 1.0$$

$$\phi \frac{GK_T\lambda}{M_z} = N(2.0) - \sinh N(2.0) + \tanh(2.0)[\cosh N(2.0) - 1]$$

Table 5-5 Torsional functions

N	ϕ	ϕ'	ϕ''	ϕ'''
0.0	0.0	0.0	.96	−1.0
.2	.06	.31	.63	−.69
.4	.24	.52	.40	−.48
.6	.47	.64	.23	−.36
.8	.74	.70	.10	−.30
1.0	1.01	.73	.00	−.27

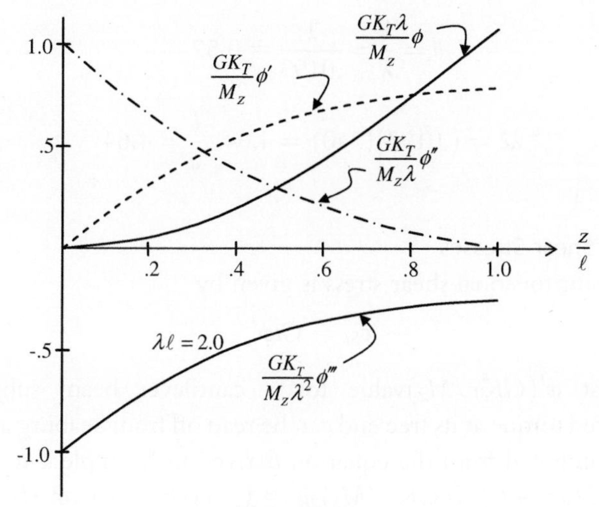

Figure 5-23

Example 2

A concentrated load of 5 kips is applied at the free end of a cantilever beam (W 12 × 50) of 20 feet long as shown in Figure 5-24. $E = 29 \times 10^3$ ksi, $G = 11.2 \times 10^3$ ksi, $P = 5$ kips, $e = 1/2$ in, $K_T = 1.82$ in^4. Find the maximum stresses.

Torsional moment is,

$$M_z = -P \times e = -5 \times \frac{1}{2} = -2.5 \text{ k-in}$$

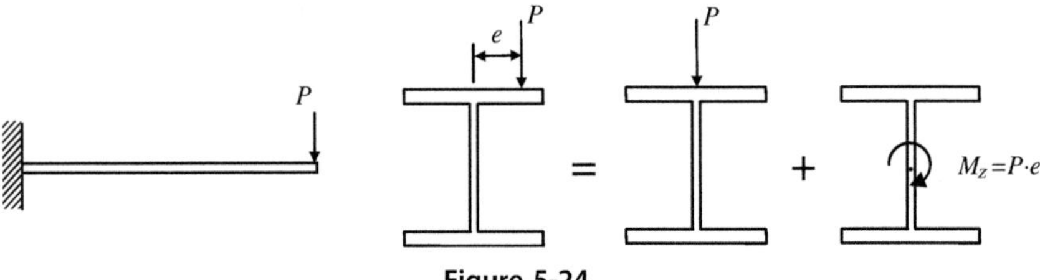

Figure 5-24

The warping constant is

$$I_w = \frac{I_y h^2}{4} = 1,881.0 \text{ in}^6 \text{ (This formula is good for W shapes only)}$$

Then

$$\lambda^2 = \frac{GK_T}{EI_W} = \frac{(11.2)(1.82)}{(29)(1881)} = 0.374 \times 10^{-3} \text{ in}^{-2}, \lambda = 0.0193 \text{ in}^{-1}$$

$$a = \frac{1}{\lambda} = \frac{1}{.0193} \doteq 0.52$$

$$\lambda\ell = (.0193)(240) = 4.64, \ \frac{\ell}{a} = 4.64$$

Stresses

1. Torsional Shear Stresses

The St. Venant torsional shear stress is given by

$$\tau_{St} = Gt\phi'$$

The greatest $\phi'(GK_T/M_z)$value for a cantilever beam subjected to a concentrated torque at its free end can be read off from Seaburg and Carter (1997) or computed from the equation derived in Example 1 to be 0.981, say 1.0, for $\ell/a = 4.64$. $((GK_T/M_z)\,\phi' = 1 - \cosh\lambda z + \tanh\lambda\ell(\sinh\lambda z))$.

$$\phi' = 1 \times \frac{M_z}{GK_T}, \ \tau_{st} = Gt \times \frac{M_z}{GK_T} = t\frac{M_z}{K_T}$$

$$\tau_{st} = \frac{M_z t}{K_T} = \frac{-2.5 \times t}{1.82} = -1.37t, \ t = (t_f \text{ or } t_w)$$

$$t_f = .641 \quad \tau_{st}^{flg} = -1.37 \times 0.641 = -0.88 \ ksi$$

$$t_w = 0.371 \ \tau_{st}^{web} = -1.37 \times .371 = -0.51 \ ksi$$

The St. Venant shear stress is equal to zero at the fixed end, and the St. Venant shear stress distribution is illustrated in Fig. 5-20. Hence, there is no net shear flow due to St. Venant torsion.

The warping shear stress is given by $\tau_{ws} = -(ES_w/t)\phi'''$ where S_w is referred to as the warping statical moment and can be calculated for simple structural shapes from formulas given by Seaburg and Carter (1997) or elsewhere. Yoo and Acra (1986) present a general method of calculating cross-sectional properties of general thin-walled sections.

$$\frac{S_w}{t} = \frac{b^2 h}{16} = 47.1 \text{ in}^3 \text{ for W12} \times 50$$

$$\tau_{ws} = -\frac{Eb^2(d-t)}{16}\phi''' = -\frac{Eb^2 h}{16}\phi'''$$

The greatest $\phi'''[GK_T/(\lambda^2 M_z)]$ ($= -\cosh \lambda z + \tanh \lambda \ell \sinh \lambda z$) value for a cantilever beam subjected to a concentrated torque at its free end can be read off from Seaburg and Carter (1997) or computed from the equation derived in Example 1 to be –1.0 at the fixed end for $\ell/a = 4.64$.

$$\phi''' = -1.0 \times \frac{M_z \lambda^2}{GK_T}$$

$$\tau_w = -\frac{Eb^2 h}{16} \times \phi''' = -\frac{Eb^2 h}{16} \times (-1)\frac{M_z \lambda^2}{GK_T}$$

$$\tau_{ws} = \frac{47.1}{1881}(-2.5) = -.0625 \text{ ksi at } \frac{z}{\ell} = 0.0 \text{ and } \tau_{ws} = 0.0 \text{ at } \frac{z}{\ell} = 1.0$$

2. Warping Normal Stress

The warping normal stress is to be computed from

$$\sigma_w = \frac{BM\, W_n}{I_w}$$

where BM is the bimoment given by $-EI_w\phi''$and W_n is the normalized warping function given by $bh/4$ for doubly-symmetric I-shape sections. A general method of evaluating these section properties is given by Yoo and Acra (1986). Hence,

$$\sigma_w = -EW_n\phi'' = -\frac{Ebh}{4}\phi''$$

in which

$$W_n = \frac{bh}{4} = 23.32 \text{ in}^2 \text{ for W12} \times 50$$

The greatest $\phi''(GK_T/M_z\lambda)$ $(= -\sinh\lambda z + \tanh \lambda\ell \cosh \lambda z)$ value for a cantilever beam subjected to a concentrated torque at its free end can be read off from Seaburg and Carter (1997) or computed from the equation derived in Example 1 to be 1.0 at the fixed end for $\ell/a = 4.64$.

$$\phi'' = \frac{M_z\lambda}{GK_T}$$

$$\sigma_w = -E\frac{(bh)}{4}\frac{M_z}{GK_T}\lambda = -EW_n \times \frac{M_z}{GK_T}\lambda$$

$$\sigma_w = \frac{-29 \times 23.3 \times 0.0193}{11.2 \times 1.82}(-2.5) = +0.64 \text{ ksi at } \frac{z}{\ell} = 0.0 \text{ and}$$

$$\sigma_w = 0.0 \text{ at } \frac{z}{\ell} = 1.0$$

3. Bending Shear Stress

The bending shear stress given by $VQ/I_x t$ is constant along the length of the cantilever beam subjected to a concentrated load at it tip. The bending statical moment, Q, is evaluated at the flange and the web for the maximum value of the shear stress.

$$\frac{Q_f}{t_f} = \frac{14.67}{.641} = 22.8\ in^2,\ \tau_{fb} = \frac{5 \times 22.8}{394.5} = .282\ ksi$$

$$\frac{Q_w}{t_w} = \frac{36.24}{.371} = 98\ in^2,\ \tau_{wb} = \frac{5 \times 98}{394.5} = 1.27\ ksi$$

4. Bending Normal Stress

$$\sigma_b = \frac{Mc}{I} = \frac{M}{S} = \frac{5 \times 240}{64.7} = \pm 18.55\ ksi$$

5. Summary of Stresses

Table 5-6 Stresses at support

z/ℓ		τ			σ	
		τ_{st}	τ_b	τ_{ws}	σ_w	σ_b
0.0	flg	0.0	0.282	−0.0625	1.60	18.55
	Σ	0.3445 or 0.2195			16.95 or 20.15	
	Web	0.0	1.27	0.0		
	Σ		1.27			

Table 5-7 Stresses at free end

z/ℓ		τ			σ	
		τ_{st}	τ_b	τ_{ws}	σ_w	σ_b
1.0	flg	−0.88	0.282	0.0	0.0	0.0
	Σ	1.162 or 0.598				
	Web	−0.51	1.27	0.0	0.0	0.0
	Σ	1.78 or 0.76				

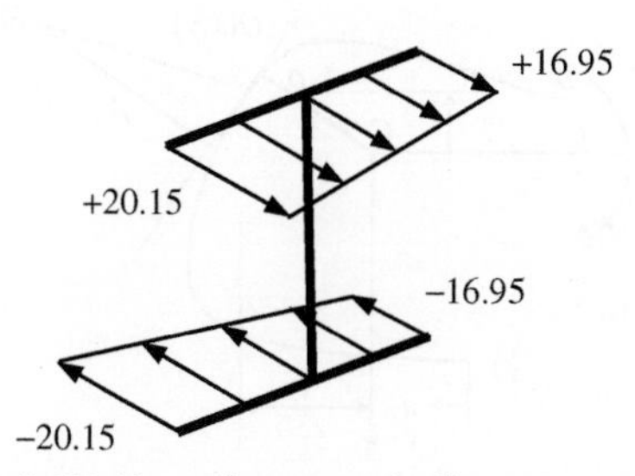

Figure 5-25 Normal stresses in flanges at support

5.8. NONUNIFORM TORSION OF THIN-WALLED OPEN CROSS SECTIONS

In the previous section, nonuniform torsion on doubly symmetrical sections was briefly considered. An approximate analysis of nonuniform torsion of a member with a general thin-walled open cross section may be developed within the confinement of assumptions employed. The literature based on the assumption that the shape of the thin-walled open cross section remains unchanged and is quite extensive. A more detailed treatment of nonuniform torsion may be found in Brush and Almroth (1975), Galambos (1968), Kollbrunner and Basler (1969), Nakai and Yoo (1988), and Timoshenko and Gere (1961). The present development of warping deformation and stress of open cross section follows, in some respects, the analysis in Timoshenko and Gere (1961), Kollbrunner and Basler (1969), Galambos (1968), and Brush and Almroth (1975).

5.8.1. Assumptions

1) Members are subjected to torsion only.
2) Members are prismatic and retain their original shapes.
3) Hooke's law holds.
4) Cross-sectional coordinates, x and y, are the principal coordinates and the z-axis is the longitudinal axis through the centroid of the cross section.
5) There is an axis parallel to the z-axis about which twisting takes place and the centroid and shear center of the cross section are denoted by C and S, respectively.
6) Deformations are small.
7) Shear at the middle line is equal to zero.

5.8.2. Symbols

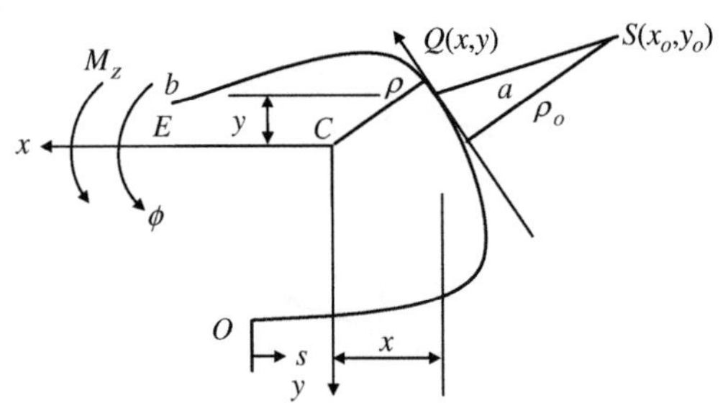

Figure 5-26 Perpendicular distances, ρ and ρ_0, to a tangent at Q

Symbols in Fig. 5-26 are defined as follows:
C: centroid of cross section ($x = 0, y = 0$)
S: shear center ($x = x_0, y = y_0$)
ϕ: angle of twist
M_z: twisting moment
Q: a point on the middle line of cross section (x, y, s)
s: perimeter coordinate measured along the middle line from point O to Q
b: total perimeter length of the middle line, O to E
a: distance between Q and S
ρ_o: perpendicular distance between S and the tangent line at Q
ρ: perpendicular distance between C and the tangent line at Q

5.8.3. Warping Torsion

Figure 5-27 shows the relationship between the angle of twist and a longitudinal displacement of a point in a member. Warping represents

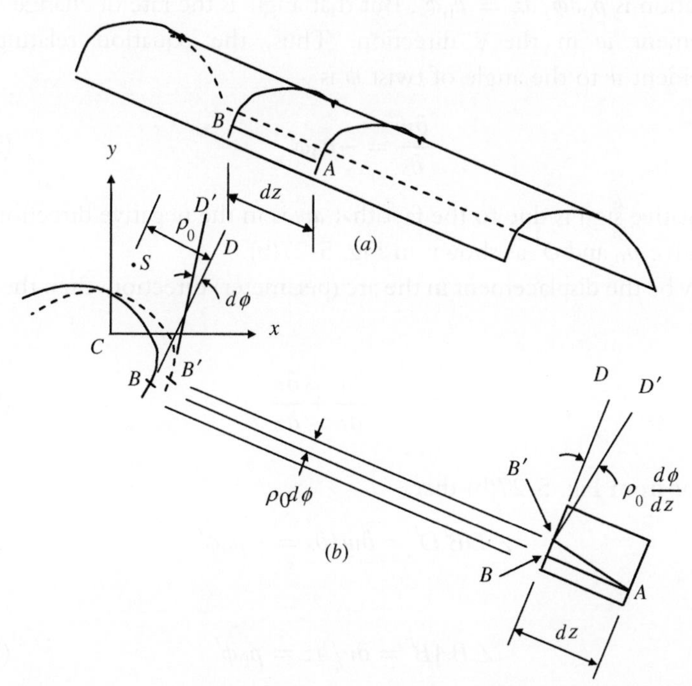

Figure 5-27 Segment of member showing warping deformation (after Brush and Almroth, *Buckling of Bars, Plates, and Shells*, New York: McGraw-Hill, 1975). Reproduced by permission.

a longitudinal displacement of points in a member due to twisting. An equation relating the longitudinal displacement component w to angle of twist ϕ may be derived. The middle surface of an element of length dz is shown in the undeformed configuration in Fig. 5-27(a). The element is shown in orthographic projections in Fig. 5-27(b), where a side view is placed on the right and a section view from the positive end of the z-axis is shown on the left. Line AB in the side view is a longitudinal line on the middle surface prior to deformation, and BD is tangent to the middle surface at B and is perpendicular to line AB. When the member is twisted, one end of the longitudinal element rotates about the shear center by a small angle $d\phi$, as shown. Then point B moves to B', and the angle BAB' in the side view is $\rho_0 d\phi/dz$, where ρ_0 is the perpendicular distance from the shear center to the tangent BD, as defined in Fig. 5-26. The variable ρ_0 is positive if a vector along the tangent in the direction of increasing s acts counterclockwise about the shear center. After the deformation, the tangent $B'D'$ remains perpendicular to AB' as shown in the side view of Fig. 5-27(b). Thus, in the side view, the angle between the tangents before and after deformation is $\rho_0 d\phi/dz \equiv \rho_0\phi'$. But that angle is the rate of change of the displacement w in the s direction. Thus, the equation relating the displacement w to the angle of twist ϕ is

$$\frac{\partial w}{\partial s} = -\rho_0\phi' \tag{5.8.1}$$

The negative sign is due to the fact that dw is in the negative direction of z for positive ρ_0 and ϕ' as shown in Fig. 5-27(b).

Let v be the displacement in the arc (perimeter) direction, then the shear strain is

$$\gamma_{sz} = \frac{\partial w}{\partial s} + \frac{\partial v}{\partial z} \tag{5.8.2}$$

It is clear from Fig. 5-27(b) that

$$\angle DB'D' = \partial w/\partial s = -\rho_0\phi' \tag{5.8.3}$$

and

$$\angle BAB' = \partial v/\partial z = \rho_0\phi' \tag{5.8.4}$$

Substituting these relations into Eq. (5.8.2) leads the shear strain to be zero. Hence, there will be no warping shear stress τ_{sz} developed, which appears to be contradictory, as will be evidenced later in Fig. 5-28. This is true under unrestrained warping as in the member that is twisted by a concentrated torque at each free end. In this case, line AB' remains straight and the original right angle of the element $\angle ABD$ remains unchanged after deformation ($\phi' =$ constant, $\phi'' = 0$). When warping is restrained, however, line AB' cannot remain straight and line AB' in Fig. 5-27(b) may be interpreted as an average. A shear strain measurement based on an average deformation is not representative of true strain.

Integration of Eq. (5.8.1) gives

$$w = w_0 - \phi' \int_0^s \rho_0 ds \tag{5.8.5}$$

where w_0 (z) is the constant of integration and equal to w at $s = 0$. If one defines

$$\omega_0 = \int_0^s \rho_0 ds \tag{5.8.6}$$

as the sectorial coordinate or the unit warping function (length^2 unit) with respect to the shear center, Eq. (5.8.5) may be rewritten as

$$w = w_0 - \phi' \omega_0 \tag{5.8.7}$$

If the warping longitudinal displacement given by Eq. (5.8.7) is introduced into one-dimensional Hooke's law, one arrives at the warping normal stress

$$\sigma_z = E\frac{dw}{dz} = E(w_0' - \phi'' \omega_0) \tag{5.8.8}$$

Since only a twisting moment M_z is applied, the resultant axial force and the bending moments due to warping normal stresses must be zero at any cross section. That is

$$N = 0 = \int_0^b \sigma_z t ds \tag{5.8.9a}$$

$$M_x = 0 = \int_0^b y \sigma_z t ds \tag{5.8.9b}$$

$$M_y = 0 = \int_0^b x \sigma_z t ds \tag{5.8.9c}$$

Equation (5.8.9a) serves to eliminate the constant of integration w_0 (z). Substituting Eq. (5.8.8) into Eq. (5.8.9a) leads to

$$w_0' \int_0^b tds - \phi'' \int_0^b \omega_0 tds = 0 \tag{5.8.10a}$$

or

$$w_0' = \frac{\phi''}{A} \int_0^b \omega_0 tds \tag{5.8.10b}$$

where

$$A = \int_0^b tds \tag{5.8.11}$$

Substituting Eq. (5.8.10b) into Eq. (5.8.8) yields

$$\sigma_z = E\phi'' \left(\frac{1}{A} \int_0^b \omega_0 tds - \omega_0 \right) \tag{5.8.12}$$

Defining a new cross-sectional property ω_n[1], the normalized unit warping (length^2 unit), as

$$\omega_n = \frac{1}{A} \int_0^b \omega_0 tds - \omega_0 \tag{5.8.13}$$

one can rewrite Eq. (5.8.12) as

$$\sigma_z = E\omega_n \phi'' \tag{5.8.14}$$

The variation of the normal stresses σ_z along the z-axis produces shearing stresses, which constitute resisting warping torque M_z^w.

To calculate the shearing stresses, consider an element *mnop* (Fig. 5-28) cut out from the wall of the member in Fig. 5-27.

From summing forces in the z-direction, one obtains

$$\frac{\partial(\tau_{sz}t)}{\partial s} dsdz + t\frac{\partial \sigma_z}{\partial z} dsdz = 0$$

[1] The first term in Eq. (5.8.12), $\left(\int_0^b \omega_0 tds\right)/A$, is replaced by $\left(\int_0^b \omega_0 ds\right)/b$ in Timoshenko and Gere (1961) and Brush and Almroth (1975). If the integration process is replaced by a summation of discrete elements as $\sum_i \omega_{0i} t_i L_i / \sum_i t_i L_i$ such that t_i is constant in each element, t_i may be replaced by $t_{avg}\left(\sum_i t_i L_i / b\right)$ as it is independent of ω_{0i} and L_i. Then $\sum_i \omega_{0i} t_i L_i / \sum_i t_i L_i = t_{avg} \sum_i \omega_{0i} L_i / t_{avg} \sum_i L_i = \sum_i \omega_{0i} L_i / \sum_i L_i$. In fact, the two expressions are identical. This has been confirmed numerically by SECP (Yoo and Acra 1986).

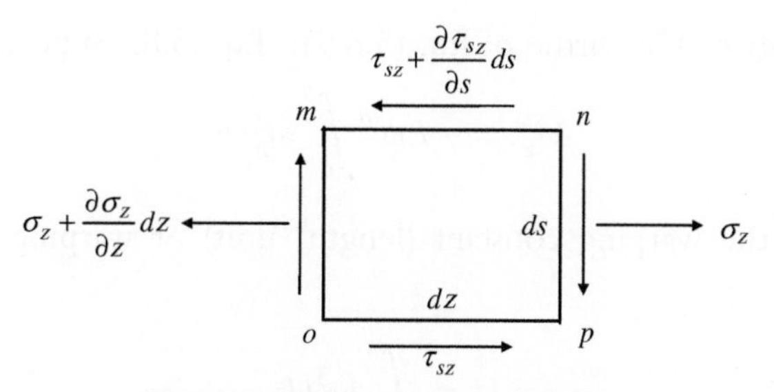

Figure 5-28 Stresses in an element

or

$$\frac{\partial(\tau_{sz}t)}{\partial s}=-t\frac{\partial\sigma_z}{\partial z}=-tE\omega_n\frac{d^3\phi}{dz^3} \tag{5.8.15}$$

Integrating Eq. (5.8.15) with respect to s and noting that ϕ is independent of s and $\tau_{sz}=0$ at $s=0$, one obtains

$$\tau_{sz}t=-E\frac{d^3\phi}{dz^3}\int_0^s\omega_n\,t\,ds \tag{5.8.16}$$

As the warping shearing stress τ_{sz} is related to the warping torque by the equation

$$M_z^w=\int_0^b\tau_{zs}\rho_0\,t\,ds=-E\frac{d^3\phi}{dz^3}\int_0^b\left[\int_0^s\omega_n\,t\,ds\right]\rho_0\,ds \tag{5.8.17}$$

Integrating Eq. (5.8.17) by parts ($\int udv=uv-\int vdu$) and letting $u=\int_0^s\omega_ntds$ and $dv=\rho_0ds$ to lead $du=\omega_ntds$ and $v=\int_0^s\rho_0ds=\omega_0$, one obtains

$$M_z^w=-E\phi'''\left(\omega_0\Big|_0^b\int_0^b\omega_ntds-\int_0^b\omega_0\omega_ntds\right) \tag{5.8.18}$$

Noting that the first term in Eq. (5.8.13), referred to as an average warping function, is a constant, one gets for the first term of Eq. (5.8.18)

$$\int_0^b\omega_ntds=\left(\frac{1}{A}\int_0^b\omega_0tds\right)\int_0^b tds-\int_0^b\omega_0tds=0$$

After substituting $\omega_0=(1/A)\int_0^b\omega_0tds-\omega_n$, the second term yields

$$\int_0^b\omega_0\omega_ntds=\left(\frac{1}{A}\int_0^b\omega_0tds\right)\int_0^b\omega_ntds-\int_0^b\omega_n^2tds$$

Since $\int_0^b \omega_n t ds = 0$ by virtue of Eq. (5.8.9a), Eq. (5.8.18) becomes

$$M_z^w = -E\phi''' \int_0^b \omega_n^2 t ds \tag{5.8.19}$$

Introducing the warping constant (length^6 unit) or warping moment of inertia, I_w

$$I_w = \int_0^b \omega_n^2 t ds \tag{5.8.20}$$

one gets

$$M_z^w = -EI_w \phi''' \tag{5.8.21}$$

The total resisting twisting moment is the sum of the warping contribution and the St. Venant contribution; that is, as per Eq. (5.7.1),

$$M_z = M_z^{St} + M_z^w$$

For concentrated torques

$$M_z = GK_T \phi' - EI_w \phi''' \tag{5.7.5}$$

For a distributed torque

$$m_z = EI_w \phi^{iv} - GK_T \phi'' \tag{5.7.6}$$

The warping shear flow is given by Eq. (5.8.16). Defining the warping statical moment (length^4 unit) as

$$S_w = \int_0^s \omega_n t ds \tag{5.8.22}$$

the warping shear flow equation is

$$\tau_w t = -ES_w \phi''' \tag{5.8.23}$$

The bimoment (force-length^2 unit) is defined by

$$BM = EI_w \phi'' \tag{5.8.24}$$

or

$$\phi'' = \frac{BM}{EI_w} \tag{5.8.25}$$

Substitution of Eq. (5.8.25) into Eq. (5.8.14) yields

$$\sigma_z = \frac{BM\omega_n}{I_w} \tag{5.8.26}$$

5.9. CROSS-SECTION PROPERTIES

5.9.1. Shear Center Location - General Method

Definition

If a general system of forces acting on a member is resolved into torsional and bending components with respect to the shear center, these cause, respectively, pure rotation and pure bending of the member (Kollbrunner and Basler 1969). That is, if a member is fixed at one end and subjected to a transverse load applied through the shear center at the other end, it undergoes bending without twisting. Conversely, a torque applied to this member induces no transverse deflection of the shear center. Hence, in such a case the shear-center axis remains straight during twisting, and the cross sections of the member rotate about the shear center during the deformation (Brush and Almroth 1975). The position of the shear center depends on the properties of the cross section only. It is therefore constant with respect to the cross section for prismatic members and the shear-center axis remains parallel to the centroidal axis. The resultant of the shear flows must be equal to the shearing forces acting on the cross section.

Development

In the theoretical development of the shear center location, the following assumptions are employed:

1) No torsion (only pure bending is considered).
2) Bending about one centroidal axis (not necessarily principal) is considered because the case of biaxial bending can be handled by repeating the same procedure.
3) Hooke's law holds.
4) Shearing stress is constant across plate thickness (i.e., thin-walled x-section).
5) Member is prismatic.
6) Cross section retains shape.
7) Small deflection.
8) Open cross section.
9) The thickness of the cross section is a function of the perimeter coordinate s, but not the longitudinal member axis.

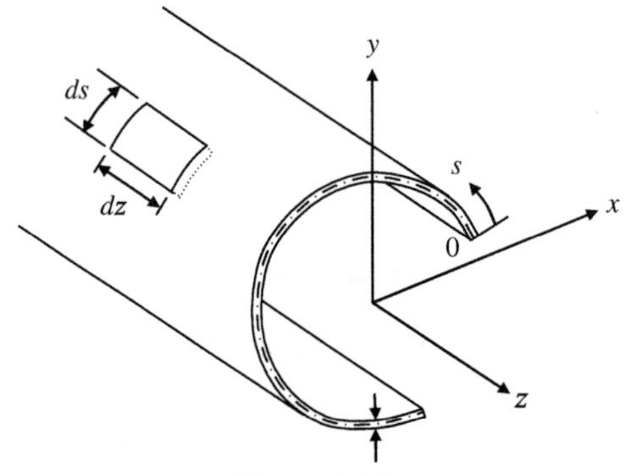

Figure 5-29

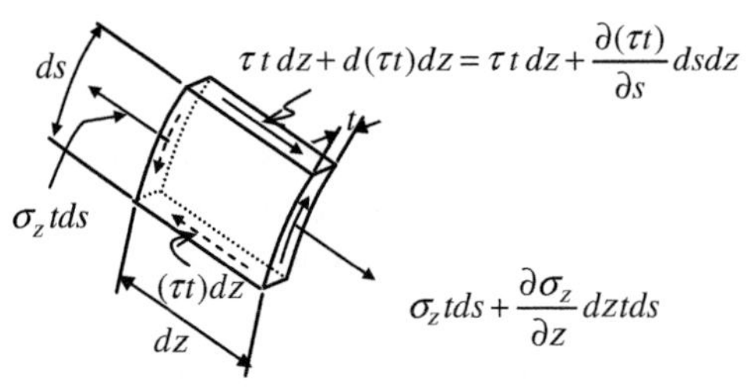

Figure 5-30

In the derivation, a general reference is made to Timoshenko (1945), Vlasov (1961), Galambos (1968), Kollbrunner and Basler (1969), and Heins (1975). An element isolated from the body in Fig. 5-29 is redrawn in Fig. 5-30, along with the stresses acting on it. Equilibrium of the forces in the z-direction gives

$$\sigma_z\, t\, ds + \tau\, t\, dz - \left[\tau t\, dz + \frac{\partial(\tau t)}{\partial s} ds\, dz\right] - \left(\sigma_z\, tds + \frac{\partial \sigma_z}{\partial z} t\, ds\, dz\right) = 0$$

From which one obtains

$$t\frac{\partial \sigma_z}{\partial z} + \frac{\partial(\tau t)}{\partial s} = 0 \tag{5.9.1}$$

From the equations of pure flexure (in the absence of axial force and $M_y = 0$), one gets

$$\sigma_z = M_x \left(\frac{I_{xy}x - I_y y}{I_{xy}^2 - I_x I_y}\right) \tag{5.9.2}$$

$$\frac{\partial \sigma_z}{\partial z} = \frac{\partial M_x}{\partial z}\left(\frac{I_{xy}x - I_y y}{I_{xy}^2 - I_x I_y}\right) \tag{5.9.3}$$

$$\frac{\partial M_x}{\partial z} = V_x \tag{5.9.4}$$

It should be noted that the concept of the shear center is meaningless in the constant-moment zone where $V_x = 0$.

Substituting Eq. (5.9.4) into Eq. (5.9.3) and Eq. (5.9.3) into Eq. (5.9.1), one gets

$$\frac{\partial(\tau t)}{\partial s} = -V_x t\left(\frac{I_{xy}x - I_y y}{I_{xy}^2 - I_x I_y}\right) \tag{5.9.5}$$

Integrating Eq. (5.9.5) with respect to s gives

$$\begin{aligned}\tau t &= -\int_0^s \left(\frac{I_{xy}x - I_y y}{I_{xy}^2 - I_x I_y}\right) V_x t\, ds \\ &= \frac{V_x}{(I_{xy}^2 - I_x I_y)}\left(I_y \int_0^s yt\, ds - I_{xy}\int_0^s xt\, ds\right)\end{aligned} \tag{5.9.6}$$

Summing the moment of the forces in Fig. 5-31 with respect to the centroid C, one gets

$$\sum M_C = -V_x x_0 + \int_0^b \rho(\tau t) ds = 0$$

From which one obtains

$$x_0 = (1/V_x)\int_0^b \rho(\tau t) ds$$

Substituting Eq. (5.9.6) for the shear flow gives

$$x_0 = \frac{1}{(I_{xy}^2 - I_x I_y)}\left(I_y \int_0^b \rho\, ds \int_0^s yt\, ds - I_{xy}\int_0^b \rho\, ds \int_0^s x\, t\, ds\right) \tag{5.9.7}$$

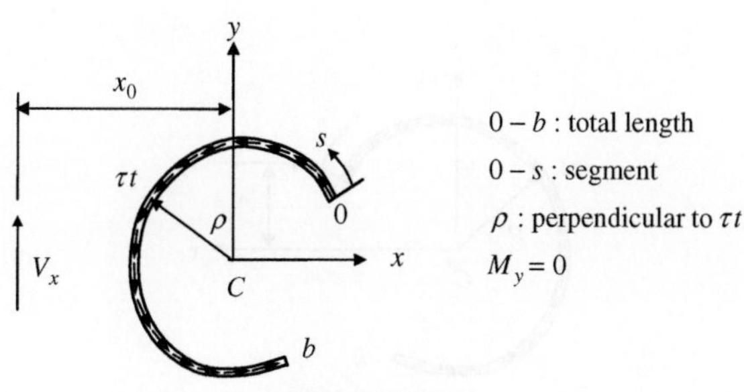

Figure 5-31

From the equations of flexure (axial force $= 0$ and $M_x = 0$), one gets

$$\sigma_z = M_y \left(\frac{I_{xy}y - I_x x}{I_{xy}^2 - I_x I_y} \right) \quad \text{and} \quad \frac{\partial \sigma_z}{\partial z} = V_y \left(\frac{I_{xy}y - I_x x}{I_{xy}^2 - I_x I_y} \right)$$

Substituting these equations into Eq. (5.9.1) yields

$$\left(\partial(\tau t)/\partial s \right) = -V_y t \left(\left(I_{xy}y - I_x x \right) / \left(I_{xy}^2 - I_x I_y \right) \right)$$

Integrating with respect to the perimeter coordinate gives

$$\tau t = -\int_0^s \left(\frac{I_{xy}y - I_x x}{I_{xy}^2 - I_x I_y} \right) V_y t \, ds = \frac{V_y}{I_{xy}^2 - I_x I_y} \left(I_x \int_0^s x \, t \, ds - I_{xy} \int_0^s y \, t \, ds \right) \tag{5.9.8}$$

Summing the moment of the forces with respect to the centroid gives

$$\sum M_C = V_y y_0 - \int_0^b \rho(\tau t) \, ds = 0$$

From which one obtains

$$y_0 = -\left(1/V_y \right) \int_0^b \rho(\tau t) ds$$

Substituting Eq. (5.9.8) for the shear flow gives

$$y_0 = \frac{1}{(I_{xy}^2 - I_x I_y)} \left(I_{xy} \int_0^b \rho \, ds \int_0^s y \, t \, ds - I_x \int_o^b \rho \, ds \int_0^s x \, t \, ds \right) \tag{5.9.9}$$

Evaluation of Integrals

Let the unit warping function (length^2 unit) with respect to the centroid C be defined as

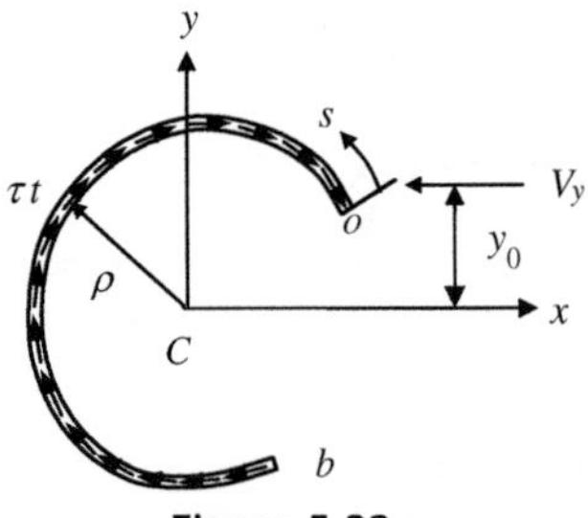

Figure 5-32

$$\omega = \int_0^s \rho ds \tag{5.9.10}$$

Then $d\omega = \rho ds$ and

$$\int_0^b \rho ds \int_0^s ytds = \int_o^b d\omega \int_0^s ytds$$

Integration by parts ($\int udv = uv - \int vdu$) and letting $u = \int_0^s ytds$ and $dv = d\omega$ to lead $du = y\,t\,ds$ and $v = \omega$, one obtains

$$\int_0^b \rho ds \int_0^s ytds = \int_o^b d\omega \int_0^s ytds = [\omega \int_o^s ytds]_0^b - \int_0^b \omega ytds$$

in which, by definition of centroid

$$[\omega \int_o^s ytds]_0^b = 0$$

Hence,

$$\int_0^b \rho ds \int_0^s ytds = -\int_0^b \omega ytds = -I_{wy} \tag{5.9.11}$$

Similarly,

$$\int_0^b \rho ds \int_0^s xtds = -\int_0^b \omega xtds = -I_{wx} \tag{5.9.12}$$

Substituting Eqs. (5.9.11) and (5.9.12) into Eqs. (5.9.7) and (5.9.9) gives

$$x_0 = \frac{I_{xy}I_{wx} - I_y I_{wy}}{I_{xy}^2 - I_x I_y} \tag{5.9.13}$$

$$y_0 = \frac{I_x I_{wx} - I_{xy} I_{wy}}{I_{xy}^2 - I_x I_y} \tag{5.9.14}$$

If x, y are principal axes, then ($I_{xy} = 0$)

$$x_0 = \frac{I_{wy}}{I_x} \tag{5.9.15}$$

$$y_0 = -\frac{I_{wx}}{I_y} \tag{5.9.16}$$

Although Eqs. (5.9.13) through (5.9.16) can be applied to simple thin-walled sections, when the section becomes complex with nonprismatic elements such as in the case of S shapes, as well as compound sections and multiply-connected cellular sections including stiffening interior cells, execution of these equations by an analytical means is simply not a viable option. When a cross section consists of n cells as in the case of a ship hull cross section, there will be n redundant shear flows. Hence, it is desirable to devise a numerical scheme that is readily programmable.

5.9.2. Numerical Computations of Section Properties

Usually thin-walled open sections are made up of a series of flat-plate elements. In the case of such sections, the numerical work can be simplified into a tabular form.

Determination of ω

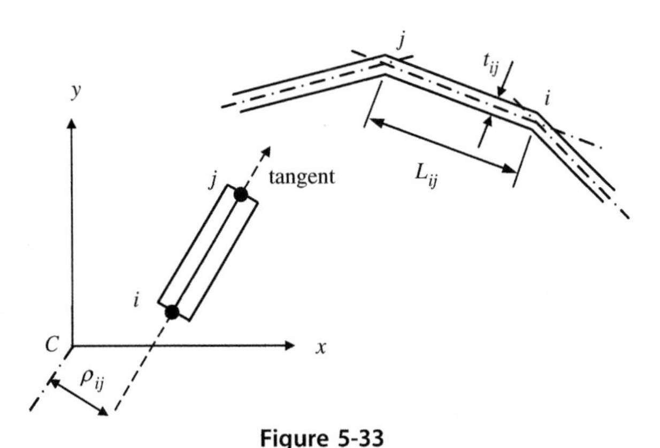

Figure 5-33

Based on Eq. (5.9.10), it becomes clear that the unit warping function with respect to the centroid C at the node j can be written as $\omega_j = \omega_i + \rho_{ij} L_{ij}$. Hence, at any node k

$$\omega_k = \sum_{i=1}^{j=k} \rho_{ij} L_{ij} \tag{5.9.17}$$

The definition adopted in the computation of ω_0 applies here likewise; that is, ρ is positive if centroid is to left when facing tangent line and ω varies linearly between two adjacent nodes i and j.

Determination of $I_{\omega x}$

From the geometry shown in Fig. 5-34, the following relations can be readily established:

$$L_{ij} = \frac{x_j - x_i}{\cos \alpha_{ij}} \tag{5.9.18a}$$

$$s = \frac{x - x_i}{\cos \alpha_{ij}} \tag{5.9.18b}$$

$$\frac{s}{L_{ij}} = \frac{x - x_i}{x_j - x_i} \tag{5.9.18c}$$

Similarly

$$ds = \frac{dx}{\cos\alpha_{ij}} \tag{5.9.18d}$$

ω is varying linearly between the two adjacent nodes and the ratio is

$$(\omega_j - \omega_i)/L_{ij} = (\omega - \omega_i)/s$$

From which

$$\omega = \omega_i + (\omega_j - \omega_i)(s/L_{ij})$$

Replacing s/L_{ij} by Eq. (5.9.18c) yields

$$\omega = \omega_i + \frac{(\omega_j - \omega_i)(x - x_i)}{(x_j - x_i)} \tag{5.9.19}$$

Substituting Eqs. (5.9.19) and (5.9.18d) into Eq. (5.9.12), one obtains

$$I_{wx} = \int_0^b \omega\, x\, t\, ds = \sum_0^b \frac{t_{ij}}{\cos \alpha_{ij}} \int_{x_i}^{x_j} \left[\omega_i + \frac{(\omega_j - \omega_i)(x - x_i)}{(x_j - x_i)} \right] x\, dx$$

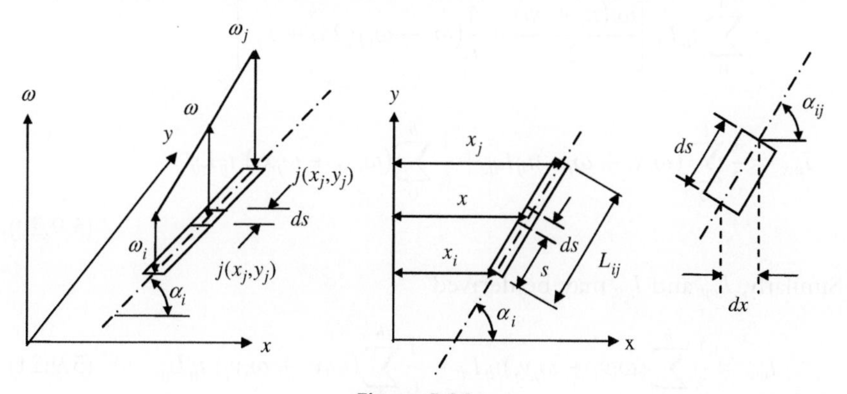

Figure 5-34

Expansion of Integral

$$\int_{x_i}^{x_j}\left[\omega_i+\frac{(\omega_j-\omega_i)(x-x_i)}{(x_j-x_i)}\right]x\,dx$$

$$=\frac{\omega_i}{2}\left(x_j^2-x_i^2\right)+\left(\frac{\omega_j-\omega_i}{x_j-x_i}\right)\left(\frac{x_j^3}{3}-\frac{x_i^3}{3}-\frac{x_ix_j^2}{2}+\frac{x_i^3}{2}\right)$$

$$=\frac{\omega_i}{2}[(x_j+x_i)(x_j-x_i)]+\frac{1}{6}\left(\frac{\omega_j-\omega_i}{x_j-x_i}\right)[2x_j^2(x_j-x_i)+x_i(x_i^2-x_j^2)]$$

Recalling $(x_j-x_i)=L_{ij}\cos\alpha_{ij}$

$$I_{wx}=\sum_0^b\left\{\frac{t_{ij}}{\cos\alpha_{ij}}\left[\frac{\omega_i}{2}(x_j+x_i)L_{ij}\cos\alpha_{ij}\right.\right.$$

$$\left.\left.+\frac{1}{6}(\omega_j-\omega_i)\frac{2x_j^2(x_j-x_i)-x_i(x_j^2-x_i^2)}{(x_j-x_i)}\right]\right\}$$

$$=\sum_0^b(t_{ij})\left[\frac{\omega_iL_{ij}(x_j+x_i)}{2}+\frac{1}{6}\frac{(\omega_j-\omega_i)(2x_j^2-x_ix_j-x_i^2)L_{ij}}{(x_j-x_i)}\right]$$

$$=\sum_0^b(t_{ij}L_{ij})\left[\frac{\omega_i(x_j+x_i)}{2}+\frac{1}{6}(\omega_j-\omega_i)\frac{(2x_j^2-2x_ix_j-x_i^2+x_ix_j)}{(x_j-x_i)}\right]$$

$$=\sum_0^b t_{ij}L_{ij}\left[\frac{\omega_i(x_j+x_i)}{2}+\frac{1}{6}(\omega_j-\omega_i)(2x_j+x_i)\right]$$

$$I_{wx}=\frac{1}{3}\sum_0^b(\omega_ix_i+\omega_jx_j)t_{ij}L_{ij}+\frac{1}{6}\sum_0^b(\omega_ix_j+\omega_jx_i)\;t_{ij}L_{ij} \tag{5.9.20}$$

Similarly, I_{wy} and I_{xy} may be derived

$$I_{wy}=\frac{1}{3}\sum_0^b(\omega_iy_i+\omega_jy_j)t_{ij}L_{ij}+\frac{1}{6}\sum_0^b(\omega_iy_j+\omega_jy_i)\;t_{ij}L_{ij} \tag{5.9.21}$$

$$I_{xy} = \frac{1}{3}\sum_{0}^{b}(x_iy_i + x_jy_j)t_{ij}L_{ij} + \frac{1}{6}\sum_{0}^{b}(x_iy_j + x_jy_i)\ t_{ij}L_{ij} \tag{5.9.22}$$

Likewise, I_x and I_y in numerical expressions are

$$I_x = \frac{1}{3}\sum_{0}^{b} t_{ij}L_{ij}(y_i^2 + y_iy_j + y_j^2) \tag{5.9.23}$$

$$I_y = \frac{1}{3}\sum_{0}^{b} t_{ij}L_{ij}(x_i^2 + x_ix_j + x_j^2) \tag{5.9.24}$$

Hence, quantities needed for Eqs.(5.9.13) through (5.9.16) are numerically evaluated in Eqs. (5.9.20) through (5.9.24).

Example

Determine the shear center of the section shown in Fig. 5-35. The thickness t is uniform ($t = 0.5$ in.).

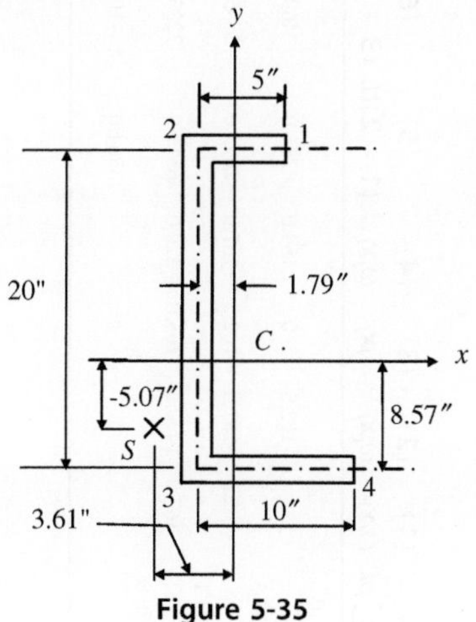

Figure 5-35

Numerical values are to be taken from Table 5-8.

$$\begin{aligned} I_{\omega x} &= \frac{1}{3}\sum(\omega_i x_i + \omega_j x_j)\ tL + \frac{1}{6}\sum(\omega_i x_j + \omega_j x_i)\ tL \\ &= \frac{1}{3}(3564) + \frac{1}{6}(-10) = 1186.3\ \text{in}^5 \end{aligned}$$

$$\begin{aligned} I_{wy} &= \frac{1}{3}\sum(\omega_i y_i + \omega_j y_j)\ tL + \frac{1}{6}\sum(\omega_i y_j + \omega_j y_i)\ tL \\ &= \frac{1}{3}(-11390) + \frac{1}{6}(-4230) = 4501.7\ \text{in}^5 \end{aligned}$$

Table 5-8 Shear center location

Node	x	y	L_{ij}	ρ_{ij}	$\rho_{ij}L_{ij}$	$\omega = \sum \rho L$	(1) $\omega_i x_i$	(2) $\omega_j x_j$	(3) $\omega_i x_j$	(4) $\omega_j x_i$	(5) $(1+2)tL$	(6) $(3+4)tL$	(7) $\omega_i y_i$	(8) $\omega_j y_j$	(9) $\omega_i y_j$	(10) $\omega_j y_i$	(11) $(7+8)tL$	(12) $(9+10)tL$
1	3.21	11.43				0												
1-2			5	11.43	57.3		0	−102.5	0	184.0	−256.0	460.0	0	655.0	0	655.0	1640.0	1640.0
2	−1.79	11.43				57.3												
2-3			20	1.79	35.8		−102.5	−167.0	−102.5	−167.0	−2695.0	−2695.0	655.0	−796.0	−490.0	1065.0	−1410.0	5750.0
3	−1.79	−8.57				93.1												
3-4			10	8.57	85.7		−167.0	1470.0	765.0	−320.0	6515.0	2225.0	−796.0	−1530.0	−798.0	−1530.0	−11620.0	−11620.0
4	8.21	−8.57				178.8												
$\sum$											3564.0	−10.0					−11390.0	−4230.0

$$I_{xx} = \frac{1}{12}(0.5)^3(10+5) + 0.5(5)(11.43)^2 + \frac{1}{12}(0.5)(20)^3 + 10(1.43)^2$$
$$= 1047.6 \text{ in}^4$$

$$I_{yy} = \frac{1}{12}(0.5)(10^3+5^3) + 5(3.21)^2 + 2.5(0.71)^2 + \frac{1}{12}(20)(0.5)^3$$
$$+ 10(1.79) = 132.5 \text{ in}^4$$

$$I_{xy} = 2.5(11.43)(0.75) + 5(-8.57)(3.25) + 10(-1.75)(1.43)$$
$$= -142.6 \text{ in}^4$$

$$x_0 = \frac{I_{xy}I_{wx} - I_yI_{wy}}{I_{xy}^2 - I_xI_y} = \frac{(-142.6)(1186.3) - (132.5)(-4501.7)}{(-142.6)^2 - (1047.6)(132.5)}$$
$$= -3.61 \text{ in}$$

$$y_0 = \frac{I_xI_{wx} - I_{xy}I_{wy}}{I_{xy}^2 - I_xI_y} = \frac{(1047.6)(1186.3) - (-142.6)(-4501.7)}{(-142.6)^2 - (1047.6)(132.5)}$$
$$= -5.07 \text{ in}$$

This is just a simple example. If a cross section consists of multiple-cellular sections combined with protruding elements, formulas to evaluate cross sectional properties for such sections are not available. Each closed cell must be made an open section by introducing a fictitious cut (Heins 1975) somewhere in the cell perimeter. Then, the section properties on this pseudo-open section are evaluated. The compatibility condition at the cut will provide a condition equation to determine the redundant shear flow or to determine other properties such as the normalized warping function to be consistent at the cut.

Although a few attempts to evaluate the cross-sectional properties by digital computers can be found in the literature, SECP (Yoo and Acra 1986) is believed to be the most comprehensive program currently (2010) available to compute cross-sectional properties, particularly multicellular sections with internal stiffening cells such as those found in an orthotropic bridge deck. This program can be downloaded from the senior author's Web pages. Access codes are available from the back flap of the book. The user documentation is included in the Fortran source code by liberal use of Comment statements. Once anyone experiences the power of SECP, a longhand computation of cross-sectional properties will not likely be attempted anymore.

REFERENCES

Brush, D. O., & Almroth, B. O. (1975). *Buckling of Bars, Plates, and Shells.* New York: McGraw-Hill.

Galambos, T. V. (1968). *Structural Members and Frames.* Englewood Cliffs, NJ: Prentice-Hall.

Heins, C. P. (1975). *Bending and Torsional Design in Structural Members.* Lexington, MA: Lexington Book.

Heins, C. P., & Seaburg, P. A. (1963). *Torsional Analysis of Rolled Steel Sections.* Bethlehem, PA: Bethlehem Steel Company.

Kollbrunner, C. F., & Basler, K. (1969). Torsion in Structure: An engineering Approach. In: E. C. Glauser (Ed.). Berlin: Springer-Verlag. translated from German.

Nadai, A. (1923). Sand-heap Analogy. *Z. angew. Math. Mechanik, Vol. 3*(No. 6), 442.

Nadai, A. (1950) (2nd ed.).*Theory of Flow and Fracture of Solids, Vol. 1.* New York: McGraw-Hill.

Nakai, H., & Yoo, C. H. (1988). *Analysis and Design of Curved Steel Bridges.* New York: McGraw-Hill.

Prandtl, L. (1903). Membrane Analogy. *Physik. Z., Vol. 4.*

Rees, D. W. A. (2000). *Mechanics of Solids and Structures.* London: Imperial College Press.

Saada, A. S. (1974). *Elasticity, Theory and Application* (2nd ed.). Malabar, FL: Krieger.

Saint-Venant, B. D. (1855). *Mém. acad. sci. savants, Vol. 14*, 233–560.

Seaburg, P. A., & Carter, C. J. (1997). *Torsional Analysis of Structural Steel Members.* Chicago, IL: American Institute of Steel Construction.

Sokolnikoff, I. S. (1956). *Mathematical Theory of Elasticity* (2nd ed.). New York: McGraw-Hill.

Timoshenko, S. P. (1945). *Theory of Bending, Torsion, and Buckling of Thin-Walled Members of Open Cross Section.* Philadelphia: J. Franklin Institute. Vol. 239, No. 3, p. 201, No. 4, pp. 248, No. 5, p. 343.

Timoshenko, S. P., & Gere, J. M. (1961). *The Theory of Elastic Stability* (2nd ed.). New York: McGraw-Hill.

Timoshenko, S. P., & Goodier, J. N. (1951). *Theory of Elasticity* (2nd ed.). New York, NY: McGraw-Hill.

Vlasov, V. Z. (1940). *Tonkostennye uprugie sterzhni (Thin-walled elastic beams).* Stroiizdat (in Russian).

Vlasov, V. Z. (1961). In *Tonkostennye uprugie sterzhni (Thin-walled elastic beams).* Moscow, 1959, (2nd ed.), translated by Schechtman. Jerusalem: Israel Program for Scientific Translation.

Yoo, C. H. (2000). Torsional and Other Properties of Prestressed Concrete Sections. *PCI Journal, Vol. 45*(No. 3), 66–72.

Yoo, C. H., & Acra, S. V. (1986). Cross Sectional Properties of Thin-Walled Multi-Cellular Section. *Computers and Structures, Vol. 112*(No. 9), 53–61.

PROBLEMS

5.1 In order to minimize the potential stress concentration at the reentrant corners at the bottom of a rectangular keyway, the sharp corners are smoothed out by a circular hole. Show that the Prandtl stress function $\phi = m(r^2 - b^2)(2a \cos \theta / r - 1)$ (Sokolnikoff 1956) leads to the solution of the circular shaft with a circular keyway, shown in Fig. P5-1. Determine the constant m and the expressions of the stresses, τ_{zx} and

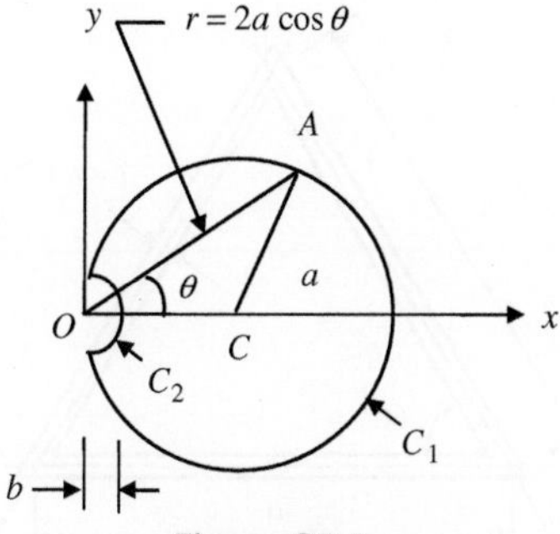

Figure P5-1

τ_{zy} on the boundaries C_1 and C_2. If $a = 1$ in. and $b = 1/8$ in., show that the ratio of the maximum shear stresses that are developed in C_2 and C_1 is approximately 2 to 1.

5.2 Three rods with solid cross sections, square, equilateral triangle, and circle, have equal cross-sectional areas and are subjected to equal twisting moments (Saada 1974). Compute the maximum shearing stresses developed and St. Venant torsional constants. Evaluate the shape factors and assess the effectiveness of each shape.

5-3 Develop $M_z - \theta'$ relationship for pure torsion over the elastic and plastic range for an angle section shown in Fig. P5-3 made of a material obeying an ideal elastic-plastic stress strain law. Neglect end effect ($t<<L$). Plot $M - \theta$ curve, nondimensionally (M_z/M_{zy}) vs θ'/θ_y. Compute the shape factor.

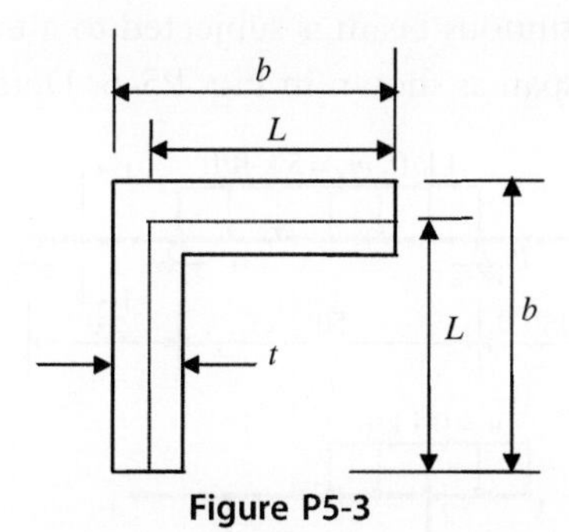

Figure P5-3

5.4 For the triangular section of side length "a" and $t = a/20$ shown in Fig. P5-4, evaluate K_{T0} and K_{Tc}. Cut at point "1." If $\tau_{max} = \tau_y$, compute the ratio T_c/T_0.

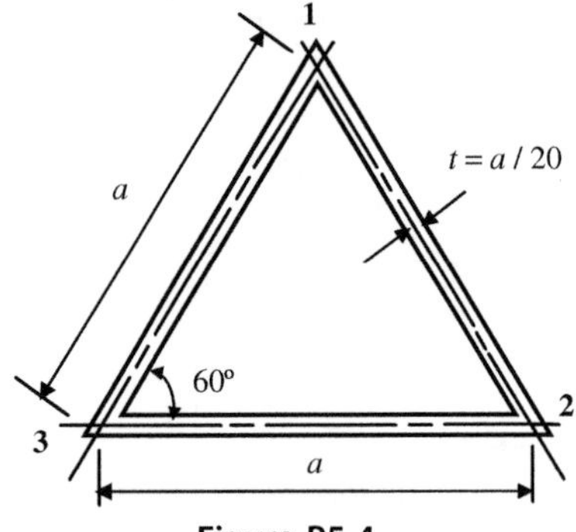

Figure P5-4

5.5 A three-cell thin-walled box section is made of steel plates of constant thickness as shown in Fig. P5-5. Determine the maximum shearing stresses in various elements if the box is subjected to a torque of 1000 kip-in. Determine the shearing stress in the central cell if a cut was made in each side cell wall. Use $G = 11.2 \times 10^3$ ksi.

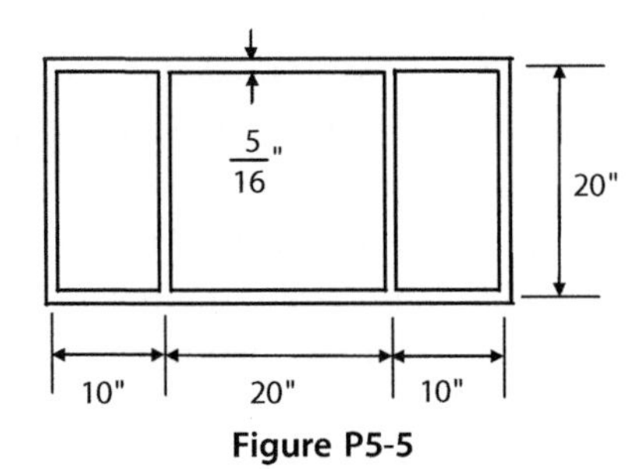

Figure P5-5

5.6 A three-span continuous beam is subjected to a uniformly distributed load at its center span as shown in Fig. P5-6. Determine the locations

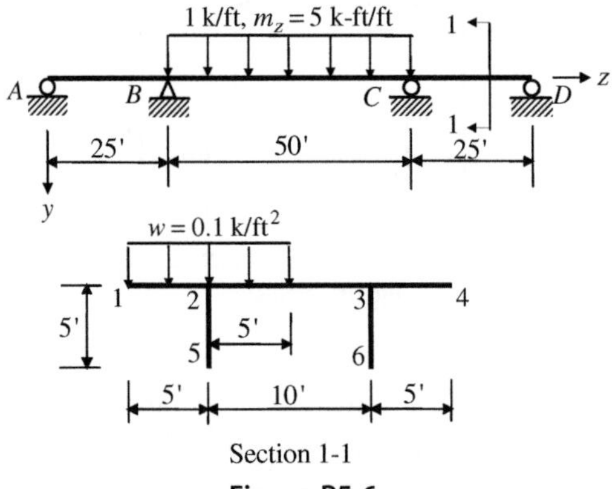

Figure P5-6

and magnitudes of the maximum normal and maximum shearing stress. Use $E = 3000$ ksi and $G = 1000$ ksi. The thickness of each element of the cross section is 5 inches.

5.7 **(a)** Using the numerical procedure, determine the location of the shear center of the shape shown in Figure P5-7 in terms of variables b, c, d, and t.

(b) What is x_0 if $b = 3''$, $c = 1''$, $d = 5''$, $t = 0.25''$?

(c) Verify the computation when $c = 0$. (Use any known value.)

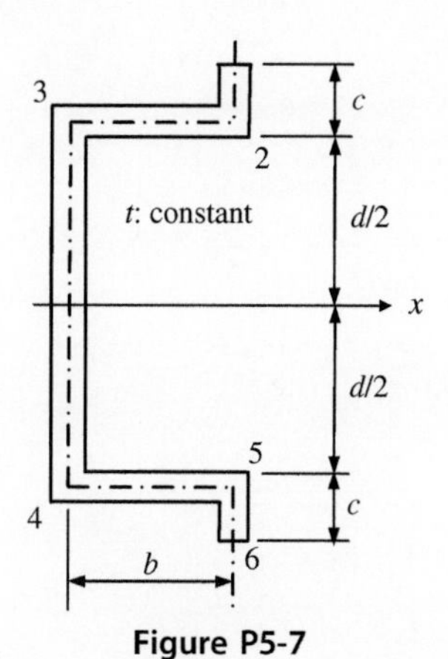

Figure P5-7

5.8 Using the numerical procedure, determine the shear center.

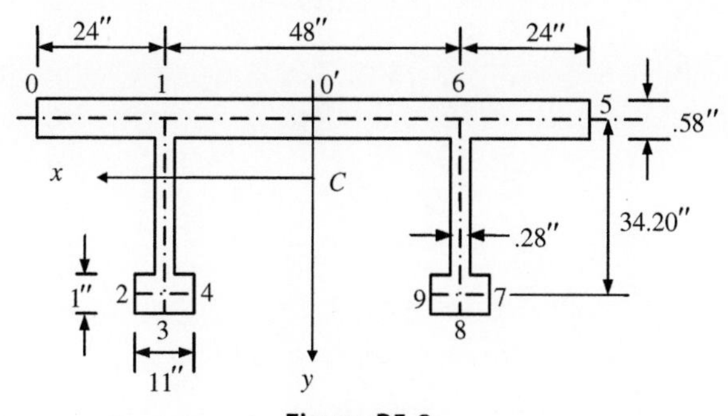

Figure P5-8

Torsional and Flexural-Torsional Buckling

Contents

6.1. INTRODUCTION

In Chapter 1, the fundamental case of buckling of centrally loaded columns is presented under the assumption that columns will buckle in the plane of a principal axis without the accompanying rotation of the cross sections. This assumption, first made by Euler (1744), appears reasonable for the doubly symmetric cross section but becomes problematic if cross sections have only one axis of symmetry or none at all. The possibility of torsional column failure had never been recognized until open thin-walled sections were used in aircraft design in the 1930s. Experience has revealed that columns having an open section with only one or no axis of symmetry show a tendency to bend and twist simultaneously under axial compression. The ominous nature of this type of failure lies in the fact that the actual critical load of such columns may be less than that predicted by the generalized Euler formula due to their small torsional rigidities. Bleich (1952) gives a fairly thorough overview of the early development of the theory on the torsional buckling.

Bleich and Bleich (1936) were among the early developers of the theory on torsional buckling along with Wagner and Pretschner (1936), Ostenfeld (1931), Kappus (1938), Lundquist and Fligg (1937), Goodier (1941), Hoff

ISBN 978-0-12-385122-2, doi:10.1016/B978-0-12-385122-2.10006-5

(1944), and Timoshenko (1945). All of these authors make the fundamental assumption that the plane cross sections of the column warp but that their geometry does not change during buckling. Thus, the theories consider primary failure (global buckling) of columns as opposed to local failure characterized by distortion of the cross sections. The dividing line between primary and local failure is not always sharp. Separate analysis of primary and local buckling based on governing differential equations, without abandoning the assumption that cross sections of the column will not deform, may yield only approximate solutions since there could be coupling of primary and local buckling. Modern finite element codes with refined modeling capabilities incorporating at least flat shell elements may be able to assess this combined buckling action.

The notion of "unit warping" or the concept of sectorial coordinate appears to have gained currency in the literature including Goodier (1941), Galambos (1968), Kollbrunner and Basler (1969), Timoshenko (1945), and Vlasov (1961). Bleich and Bleich (1936) developed their differential equations governing the torsional buckling behavior of columns with thin-walled open sections based on the principle of minimum potential energy without invoking the notion of "unit warping" or the concept of the sectorial coordinate. Although they claim that their equations are practically the same as those developed under the concept of the sectorial coordinate, they differ in a significant aspect. The warping constant I_w (or Γ in their notation) for the cross section consisting of thin rectangular elements does not vanish according to their theory, in which the axial strain and the curvature are considered to be coupled. This consideration appears to be odd since in the linearized bifurcation-type buckling analysis, an adjacent equilibrium configuration is examined after all static deformations have taken place. As a consequence of their theory of nonvanishing warping constant, a beam having a cross section consisting of a series of narrow rectangular elements that meet at the shear center will become warped. This is a direct contradiction to Timoshenko and Gere (1961)[1] and Vlasov (1961).[2] According to the definition of unit warping, the warping constant I_w must be equal to zero for such sections where the perpendicular distance from the shear center to each element wall becomes zero. Therefore, the differential equations to be developed for torsional and flexural-tosional

[1] See page 217.

[2] See page 27: a thin-walled beam consisting of a single bundle of very thin rectangular plates does not become warped.

buckling here in this chapter and for lateral-torsional buckling in the next chapter will be based on the concept of unit warping.

6.2. STRAIN ENERGY OF TORSION

Recall that the concept of the stress tensor arises from equilibrium considerations and that the concept of the strain tensor arises from kinematic (deformation) considerations. These tensors are related to each other by laws that are called constitutive laws. The constitutive laws relating stresses and strains directly and uniquely can be expressed mathematically as

$$\tau_{ij} = \tau_{ij}(\varepsilon_{11}, \varepsilon_{12}, ..., \varepsilon_{33}) \tag{6.2.1}$$

where τ_{ij} = stresses and ε_{ij} = strains.

Consider a quantity, U_0, strain energy density function which is also known to be a point function that is independent from the integral path taken:

$$U_0 = \int_0^{\varepsilon_{ij}} \tau_{ij} d\varepsilon_{ij}$$

A strain energy density function is measured in energy per volume and is a scalar quantity. For linearly elastic materials, it becomes

$$U_0 = \frac{1}{2} \tau_{ij}\, \varepsilon_{ij} \tag{6.2.2}$$

Then, the strain energy stored in a body is

$$U = \int_V U_o dv = \frac{1}{2} \int_V \tau_{ij}\, \varepsilon_{ij} dv \tag{6.2.3}$$

The strain energy stored in a twisted member is broken down into two parts, one due to St. Venant torsion and the other due to warping torsion. In order to maintain a generality, torsional stresses and corresponding strain expressions are explicitly developed and substituted in Eq. (6.2.3). As columns of a closed cross section are not likely to develop torsional or flexural-torsional buckling, such columns are not considered here.

6.2.1. St. Venant Torsion

Recall that the St. Venant torsional moment is given by Eq. (5.4.1) as

$$M_z^{St} = GK_T \phi'$$

The corresponding shear stress in an open cross section shown in Fig. 5-20 is given by

$$\tau_{zs} = \tau_z^{St} = \frac{2M_z^{St}h}{K_T} = 2Gh\phi' \tag{6.2.4a}$$

and the corresponding shear strain is

$$\varepsilon_{zs} = \frac{\gamma_{zs}}{2} = \frac{\tau_{zs}}{2G} = h\phi' \tag{6.2.4b}$$

Then, the strain energy due to St. Venant torsion is

$$U_T^{St} = \int_V U_o dv = \frac{1}{2}\int_V \tau_{ij}\varepsilon_{ij} dv = \frac{1}{2}\int_V (\tau_{zs}\varepsilon_{zs} + \tau_{sz}\varepsilon_{sz}) dv$$

Substituting (6.2.4a) and (6.2.4b) gives

$$U_z^{St} = 2G(\phi')^2 \int_0^\ell \int_0^b \int_{-t/2}^{t/2} h^2 dh ds dv = 2G(\phi')^2 \int_0^\ell \int_0^b h^3 \Big|_{-\frac{t}{2}}^{\frac{t}{2}} ds dz$$

$$= \frac{1}{2}\int_0^\ell GK_T(\phi')^2 dz \tag{6.2.4c}$$

where b is the width of a thin-walled element and h is measured from the centerline of a thin-walled element thickness so that $h_{\max} = t/2$.

6.2.2. Warping Torsion

For a member subjected to warping torsion, the dominant strain energy stored in the member due to its resistance to warping is assumed to be the strain energy due to warping normal stresses. Even though warping shear stresses produce strain energy, it is usually considered to be negligibly small and is neglected as in the case of not including the shear deformation effect in ordinary flexural analysis. Warping normal stresses and corresponding strains are evaluated by

$$\sigma_z^w = \frac{BM\omega_n}{I_w} = \frac{EI_w\phi''\omega_n}{I_w} = E\omega_n\phi'' \tag{5.8.26}$$

$$\varepsilon_z^w = \frac{\sigma_z^w}{E} = \omega_n\phi''$$

Hence,

$$U_T^w = \frac{1}{2}\int_V \sigma_z^w \varepsilon_z^w dv = \frac{1}{2}E(\phi'')^2 \int_0^\ell \int_A (\omega_n)^2 dA dz = \frac{1}{2}\int_0^\ell EI_w(\phi'')^2 dz$$
$$= \frac{1}{2}\int_0^\ell EI_w(\phi'')^2 dz \tag{6.2.5}$$

where $\int_A (\omega_n)^2 dA = I_w$ as per Eq. (5.8.20).

6.3. TORSIONAL AND FLEXURAL-TORSIONAL BUCKLING OF COLUMNS

It is assumed that the cross section retains its original shape during buckling. For prismatic members having thin-walled open sections, there are two parallel longitudinal reference axes: One is the centroidal axis, and the other is the shear center axis. The column load P must be placed at the centroid to induce a uniform compressive stress over the entire cross section. Transverse loads for pure bending must be placed along the shear center axis in order to not induce unintended torsional response. Since the cross sectional rotation is measured by the rotation about the shear center axis, the only way not to generate unintended torsional moment by the transverse load is to place the transverse load directly on the shear center axis so as to eliminate the moment arm.

It is assumed that the member ends are simply supported for simplicity so that displacements in the x- and y-directions and the moments about these axes vanish at the ends of the member. Hence,

$$u = u'' = v = v'' = 0 \quad \text{at } z = 0 \text{ and } \ell \tag{6.3.1}$$

The member ends are assumed to be simply supported for torsion so that the rotation with respect to the shear center axis and warping restraint are equal to zero at the ends of the member. Thus

$$\phi = \phi'' = 0 \tag{6.3.2}$$

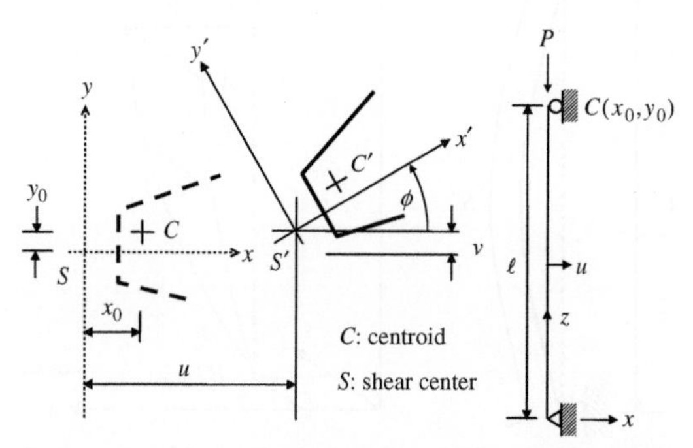

Figure 6-1 Flexural-torsional buckling deformation

In order to consider a meaningful warping restraint, the member ends must be welded (not bolted) thoroughly with thick end plates or embedded into heavy bulkhead with no gap at the ends. These types of torsional boundary conditions are not expected to be encountered in ordinary construction practice.

Strain energy stored in the member in the adjacent equilibrium configuration consists of four parts, ignoring the small contribution of the bending shear strain energy and the warping shear strain energy: the energies due to bending in the x- and y-directions; the energy due to St. Venant shear stress; and the energy due to warping torsion. Thus

$$U = \frac{1}{2}\int_0^{\ell} EI_y(u'')^2\,dz + \frac{1}{2}\int_0^{\ell} EI_x(v'')^2\,dz + \frac{1}{2}\int_0^{\ell} GK_T(\phi')^2\,dz + \frac{1}{2}\int_0^{\ell} EI_w(\phi'')^2\,dz \tag{6.3.3}$$

The loss of potential energy of external loads is equal to the negative of the product of the loads and the distances they travel as the column takes an adjacent equilibrium position. Figure 6-2 shows a longitudinal fiber whose ends get close to one another by an amount Δ_b. The distance Δ_b is equal to the difference between the arc length S and the chord length ℓ of the fiber. Thus

$$V = -\int_A \Delta_b \sigma dA \tag{6.3.4}$$

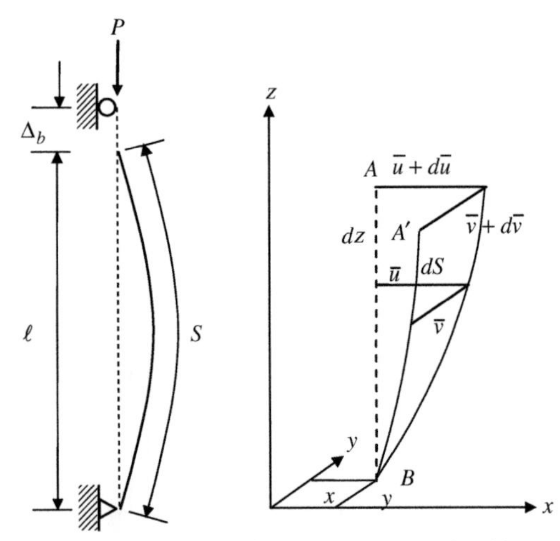

Figure 6-2 Fiber deformations due to buckling

As shown in Fig. 6-2 when the x and y displacements of the lower end of a differential element dz of the column are designated as $\bar{u}$ and $\bar{v}$, then the corresponding displacements at the upper end are $\bar{u} + d\bar{u}$ and $\bar{v} + d\bar{v}$. From the Pythagorean theorem, the length of the deformed element is

$$dS = \sqrt{(d\bar{u})^2 + (d\bar{v})^2 + (dz)^2} = \sqrt{\left(\frac{d\bar{u}}{dz}\right)^2 + \left(\frac{d\bar{v}}{dz}\right)^2 + 1}\; dz \quad (6.3.5)$$

In Section 1.6 it was shown that the binomial expansion can be applied to Eq. (6.3.5) if the magnitude of the derivatives is small compared to unity. Hence,

$$\sqrt{\left(\frac{d\bar{u}}{dz}\right)^2 + \left(\frac{d\bar{v}}{dz}\right)^2 + 1}\; dz \doteq \left[\frac{1}{2}\left(\frac{d\bar{u}}{dz}\right)^2 + \frac{1}{2}\left(\frac{d\bar{v}}{dz}\right)^2 + 1\right] dz \quad (6.3.6)$$

Integrating Eq. (6.3.6) gives

$$S = \int_0^\ell \left[\frac{1}{2}\left(\frac{d\bar{u}}{dz}\right)^2 + \frac{1}{2}\left(\frac{d\bar{v}}{dz}\right)^2 + 1\right] dz \quad (6.3.7)$$

from which

$$\Delta_b = S - \ell = \frac{1}{2}\int_0^\ell \left[\left(\frac{d\bar{u}}{dz}\right)^2 + \left(\frac{d\bar{v}}{dz}\right)^2\right] dz \quad (6.3.8)$$

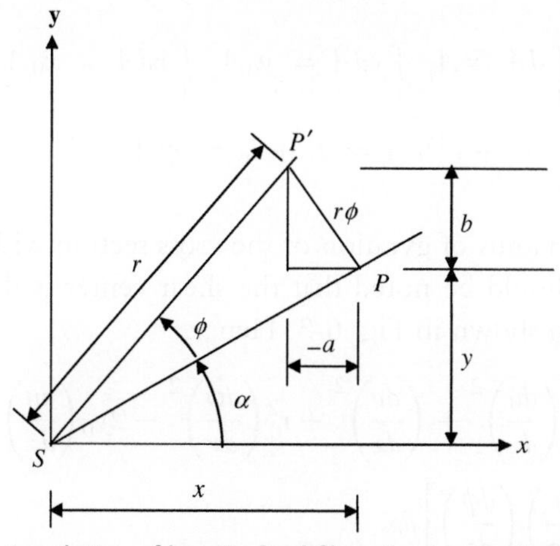

Figure 6-3 Lateral translation of longitudinal fiber due to rotation about shear center

where $\overline{u}$ *and* $\overline{v}$ are the translation of the shear center u *and* v plus additional translation due to rotation of the cross section about the shear center. The additional translations $d\overline{u}$ *and* $d\overline{v}$, in the x- *and* y-directions, are denoted, as shown in Fig. 6-3 by $-a$ and b. From the geometry of the figure, it is evident that $PP' = r\phi$, $a = r\phi \sin\alpha$, and $b = r\phi \cos\alpha$.

Since $x = r\cos\alpha$ and $y = r\sin\alpha$, one may also write $-a = -y\phi$ and $b = x\phi$. Hence, the total displacements of the fiber are

$$\begin{aligned}\overline{u} &= u - y\phi \\ \overline{v} &= v + x\phi\end{aligned} \tag{6.3.9}$$

Substituting Eq. (6.3.9) into Eq. (6.3.8) yields

$$\begin{aligned}\Delta_b = \frac{1}{2}\int_0^\ell \bigg[\left(\frac{du}{dz}\right)^2 + \left(\frac{dv}{dz}\right)^2 + (x^2+y^2)\left(\frac{d\theta}{dz}\right)^2 \\ - 2y\left(\frac{du}{dz}\right)\left(\frac{d\theta}{dz}\right) + 2x\left(\frac{dv}{dz}\right)\left(\frac{d\theta}{dz}\right)\bigg]dz\end{aligned} \tag{6.3.10}$$

$$\begin{aligned}V = -\frac{1}{2}\int_0^\ell \int_A \sigma\bigg[\left(\frac{du}{dz}\right)^2 + \left(\frac{dv}{dz}\right)^2 + (x^2+y^2)\left(\frac{d\theta}{dz}\right)^2 \\ -2y\left(\frac{du}{dz}\right)\left(\frac{d\theta}{dz}\right) + 2x\left(\frac{dv}{dz}\right)\left(\frac{d\theta}{dz}\right)\bigg]dA\,dz\end{aligned} \tag{6.3.11}$$

In order to simplify Eq. (6.3.11), the following geometric relations can be used:

$$\begin{aligned}&\int_A dA = A, \ \int_A y\,dA = y_0 A, \ \int_A x\,dA = x_0 A \\ &\int_A (x^2+y^2)dA = I_x + I_y = r_0^2 A\end{aligned} \tag{6.3.12}$$

where r_0 is polar radius of gyration of the cross section with respect to the shear center. It should be noted that the shear center is the origin of the coordinate system shown in Fig. 6-3. Hence,

$$\begin{aligned}V = -\frac{P}{2}\int_0^\ell \bigg[\left(\frac{du}{dz}\right)^2 + \left(\frac{dv}{dz}\right)^2 + r_0^2\left(\frac{d\phi}{dz}\right)^2 - 2y_0\left(\frac{du}{dz}\right)\left(\frac{d\phi}{dz}\right) \\ + 2x_0\left(\frac{dv}{dz}\right)\left(\frac{d\phi}{dz}\right)\bigg]dz\end{aligned} \tag{6.3.13}$$

The total potential energy functional π is given by the sum of Eq. (6.3.3) and Eq. (6.3.13) as

$$\Pi = U(\text{Eq.}(6.3.3)) + V(\text{Eq.}(6.3.13)) = \int_{\ell} F(z, u', v', \phi', u'', v'', \phi'')dz \tag{6.3.14}$$

According to the rules of the calculus of variations, Π will be stationary (minimum) if the following three Euler-Lagrange differential equations are satisfied:

$$\begin{aligned} \frac{\partial F}{\partial u} - \frac{d}{dz}\frac{\partial F}{\partial u'} + \frac{d^2}{dz^2}\frac{\partial F}{\partial u''} &= 0 \\ \frac{\partial F}{\partial v} - \frac{d}{dz}\frac{\partial F}{\partial v'} + \frac{d^2}{dz^2}\frac{\partial F}{\partial v''} &= 0 \\ \frac{\partial F}{\partial \phi} - \frac{d}{dz}\frac{\partial F}{\partial \phi'} + \frac{d^2}{dz^2}\frac{\partial F}{\partial \phi''} &= 0 \end{aligned} \tag{6.3.15}$$

Execution of Eq. (6.3.15) gives

$$EI_y u^{iv} + Pu'' - Py_0\phi'' = 0 \tag{6.3.16a}$$

$$EI_x v^{iv} + Pv'' + Px_0\phi'' = 0 \tag{6.3.16b}$$

$$EI_w\phi^{iv} + (r_0^2 P - GK_T)\phi'' - y_0 Pu'' + x_0 Pv'' = 0 \tag{6.3.16c}$$

These three differential equations are the simultaneous differential equations of torsional and flexural-torsional buckling for centrally applied loads only. Each of the three equations in Eq. (6.3.16) is a fourth-order differential equation. Hence, the system must have 12 (4 × 3) boundary conditions to determine uniquely the integral constants.

Equations (6.3.16) are linear and homogeneous, and have constant coefficients. Their general solution in the most general case can be obtained by means of the characteristic polynomial approach. Assume the solution to be of the form,[3]

$$u = A\sin\frac{\pi z}{\ell}, \quad v = B\sin\frac{\pi z}{\ell}, \quad \phi = C\sin\frac{\pi z}{\ell}$$

where A, B, and C are arbitrary constants. Substituting derivatives of these functions into the differential equations (6.3.16) and reducing by the common factor $\sin(\pi z/\ell)$, one obtains

[3] Vlasov (1961) shows this is indeed the solution of the eigenfunctions for a simply supported column; see p. 271.

$$
\begin{aligned}
(EI_y k^2 - P)A + y_0 PC &= 0 \\
(EI_x k^2 - P)B - x_0 PC &= 0 \\
y_0 PA - x_0 PB + (EI_w k^2 + GK_T - r_0^2 P)C &= 0
\end{aligned} \tag{6.3.17}
$$

where $k^2 = \pi^2/\ell^2$.

For a nontrivial solution for A, B, and C, the determinant of the system of homogeneous equations must vanish. Thus

$$
\begin{vmatrix} P_y - P & 0 & y_0 P \\ 0 & P_x - P & -x_0 P \\ y_0 P & -x_0 P & r_0^2 (P_\phi - P) \end{vmatrix} = 0 \tag{6.3.18}
$$

where

$$
P_x = \frac{\pi^2 EI_x}{\ell^2},\ P_y = \frac{\pi^2 EI_y}{\ell^2},\ P_\phi = \frac{1}{r_0^2}\left(EI_w \frac{\pi^2}{\ell^2} + GK_T \right) \tag{6.3.19}
$$

Expanding Eq. (6.3.18) gives

$$
(P_y - P)(P_x - P)(P_\phi - P) - (P_y - P)\frac{P^2 x_0^2}{r_0^2} - (P_x - P)\frac{P^2 y_0^2}{r_0^2} = 0 \tag{6.3.20}
$$

The solution of the above cubic equation gives the critical load of the column.

Case 1: If the cross section is doubly symmetrical, then $x_0 = y_0 = 0$, and Eq. (6.3.20) reduces to

$$
(P_y - P)(P_x - P)(P_\phi - P) = 0
$$

The three roots and corresponding mode shapes are:

$$
P_{cr} = P_y = \frac{\pi^2 EI_y}{\ell^2}: \quad A \neq 0, B = C = 0 \Rightarrow \text{pure flexural buckling}
$$

$$
P_{cr} = P_x = \frac{\pi^2 EI_x}{\ell^2}: \quad B \neq 0, A = C = 0 \Rightarrow \text{pure flexural buckling}
$$

$$
P_{cr} = P_\phi = \frac{1}{r_0^2}\left(\frac{\pi^2 EI_w}{\ell^2} + GK_T \right):
$$

$$
C \neq 0, A = B = 0 \Rightarrow \text{pure torsional buckling}
$$

Coupled flexural-torsional buckling does not occur in a column with a cross section where the shear center coincides with the center of gravity. Doubly symmetric sections and the Z purlin section have the shear center and the center of gravity at the same location.

Example Consider a pinned column (W14 × 43) of length $\ell = 280$ inches. Use $E = 29 \times 10^3$ ksi and $G = 11.2 \times 10^3$ ksi
For W14 × 43, $I_x = 428$ in^4, $I_y = 45.2$ in^4, $K_T = 1.05$ in^4, $I_w = 1,950$ in^6

$$P_y = \frac{\pi^2 EI_y}{\ell^2} = \frac{\pi^2 \times 29 \times 10^3 \times 45.2}{280^2} = 165 \text{ kips}$$

$$P_x = \frac{\pi^2 EI_y}{\ell^2} = \frac{\pi^2 \times 29 \times 10^3 \times 428}{280^2} = 1,563 \text{ kips}$$

$$r_0 = [(I_x + I_y)/A]^{1/2} = [(428 + 45.2)/12.6]^{1/2} = 6.13 \text{ in}$$

$$P_\phi = \frac{1}{6.13^2}\left(11200 \times 1.05 + \frac{\pi^2 \times 29000 \times 1950}{280^2}\right) = 505 \text{ kips}$$

Usually in a column fabricated of a W section, torsional buckling is not checked as it is likely to buckle with respect to the weak axis.

Case 2: If there is only one axis of symmetry as shown in Fig. 6-4, say the x axis, then shear center lies on the x axis and $y_0 = 0$. Then Eq. (6.3.20) reduces to

$$(P_y - P)\left[(P_x - P)(P_\phi - P) - \frac{P^2 x_0^2}{r_0^2}\right] = 0$$

This equation is satisfied either if

$$P_{cr} = P_y$$

or if

$$(P_x - P)(P_\phi - P) - \frac{P^2 x_0^2}{r_0^2} = 0$$

The first expression corresponds to pure flexural buckling with respect to the y axis. The second is a quadratic equation in P and its solutions correspond to buckling by a combination of flexure and twisting, that is, flexural-torsional buckling. The smaller root of the second equation is

$$P_{F-T} = \frac{1}{2K}\left[P_\phi + P_x - \sqrt{(P_\phi + P_x)^2 - 4KP_x P_\phi}\right]$$

where

$$K = \left[1 - \left(\frac{x_0}{r_0}\right)^2\right]$$

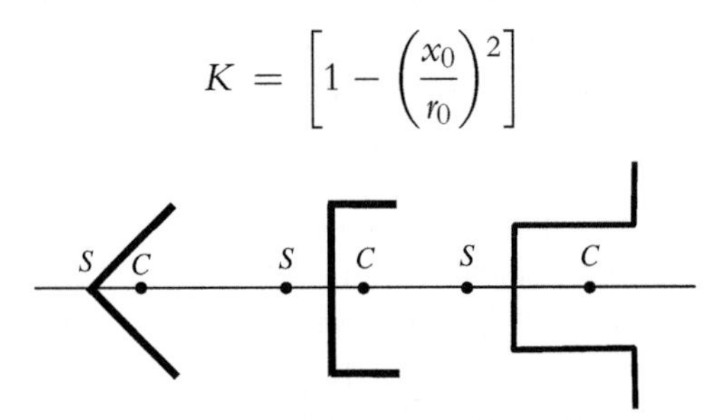

Figure 6-4 Singly symmetric sections

Example For the simply supported column of a singly symmetric hat section shown in Fig. 6-5, develop the elastic buckling strength envelope for the data given. Section and material properties are:

$$A = 6bt,\ I_x = 7.333b^3t,\ I_y = 1.167b^3t,\ I_w = 0.591b^5t,$$

$$K_T = 2bt^3,\ r_0^2 = 2.086b^2,\ t = 0.1 \text{ in},\ b/t = 10$$

$$\sigma_y = 32 \text{ ksi},\ E = 10,300 \text{ ksi},\ G = 3,850 \text{ ksi}$$

where terms having t^3 are neglected in the computation of I_x and I_y.

$$P_0 = \sigma_y A$$

$$\frac{P_x}{P_0} = 12.062\left(\frac{E}{\sigma_y}\right)\left(\frac{b}{\ell}\right)^2$$

$$\frac{P_y}{P_0} = 1.9196\left(\frac{E}{\sigma_y}\right)\left(\frac{b}{\ell}\right)^2$$

$$\frac{P_\phi}{P_0} = 0.160\left(\frac{G}{\sigma_y}\right)\left(\frac{t}{b}\right)^2 + 0.466\left(\frac{E}{\sigma_y}\right)\left(\frac{b}{\ell}\right)^2$$

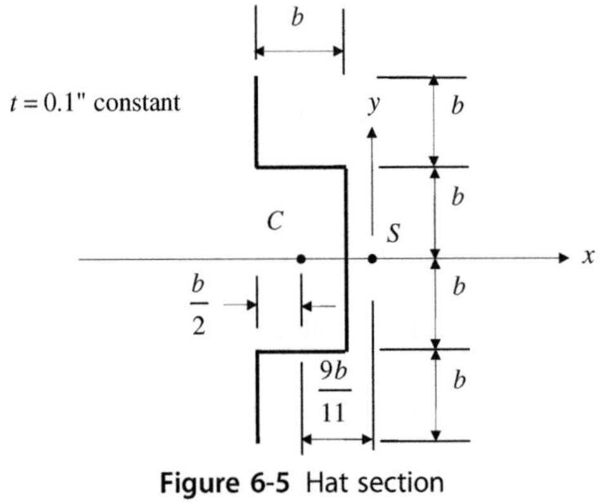

$t = 0.1$" constant

Figure 6-5 Hat section

Results are tabulated in Table 6-1 and plotted in Fig. 6-6. As can be seen from Fig. 6-6, the flexural-torsional buckling strength controls the lowest critical load for short columns until the Euler buckling load takes over at a longer column length. It is particularly ominous for cross sections where warping constants vanish, for which the pure torsional buckling load is independent from the column length.

Case 3: If there is no axis of symmetry, then $x_0 \neq 0, y_0 \neq 0$ and Eq. (6.3.20) cannot be simplified.

In such cases, bending about either principal axis is coupled with both twisting and bending about the other principal axis. All the three roots to Eq. (6.3.20) correspond to torsional-flexural buckling and are lower than all the separable critical loads. Hence, if $P_y < P_x < P_\phi$, then

$$P_{cr} < P_y < P_x < P_\phi$$

Table 6-1 Comparison of flexural-torsional buckling analysis

	$\ell = 1'$			$\ell = 10'$		
	Exact	**STSTB**	**% Error**	**Exact**	**STSTB**	**% Error**
P_ϕ (kips)	23.67	23.65	0.084	3.890	3.916	0.690
P_x (kips)	517.80	518.00	0.039	5.178	5.180	0.035
P_y (kips)	82.60	82.64	0.048	0.826	0.826	0.051
P_{T-F} (kips)	23.30	23.32	0.085	2.815	2.828	0.450

Note: Only four elements were used in STSTB (Yoo 1980). For $t = 0.1$ in.

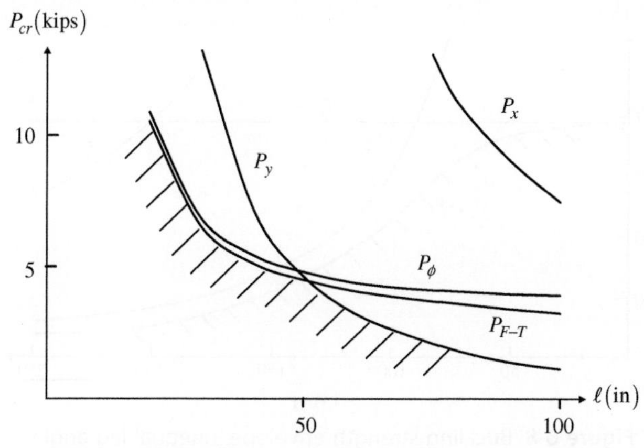

Figure 6-6 Buckling strength envelope, hat section

Example For the simply supported column of an unequal leg angle, $L8 \times 6 \times 1/2$, shown in Fig. 6-7, develop the elastic buckling strength envelope. Use $E = 29 \times 10^3$ ksi and $G = 11.2 \times 10^3$ ksi.

Neglecting the fillet or corner effect, SECP gives the following section properties:

$$I_x = 54.52 \text{ in}^4, I_y = 11.36 \text{ in}^4, K_T = 0.563 \text{ in}^4, I_w = 0,$$

$$r_0 = 4.02 \text{ in}, x_0 = 2.15 \text{ in}, y_0 = -1.33 \text{ in}.$$

In this case the cubic equation, Eq. (6.3.23), must be solved for each set of critical loads, P_x, P_y, P_ϕ. This can be best accomplished by utilizing **Maple**® with Exel. As expected, the lowest elastic buckling load is controlled by the flexural-torsional buckling as shown in Fig. 6-8.

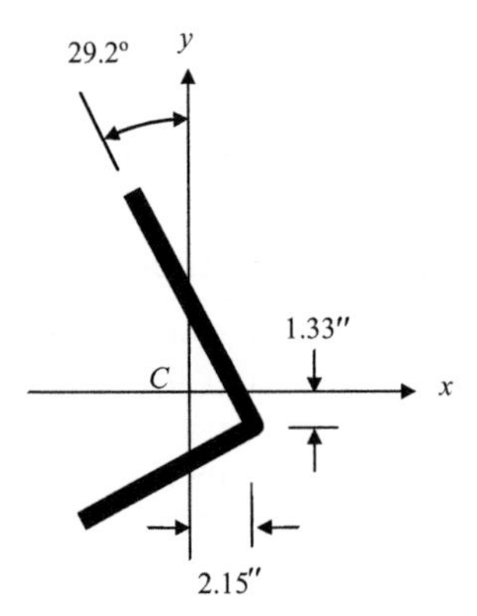

Figure 6-7 Unequal leg angle

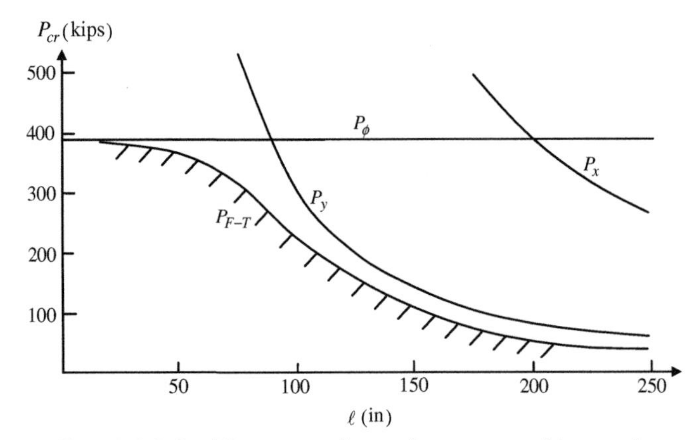

Figure 6-8 Buckling strength envelope, unequal leg angle

Torsional buckling and flexural-torsional buckling are particularly onerous in cold-formed steel design because of the thin gages utilized, which in turn yield lower torsional rigidities. A series of research projects sponsored by the American Iron and Steel Institute have produced rich research results in the area. Chajes and Winter (1965) is a good example of this effort.

6.4. TORSIONAL AND FLEXURAL-TORSIONAL BUCKLING UNDER THRUST AND END MOMENTS

In the previous section, the total potential energy functional of a column for torsional and flexural-torsional buckling expressed with respect to the shear center was derived. When it is desired to express the same in the centroidal coordinate system, it can be done readily, provided that the sign of x_0 and y_0 needs to be reversed as they are defined in two separate coordinate systems (this time they are measured from the centroid). Hence,

$$U = \frac{1}{2}\int_0^\ell EI_y\left(\frac{d^2u}{dz^2}\right)^2 dz + \frac{1}{2}\int_0^\ell EI_x\left(\frac{d^2v}{dz^2}\right)^2 dz + \frac{1}{2}\int_0^\ell GK_T\left(\frac{d\phi}{dz}\right)^2 dz + \frac{1}{2}\int_0^\ell EI_w\left(\frac{d^2\phi}{dz^2}\right)^2 dz$$

and

$$V = -\frac{P}{2}\int_0^\ell \left[\left(\frac{du}{dz}\right)^2 + \left(\frac{dv}{dz}\right)^2 + r_0^2\left(\frac{d\phi}{dz}\right)^2 + 2y_0\left(\frac{du}{dz}\right)\left(\frac{d\phi}{dz}\right) - 2x_0\left(\frac{dv}{dz}\right)\left(\frac{d\phi}{dz}\right)\right] dz \tag{6.4.1}$$

It can be seen from Eq. (6.4.1) that the sign of x_0 and y_0 is reversed from that in Eq. (6.3.13). It should be noted that r_0 is the polar radius of gyration of the cross section with respect to the shear center.

Although the differential equations for torsional and flexural-torsional buckling have been successfully derived in the previous section with the coordinate center located at the shear center as was done by Goodier (1941) and Hoff (1944), the same equations, except the sign of x_0 and y_0, can be derived with the coordinate center located at the centroid. Also, it may be informative to examine the detailed mechanics in the neighboring equilibrium position instead of relying blindly on the calculus of variations procedure for the derivation of the differential equations.

6.4.1. Pure Torsional Buckling

In order to show how a compressive load may cause purely torsional buckling, consider a column of a cruciform with four identical thin-walled flanges of width b and thickness t as shown in Fig. 6-9. As demonstrated by Case 1 in the previous section, the torsional buckling load will be the lowest for the cruciform column unless the column length is longer than 40 times the flange width where the thickness is 5% of the width.

It is imperative to draw a slightly deformed configuration of the column corresponding to the type of buckling to be examined (in this case, torsional buckling). The controidal axis z (which coincides with the shear center axis in this case) does not bend but twists slightly such that mn becomes part of a curve with a displacement component of v in the y-direction. As has been illustrated in Fig. 5-4, the membrane force times the second derivative produces a fictitious lateral load of intensity $-Pd^2v/dz^2$. Consider an element mn shown in Fig. 6-9 in the form of a strip of length dz located at a distance r from the z-axis and having a cross sectional area tdr. The displacement of this element in the y-direction becomes

$$v = r\phi \tag{6.4.2}$$

The compressive forces acting on the ends of the element mn are $-\sigma tdr$, where $\sigma = P/A$. The statically equivalent fictitious lateral load is then $-(\sigma tdr)(d^2v/dz^2)$ or $-(\sigma trdr)(d^2\phi/dz^2)$. The twisting moment

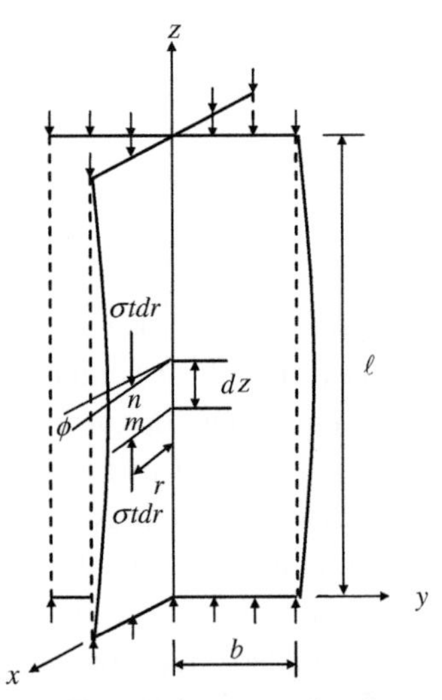

Figure 6-9 Pure torsional buckling (after Timoshenko and Gere, *Theory of Elastic Stability*, 2nd ed., McGraw-Hill, 1961). Reproduced by permission.

about the z-axis due to this fictitious lateral load acting on the element mn is then $(-\sigma)(d^2\phi/dz^2)(dz)(t)(r^2dr)$. Summing up the twisting moments for the entire cross section yields

$$-\sigma\frac{d^2\phi}{dz^2}dz\int_A tr^2dr = -\sigma\frac{d^2\phi}{dz^2}dz\int_A r^2dA = -\sigma\frac{d^2\phi}{dz^2}dzI_0 \qquad (6.4.3)$$

where I_0 is the polar moment of inertia of the cross section with respect to the shear center S, coinciding in this case with the centroid. Recalling the notation for the distributed torque, one obtains

$$m_z = -\sigma\frac{d^2\phi}{dz^2}I_0 \qquad (6.4.4)$$

Substituting Eq. (6.4.4) into Eq. (5.8.23) yields

$$EI_w\phi^v - (GK_T - \sigma I_0)\phi'' = 0 \qquad (6.4.5)$$

For column cross sections in which all elements meet at a point such as that shown in Fig. 6-9, angles and tees, the warping constant vanishes. Hence, in the case of torsional buckling, Eq. (6.4.5) is satisfied if

$$GK_T - \sigma I_0 = 0$$

which yields

$$\sigma_{cr} = \frac{GK_T}{I_0} = \frac{(4/3)bt^3G}{(4/3)tb^3} = \frac{Gt^2}{b^2} \qquad (6.4.6)$$

For cases in which the warping constant does not vanish, the critical compressive stress can also be obtained form Eq. (6.4.5). Introduce $k^2 = (\sigma I_0 - GK_T)/(EI_w)$ into Eq. (6.4.5) to transform it into $\phi^{iv} + k^2\phi'' = 0$, a similar form to a beam-column equation. The general solution of this equation is given by $\phi = a\sin kz + b\cos kz + cz + d$. Applying boundary conditions of a simply supported column gives $\phi = a\sin k\ell = 0$, from which $k\ell = n\pi$. Substituting for k yields

$$\sigma_{cr} = \frac{1}{I_0}\left(GK_T + \frac{n^2\pi^2}{\ell^2}EI_w\right) \qquad (6.4.7)$$

which is identical to Eq. (6.3.19) obtained by the calculus of variations procedure in the previous Section as it should be.

6.4.2. Flexural-Torsional Buckling

Consider an unsymmetrical section shown in Fig. 6-10. The x and y are principal axes, and x_0 and y_0 are the coordinate of the shear center

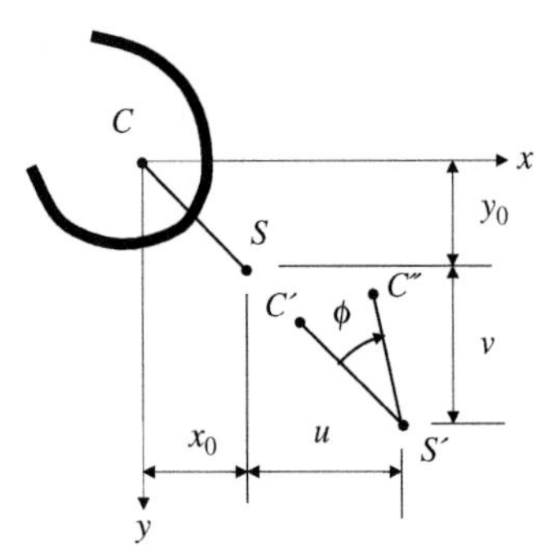

Figure 6-10

S measured from the centroid C. During buckling the centroid translates to C' and then rotates to C''. Therefore, the final position of the centroid is $u + y_0\phi$ and $v - x_0\phi$. If only central load P is applied on a simply supported column, the bending moments with respect to the principal axes are

$$M_x = -P(v - x_0\phi) \quad \text{and} \quad M_y = -P(u + y_0\phi)$$

The sign convention for M_x and M_y is such that they are considered positive when they create positive curvature

$$EI_x v'' = +M_x = -P(v - x_0\phi) \tag{6.4.8}$$

$$EI_y u'' = +M_y = -P(u + y_0\phi) \tag{6.4.9}$$

Consider a small longitudinal strip of cross section tds defined by coordinate x and y as was done in the case of pure torsional buckling. The components of its displacements in the - x and y directions during buckling are $u + (y_0 - y)\phi$ and $v - (x_0 - x)\phi$, respectively. Recalling the procedure illustrated in Fig. 5-4, the products of the compressive force acting on the slightly bend element, σtds and the second derivative of the displacements give a fictitious lateral load in the x- and y-directions of intensity

$$-(\sigma tds)\frac{d^2}{dz^2}[u + (y_0 - y)\phi]$$

$$-(\sigma tds)\frac{d^2}{dz^2}[v - (x_0 - x)\phi]$$

These fictitious lateral loads produce twisting moment about the shear center per unit length of the column of intensity

$$dm_z = -(\sigma tds)\frac{d^2}{dz^2}[u + (y_0 - y)\phi](y_0 - y)$$
$$+(\sigma tds)\frac{d^2}{dz^2}[v - (x_0 - x)\phi](x_0 - x)$$

Integrating over the entire cross-sectional area and realizing that

$$\sigma\int_A tds = P, \quad \int_A xtds = \int_A ytds = 0, \quad \int_A y^2 tds = I_x,$$
$$\int_A x^2 tds = I_y, \quad I_0 = I_x + I_y + A(x_0^2 + y_0^2)$$

one obtains

$$m_z = \int_A dm_z = P(x_0 v'' - y_0 u'') - r_0^2 P\phi'' \tag{6.4.10}$$

where $r_0^2 = I_0/A$.

Substituting Eq. (6.4.10) into Eq. (5.8.23) yields

$$EI_w\phi^{iv} - (GK_T - r_0^2)\phi'' - x_0 Pv'' + y_0 Pu'' = 0 \tag{6.4.11}$$

Equations (6.4.8), (6.4.9), and (6.4.11) are the three simultaneous differential equations for torsional, flexural-torsional buckling of columns with arbitrary thin-walled cross sections. They are identical to Eq. (6.3.16) derived in the previous section as expected, except the signs of x_0 and y_0 are reversed as they are measured from the opposite reference point.

6.4.3. Torsional and Flexural-Torsional Buckling under Thrust and End Moments

Consider the case when the column is subjected to bending moments M_x and M_y applied at the ends in addition to the concentric load P. The bending moments M_x and M_y are taken positive when they produce positive curvatures in the plane of bending. It is assumed that the effect of P on the bending stresses can be neglected and the initial deflection of the column due to the moments is considered to be small. Under this assumption, the normal stress at any point on the cross section of the column is independent of z and is given by

$$\sigma = -\frac{P}{A} - \frac{M_x y}{I_x} - \frac{M_y x}{I_y} \tag{6.4.12}$$

As is customarily done in the elastic buckling analysis, any prebuckling deformations are not considered in the adjacent equilibrium condition. Additional deflections u and v of the shear center and rotation ϕ with respect to the shear center axis are produced during buckling, and examination is being conducted on this new slightly deformed configuration. Thus, the components of deflection of any longitudinal fiber of the column are $u + (y_0 - y)\phi$ and $v - (x_0 - x)\phi$. Hence, the fictitious lateral loads and distributed twisting moment resulting from the initial compressive force in the fibers acting on their slightly bent and rotated cross sections are obtained in a manner used earlier.

$$q_x = -\int_A (\sigma t ds)\frac{d^2}{dz^2}[u + (y_0 - y)\phi]$$

$$q_y = -\int_A (\sigma t ds)\frac{d^2}{dz^2}[v - (x_0 - x)\phi]$$

$$\begin{aligned} m_z = & -\int_A (\sigma t ds)\frac{d^2}{dz^2}[u + (y_0 - y)\phi](y_0 - y) \\ & + \int_A (\sigma t ds)\frac{d^2}{dz^2}[v - (x_0 - x)\phi](x_0 - x) \end{aligned}$$

Substituting Eq. (6.4.12) into the above equations and integrating yields

$$q_x = -P\frac{d^2 u}{dz^2} - (Py_0 - M_x)\frac{d^2\phi}{dz^2}$$

$$q_y = -P\frac{d^2 v}{dz^2} + (Px_0 - M_y)\frac{d^2\phi}{dz^2}$$

$$m_z = -(Py_0 - M_x)\frac{d^2 u}{dz^2} + (Px_0 - M_y)\frac{d^2 v}{dz^2} - (M_x\beta_x + M_y\beta_y + r_0^2 P)\frac{d^2\phi}{dz^2}$$

where the following new cross-sectional properties are introduced:

$$\beta_x = \frac{1}{I_x}\left(\int_A y^3 dA + \int_A x^2 y dA\right) - 2y_0 \qquad (6.4.13$$

$$\beta_y = \frac{1}{I_y}\left(\int_A x^3 dA + \int_A xy^2 dA\right) - 2x_0$$

The three equations for bending and torsion of the column are

$$EI_y u^{iv} + Pu'' + (Py_0 - M_x)\phi'' = 0 \tag{6.4.14}$$

$$EI_x v^{iv} + Pv'' - (Px_0 - M_y)\phi'' = 0 \tag{6.4.15}$$

$$EI_w \phi^{iv} - (GK_T - M_x\beta_x - M_y\beta_y - r_0^2 P)\phi'' + (Py_0 - M_x)u'' - (Px_0 - M_y)v'' = 0 \tag{6.4.16}$$

These are three simultaneous differential equations with constant coefficients. Hence, the critical values of the external forces can be computed for any combinations of end conditions.

If the load P is applied eccentrically with the coordinate of the point of application of P by e_x and e_y measured from the centroid, the end moments become $M_x = Pe_y$ and $M_y = Pe_x$. Equations (6.4.14) through (6.4.16) take the form

$$EI_y u^{iv} + Pu'' + P(y_0 - e_y)\phi'' = 0 \tag{6.4.17}$$

$$EI_x v^{iv} + Pv'' - P(x_0 - e_x)\phi'' = 0 \tag{6.4.18}$$

$$EI_w \phi^{iv} - (GK_T - Pe_y\beta_x - Pe_x\beta_y - r_0^2 P)\phi'' + P(y_0 - e_y)u'' - P(x_0 - e_x)v'' = 0 \tag{6.4.19}$$

If the thrust P acts along the shear center axis ($x_0 = e_x$ and $y_0 = e_y$), Eqs. (6.4.17) through (6.4.19) become very simple as they become independent of each other. The first two equations yield the Euler loads, and the third equation gives the critical load corresponding to pure torsional buckling of the column.

If the thrust becomes zero, one obtains the case of pure bending of a beam by couples M_x and M_y at the ends. Equations (6.4.17) through (6.4.19) take the form

$$EI_y u^{iv} - M_x\phi'' = 0 \tag{6.4.20}$$

$$EI_x v^{iv} + M_y\phi'' = 0 \tag{6.4.21}$$

$$EI_w \phi^{iv} - (GK_T - M_x\beta_x - M_y\beta_y)\phi'' - M_x u'' + M_y v'' = 0 \tag{6.4.22}$$

Assume the x-axis is the strong axis. If $M_y = 0$, then the critical lateral-torsional buckling moment can be computed from

$$EI_y u^{iv} - M_x \phi'' = 0 \tag{6.4.23}$$

$$EI_w \phi^{iv} - (GK_T - M_x \beta_x)\phi'' - M_x u'' = 0 \tag{6.4.24}$$

If the ends of the beam are simply supported, the displacement functions for u and ϕ can be taken in the form

$$u = A \sin \frac{\pi z}{\ell} \quad \phi = B \sin \frac{\pi z}{\ell}$$

Substituting derivatives of the displacement functions, one obtains the following characteristic polynomial for the critical moment:

$$\frac{\pi^2 EI_y}{\ell^2}\left(GK_T + EI_w \frac{\pi^2}{\ell^2} - M_x \beta_x \right) - M_x^2 = 0 \tag{6.4.25}$$

Incorporating the following notations

$$P_y = \frac{\pi^2 EI_y}{\ell^2}, \quad P_\phi = \frac{1}{r_0^2}\left(GK_T + EI_w \frac{\pi^2}{\ell^2} \right)$$

Eq. (6.4.25) becomes

$$M_x^2 + P_y \beta_x M_x - r_0^2 P_y P_\phi = 0 \tag{6.4.26}$$

The roots of Eq. (6.4.26) are

$$M_{xcr} = -\frac{P_y \beta_x}{2} \pm \sqrt{\left(\frac{P_y \beta_x}{2} \right)^2 + r_0^2 P_y P_\phi} \tag{6.4.27}$$

If the beam has two axes of symmetry, β_x vanishes and the critical moment becomes

$$\begin{aligned} M_{xcr} &= \pm \sqrt{r_0^2 P_y P_\phi} = \pm \sqrt{r_0^2 \frac{EI_y \pi^2}{\ell^2} \frac{1}{r_0^2}\left(GK_T + EI_w \frac{\pi^2}{\ell^2} \right)} \\ &= \pm \frac{\pi}{\ell}\sqrt{EI_y \left(GK_T + EI_w \frac{\pi^2}{\ell^2} \right)} \end{aligned} \tag{6.4.28}$$

where $\pm$ sign in Eq. (6.4.28) implies that a pair of end moments equal in magnitude but opposite in direction can cause lateral-torsional buckling of a doubly symmetrical beam.

In this discussion, considerations have been given for the bending of a beam by couples applied at the ends so that the normal stresses caused by these moments remain constant, thereby maintaining the governing differential equations with constant coefficients. If a beam is subjected to lateral loads, the bending stresses vary with z and the resulting differential equations will have variable coefficients, for which there are no general closed-form solutions available and a variety of numerical integration schemes are used. The computation of critical loads of lateral-torsional buckling is discussed in the next chapter.

REFERENCES

Bleich, F. (1952). *Buckling Strength of Metal Structures.* New York: McGraw-Hill.

Bleich, F. & Bleich, H. (1936). Bending Torsion and Buckling of Bars Composed of Thin Walls, Prelim. Pub. 2nd Cong. International Association for Bridge and Structural Engineering, English ed., p. 871, Berlin.

Euler, L. (1744). *Methodus Inveniendi Lineas Curvas Maximi Minimive Propreietate Gaudentes* (Appendix, De Curvis Elasticis). Lausanne and Geneva: Marcum Michaelem Bousquet.

Chajes, A., & Winter, G. (1965). Torsional-Flexural Buckling of Thin-Walled Members. *Journal of the Structural Div., ASCE, Vol. 91*(No. ST4), 103–124.

Galambos, T. V. (1968). *Structural Members and Frames.* Englewood Cliffs, NJ: Prentice-Hall.

Goodier, J. N. (1941). *The Buckling of Compressed Bars by Torsion and Flexure.* Cornell University Engineering Experiment Station. Bulletin No. 27.

Hoff, N. J. (1944). A Strain Energy Derivation of the Torsional-Flexural Buckling Loads of Straight Columns of Thin-Walled Open Sections. *Quarterly of Applied Mathematics, Vol. 1*(No. 4), 341–345.

Kappus, R. (1938). Drillknicken zentrisch gedrhckter Stäbe emit offenem Profile im elastischen Bereich. *Lufthahrt-Forschung, Vol. 14*(No. 9), 444–457, English translation, Twisting Failure of Centrally Loaded Open-Sections Columns in the Elastic Range, NACA Tech. Mem. No. 851.

Kollbrunner, C. F., & Basler, K. (1969). E.C. Glauser (ed.) *Torsion in Structures; An Engineering Approach,* translated from German Berlin: Springer-Verlag.

Lundquist, E.F., & Fligg, C.M. (1937). A Theory for Primary Failure of Straight Centrally Loaded Columns, NACA Tech: Report No. 582.

Ostenfeld, A. (1931). *Politechnisk Laereanstalts Laboratorium for Bygningsstatik.* Copenhagen: Meddelelse No. 5.

Timoshenko, S.P. (1945). Theory of Bending, Torsion and Buckling of Thin-Walled Members of Open Cross Section, J. Franklin Institute (Philadelphia), *Vol. 239,* No. 3, p. 201, No. 4, p. 248, No. 5, p. 343.

Vlasov, V.Z. (1961). *Tonkostennye uprugie sterzhni* (*Thin-walled elastic beams*), (2nd ed.), Moscow, 1959, translated by Schechtman, Israel Program for Scientific Translation, Jerusalem.

Wagner, H. & Pretschner, W. (1936). Torsion and Buckling of Open Sections, NACA Tech. mem. No. 784.

Yoo, C. H. (1980). Bimoment Contribution to Stability of Thin-Walled Assemblages. *Computers and Structures, Vol. 11*(No. 5), 465–471.

PROBLEMS

6.1 Develop the buckling strength envelope for a simply supported column of light gage channel section shown in Fig. P6-1. Material properties and dimensions of the cross section are as follows:

$E = 29 \times 10^3$ ksi, $G = 11.2 \times 10^3$ ksi, $a = b = 5$ in, $c = 1$ in , $t = 0.1$ in.

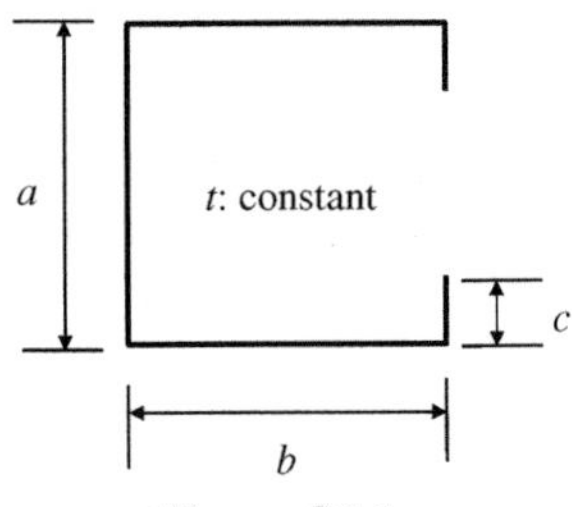

Figure P6-1

6.2 Develop the buckling strength envelope for a simply supported column of an unequal leg angle shown in Fig. P6-2. Material properties and dimensions of the cross section are as follows: $E = 29 \times 10^3$ ksi, $G = 11.2 \times 10^3$ ksi, $L_1 = 6$ in, $L_2 = 4$ in, $t = 1/2$ in.

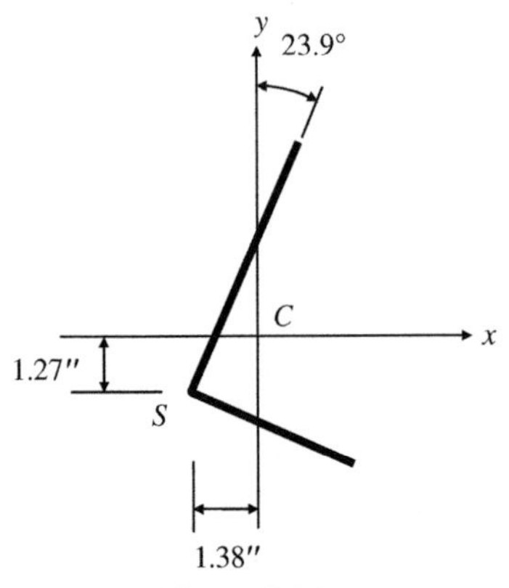

Figure P6-2

6.3 For the coupled system of differential equations given by Eqs. (6.4.23) and (6.4.24), prove why the solution eigenfunctions are sine functions for a simply supported beam of a doubly symmetric section. In other words, state the reason why the eigenfunctions do not include hyperbolic functions or polynomials.

CHAPTER 7

Lateral-Torsional Buckling

Contents

7.1. INTRODUCTION

A transversely (or combined transversely and axially) loaded member that is bent with respect to its major axis may buckle laterally if its compression flange is not sufficiently supported laterally. The reason buckling occurs in a beam at all is that the compression flange or the extreme edge of the compression side of a narrow rectangular beam, which behaves like a column resting on an elastic foundation, becomes unstable. If the flexural rigidity of the beam with respect to the plane of the bending is many times greater than the rigidity of the lateral bending, the beam may buckle and collapse long before the bending stresses reach the yield point. As long as the applied loads remain below the limit value, the beam remains stable; that is, the beam that is slightly twisted and/or bent laterally returns to its original configuration upon the removal of the disturbing force. With increasing load intensity, the restoring forces become smaller and smaller, until a loading is reached at which, in addition to the plane bending equilibrium configuration, an adjacent, deflected, and twisted, equilibrium position becomes equally possible. The original bending configuration is no longer stable, and the lowest load at which such an alternative equilibrium configuration becomes possible is the critical load of the beam. At the critical load, the compression flange tends to bend laterally, exceeding the

Stability of Structures
ISBN 978-0-12-385122-2, doi:10.1016/B978-0-12-385122-2.10007-7

restoring force provided by the remaining portion of the cross section to cause the section to twist. Lateral buckling is a misnomer, for no lateral deflection is possible without concurrent twisting of the section.

Bleich (1952) gives credit to Prandtl (1899) and Michell (1899) for producing the first theoretical studies on the lateral buckling of beams with long narrow rectangular sections. Similar credit is also extended to Timoshenko (1910) for deriving the fundamental differential equation of torsion of symmetrical I-beams and investigating the lateral buckling of transversely loaded deep I-beams with the derived equation. Since then, many investigators, including Vlasov (1940), Winter (1943), Hill (1954), Clark and Hill (1960), and Galambos (1963), have contributed on both elastic and inelastic lateral-torsional buckling of various shapes. Some of the early developments of the resisting capacities of steel structural members leading to the Load and Resistance Factor Design (LRFD) are summarized by Vincent (1969).

7.2. DIFFERENTIAL EQUATIONS FOR LATERAL-TORSIONAL BUCKLING

If transverse loads do not pass through the shear center, they will induce torsion. In order to avoid this additional torsional moment (thereby weakening the flexural capacity) in the flexural members, it is customary to use flexural members of at least singly symmetric sections so that the transverse loads will pass through the plane of the web as shown in Fig. 7-1. The section is symmetric about the y-axis, and u and v are the components of the displacement of the shear center parallel to the axes ξ and η. The rotation of the shear center ϕ is taken positive about the z-axis according to the right-hand screw rule, and the z-axis is perpendicular to the $\xi\eta$ plane. The following assumptions are employed:

1. The beam is prismatic.
2. The member cross section retains its original shape during buckling.
3. The externally applied loads are conservative.
4. The analysis is limited within the elastic limit.
5. The transverse load passes through the axis of symmetry in the plane of bending.

In the derivation of the governing differential equations of the lateral-torsional buckling of beams, it is necessary to define two coordinate systems: one for the undeformed configuration, x, y, z, and the other for the deformed configuration, ξ, η, ς as shown in Fig. 7-1. Hence, the fixed

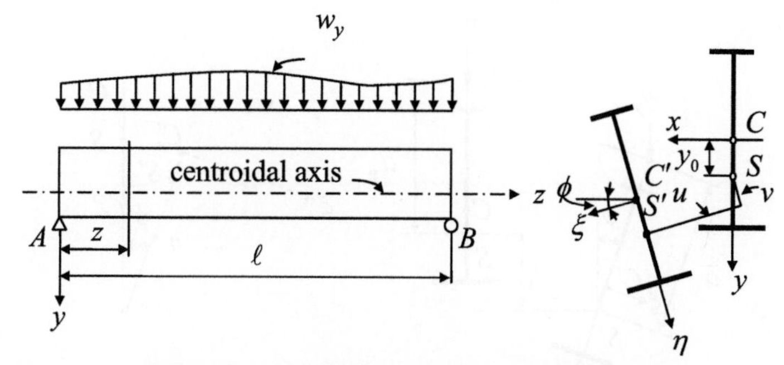

Figure 7-1 Coordinate systems and loading w_y

coordinate axes, x, y, z, constitute a right-hand rectangular coordinate system, while the coordinate axes ξ, η, ς make a pointwise rectangular coordinate system as the ς axis is tangent to the centroidal axis of the deformed configuration. As the loading will constitute the conservative force system, it will become necessary to relate the applied load in the fixed coordinate system to those in the deformed configuration. This can be readily accomplished by considering the direction cosines of the angles between the axes shown in Fig. 7-1. These cosines are summarized in Table 7-1. The curvatures of the deflected axis of the beam in the xz and yz planes can be taken as d^2u/dz^2 and d^2v/dz^2, respectively for small deflections. M_x and M_y are assumed positive when they create positive curvatures; $EI_x\eta'' = M_x$ and $EI_y\xi'' = M_y$.

Since column buckling due to the axial load and the lateral-torsional buckling of beams under the transverse loading are uncoupled in the linear elastic first-order analysis, only the transverse loading will be considered in the derivation of the governing differential equations. Excluding the strain energy of vertical bending prior to buckling, the strain energy in the neighboring equilibrium configuration is

$$U = \frac{1}{2}\int_0^\ell \left[EI_y(u'')^2 + EI_w(\phi'')^2 + GK_T(\phi')^2\right] dz \qquad (7.2.1)$$

The load w_y, is lowered by a net distance of $y_s + |\bar{a}_y|(1 - \cos\phi)$. Since ϕ is small, $1 - \cos\phi = \phi^2/2$. The vector distance $\bar{a}_y$ is measured from the

Table 7-1 Cosines of angles between axes in Fig. 7-1

	x	y	z
ξ	1	ϕ	$-du/dz$
η	$-\phi$	1	$-dv/dz$
ς	du/dz	dv/dz	1

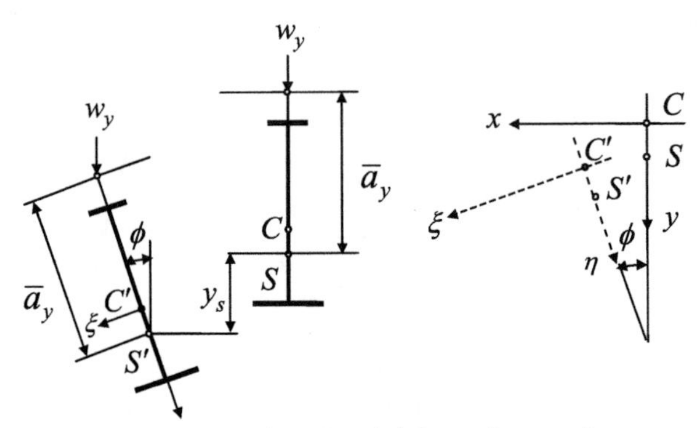

Figure 7-2 Lateral-torsional deformations under w_y

shear center to the transverse load application point. Hence, the loss of the potential energy of the transverse load w_y is

$$V_{wy} = -\int_0^{\ell} w_y y_s dz + \frac{\bar{a}_y}{2}\int_0^{\ell} w_y \phi^2 dz \tag{7.2.2}$$

It is noted that the sign of y_s is positive and $\bar{a}_y$ is negative as shown in Fig. 7-2. It should be noted that the position of the transverse load $\bar{a}_y$ affects the lateral-torsional buckling strength significantly. When the load is applied at the upper flange, it tends to increase the positive rotation of the cross section as shown in Fig. 7-2, thereby lowering the critical load. This could result in a significantly lower critical value than that when the load is applied at or below the shear center. Although the difference in the critical values is gradually decreasing following the increase of the span length, the position of the transverse load should be properly reflected whenever it is not negligibly small.

The first term of Eq. (7.2.2) can be expanded by integration by parts using the relationships that can be derived from Fig. 7-3.

$$\sum F_y = 0 = -Q_{wy} + Q_{wy} + dQ_{wy} + w_y dz$$

$$\frac{dQ_{wy}}{dz} = -w_y \tag{7.2.3}$$

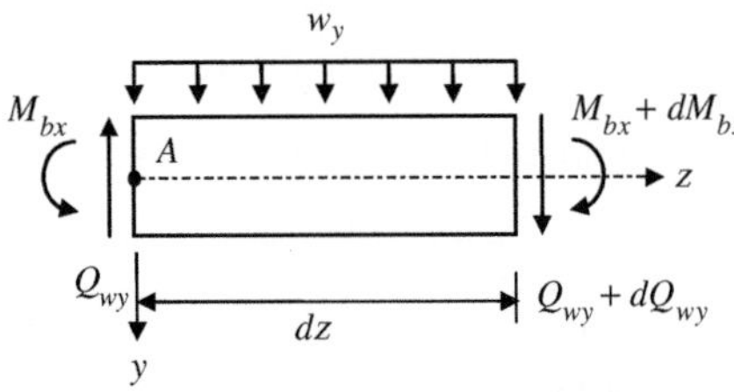

Figure 7-3 Free body of a differential element with w_y

$$\sum M_A = 0 = +M_{bx} - w_y dz\frac{dz}{2} - (Q_{wy} + dQ_{wy})dz - M_{bx} - dM_{bx}$$

$$\frac{dM_{bx}}{dz} = -Q_{wy} \tag{7.2.4}$$

Hence,

$$-\int_0^\ell w_y y_s dz = \int_0^\ell \frac{dQ_{wy}}{dz} y_s dz = \cancel{[Q_{wy} y_s]_0^\ell} - \int_0^\ell Q_{wy}\frac{dy_s}{dz}dz$$

$$= +\cancel{\left[M_{bx}\frac{dy_s}{dz}\right]_0^\ell} - \int_0^\ell M_{bx}\frac{d^2y_s}{dz^2}dz \tag{7.2.5}$$

Reflecting any combination of the geometric and natural boundary conditions at the ends of the beam, the two terms in the above equation indicated by slashes must vanish. Therefore,

$$V_{wy} = -\int_0^\ell M_{bx}\frac{d^2y_s}{dz^2}dz + \frac{\bar{a}_y}{2}\int_0^\ell w_y\phi^2 dz \tag{7.2.6}$$

The term d^2y_s/dz^2 represents the curvature in the yz plane; all deformations being small, the curvatures in other planes may be related as a vectorial sum indicated in Fig. 7-4 (it can also be seen from Fig. 7-1, $y_s = v\cos\phi + u\sin\phi$)

$$\frac{d^2y_s}{dz^2} = v''\cos\phi + u''\sin\phi \cong v'' + \phi u'' \tag{7.2.7}$$

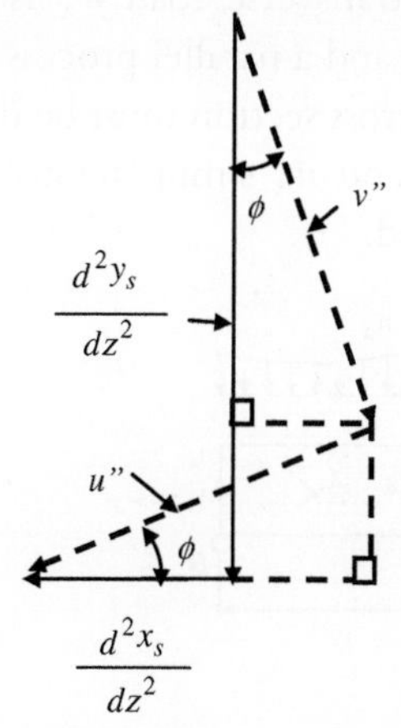

Figure 7-4 Relationship between u'' and v''

Therefore, the loss of potential energy is

$$
\begin{aligned}
V_{wy} &= -\int_0^\ell M_{bx}(v'' + \phi u'')dz + \frac{\overline{a}_y}{2}\int_0^\ell w_y\phi^2 dz \\
&= -\int_0^\ell M_{bx}v''dz - \int_0^\ell M_{bx}\phi u''dz + \frac{\overline{a}_y}{2}\int_0^\ell w_y\phi^2 dz \qquad (7.2.8)
\end{aligned}
$$

The above equation is the change of potential energy from unloaded to the buckled state. Just prior to buckling, $\phi = u'' = 0$ and the static potential energy is

$$
-\int_0^\ell M_{bx}v''dz \qquad (7.2.9)
$$

Hence, the loss of potential energy due to buckling (in the neighboring equilibrium) is

$$
V_{wy} = -\int_0^\ell M_{bx}\phi u''dz + \frac{\overline{a}_y}{2}\int_0^\ell w_y\phi^2 dz \qquad (7.2.10)
$$

The total potential energy functional becomes

$$
\begin{aligned}
\Pi &= U + V \\
&= \frac{1}{2}\int_0^\ell \left[EI_y(u'')^2 + EI_w(\phi'')^2 + GK_T(\phi')^2\right]dz \\
&\quad - \int_0^\ell M_{bx}\phi u'' + \frac{\overline{a}_y}{2}\int_0^\ell w_y\phi^2 dz \qquad (7.2.11)
\end{aligned}
$$

In the case when the transverse load w_x is considered for a similar derivation, Fig. 7-5 is used, and a parallel process can be applied. By virtue of assumption 5, the beam cross section must be doubly symmetric in order to accommodate both w_x and w_x simultaneously, and as a consequence, biaxial bending is uncoupled.

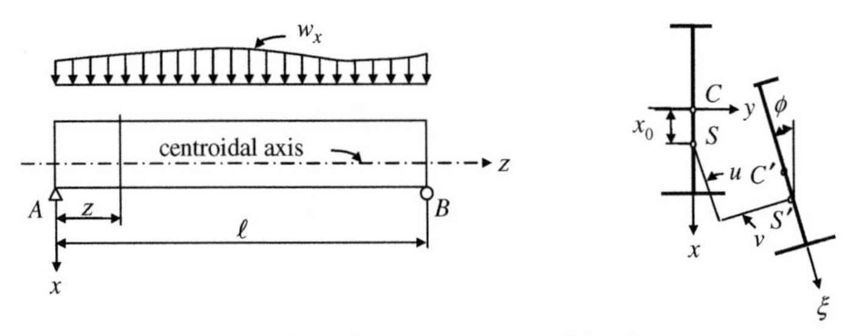

Figure 7-5 Coordinate systems and loading w_x

The load w_x is lowered by a distance $x_s + |\bar{a}_x|(1 - \cos\phi)$ as shown in Fig. 7-6. Since ϕ is small, $1 - \cos\phi = \phi^2/2$. The vector distance $\bar{a}_x$ is measured from the shear center to the transverse load point.
Hence,

$$V_{wx} = -\int_0^\ell w_x x_s dz + \frac{\bar{a}_x}{2}\int_0^\ell w_x \phi^2 dz \tag{7.2.12}$$

It is noted that the sign of x_s is positive and $\bar{a}_x$ is negative as is shown in Fig. 7-6.
The first term of Eq. (7.2.12) can be expanded by integration by parts using the relationships that can be derived from Fig. 7-7.

$$\sum F_x = -Q_{wx} + Q_{wx} + dQ_{wx} + w_x dz = 0$$

$$\frac{dQ_{wx}}{dz} = -w_x \tag{7.2.13}$$

$$\sum M_A = -M_{by} + w_x dz\frac{dz}{2} + (Q_{wx} + dQ_{wx})dz + M_{by} + dM_{by} = 0$$

$$\frac{dM_{by}}{dz} = -Q_{wy} \tag{7.2.14}$$

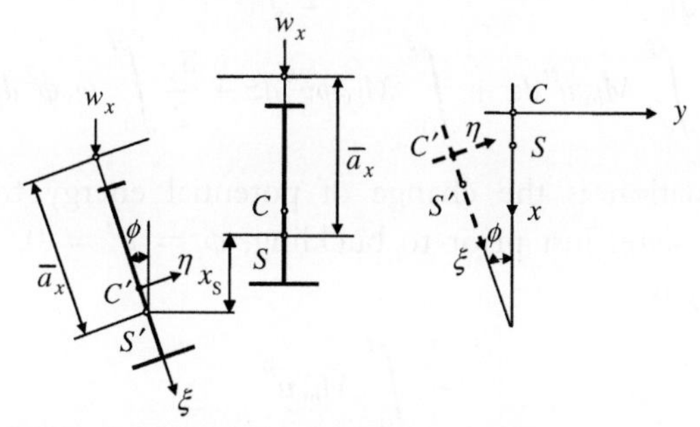

Figure 7-6 Lateral-torsional deformations under w_x

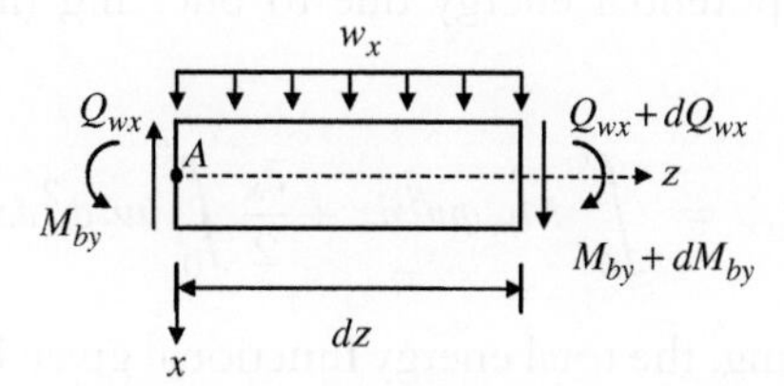

Figure 7-7 Free body of a differential element with w_x

Hence,

$$-\int_0^\ell w_x x_s dz = \int_0^\ell \frac{dQ_{wx}}{dz} x_s dz = [Q_{wx} x_s]_0^\ell - \int_0^\ell Q_{wx} \frac{dx_s}{dz} dz$$
$$= +\left[M_{by}\frac{dx_s}{dz}\right]_0^\ell - \int_0^\ell M_{by}\frac{d^2x_s}{dz^2} dz \tag{7.2.15}$$

Reflecting any combination of the geometric and natural boundary conditions at the ends of the beam, the two terms in the above equation indicated by slashes must vanish. Therefore,

$$V_{wx} = -\int_0^\ell M_{by}\frac{d^2x_s}{dz^2}\,dz + \frac{\overline{a}_x}{2}\int_0^\ell w_x \phi^2 dz \tag{7.2.16}$$

The term d^2x_s/dz^2 represents the curvature in the xz plane; all deformations being small, the curvatures in other planes may be related as a vectorial sum as indicated in Fig. 7-4.

$$\frac{d^2x_s}{dz^2} = u'' \cos\phi - v'' \sin\phi \cong u'' - \phi v'' \tag{7.2.17}$$

Therefore, the loss of potential energy is

$$V_{wx} = -\int_0^\ell M_{by}(u'' - \phi v'')dz + \frac{\overline{a}_x}{2}\int_0^\ell w_x \phi^2 dz$$
$$= -\int_0^\ell M_{by} u'' dz + \int_0^\ell M_{by}\phi v'' dz + \frac{\overline{a}_x}{2}\int_0^\ell w_x \phi^2 dz \tag{7.2.18}$$

The above equation is the change of potential energy from unloaded to the buckled state. Just prior to buckling, $\phi = v'' = 0$, and the static potential energy is

$$-\int_0^\ell M_{by} u''$$

Hence, the loss of potential energy due to buckling (in the neighboring equilibrium) is

$$V_{wx} = \int_0^\ell M_{by}\phi v'' dz + \frac{\overline{a}_x}{2}\int_0^\ell w_x \phi^2 dz \tag{7.2.19}$$

For biaxial bending, the total energy functional given by Eq. (7.2.11) can be extended as

$$\Pi = \frac{1}{2}\int_0^\ell \left[EI_y\left(u''\right)^2 + EI_x\left(v''\right)^2 + EI_w\left(\phi''\right)^2 + GK_T\left(\phi'\right)^2 \right.$$
$$\left. - \int_0^\ell M_{bx}\phi u'' dz + \int_0^\ell M_{by}\phi v'' dz + \frac{1}{2}\int_0^\ell \left(\bar{a}_x w_x + \bar{a}_y w_y\right)\phi^2 \right] dz$$
$$= \int_0^\ell F\left(u'', v'', \phi, \phi', \phi''\right) dz$$

(7.2.20)

It should be noted that biaxial bending can only be considered for doubly symmetric sections by virtue of assumption 5. Π will be stationary (minimum) if the following Euler-Lagrange equations are satisfied:

$$\frac{\partial F}{\partial u} - \frac{d}{dz}\frac{\partial F}{\partial u'} + \frac{d^2}{dz^2}\frac{\partial F}{\partial u''} = 0 \qquad (7.2.21a)$$

$$\frac{\partial F}{\partial v} - \frac{d}{dz}\frac{\partial F}{\partial v'} + \frac{d^2}{dz^2}\frac{\partial F}{\partial v''} = 0 \qquad (7.2.21b)$$

$$\frac{\partial F}{\partial \phi} - \frac{d}{dz}\frac{\partial F}{\partial \phi'} + \frac{d^2}{dz^2}\frac{\partial F}{\partial \phi''} = 0 \qquad (7.2.21c)$$

Noting that

$$\frac{\partial F}{\partial u} = 0, \quad \frac{\partial F}{\partial u'} = 0, \quad \frac{\partial F}{\partial u''} = EI_y u'' - M_{bx}\phi$$

Eq. (7.2.21a) becomes

$$EI_y u^{iv} - \frac{d^2}{dz^2}\left(M_{bx}\phi\right) = 0 \qquad (7.2.22)$$

Similarly, Eq. (7.2.21b) becomes

$$EI_x v^{iv} + \frac{d^2}{dz^2}\left(M_{by}\phi\right) = 0 \qquad (7.2.23)$$

Substituting the followings into Eq. (7.2.21c)

$$\frac{\partial F}{\partial \phi} = -M_{bx}u'' + M_{by}v'' + \left(\bar{a}_x w_x + \bar{a}_y w_y\right)\phi$$

$$\frac{\partial F}{\partial \phi'} = GK_T\phi'$$

$$\frac{\partial F}{\partial \phi''} = EI_w \phi''$$

one obtains

$$EI_w \phi^{iv} - GK_T \phi'' - M_{bx} u'' + M_{by} v'' + \left(\bar{a}_x w_x + \bar{a}_y w_y \right) \phi = 0 \quad (7.2.24)$$

Equations (7.2.22), (7.2.23), and (7.2.24) are general differential equations describing the lateral-torsional buckling behavior of prismatic straight beams. The total potential energy functional given by Eq. (7.2.20) can be readily transformed into matrix eigenvalue problems. When the beam is subjected to varying loads, in order to make the analysis simple it can be subdivided into a series of elements subjected to an equivalent uniform load determined by a stepwise uniform load. Experience has shown that no more than three subdivisions are satisfactory for most practical engineering problems. These equations check well with those given by Timoshenko and Gere (1961)[1] and Bleich (1952).[2] It is noted that the sign adopted herein for positive values of $\bar{a}_y$ and M_{bx} is reversed from that in Bleich (1952). If the beam is subjected to a transverse load, the resulting bending moment will become a function of the longitudinal axis, thereby rendering these differential equations to contain variable coefficients. Hence, no analytical solution for the critical load, in general, appears possible, and a variety of numerical integration schemes have been proposed. An approximate energy method based on an assumed displacement function is always possible.

7.3. GENERALIZATION OF GOVERNING DIFFERENTIAL EQUATIONS

If a wide flange beam is subjected to constant bending moment M_{bx} only, the three general governing differential equations (7.2.22 to 7.2.24) are reduced to

$$\begin{aligned} EI_y u^{iv} - \frac{d^2}{dz^2}(M_{bx}\phi) &= 0 \\ EI_w \phi^{iv} - GK_T \phi'' - M_{bx} u'' &= 0 \end{aligned} \quad (7.3.1)$$

[1] Page 245.
[2] Page 158.

Vlasov (1961)[3] pointed out a potential limitation of the governing differential equations on the lateral-torsional buckling of wide flange beams in some of the references, including Bleich (1952) and Timoshenko and Gere (1961). The equations discussed by Vlasov have the form:

$$\begin{aligned} EI_y u'' - M_{bx}\phi &= 0 \\ EI_w\phi''' - GK_T\phi' + M_{bx}u' - M_{bx}{}'u + \int_0^\ell M_{bx}{}''udz &= 0 \end{aligned} \tag{7.3.2}$$

Integrating the first equation of Eqs. (7.3.1) twice, the second equation once, and applying in the second equation integration by parts ($\int M_{bx}u''dz = M_{bx}u' - \int u'M_{bx}{}'dz = M_{bx}u' - M_{bx}{}'u + \int M_{bx}{}''udz$), one obtains

$$\begin{aligned} EI_y u'' + M_{bx}\phi &= Az + B \\ EI_w\phi''' - GK_T\phi' + M_{bx}u' - M_{bx}{}'u + \int_0^\ell M_{bx}{}''udz &= C \end{aligned} \tag{7.3.3}$$

where A, B, and C are arbitrary integral constants. These integral constants, as evident from the statical meaning of the transformation of Eqs. (7.3.1) into Eqs. (7.3.3), are respectively equal to the variations of the transverse shear force Q_x acting in the initial section $z = 0$ in the direction of the axis x, of the bending moment M_y with respect to the axis y, and of the torsional moment M_z with respect to the axis z. If the variations of the statical factors, Q_x, M_y, and M_z vanish in the initial section $z = 0$, which is the case in a cantilever at the free end, then the integration constants, A, B, and C are equal to zero and Eqs. (7.3.3) reduce to Eqs. (7.3.2).

If the beam has at the ends a rigid or elastic fixing to restrain translation and rotation, the integration constants, A, B, and C will not vanish and the general Eqs. (7.3.1) must be used.

7.4. LATERAL-TORSIONAL BUCKLING FOR VARIOUS LOADING AND BOUNDARY CONDITIONS

If the external load consists of a couple of end moments so that the moment remains constant along the beam length, then Eqs. (7.3.1) become

$$\begin{aligned} EI_y u^{iv} - M\phi'' &= 0 \\ EI_w\phi^{iv} - GK_T\phi'' - Mu'' &= 0 \end{aligned} \tag{7.4.1}$$

[3] Pages 326–328.

Equations (7.4.1) are a pair of differential equations with constant coefficients. Assume $u = A \sin \pi z/\ell$ and $\phi = B \sin \pi z/\ell$. It should be noted that the assumed displacement functions are indeed the correct eigenfunctions.[4] Therefore, one expects to have the exact solution.
Differentiating the assumed functions, one obtains

$$u' = A\frac{\pi}{\ell}\cos\frac{\pi z}{\ell},\ u'' = -A\left(\frac{\pi}{\ell}\right)^2 \sin\frac{\pi z}{\ell},\ u''' = -A\left(\frac{\pi}{\ell}\right)^3 \cos\frac{\pi z}{\ell},$$

$$u^{iv} = A\left(\frac{\pi}{\ell}\right)^4 \sin\frac{\pi z}{\ell}$$

$$\phi' = B\frac{\pi}{\ell}\cos\frac{\pi z}{\ell},\ \varepsilon'' = -B\left(\frac{\pi}{\ell}\right)^2 \sin\frac{\pi z}{\ell},\ \phi''' = -B\left(\frac{\pi}{\ell}\right)^3 \cos\frac{\pi z}{\ell},$$

$$\phi^{iv} = B\left(\frac{\pi}{\ell}\right)^4 \sin\frac{\pi z}{\ell}$$

Substituting these derivatives into Equations (7.4.1) yields

$$\begin{vmatrix} \left(\frac{\pi}{\ell}\right)^2 EI_y & M \\ -M & \left[\left(\frac{\pi}{\ell}\right)^2 EI_w + GK_T\right] \end{vmatrix} = 0$$

Solving this characteristic equation for the critical moment gives

$$M_{cr} = \frac{\pi}{\ell}\sqrt{EI_y(EI_w\pi^2/\ell^2 + GK_T)} \tag{7.4.2}$$

In the case of a uniformly distributed load w_y, the bending moment in a simple beam as shown in Fig. 7-8 becomes $M_x(z) = w_y z(\ell - z)/2$. For this load, Eqs. (7.3.1) become

$$EI_y u^{iv} + \frac{w_y}{2}[z(\ell - z)\phi]'' = 0$$

$$EI_w \phi^{iv} - GK_T\phi'' + \frac{w_y}{2}z(\ell - z)u'' = 0 \tag{7.4.3}$$

[4] See Vlasov (1961), page 272. As shown earlier in the solution of Problem 6.3, the correct eigenfunction is indeed a sine function.

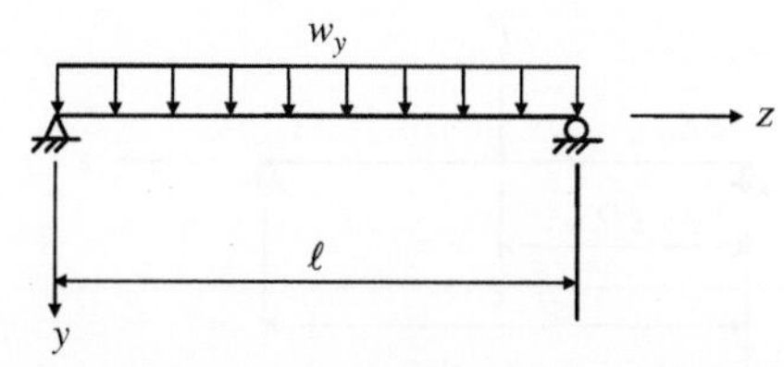

Figure 7-8 Simple beam subjected to a uniform load w_y

Equations (7.4.3) are coupled differential equations with variable coefficients. Timoshenko (1910) integrated Eqs. (7.4.3) by the method of infinite series. The critical load $(w_y\ell)_{cr}$ is given by

$$(w_y\ell)_{cr} = \frac{\gamma_1\sqrt{EI_yGK_T}}{\ell^2} \tag{7.4.4}$$

The coefficient γ_1 depends on the parameter

$$m = \frac{GK_T\ell^2}{EI_w} \tag{7.4.5}$$

Table 7-2 gives a series of values of γ_1 for a wide range of combination of the load positions and m for beams with doubly symmetric sections.

If the beam is loaded by a concentrated load at its midspan as shown in Fig. 7-9, the bending moment becomes $M_x(z) = Pz\ /\ 2$ For this load, Eqs. (7.3.1) become

$$\begin{aligned} EI_yu^{iv} + \frac{P}{2}(z\phi'') &= 0 \\ EI_w\phi^{iv} - GK_T\phi'' + \frac{P}{2}zu'' &= 0 \end{aligned} \tag{7.4.6}$$

Table 7-2 Values of γ_1 for simply supported I-beam under uniformly distributed load

	m								
Load at	**0.4**	**4**	**8**	**16**	**32**	**64**	**128**	**256**	**512**
TF	92.1	35.9	30.1	27.1	25.9	25.7	26.0	26.4	26.9
SC	144.2	52.9	42.5	36.1	32.5	30.5	29.4	28.9	28.6
BF	226.0	78.2	60.0	48.2	40.8	36.3	33.4	31.6	30.5

Notes: TF = Top flange, SC = Shear center, BF = Bottom flange.

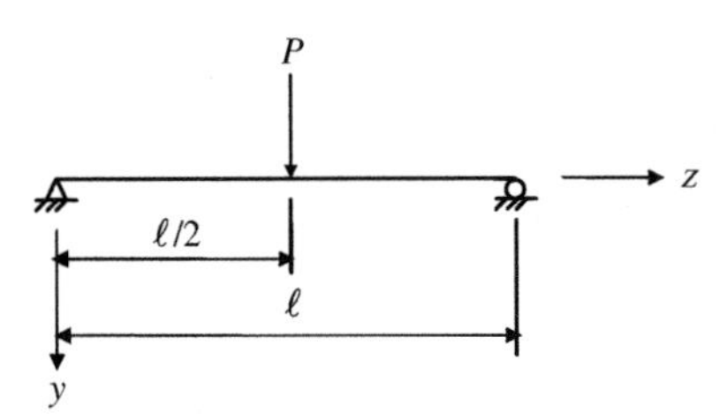

Figure 7-9 Simple beam subjected to a concentrated load P

Equations (7.4.6) are coupled differential equations with variable coefficients. Timoshenko (1910) integrated Eqs. (7.4.6) by the method of infinite series. The critical load P_{cr} is given by

$$P_{cr} = \frac{\gamma_2 \sqrt{EI_y GK_T}}{\ell^2} \tag{7.4.7}$$

The stability coefficient γ_2 depends on the parameter m defined by Eq. (7.4.5). Table 7-3 gives a series of values for a wide range of combination of γ_2 and m for beams with doubly symmetric section.

If both ends fixed beams are subjected to a uniformly distributed load, the critical loads may be expressed by Eq. (7.4.8).

$$(w_y \ell)_{cr} = \frac{\gamma_3 \sqrt{EI_y GK_T}}{\ell^2} \tag{7.4.8}$$

The stability coefficient γ_3 depends on the parameter m defined by Eq. (7.4.5). Table 7-4 gives a series of values for a wide range of combinations of γ_3 and m for beams with doubly symmetric sections.

If both ends fixed beams are loaded by a concentrated load, the critical load may be expressed by Eq. (7.4.9).

$$P_{cr} = \frac{\gamma_4 \sqrt{EI_y GK_T}}{\ell^2} \tag{7.4.9}$$

Table 7-3 Values of γ_2 for simply supported I-beam under concentrated load at the midspan

Load	*m*								
at	**0.4**	**4**	**8**	**16**	**32**	**64**	**128**	**256**	**512**
TF	50.7	19.9	16.8	15.3	14.7	14.8	15.0	15.4	15.7
SC	86.8	31.9	25.6	21.8	19.5	18.3	17.7	17.3	17.1
BF	148.8	50.9	38.7	30.8	25.8	22.7	20.7	19.4	18.6

Notes: TF = Top flange, SC = Shear center, BF = Bottom flange.

Table 7-4 Values of γ_3 for both ends fixed I-beam under uniformly distributed load

	m								
Load at	**0.4**	**4**	**8**	**16**	**32**	**64**	**128**	**256**	**512**
TF	610.6	206.8	156.7	125.0	107.0	98.9	97.1	98.7	101.6
SC	1316.8	434.1	320.4	244.4	195.4	165.1	146.8	135.8	128.8
BF	2802.0	900.3	647.2	482.0	352.6	272.7	220.0	185.4	162.4

Notes: TF = Top flange, SC = Shear center, BF = Bottom flange.

The stability coefficient γ_4 depends on the parameter m defined by Eq. (7.4.5). Table 7-5 gives a series of values for a wide range of combinations of γ_4 and m for beams with doubly symmetric sections.

For beams with simple-fixed end conditions subjected to a uniformly distributed load, the critical load may be expressed by Eq. (7.4.10).

$$(w_y\ell)_{cr} = \frac{\gamma_5\sqrt{EI_yGK_T}}{\ell^2} \tag{7.4.10}$$

The stability coefficient γ_5 depends on the parameter m defined by Eq. (7.4.5). Table 7-6 gives a series of values for a wide range of combinations of γ_5 and m for beams with doubly symmetric sections.

If beams with simple-fixed end conditions are loaded by a concentrated load, the critical load may be expressed by Eq. (7.4.11).

$$P_{cr} = \frac{\gamma_6\sqrt{EI_yGK_T}}{\ell^2} \tag{7.4.11}$$

The stability coefficient γ_6 depends on the parameter m defined by Eq. (7.4.5). Table 7-7 gives a series of values for a wide range of combinations of γ_6 and m for beams with doubly symmetric sections.

The stability coefficients γ_1 through γ_6 given in Tables 7-2 through 7-7 have been generated by STSTB (Yoo, 1980). For other combinations of loading conditions and boundary conditions not listed in these tables,

Table 7-5 Values of γ_4 for both ends fixed I-beam under concentrated load at the midspan

	m								
Load at	**0.4**	**4**	**8**	**16**	**32**	**64**	**128**	**256**	**512**
TF	238.4	80.9	61.5	52.5	42.3	39.3	39.0	39.9	42.9
SC	530.9	175.2	129.3	98.7	78.9	66.8	59.5	55.2	52.5
BF	1210.2	371.4	278.9	194.1	144.5	111.3	89.5	75.4	66.2

Notes: TF = Top flange, SC = Shear center, BF = Bottom flange.

Table 7-6 Values of γ_5 for simple-fixed I-beam under a uniformly distributed load

	m								
Load at	**0.4**	**4**	**8**	**16**	**32**	**64**	**128**	**256**	**512**
TF	259.0	92.4	73.0	61.6	56.0	54.2	54.3	55.3	56.5
SC	468.3	160.4	122.3	97.8	82.8	74.0	69.0	66.1	64.3
BF	838.8	275.9	203.0	153.8	121.4	100.6	87.3	78.7	73.0

Notes: TF = Top flange, SC = Shear center, BF = Bottom flange.

Table 7-7 Values of γ_6 for simple-fixed I-beam under concentrated load at the midspan

	m								
Load at	**0.4**	**4**	**8**	**16**	**32**	**64**	**128**	**256**	**512**
TF	129.1	46.1	36.5	30.9	28.2	27.4	27.7	28.5	29.4
SC	257.4	88.0	67.0	53.5	45.1	40.2	37.3	35.6	34.5
BF	499.6	160.6	118.1	89.2	70.0	57.4	49.2	43.8	40.2

Notes: TF = Top flange, SC = Shear center, BF = Bottom flange.

reasonably accurate (depending on the number of elements modeled) elastic lateral-torsional buckling loads can be determined by STSTB that can be downloaded from the senior author's Web pages. Access codes are available at the back flap of the book.

Similar tables are given for γ_1 and γ_2 in Timoshenko and Gere (1961).[5] The values of γ_1 and γ_6 in Tables 7-2 and 7-3 are very close to those given by Timoshenko and Gere (1961). It is of interest to note that the transverse load point has a significant impact on the critical lateral-torsional buckling load in beams with very short spans or unbraced lengths, but it tapers off for long and slender beams. Perhaps Australia is the only nation that has a design code that reflects the effect of the transverse load application point in computing the critical load. AISC (2005)[6] directs the designers' attention to the Commentary to Chapter 5 of the SSRC Guide[7] (Galambos 1998) when the loads are not applied at the shear center of the member. Its adverse effect is particularly onerous when the loads are applied at the top flange of the long unbraced cantilever. In practical design, however, determining the transverse load application point is problematic as beams are subjected to some combination of dead and live loads. For beams with very short spans or unbraced lengths, the

[5] See pages 264 and 268.
[6] See page 16.1–274.
[7] See page 207.

implication of this significant difference due to the transverse load points may become a mute issue because the elastic lateral-torsional buckling moment is likely to be greater than the full-plastic moment.

7.5. APPLICATION OF BESSEL FUNCTION TO LATERAL-TORSIONAL BUCKLING PROBLEMS

For a uniaxial bending problem, Eq. (7.2.24) takes the form

$$EI_w\phi^{iv} - GK_T\phi'' - M_{bx}u'' + \bar{a}_y w_y \phi = 0 \tag{7.5.1}$$

There is no closed-form solution available for the coupled equations of Eqs. (7.2.22) and (7.5.1) if the moment is not constant, and appropriate numerical solution techniques must be used. For a narrow rectangular section, or any section of which warping constant, I_w, is equal to zero, it is only necessary to omit the term in the equation containing the warping constant.

Consider as the first example lateral buckling of a cantilever beam subjected to a concentrated load P applied at its free end at the centroid as shown in Fig. 7-10.

Integrating Eq. (7.2.22) twice with respect to z, results in the following form:

$$EI_y u'' - M_{bx}\phi = Az + B \tag{7.5.2}$$

The integral constants A and B vanish for the reasons discussed in Section 7.3. The moment of the vertical load P with respect to axes through the centroid parallel to the x, y, and z axes are

$$M_x = P(\ell - z) \quad M_y = 0 \quad M_z = P[u(\ell) - u(z)] \tag{7.5.3}$$

Taking the components of moments in Eqs. (7.5.3) about the ξ, η, ς axes by using Table 7-1 for the cosines of the angles between the axes yields

$$M_\xi = P(\ell - z) \quad M_\eta = \phi P(\ell - z) \quad M_\varsigma = -P(\ell - z)\frac{du}{dz} + P[u(\ell) - u(z)] \tag{7.5.4}$$

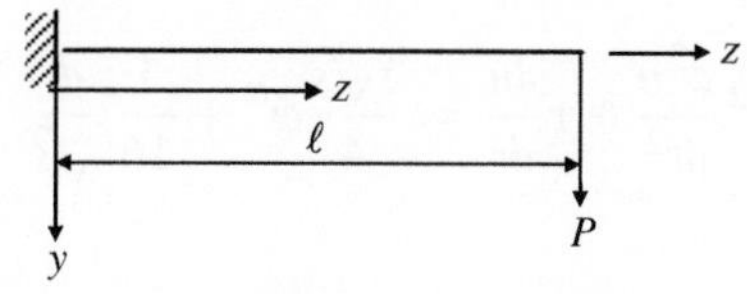

Figure 7-10 Cantilever beam subjected to a concentrated load at its free end

Substituting these values into Eqs. (7.5.2) and (5.8.22) gives

$$
\begin{aligned}
EI_y \frac{d^2u}{dz^2} - P(\ell - z)\phi &= 0 \\
-GK_T\phi' - P(\ell - z)\frac{du}{dz} + P[u(\ell) - u(z)] &= 0
\end{aligned}
\tag{7.5.5}
$$

Differentiating the second equation of Eq. (7.5.5) and eliminating d^2u/dz^2 gives

$$GK_T\phi'' + \frac{P^2}{EI_y}(\ell - z)^2\phi = 0 \tag{7.5.6}$$

Introduce new variables

$$s = \ell - z \tag{7.5.7}$$

and

$$k^2 = \frac{P^2}{EI_y GK_T} \tag{7.5.8}$$

to give

$$\frac{d^2\phi}{ds^2} + k^2 s^2 \phi = 0 \tag{7.5.9}$$

Equation (7.5.9) has a variable coefficient. This homogeneous differential equation has nontrivial solutions only for discrete values of the parameter P. The smallest such value is P_{cr}. Equation (7.5.9) is a typical Bessel differential equation classified by Bowman (1938)[8] as class (iv). The following substitution will reduce Eq. (7.5.9) to a Bessel equation as per Bowman (1938) and Grossman and Derrick (1988). Let $u = \phi/s$ and $r = ks^2/2$. Then

$$\frac{du}{dr} = \frac{du/ds}{dr/ds} = \frac{\phi'}{ks\sqrt{s}} - \frac{\phi}{2ks^2\sqrt{s}}$$

$$\frac{d^2u}{dr^2} = \frac{\frac{d}{ds}\left(\frac{du}{dr}\right)}{dr/ds} = \frac{\phi''}{k^2s^2\sqrt{s}} - \frac{2\phi'}{k^2s^3\sqrt{s}} + \frac{5}{4}\frac{\phi}{k^2s^4\sqrt{s}}$$

Hence

$$r^2\frac{d^2u}{dr^2} + r\frac{du}{dr} = \frac{s\sqrt{s}}{4}\phi'' + \frac{1}{16}\frac{\phi}{\sqrt{s}}$$

[8] See page 118.

and substituting Eq. (7.5.9) for ϕ'', one obtains

$$r^2\frac{d^2u}{dr^2}+r\frac{du}{dr} = \frac{s\sqrt{s}}{4}(-k^2s^2\phi)+\frac{1}{16}\frac{\phi}{\sqrt{s}} = -k^2\frac{s^4}{4}\frac{\phi}{\sqrt{s}}+\frac{1}{16}\frac{\phi}{\sqrt{s}}$$
$$= -\left(r^2-\frac{1}{16}\right)u$$

or

$$r^2\frac{d^2u}{dr^2}+r\frac{du}{dr}+\left(r^2-\frac{1}{16}\right)u = 0 \tag{7.5.10}$$

which is the Bessel equation of order 1/4. Since Eq. (7.5.10) has the general solution

$$u(r) = AJ_{1/4}(r)+BJ_{-1/4}(r),$$

Equation (7.5.10) has the solution

$$\phi(s) = \sqrt{s}\left[A_1J_{1/4}\left(\frac{k}{2}s^2\right)+A_2J_{-1/4}\left(\frac{k}{2}s^2\right)\right] \tag{7.5.11}$$

The constants of integration A_1 and A_2 in the general solution (7.5.11) are determined from the proper end boundary conditions. From Bowman (1938) and Grossman and Derrick (1988), one may extract useful relationships

$$\frac{d}{ds}\left[\left(\frac{k}{2}s^2\right)^{1/4}J_{1/4}\left(\frac{k}{2}s^2\right)\right] = \left[\left(\frac{k}{2}s^2\right)^{1/4}J_{-3/4}\left(\frac{k}{2}s^2\right)^{1/4}\right]ks$$
$$\frac{d}{ds}\left[\left(\frac{k}{2}s^2\right)^{1/4}J_{-1/4}\left(\frac{k}{2}s^2\right)\right] = \left[-\left(\frac{k}{2}s^2\right)^{1/4}J_{3/4}\left(\frac{k}{2}s^2\right)^{1/4}\right]ks \tag{7.5.12}$$

For the cantilever beam shown in Fig. 7-10, the boundary conditions are

$$\phi = 0 \text{ (no twisting) at } s = \ell \text{ (built-in end) and}$$

$$\frac{d\phi}{ds} = 0 \text{ (no torque) at } s = 0 \text{ (free end)} \tag{7.5.13}$$

From Eqs. (7.5.11) and (7.5.12),

$$\frac{d\phi}{ds} = 2\left(\frac{k}{2}\right)^{1/4}\left[A_1\left(\frac{k}{2}s^2\right)^{3/4}J_{-3/4}\left(\frac{k}{2}s^2\right)-A_2\left(\frac{k}{2}s^2\right)^{3/4}J_{3/4}\left(\frac{k}{2}s^2\right)\right]$$

For $s = 0, J_{-3/4}(0)\neq 0$ and $J_{3/4}(0) = 0$ according to **Maple®**. Hence A_1 must be equal to zero to satisfy the second boundary condition given in Eq. (7.5.13). Then, from Eq. (7.5.11),

$$\phi(s) = \sqrt{s}A_2 J_{-1/4}\left(\frac{k}{2}s^2\right) \tag{7.5.14}$$

The first boundary condition of Eq. (7.5.13) applied to Eq. (7.5.14) gives

$$0 = J_{-1/4}\left(\frac{k}{2}\ell^2\right) \tag{7.5.15}$$

The smallest value to satisfy Eq. (7.5.15) according to **Maple**® is $k\,\ell^2/2 = 2.0063$. Then $k = 4.0126/\ell^2$. From which

$$P_{cr} = \frac{4.0126}{\ell^2}\sqrt{EI_y GK_T} \tag{7.5.16}$$

This result, Eq. (7.5.16), was obtained by Prandtl (1899).

Consider, as another example of applying the Bessel equation, a simply supported beam of narrow rectangular section subjected to a concentrated load applied at the centroid at the midspan as shown in Fig. 7-9. For convenience, the origin of the coordinate system is moved to the midspan. The moments with respect to axes through the centroid of the cross section parallel to the x, y, and z axes are

$$M_x = -\frac{P}{2}\left(\frac{\ell}{2} - z\right) \quad M_y = 0 \quad M_z = -\frac{P}{2}[u(0) - u(z)] \tag{7.5.17}$$

Taking the components of moments in Eqs. (7.5.17) about the ξ, η, ς axes by using Table 7-1 for the cosines of the angles between the axes yields

$$M_\xi = -\frac{P}{2}\left(\frac{\ell}{2} - z\right) \quad M_\eta = -\phi\frac{P}{2}\left(\frac{\ell}{2} - z\right)$$

$$M_\varsigma = \frac{P}{2}\left(\frac{\ell}{2} - z\right)\frac{du}{dz} - P[u(0) - u(z)] \tag{7.5.18}$$

Substituting these values in to Eqs. (7.5.2) and (5.8.22) gives

$$\begin{aligned} EI_y\frac{d^2u}{dz^2} + \frac{P}{2}\left(\frac{\ell}{2} - z\right)\phi &= 0 \\ -GK_T\phi' + \frac{P}{2}\left(\frac{\ell}{2} - z\right)\frac{du}{dz} - \frac{P}{2}[u(0) - u(z)] &= 0 \end{aligned} \tag{7.5.19}$$

Eliminating d^2u/dz^2 in Eqs. (7.5.19) gives

$$GK_T\phi'' + \frac{P^2}{4EI_y}\left(\frac{\ell}{2} - z\right)^2\phi = 0 \tag{7.5.20}$$

Introducing the variable $t = \ell/2 - z$ and the notation

$$k^2 = \frac{P^2}{4EI_y GK_T} \tag{7.5.21}$$

to give

$$\frac{d^2\phi}{dt^2} + k^2 t^2 \phi = 0 \tag{7.5.22}$$

Equation (7.5.22) is identical to Eq. (7.5.9). The general solution of Eq. (7.5.22) is

$$\phi = \sqrt{t}\left[A_1 J_{1/4}(kt^2) + A_2 J_{-1/4}(kt^2)\right] \tag{7.5.23}$$

For a simply supported beam, the proper boundary conditions are

$$\phi = 0 \quad \text{at } t = 0 \quad \frac{d\phi}{dt} = 0 \quad \text{at } t = \frac{\ell}{2} \tag{7.5.24}$$

In order to satisfy the first condition of Eq. (7.5.24) ($J_{-1/4}(0) \neq 0$, $J_{1/4}(0) = 0$), $A_2 = 0$. Then,

$$\frac{d\phi}{dt} = 2\left(\frac{k}{2}\right)^{1/4} A_1 \left(\frac{k}{2}t^2\right)^{3/4} J_{-3/4}\left(\frac{k}{2}t^2\right) = 0 \quad \text{at } t = \frac{\ell}{2}$$

Hence, $J_{-3/4}((k/8)\,\ell^2) = 0$.
The parameter for the first zero of the Bessel function of order $-3/4$ is found from **Maple**® to be 1.0585, which leads to

$$P_{cr} = \frac{16.94\sqrt{EI_y GK_T}}{\ell^2} \tag{7.5.25}$$

It can be readily recognized that the computer-based modern matrix structural analysis (such as STSTB) and/or finite element analysis would be superior to the longhand classical solution techniques with regard to speed of analysis as well as the versatility on the loadings and boundary conditions that can be accommodated. Some of the really old classical methods of analysis are to be viewed as historical interest.

7.6. LATERAL-TORSIONAL BUCKLING BY ENERGY METHOD

The determination of the critical lateral-torsional buckling loads by longhand classical methods is very complex and tedious, particularly for nonuniform bending, as this will result in a system of differential equations with variable coefficients. In this section, the Rayleigh-Ritz method will be used to determine approximately the critical lateral-torsional buckling loads of beams following the general procedures presented by Winter (1941) and Chajes (1974). In any energy method, it is required to establish expressions for the strain energy stored in the elastic body and the loss of potential energy of the externally applied loads. It is relatively simple to come up with the expression for the strain energy by

$$U = (1/2) \int_v \sigma^T \varepsilon dv$$

where σ^T = transpose of the stress vector, ε = strain vector, and v = volume of the body. Although the loss of the potential energy of the applied loads is simple in concept as being the negative product of the generalized force and the corresponding deformation during buckling, the expression for the corresponding deformation usually requires considerable geometric analyses.

7.6.1. Uniform Bending

Consider a prismatic, simply supported doubly symmetric (for simplicity)[9] I-beam subjected to a uniform bending moment M_x as shown in Fig. 7-11. The bending moment M_x shown is negative as it produces negative curvature. The notion of the buckling analysis of the beam is to examine equilibrium in the slightly buckled (lateral-torsional deformations of the beam) configuration. Therefore, the strain energy associated with vertical bending or pre-buckling static equilibrium should be excluded from Eq. (6.3.3) in the buckling analysis because it belongs to a totally different equilibrium configuration. The strain energy stored in the beam during buckling consists of two parts: the energy associated with bending about the y-axis and the energy due to twisting about the z-axis. Thus the strain energy is

$$U = \frac{1}{2}\int_0^\ell EI_y(u'')^2 dz + \frac{1}{2}\int_0^\ell GK_T(\phi')^2 dz + \frac{1}{2}\int_0^\ell EI_w(\phi'')^2 dz \quad (7.6.1)$$

To form the total potential energy, the potential energy V of the externally applied loads must be added to the strain energy, Eq. (7.6.1). For a beam subjected to pure bending, the loss of potential energy V is equal to the negative product of the applied moments and the corresponding angles due to buckling. Hence,

$$V = -2M_x\theta \quad (7.6.2)$$

[9] Winter (1941) considered a singly symmetric cross section.

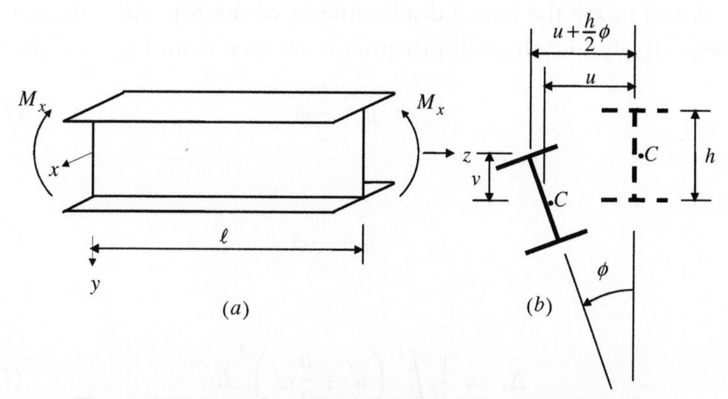

Figure 7-11 Lateral-torsional deformations of simple beam

where θ is the angle of rotation about the x-axis at each end of the beam as shown in Fig. 7-12.

By the definition of the simple support, neither twisting of the beam nor lateral deformations of the flanges is allowed at the support. Hence, the top flange deflects more than the bottom flange, as illustrated in Fig. 7-11(b). Thus, the angle θ is

$$\theta = \frac{\Delta_t - \Delta_b}{h} \tag{7.6.3}$$

where h is the depth of the cross section. Recalling Eq. (1.6.3),

$$\Delta_t = \frac{1}{4}\int_0^\ell (u_t')^2 dz \tag{7.6.4}$$

and

$$\Delta_b = \frac{1}{4}\int_0^\ell (u_b')^2 dz \tag{7.6.5}$$

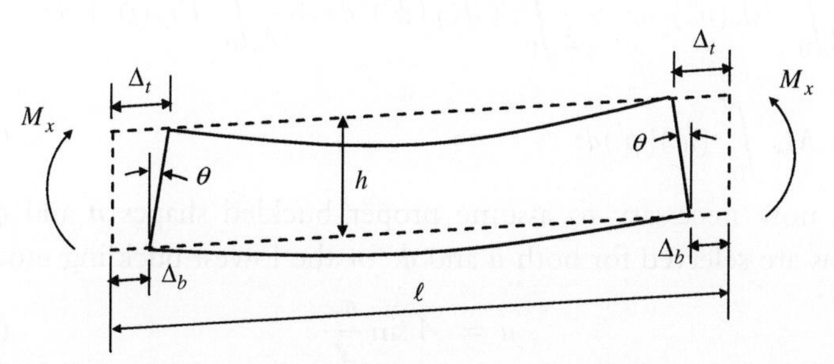

Figure 7-12 End rotations due to lateral-torsional buckling (after Winter, "Lateral Stability of Unsymmetrical I-Beams and Trusses in Bending," *Proceedings, ASCE*, Vol. 67, 1941). Reproduced by permission.

where u_t and u_b are the lateral displacements of the top and bottom of the web, respectively, and these displacements are seen from Fig. 7-11(b) to be

$$u_t = u + \frac{h}{2}\phi \tag{7.6.6}$$

and

$$u_b = u - \frac{h}{2}\phi \tag{7.6.7}$$

Thus

$$\Delta_t = \frac{1}{4}\int_0^\ell \left(u' + \frac{h}{2}\phi'\right)^2 dz \tag{7.6.8}$$

and

$$\Delta_b = \frac{1}{4}\int_0^\ell \left(u' - \frac{h}{2}\phi'\right)^2 dz \tag{7.6.9}$$

Substituting Eqs. (7.6.8) and (7.6.9) into Eq. (7.6.3) gives

$$\theta = \frac{1}{2}\int_0^\ell (u')(\phi')dz \tag{7.6.10}$$

Thus, Eq. (7.6.2) becomes

$$V = -M_x\int_0^\ell (u')(\phi')dz \tag{7.6.11}$$

Finally, the total potential energy is

$$\begin{aligned} \Pi &= U + V \\ &= \frac{1}{2}\int_0^\ell EI_y(u'')^2 dz + \frac{1}{2}\int_0^\ell GK_T(\phi')^2 dz + \frac{1}{2}\int_0^\ell EI_w(\phi'')^2 dz \\ &\quad - M_x\int_0^\ell (u')(\phi')dz \end{aligned} \tag{7.6.12}$$

It is now necessary to assume proper buckled shapes u and ϕ. Sine functions are selected for both u and ϕ for the lowest buckling mode as

$$u = A\sin\frac{\pi z}{\ell} \tag{7.6.13}$$

$$\phi = B\sin\frac{\pi z}{\ell} \tag{7.6.14}$$

As Eqs. (7.6.13) and (7.6.14) satisfy both geometric and natural boundary conditions, it is expected that the approximate solution will be very close to the exact solution. When these expressions are substituted into Eq. (7.6.12), the total potential energy becomes a function of two variables A and B. Invoking the principle of minimum total potential energy, one can determine the critical moment by solving the two equations that result if the first variation of π is made to vanish with respect to both A and B. An alternative approach is to express A in terms of B. Although the alternative approach involves fewer computations than the first, the first procedure must be used if a relation between u and ϕ is not available. Since M_x and M_y are defined to be positive when they produce positive curvature, $M_x = EI_x v''$ and $M_y = EI_y u''$. From Table 7-1, $M_y = M_x \phi$. Thus

$$\phi = \frac{EI_y}{M_x} u'' \tag{7.6.15}$$

$$A = -B\frac{\ell^2}{\pi^2}\frac{M_x}{EI_y} \tag{7.6.16}$$

The assumed function for u can now be written as

$$u = -\frac{B\ell^2}{\pi^2}\frac{M_x}{EI_y}\sin\frac{\pi z}{\ell} \tag{7.6.17}$$

Using Eqs. (7.6.14) and (7.6.17), the total potential energy becomes

$$\begin{aligned}\Pi &= U + V \\ &= \frac{1}{2}\frac{B^2 M_x^2}{EI_y}\int_0^\ell \sin^2\frac{\pi z}{\ell}dz + \frac{1}{2}GK_T B^2\frac{\pi^2}{\ell^2}\int_0^\ell \cos^2\frac{\pi z}{\ell}dz \\ &\quad + \frac{1}{2}EI_w B^2\frac{\pi^4}{\ell^4}\int_0^\ell \sin^2\frac{\pi z}{\ell}dz - \frac{M_x^2 B^2}{EI_y}\int_0^\ell \cos^2\frac{\pi z}{\ell}dz\end{aligned} \tag{7.6.18}$$

Since

$$\int_0^\ell \sin^2\frac{\pi z}{\ell}dz = \int_0^\ell \cos^2\frac{\pi z}{\ell}dz = \frac{\ell}{2}$$

Equation (7.6.18) reduces to

$$\Pi = U + V = \frac{1}{4}\left(\frac{GK_T B^2\pi^2}{\ell} + \frac{EI_w B^2\pi^4}{\ell^3} - \frac{M_x^2 B^2 \ell}{EI_y}\right) \tag{7.6.19}$$

The critical moment is reached when neutral equilibrium (or neighboring equilibrium) is possible, and the requirement for neutral equilibrium is that the derivative of Π with respect to B vanish. Hence,

$$\frac{d\Pi}{dB} = \frac{d(U+V)}{dB} = \frac{B}{2}\left(\frac{GK_T\pi^2}{\ell} + \frac{EI_w\pi^4}{\ell^3} - \frac{M_x^2\ell}{EI_y}\right) = 0 \qquad (7.6.20)$$

If neutral equilibrium is to correspond to a buckled configuration, B cannot be zero. In order to satisfy Eq. (7.6.20), the quantity inside the parentheses must be equal to zero. Thus,

$$\frac{GK_T\pi^2}{\ell} + \frac{EI_w\pi^4}{\ell^3} - \frac{M_x^2\ell}{EI_y} = 0 \qquad (7.6.21)$$

From which

$$M_{x\ cr} = \pm\frac{\pi}{\ell}\sqrt{EI_y(GK_T + \pi^2 EI_w/\ell^2)} \qquad (7.6.22)$$

Equation (7.6.22) gives the critical moment for a simply supported I-beam subjected to pure bending, and it is identical to Eq. (7.4.2). The $\pm$ sign in Eq. (7.6.22) indicates that an identical critical moment will result if the sign of pure bending is reversed from that shown in Fig. 7-12. It should also be noticed that the critical moment given by Eq. (7.6.22) is exact since the assumed displacement functions of Eqs. (7.6.13) and (7.6.14) happen to be exact eigenfunctions. This can be proved (for example, see Problem 6.3).

7.6.2. One Concentrated Load at Midspan

Consider a simply supported prismatic I-beam subjected to a concentrated load at midspan. The cross section is assumed to be doubly symmetric, and the load is applied at the centroid (the shear center) for simplicity. The case of a concentrated load applied at a point other than the shear center in a singly symmetric cross section can be handled likewise.

The strain energy stored in the beam during buckling has the same form given by Eq. (7.6.1). The potential energy of the externally applied load is, of course, the negative product of the applied load P and the vertical displacement v_0 of P that takes place during buckling. To determine v_0, it is useful to draw the lateral deflection of the shear center (the load point) of the beam as shown in Fig. 7-13 during buckling, as the vertical displacement component of the beam during buckling is equal to the product of the lateral displacement and the twisting angle.

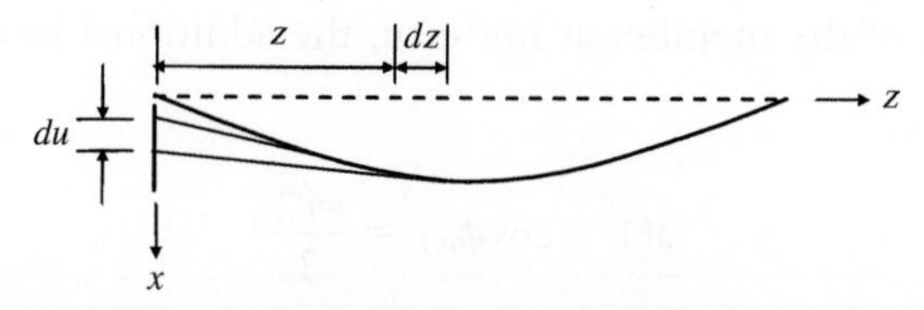

Figure 7-13 Lateral displacement of shear center of I-beam

Consider an element dz of the beam at a distance z from the left support as shown in Fig. 7-13. Due to lateral bending, there is a small vertical translation du at the support between the tangents drawn to the elastic curve at the two end points of the element. The value of the translation is, according to the moment-area theorem, given by

$$du = \frac{M_y}{EI_y} z dz \tag{7.6.23}$$

For small deformations, the increment in the vertical displacements dv corresponding to du is

$$dv = \phi du = \frac{M_y}{EI_y} \phi z dz \tag{7.6.24}$$

Thus the vertical displacement v_0 at the shear center at midspan is

$$v_0 = \int_0^{\ell/2} dv = \int_0^{\ell/2} \frac{M_y}{EI_y} \phi z dz \tag{7.6.25}$$

According to Table 7-1, the lateral bending moment at the buckled configuration is

$$M_y = M_x \phi = \frac{P}{2} z \phi \tag{7.6.26}$$

Thus

$$v_0 = \int_0^{\ell/2} \frac{Pz^2 \phi^2}{2EI_y} dz \tag{7.6.27}$$

and the potential energy of the applied load P is

$$V = -Pv_0 = -\int_0^{\ell/2} \frac{P^2 z^2 \phi^2}{2EI_y} dz \tag{7.6.28}$$

If the load P is applied at a distance "a" above the shear center, an additional lowering of the load must be considered. If ϕ_0 is the twisting angle of the member at midspan, the additional lowering of the load is

$$a(1 - \cos\phi_0) \approx \frac{a\phi_0^2}{2} \tag{7.6.29}$$

and an additional loss of the potential energy is

$$\Delta V = -\frac{Pa\phi_0^2}{2} \tag{7.6.30}$$

Combining Eqs. (7.6.1) and (7.6.28), the total potential energy of the system becomes

$$\begin{aligned}\Pi &= U + V\\ &= \frac{1}{2}\int_0^\ell EI_y(u'')^2 dz + \frac{1}{2}\int_0^\ell GK_T(\phi')^2 dz + \frac{1}{2}\int_0^\ell EI_w(\phi'')^2 dz\\ &\quad - \frac{P^2}{2EI_y}\int_0^{\ell/2} \phi^2 z^2 dz\end{aligned} \tag{7.6.31}$$

As before, it is desirable to reduce the number of variables by expressing u in terms of ϕ. From Eq. (7.6.15)

$$\phi = \frac{EI_y}{M_x}u'' = -\frac{2EI_y u''}{Pz}$$

Hence

$$u'' = -\frac{Pz\phi}{2EI_y} \tag{7.6.32}$$

Substituting Eq. (7.6.32) into Eq. (7.6.31) gives

$$\begin{aligned}\Pi &= U + V\\ &= \frac{1}{2}\int_0^\ell GK_T(\phi')^2 dz + \frac{1}{2}\int_0^\ell EI_w(\phi'')^2 dz - \frac{P^2}{4EI_y}\int_0^{\ell/2} \phi^2 z^2 dz\end{aligned} \tag{7.6.33}$$

Assume ϕ to be of the form

$$\phi = B\sin\frac{\pi z}{\ell} \tag{7.6.34}$$

Substituting ϕ and its derivatives into Eq. (7.6.33) yields

$$U + V = -\frac{P^2B^2}{4EI_y}\int_0^{\ell/2} z^2 \sin^2\frac{\pi z}{\ell}dz + \frac{GK_TB^2\pi^2}{2\ell^2}\int_0^{\ell}\cos^2\frac{\pi z}{\ell}dz + \frac{EI_wB^2\pi^4}{2\ell^4}\int_0^{\ell}\sin^2\frac{\pi z}{\ell}dz \tag{7.6.35}$$

Substituting the definite integrals

$$\int_0^{\ell/2} z^2\sin^2\frac{\pi z}{\ell}dz = \frac{\ell^3}{48\pi^2}(\pi^2+6)$$

$$\int_0^{\ell}\sin^2\frac{\pi z}{\ell}dz = \int_0^{\ell}\cos^2\frac{\pi z}{\ell}dz = \frac{\ell}{2} \tag{7.6.36}$$

into Eq. (7.6.35) gives

$$U + V = -\frac{P^2B^2\ell^3}{192EI_y\pi^2}(\pi^2+6) + \frac{GK_TB^2\pi^2}{4\ell} + \frac{EI_wB^2\pi^4}{4\ell^3} \tag{7.6.37}$$

At the critical load, the first variation of $U + V$ with respect to B must vanish. Thus,

$$\frac{d}{dB}(U+V) = \frac{B}{2}\left[-\frac{P^2\ell^3}{48EI_y\pi^2}(\pi^2+6) + \frac{GK_T\pi^2}{\ell} + \frac{EI_w\pi^4}{\ell^3}\right] = 0$$

which leads to

$$P_{cr} = \pm\frac{4\pi^2}{\ell^2}\sqrt{\frac{3}{\pi^2+6}EI_y\left(GK_T + \frac{EI_w\pi^2}{\ell^2}\right)} \tag{7.6.38}$$

Equation (7.5.38) gives the critical load for a simply supported I-beam subjected to a concentrated load at midspan. The $\pm$ sign in Eq. (7.5.38) indicates that an identical critical load will result if the direction of the load is reversed from that shown in Fig. 7-9.

7.6.3. Uniformly Distributed Load

The procedure described above for the case of a concentrated load at midspan can also be used when the I-beam (Fig. 7-8) carries a uniformly distributed load. The strain energy given by Eq. (7.6.1) remains unchanged. However, the expression for the loss of potential energy of the externally applied load must be determined.

Assume ϕ to be of the form

$$\phi = B \sin \frac{\pi z}{\ell}$$

The vertical displacement of the shear center at midspan of the I-beam as shown in Fig. 7-14 is, according to the moment-area theorem, given by

$$v_0 = \int_{\ell/2}^{0} \frac{M_y}{EI_y} \varsigma \phi \, d\varsigma$$

where $M_y = M_x \phi$ as per Table 7-1. The relationships between the lateral deflections and the vertical deflections are given by

$$v_0 = u_0 \phi, \; v_1 = u_1 \phi, \text{ and } v_2 = u_2 \phi$$

where u_0 and u_1 are the lateral displacements of the beam at midspan and at a distance z from the support, respectively, and u_2 is equal to u_0 subtracted by u_1 as shown in Fig. 7-14.

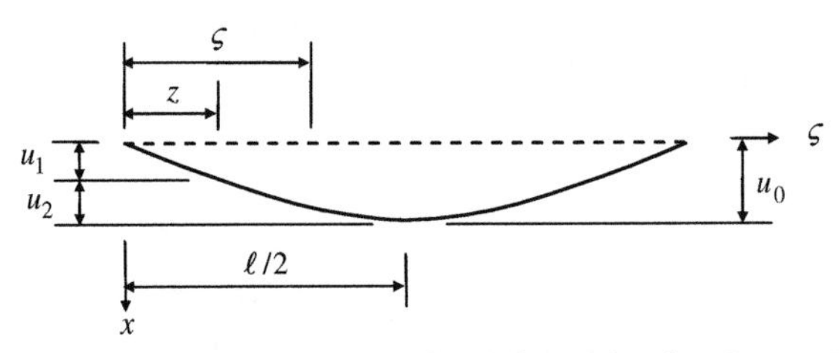

Figure 7-14 Lateral displacements in ξς plan (after Schrader, Discussion on "Lateral Stability of Unsymmetrical I-Beams and Trusses in Bending," by Winter, *Proceedings, ASCE*, 1943). Reproduced by permission.

Substituting the expression for the moment M_y and the rotation ϕ, the vertical displacement of the beam at midspan takes the following form:

$$v_0 = \frac{-1}{EI_y} \int_{\ell/2}^{0} \frac{w_y}{2} (\ell \varsigma - \varsigma^2) \varsigma \left(B \sin \frac{\pi \varsigma}{\ell} \right)^2 d\varsigma = \frac{w_y B^2}{2EI_y} \int_{0}^{\ell/2} (\ell \varsigma^2 - \varsigma^3) \sin^2 \frac{\pi \varsigma}{\ell} d\varsigma$$

Expanding the definite integral by **Maple®** gives

$$v_0 = \frac{w_y B^2 \ell^4}{768 \pi^4 EI_y} (5\pi^4 + 12\pi^2 + 144) \tag{7.6.39}$$

Similarly,

$$v_2 = \int_{\ell/2}^{z} \frac{M_y}{EI_y} (\varsigma - z) \phi d\varsigma = \frac{w_y B^2}{2EI_y} \int_{\ell/2}^{z} (\varsigma^2 - \ell \varsigma)(\varsigma - z) (\sin \frac{\pi \varsigma}{\ell})^2 d\varsigma$$

$$v_2 = \frac{w_y B^2}{768\pi^4 EI_y}$$

$$\times \left[\begin{aligned} &5\pi^4\ell^4 - 48\pi^2\ell^3 z - 16\ell^3\pi^4 z + 12\ell^4\pi^2 - 96\ell^2\pi^2 z^2 \cos^2\frac{\pi z}{\ell} \\ &+ 48\ell^2\pi^2 z^2 + 144\ell^4 \cos^2\frac{\pi z}{\ell} - 96\ell^4\pi \cos\frac{\pi z}{\ell}\sin\frac{\pi z}{\ell} \\ &+ 96\ell^3\pi^2 z \cos^2\frac{\pi z}{\ell} - 16\pi^4 z^4 + 32\ell\pi^4 z^3 + 192\pi\ell^3 z \cos\frac{\pi z}{\ell}\sin\frac{\pi z}{\ell} \end{aligned} \right] \tag{7.6.40}$$

From Fig. 7-14, it is seen that

$$v_1 = v_0 - v_2 \tag{7.6.41}$$

Hence, the loss of the potential energy of the uniformly distributed load, w_y, is

$$V = -2w_y \int_0^{\ell/2} (v_0 - v_2)dz = -\frac{w_y^2 B^2 \ell^5}{240\pi^4 EI_y}(\pi^4 + 45) \tag{7.6.42}$$

Expanding the first term of the strain energy in Eq. (7.6.1) gives

$$\begin{aligned} \frac{EI_y}{2}\int_0^\ell (u'')^2 dz &= \frac{EI_y}{2}\int_0^\ell \left(-\frac{M_y}{EI_y}\right)^2 dz = \frac{1}{2EI_y}\int_0^\ell (-M_x\phi)^2 dz \\ &= \frac{w_y^2 B^2}{8EI_y}\int_0^\ell (\ell z - z^2)^2 \sin^2\frac{\pi z}{\ell} dz \\ &= \frac{w_y^2 B^2 \ell^5}{480\pi^4 EI_y}(\pi^4 + 45) \end{aligned}$$

Hence,

$$U = \frac{w_y^2 B^2 \ell^5}{480\pi^4 EI_y}(\pi^4 + 45) + \frac{GK_T B^2 \pi^2}{4\ell} + \frac{EI_w B^2 \pi^4}{4\ell^3} \tag{7.6.43}$$

and

$$U + V = -\frac{w_y^2 B^2 \ell^5}{480\pi^4 EI_y}(\pi^4 + 45) + \frac{GK_T B^2 \pi^2}{4\ell} + \frac{EI_w B^2 \pi^4}{4\ell^3} \tag{7.6.44}$$

At the critical load, the first variation of $U + V$ with respect to B must vanish. Thus,

$$\frac{d}{dB}(U+V) = \frac{B}{2}\left[-\frac{w_y^2\ell^5}{120EI_y\pi^4}(\pi^4+45)+\frac{GK_T\pi^2}{\ell}+\frac{EI_w\pi^4}{\ell^3}\right] = 0$$

which leads to

$$(w_y\ell)_{cr} = \pm\frac{2\pi^3}{\ell^2}\sqrt{\frac{30}{\pi^4+45}EI_y\left(GK_T+\frac{EI_w\pi^2}{\ell^2}\right)} \qquad (7.6.45)$$

Schrader (1943) obtained an expression for the critical uniformly distributed load of a simply supported prismatic beam of a singly symmetric cross section based on the energy method. In the formula, he allowed that the load could be applied at any point along the web axis. He extended the approach to include two concentrated loads applied symmetrically on the span.

$$I_x = 307 \text{ in}^4, I_y = 44.1 \text{ in}^4, K_T = 0.906 \text{ in}^4, I_w = 1440 \text{ in}^6,$$

$$\ell = 180 \text{ in}, E = 29000 \text{ ksi, and } G = 11200 \text{ ksi}$$

For $m = 8$,

$$\ell = \sqrt{8EI_w/GK_T} = \sqrt{8\times 29000\times 1440/(11200\times 0.906)}$$
$$= 181.45 \text{ in.}$$

From Tables 7-2 and 7-3, γ_1 and γ_2 are read to be 42.5 and 25.6, respectively. Hence,

$$(w_y\ell)_{cr} = \frac{\gamma_1\sqrt{EI_yGK_T}}{\ell^2} = \frac{42.5}{181.45^2}\times 113917.75 = 147.05 \text{ kips}$$

$$(w_y\ell)_{cr} = \frac{2\pi^3}{\ell^2}\sqrt{\frac{30}{\pi^4+45}EI_y\left(GK_T+\frac{EI_w\pi^2}{\ell^2}\right)} = 147.2 \text{ kips} \qquad (7.6.45)$$

and

$$P_{cr} = \frac{\gamma_2\sqrt{EI_yGK_T}}{\ell^2} = \frac{25.6}{181.45^2}\sqrt{29000\times 44.1\times 11200\times 0.906}$$
$$= 88.58 \text{ kips}$$

$$P_{cr} = \frac{4\pi^2}{\ell^2}\sqrt{\frac{3}{\pi^2+6}EI_y\left(GK_T + \frac{EI_w\pi^2}{\ell^2}\right)} = 88.76 \text{ kips} \qquad (7.6.38)$$

7.6.4. Two Concentrated Loads Applied Symmetrically

Consider the case of two concentrated loads applied symmetrically as shown in Fig. 7-15. From Table 1, $M_y = M_x\phi$. Assume $\phi = B\sin\frac{\pi z}{\ell}$

The bending moment is given by

$$M_x = Pz \text{ for } 0 \le z \le a \text{ and } M_x = Pa \text{ for } a < z \le \ell/2$$

Expanding the terms of the strain energy in Eq. (7.6.1), one obtains

$$\begin{aligned}
\frac{1}{2}EI_y\int_0^\ell \left(u''\right)^2 dz &= EI_y\int_0^{\ell/2}\left(u''\right)^2 dz = EI_y\int_0^{\ell/2}\left(\frac{M_x\phi}{EI_y}\right)^2 dz \\
&= \frac{B^2P^2}{EI_y}\left(\int_0^a z^2\sin^2\frac{\pi z}{\ell}dz + a^2\int_a^{\ell/2}\sin^2\frac{\pi z}{\ell}dz\right) \\
&= \frac{B^2P^2}{EI_y}\left(\int_0^a z^2\sin^2\frac{\pi z}{\ell}dz + a^2\int_a^{\ell/2}\sin^2\frac{\pi z}{\ell}dz\right) \\
&= \frac{P^2B^2EI_y}{12\pi^3}\left[-4\pi^3a^3 - 6\pi a\ell^2\cos^2\frac{\pi a}{\ell}\right. \\
&\quad \left. + 3\ell^3\cos\frac{\pi a}{\ell}\sin\frac{\pi a}{\ell} + 3\pi a\ell^2 + 3\pi^3a^2\ell\right]
\end{aligned} \qquad (7.6.46a)$$

$$\frac{1}{2}GK_T\int_0^\ell\left(\phi'\right)^2 dz = \frac{B^2\ell}{4}\left(\frac{\pi}{\ell}\right)^2 GK_T \qquad (7.6.46b)$$

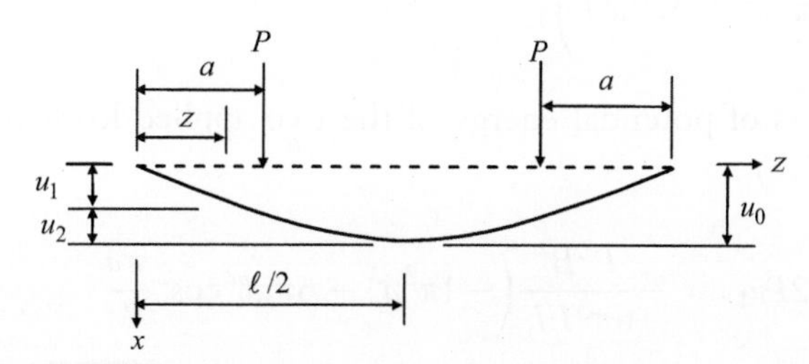

Figure 7-15 Two concentrated loads (after Schrader, Discussion on "Lateral Stability of Unsymmetrical I-Beams and Trusses in Bending," by Winter, *Proceedings, ASCE*, 1943). Reproduced by permission.

$$\frac{1}{2}EI_w \int_0^\ell \left(\phi''\right)^2 dz = \frac{B^2\ell}{4}\left(\frac{\pi}{\ell}\right)^4 EI_w \tag{7.6.46c}$$

The loss of potential energy of the applied load P is negative of the product of P and v_1. From Fig. 7-15, it is seen that $v_1 = v_0 - v_2$ as the vertical displacement v is obtained by the lateral displacement u multiplied by ϕ as per Table 7-1.

$$v_0 = \int_{\ell/2}^{0} \frac{M_y}{EI_y}\phi z\, dz$$

where $M_y = M_x\phi$. Hence,

$$\begin{aligned} v_0 &= -\frac{1}{EI_y}\left[\int_{\ell/2}^{a} Paz\left(B\sin\frac{\pi z}{\ell}\right)^2 dz + \int_a^0 Pz^2\left(B\sin\frac{\pi z}{\ell}\right)^2 dz\right] \\ &= -\frac{PB^2}{48\pi^3 EI_y}\left(3\pi^3 a\ell^2 + 12\pi a\ell^2 - 4\pi^3 a^3 - 12\pi a\ell^2\cos^2\frac{\pi a}{\ell}\right. \\ &\quad \left. + 12\ell^2\cos\frac{\pi a}{\ell}\sin\frac{\pi a}{\ell}\right) \end{aligned} \tag{7.6.47}$$

$$\begin{aligned} v_2 &= \int_{\ell/2}^{a} \frac{M_x}{EI_y}(z-a)\left(B\sin\frac{\pi z}{\ell}\right)^2 dz \\ &= -\frac{PB^2}{16\pi^2}\left(\pi^2 a\ell^2 - 4\pi^2 a^2\ell + 4a\ell^2\cos^2\frac{\pi a}{\ell} + 4\pi^2 a^3\right) \end{aligned} \tag{7.6.48}$$

$$\begin{aligned} v_1 &= -\frac{PB^2}{12\pi^3 EI_y}\left(-4\pi^3 a^3 - 6\pi a\ell^2\cos^2\frac{\pi a}{\ell} + 3\ell^3\cos\frac{\pi a}{\ell}\sin\frac{\pi a}{\ell}\right. \\ &\quad \left. + 3\pi a\ell^2 + 3\pi^3 a^2\ell\right) \end{aligned} \tag{7.6.49}$$

Hence, the loss of potential energy of the two applied loads is (note v_1 is already negative)

$$\begin{aligned} 2Pv_1 &= -\frac{P^2B^2}{6\pi^3 EI_y}\left(-4\pi^3 a^3 - 6\pi a\ell^2\cos^2\frac{\pi a}{\ell}\right. \\ &\quad \left. + 3\ell^3\cos\frac{\pi a}{\ell}\sin\frac{\pi a}{\ell} + 3\pi a\ell^2 + 3\pi^3 a^2\ell\right) \end{aligned}$$

$$
\begin{aligned}
U + V = {} & \frac{P^2 B^2 EI_y}{12\pi^3}\left[-4\pi^3 a^3 - 6\pi a\ell^2 \cos^2\frac{\pi a}{\ell} + 3\ell^3 \cos\frac{\pi a}{\ell}\sin\frac{\pi a}{\ell}\right. \\
& \left. + 3\pi a\ell^2 + 3\pi^3 a^2 \ell\right] + \frac{B^2\ell}{4}\left(\frac{\pi}{\ell}\right)^2 GK_T + \frac{B^2\ell}{4}\left(\frac{\pi}{\ell}\right)^4 EI_w \\
& - \frac{P^2 B^2}{6\pi^3 EI_y}\left(-4\pi^3 a^3 - 6\pi a\ell^2 \cos^2\frac{\pi a}{\ell} + 3\ell^3 \cos\frac{\pi a}{\ell}\sin\frac{\pi a}{\ell}\right. \\
& \left. + 3\pi a\ell^2 + 3\pi^3 a^2 \ell\right) \\
= {} & \frac{B^2\ell}{4}\left(\frac{\pi}{\ell}\right)^2 GK_T + \frac{B^2\ell}{4}\left(\frac{\pi}{\ell}\right)^4 EI_w \\
& - \frac{P^2 B^2}{12\pi^3 EI_y}\left(-4\pi^3 a^3 - 6\pi a\ell^2 \cos^2\frac{\pi a}{\ell} + 3\ell^3 \cos\frac{\pi a}{\ell}\sin\frac{\pi a}{\ell}\right. \\
& \left. + 3\pi a\ell^2 + 3\pi^3 a^2 \ell\right)
\end{aligned}
\tag{7.6.50}
$$

The stability condition equation is

$$
\begin{aligned}
& \frac{\partial(U+V)}{\partial B} = \frac{B\ell}{2}\left(\frac{\pi}{\ell}\right)^2 GK_T + \frac{B\ell}{2}\left(\frac{\pi}{\ell}\right)^4 EI_w \\
& -\frac{P^2 B}{6\pi^3 EI_y}\left(-4\pi^3 a^3 - 6\pi a\ell^2 \cos^2\frac{\pi a}{\ell} + 3\ell^3 \cos\frac{\pi a}{\ell}\sin\frac{\pi a}{\ell} + 3\pi a\ell^2 + 3\pi^3 a^2 \ell\right) \\
& = 0
\end{aligned}
$$

or

$$
\begin{aligned}
& \frac{\ell}{2}\left(\frac{\pi}{\ell}\right)^2 GK_T + \frac{\ell}{2}\left(\frac{\pi}{\ell}\right)^4 EI_w - \frac{P^2}{6\pi^3 EI_y}\left(-4\pi^3 a^3 - 6\pi a\ell^2 \cos^2\frac{\pi a}{\ell}\right. \\
& \left. + 3\ell^3 \cos\frac{\pi a}{\ell}\sin\frac{\pi a}{\ell} + 3\pi a\ell^2 + 3\pi^3 a^2 \ell\right) = 0
\end{aligned}
$$

$$EI_y\left[GK_T + EI_w\left(\frac{\pi}{\ell}\right)^2\right] = \frac{P^2\ell}{3\pi^5}\left(-4\pi^3a^3 - 6\pi a\ell^2\cos^2\frac{\pi a}{\ell} + 3\ell^3\cos\frac{\pi a}{\ell}\sin\frac{\pi a}{\ell} + 3\pi a\ell^2 + 3\pi^3a^2\ell\right)$$

$$= P^2\ell\left(-\frac{4a^3}{3\pi^2} - \frac{2a\ell^2}{\pi^4}\cos^2\frac{\pi a}{\ell} + \frac{\ell^3}{\pi^5}\cos\frac{\pi a}{\ell}\sin\frac{\pi a}{\ell} + \frac{a\ell^2}{\pi^4} + \frac{a^2\ell}{\pi^2}\right)$$

$$= P^2\ell\left(-\frac{4a^3}{3\pi^2} + \frac{a^2\ell}{\pi^2} - \frac{a\ell^2}{\pi^4}\cos\frac{2\pi a}{\ell} + \frac{\ell^3}{2\pi^5}\sin\frac{2\pi a}{\ell}\right)$$

$$P_{cr} = \sqrt{EI_y(GK_T + \pi^2EI_w/\ell^2)}\bigg/\sqrt{\ell\left(-\frac{4a^3}{3\pi^2} + \frac{a^2\ell}{\pi^2} - \frac{a\ell^2}{\pi^4}\cos\frac{2\pi a}{\ell} + \frac{\ell^3}{2\pi^5}\sin\frac{2\pi a}{\ell}\right)} \tag{7.6.51}$$

Although the approximate values of the critical load obtained by the energy method based on the principle of the minimum total potential energy are supposed to be larger than the exact values, the answers herein are very close to the exact values owing to the fact that the assumed displacement functions happen to be very close to the exact solution functions. As in all other approximate methods of analysis based on the energy principle, the accuracy of the solution depends greatly on the proper choice of the assumed displacement function. Although use of a function consisting of many terms would improve the accuracy of the solution, frequently the arithmetic operations involved could be prohibitively complex. In such a case, one ought to be able to take advantage of a computer-aided method of analysis.

7.7. DESIGN SIMPLIFICATION FOR LATERAL-TORSIONAL BUCKLING

The preceding sections determined the critical loading for beams with several different boundary conditions and loading configurations. A simply supported wide flange beam subjected to uniform bending has been shown to be in neutral equilibrium (unstable) when the applied moment reaches the value

$$M_{cr} = \frac{\pi}{\ell}\sqrt{EI_y\left(GK_T + \frac{EI_w\pi^2}{\ell^2}\right)} \tag{7.7.1}$$

The critical concentrated load applied at midspan of the same beam has been found by the energy method to be

$$P_{cr} = \frac{4\pi^2}{\ell^2}\sqrt{\frac{3}{\pi^2+6}EI_y\left(GK_T + \frac{EI_w\pi^2}{\ell^2}\right)} \tag{7.7.2}$$

Likewise, the critical uniformly distributed load on the same beam has been found to be

$$(w_y\ell)_{cr} = \frac{2\pi^3}{\ell^2}\sqrt{\frac{30}{\pi^4+45}EI_y\left(GK_T + \frac{EI_w\pi^2}{\ell^2}\right)} \tag{7.7.3}$$

Converting Eqs. (7.7.2) and (7.7.3) to the form of Eq. (7.7.1) yields

$$M_{cr} = \frac{P_{cr}\ell}{4} = 1.36\frac{\pi}{\ell}\sqrt{EI_y\left(GK_T + \frac{EI_w\pi^2}{\ell^2}\right)} \tag{7.7.4}$$

and

$$M_{cr} = \frac{(w_y\ell)_{cr}\ell}{8} = 1.13\frac{\pi}{\ell}\sqrt{EI_y\left(GK_T + \frac{EI_w\pi^2}{\ell^2}\right)} \tag{7.7.5}$$

Examination of these equations reveals that it may be possible to express the critical moment in the form

$$M_{cr} = \alpha\frac{\pi}{\ell}\sqrt{EI_y\left(GK_T + \frac{EI_w\pi^2}{\ell^2}\right)} \tag{7.7.6}$$

where the coefficient α is equal to 1.0 for uniform bending, 1.13 for a uniformly distributed load, and 1.36 for a concentrated load at applied at midspan. According to Schrader (1943) and Clark and Hill (1960), α is 1.04 for concentrated loads applied at the third points. The difference in α may be explainable from the fact that the critical bending moment diagrams of a simply supported beam are a rectangle, a triangle, and a parabola, respectively, for uniform bending, a concentrated load at midspan, and a uniformly distributed load. The area of the critical bending moment diagram for uniform bending is $M_{cr}\ell$, which is the largest. Understandably, the larger the area of the bending moment diagram, the smaller becomes the coefficient α. It is of interest to note that concentrated loads applied at the third points and a uniformly distributed load result in the same area of $2M_{cr}\ell/3$. However, the critical moment at the middle is spread wider under two third point loads than under a uniformly distributed load. This may explain the smaller value of α (1.04) for the former than that (1.13) of the latter.

Having illustrated that the equation for the critical moment of a simply supported wide flange beam subjected to uniform bending can be made applicable for other loadings by means of adjusting the factor α, the next step is to show that this equation can be made valid for different boundary conditions as well. The idea here is whether an effective-length concept analogous to that used in columns can be extended to beam buckling. Indeed it can be. Numerous researchers including Salvadori (1953, 1955), Lee (1960), and Vlasov (1961) have shown that the effective-length factor concept is also applicable to lateral-torsional buckling of beams. Based on the results given by Vlasov (1961),[10] Galambos (1968) lists values of the effective-length factor for several combinations of end conditions. Salvadori (1953) found that Eq. (7.7.6) can be made to account for the effect of moment gradient between the lateral brace points. Various lower-bound formulas have been proposed for α, but the most commonly accepted are the following:

$$C_b = 1.75 + 1.05\left(\frac{M_1}{M_2}\right) + 0.3\left(\frac{M_1}{M_2}\right)^2 \leq 2.3 \tag{7.7.7}$$

Equation (7.7.7) had been used in AISC Specifications since 1961–1993. Although Eq. (7.7.7) works well when the moment varies linearly between two adjacent brace points, it was often inadvertently used for nonlinear moment diagrams. Kirby and Nethercot (1979) present an equation that applies to various shapes of moment diagrams within the unbraced segment. Their original equation has been modified slightly to give the following:

$$C_b = \frac{12.5M_{\max}}{2.5M_{\max} + 3M_A + 4M_B + 3M_C} \leq 3.0 \tag{7.7.8}$$

Equation (7.7.8) replaces Eq. (7.7.7) in the 1993 AISC Specifications. M_B is the absolute value of the moment at the centerline, M_A and M_C are the absolute values of the quarter point and three quarter-point moments,

[10] See pages 292–297.

respectively, and M_{max} is the maximum moment regardless of its location within the brace points.

The nominal flexural strength of a beam is limited by the lateral-torsional buckling strength controlled by the unbraced length L_b of the compression flange. The critical moment equation (Eq. 7.7.6) was derived under the assumption that the material obeys Hooke's law. This means that it cannot be directly applicable to inelastic lateral-torsional buckling.

The credit goes to Lay and Galambos (1965, 1967) for determining the unbraced length L_p required for compact sections to reach the plastic bending moment M_p.

$$L_p = 2.7r_y\sqrt{\frac{E}{\sigma_y}} \tag{7.7.9}$$

where E = elastic modulus, r_y = radius of gyration with respect to the weak axis, and σ_y = mill specified minimum yield stress.

Later on, their simplified design equation was calibrated using experimental data (Bansal, 1971) in order to give compact-section beams adequate rotational capacity after reaching the plastic moment.

$$L_p = 1.76r_y\sqrt{\frac{E}{\sigma_y}} \tag{7.7.10}$$

Equation (7.7.10) is identical to Eq. (F2-5) in AISC (2005) Specifications. In plastic analysis, larger rotation capacities are required to ensure that successive plastic hinges are formed without inducing excessive lateral-torsional deformations. Bansal (1971) suggested the following equation from tests of three-span continuous beams to ensure rotation capacity greater than or equal to 3:

$$L_{pd} = \left[0.12 + 0.076\left(\frac{M_1}{M_2}\right)\right]\left(\frac{E}{\sigma_y}\right)r_y \tag{7.7.11}$$

where M_1 is the smaller moment at the ends of a laterally unbraced segment (taken positive when moments cause reverse curvature). Equation (7.7.11) is identical to Eq. (F1-17) in AISC (2001) Specifications.

The limiting value of the unbraced length for girders of compact sections to buckle in the elastic range is given by L_r. In the presence of residual stress, the maximum elastic critical moment is defined by

$$M_{cr} = S_x(\sigma_y - \sigma_r) = 0.7S_x\sigma_y = \sigma_{cr}S_x \tag{7.7.12}$$

where S_x = elastic section modulus about the x-axis, σ_r = residual stress $0.3\ \sigma_y$ for both rolled and welded shapes. From Eqs. (7.7.6) and (7.7.12),

$$\sigma_{cr} = \frac{M_{cr}}{S_x} = \frac{C_b \pi}{L_b S_x}\sqrt{EI_y GK_T + \left(\frac{\pi E}{L_b}\right)^2 I_y EI_w} \tag{7.7.13}$$

Equation (7.7.13) is identical to Eq. (F1-13) in the AISC (1986) LRFD Specifications. Equation (7.7.13) is rewritten as

$$\sigma_{cr} = \frac{C_b \pi^2 E}{\left(\dfrac{L_b}{r_{ts}}\right)^2}\sqrt{\frac{I_y GK_T L_b^2}{\pi^2 r_{ts}^4 E S_x^2} + \frac{I_y I_w}{r_{ts}^4 S_x^2}} \tag{7.7.14}$$

where

$$r_{ts}{}^2 = \frac{\sqrt{I_y E I_w}}{S_x} \tag{7.7.15}$$

Letting h_0 be the distance between the flange centroids and substituting $2G/\pi^2 E = 0.0779$ and Eq. (7.7.14), Eq. (7.7.13) becomes

$$\sigma_{cr} = \frac{C_b \pi^2 E}{\left(\dfrac{L_b}{r_{ts}}\right)^2}\sqrt{\left(\frac{L_b}{r_{ts}}\right)^2 \frac{I_y GK_T}{\pi^2 E \dfrac{\sqrt{I_y I_w}}{S_x} S_x^2} + 1}$$

$$= \frac{C_b \pi^2 E}{\left(\dfrac{L_b}{r_{ts}}\right)^2}\sqrt{1 + 0.0779 \frac{K_T c}{S_x h_0}\left(\frac{L_b}{r_{ts}}\right)^2} \tag{7.7.16}$$

where $I_w = I_y h_0^2/4$ for doubly symmetric I-beams with rectangular flanges and $c = h_0\sqrt{I_w/I_w}/2$ and hence, $c = 1.0$ for a doubly symmetric I-beam. Equation (7.7.16) is identical to Eq. (F2-4) in the AISC (2005) Specifications. Limiting the maximum critical stress in Eq. (7.7.16) to $0.7\sigma_y$ to account for residual stress σ_r and solving Eq. (7.7.16) for L_r gives

$$L_r = 1.95 r_{ts} \frac{E}{0.7\sigma_y}\sqrt{\frac{K_T c}{S_x h_0}}\sqrt{1 + \sqrt{1 + 6.767\left(\frac{0.7\sigma_y}{E}\frac{S_x h_0}{K_T c}\right)^2}} \tag{7.7.17}$$

Equation (7.7.17) is identical to Eq. (F2-6) in AISC (2005) Specifications.

The calculation of the inelastic critical moment is fairly complex. Galambos (1963, 1998) has contributed greatly to this subject. In 2005

AISC Specifications, when $L_p < L_b \leq L_r$, the nominal flexural strength M_n of compact sections is linearly interpolated between the plastic moment M_p and the elastic critical moment $M_r = 0.7S_x\sigma_y$ as

$$M_n = C_b\left[M_p - (M_p - 0.7\sigma_y S_x)\left(\frac{L_b - L_p}{L_r - L_p}\right)\right] \leq M_p \qquad (7.7.18)$$

It should be remembered that local buckling of compression flanges and web is precluded in the derivation of the unbraced lengths. The mathematical procedure for the solution of local buckling is identical to that of the lateral-torsional buckling phenomenon, except that the governing differential equations are now partial differential equations, and so the details of the solution process are different and complicated, as will be shown in the next chapter. In order to systematically reflect the effects of local buckling on the nominal flexural strength, AISC (2005) Specifications categorize sections into several types depending on the compactness of the flanges and web (compact, noncompact, or slender).

Also, it needs to be noted that the limiting values for the unbraced length given by Eqs. (7.7.10), (7.7.11), and (7.7.17) are valid only for bare-steel members. Composite systems are often utilized to maximize the efficiency of structural members. In composite girders, the top flange and concrete slab are connected with shear studs. Lateral-torsional buckling is not likely to take place when subjected to positive flexure, as the top compression flange is continuously braced by the concrete slab. However, the loss of stability should be checked when designing composite girders in negative moment zone. The steel section of a composite girder will necessarily undergo the deformation depicted in Fig. 7-16 during buckling due to the restraint provided by the concrete slab. This type of buckling is referred to as lateral-distortional buckling. The classical assumption that the member cross section retains its original shape during buckling is no longer valid in this case. Lateral-distortional buckling is basically a combined mode of lateral-torsional buckling (global buckling) and local buckling, and the derivation of a closed-form solution is, therefore, not straightforward. Limited research, including Hancock et al. (1980), Bradford and Gao (1992), and Hong et al. (2002), has shown that the unbraced length requirements for noncomposite girders give too conservative results for composite girders. A general design rule has yet to be developed due to lack of comprehensive study. With the advancement of digital computers, along with highly sophisticated computer programs such as ABAQUS, NASTRAN, and ADINA just to name a few, performing the lateral-distortional buckling analysis should present no problem.

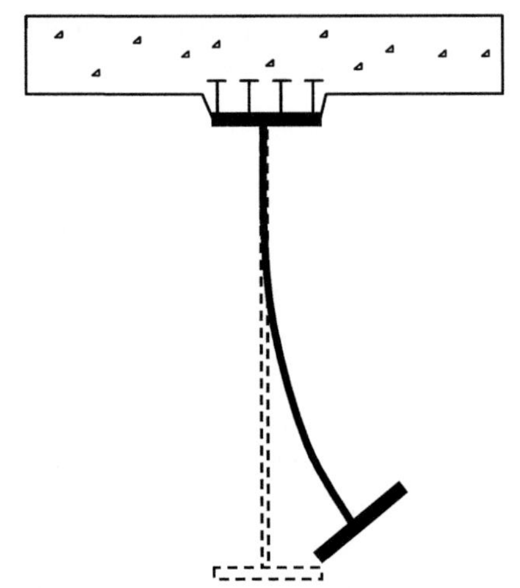

Figure 7-16 Lateral-distortional buckling

REFERENCES

AISC. (1986). *LRFD Manual of Steel Construction* (1st ed.). Chicago, IL: American Institute of Steel Construction.

AISC. (2001). *LRFD Manual of Steel Construction* (3rd ed.). Chicago, IL: American Institute of Steel Construction.

AISC. (2005). *Specification for Structural Steel Buildings*. Chicago, IL: American Institute of Steel Construction.

Bansal, J. (1971). *The Lateral Instability of Continuous Beams, AISI Report No. 3 prepared from the tests at the University of Texas*. New York: American Iron and Steel Institute.

Bleich, F. (1952). *Buckling Strength of Metal Structures*. New York: McGraw-Hill.

Bowman, F. (1938). *Introduction to Bessel Functions*. London, UK: Longmans, Green and Co.

Bradford, M. A., & Gao, Z. (1992). Distorsional Buckling Solutions for Continuous Composite Beams. *Journal of Structural Engineering, ASCE, Vol. 118*(No. 1), 73–86.

Brush, D. O., & Almroth, B. O. (1975). *Buckling of Bars, Plates, and Shells*. New York: McGraw-Hill.

Chajes, A. (1974). *Principles of Structural Stability Theory*. Englewood Cliffs, NJ: Prentice-Hall.

Clark, J. W., & Hill, H. N. (1960). Lateral Buckling of Beams. *Journal of the Structural Div., ASCE, Vol. 86*(No. ST7), 175–196.

Errera, S. J. (1964). Materials. In L. Tall (Ed.), *Structural Steel Design*. New York: Ronald Press Co.

Galambos, T. V. (1963). Inelastic Lateral Buckling of Beams. *Journal of the Structural Div., ASCE, Vol. 89*(No. ST5), 217–244.

Galambos, T. V. (1968). *Structural Members and Frames*. Englewood Cliffs, NJ: Prentice-Hall.

Galambos, T. V. (Ed.). (1998). Guide to Stability Design Criteria for Metal Structures, *Structural Stability Research Council* (5th ed.). New York: John Wiley and Sons.

Grossman, S. I., & Derrick, W. R. (1988). *Advanced Engineering Mathematics*. New York: Harper and Row.

Hill, H. N. (1954). Lateral Buckling of Channels and Z-Beams. *Transactions, ASCE, Vol. 119*, pp. 829–841.
Hong, A.R., Lee, D.S., & Lee, S.C. (2002). A Study on Unbraced Length Requirement for Composite Steel I Girder in Negative Bending, *Proceedings of 2002 Annual Conference, KSCE, Korea*, pp. 70–73 (in Korean).
Kirby, P. A., & Nethercot, D. A. (1979). *Design for Structural Stability*. New York: John Wiley and Sons.
Lay, M. G., & Galambos, T. V. (1965). Inelastic Steel Beams Under Moment Gradient. *Journal of the Structural Div., ASCE, Vol. 91*(No. ST6), 67–93.
Lay, M. G., & Galambos, T. V. (1967). Inelastic Beams Under Moment Gradient. *Journal of the Structural Div., ASCE, Vol. 93*(No. ST1), 381–399.
Lee, G.C. (1960). A Survey of the Literature on the Lateral Instability of Beams, Welding Research Council Bulletin No. 63, August.
Michell, A. G. M. (1899). Elastic Stability of Long Beams Under Transverse Forces. *Philosophic Magazine, Vol. 48*, 298.
Prandtl, L. (1899). Kipperscheinungen, Dissertation, Munich Polytechnical Institute.
Salmon, C. G., & Johnson, J. E. (1996). *Steel Structures, Design and Behavior* (4th ed.). New York: HarperCollins.
Salvadori, M.G. (1953). Lateral Buckling of I Beams Under Thrust and Unequal End Moments, *Proceedings, ASCE, Vol. 79*(No.291), pp. 1–25, also a transaction paper No 2773, Lateral Buckling of I-Beams, *Transactions, ASCE, Vol. 120*, 1955, pp. 1165–1182.
Salvadori, M.G. (1955). Lateral Buckling of Eccentrically Loaded I-Columns, *Proceedings, ASCE, Vol. 81*, pp. 607–622, also a transaction paper No. 2836 with the same title, *Transactions, ASCE, Vol. 121*, 1956, pp. 1163–1178.
Schrader, R.K. (1943). Discussion on Lateral Stability of Unsymmetrical I-Beams and Trusses in Bending, by G. Winter, Paper No. 2178, *Transactions, ASCE, Vol. 108*, 1943, pp. 247–268.
Timoshenko, S. P. (1910). Einige Stabilitäts-problme der Elasticitäts-theorie. *Zeitschrift für Mathematik und Phisik, Vol. 58*, 337–385.
Timoshenko, S. P., & Gere, J. M. (1961). *Theory of Elastic Stability* (2nd ed.). New York: McGraw-Hill.
Vincent, G. S. (1969). *Tentative Criteria for Load Factor Design of Steel Highway Bridges. AISI Report No. 15*. New York: American Iron and Steel Institute.
Vlasov, V. Z. (1940). *Tonkostennye uprugie sterzhni* (*Thin-walled elastic beams*), Stroiizdat (in Russian).
Vlasov, V. Z. (1961). *Tonkostennye uprugie sterzhni* (*Thin-walled elastic beams*), (2nd ed.), Moscow 1959, translated by Schechtman, Israel Program for Scientific Translation, Jerusalem.
Winter, G. (1941). Lateral Stability of Unsymmetrical I-Beams and Trusses in Bending, *Proceedings, ASCE. Vol. 67*, pp. 1851–1864, also a transaction paper No. 2178 with the same title, *Transactions, ASCE, Vol. 108*, 1943, pp. 247–268.
Yoo, C. H. (1980). Bimoment Contribution to Stability of Thin-Walled Assemblages. *Computers and Structures, Vol. 11*(No.5), 465–471.

PROBLEMS

7.1 Equation (7.4.2) can be nondimensionalized to investigate the influence of the various parameters affecting the lateral-torsional buckling strength.

$$\frac{M_{cr}}{M_y} = \sqrt{\frac{\pi^2 EI_y GK_T}{\ell^2 S_x^2 \sigma_y^2}\left(1 + \frac{\pi^2 EI_w}{GK_T \ell^2}\right)} \tag{P7.1.1}$$

where $M_y = S_x \sigma_y$. After introducing the following identities:

$$\varepsilon_y = \frac{\sigma_y}{E},\ I_w = \frac{I_y (d - t_f)^2}{4},\ I_x = Ar_x^2,\ I_y = Ar_y^2,\ S_x = \frac{2I_x}{d} \tag{P7.1.2}$$

and the abbreviation

$$D_T = \frac{K_T}{Ad^2} \tag{P7.1.3}$$

one gets the following nondimensionalized buckling moment:

$$\frac{M_{cr}}{M_y} = \frac{\pi (d/r_x)^2}{2\varepsilon_y}\left(\frac{\sqrt{D_T G/E}}{\ell / r_y}\right)\sqrt{1 + \frac{\pi^2 E (1 - t_f/d)^2}{4 G D_T (\ell / r_y)^2}} \tag{P7.1.4}$$

For most wide flange shapes,

$$d/r_x \cong 2.38,\ (1 - t_f)/d \cong 0.95,\ \text{and}$$

$$G/E = 1/2(1 + \mu) = 0.385 \quad \text{for } \mu = 0.3$$

Substituting these values into Eq. (P7.1.4) yields

$$\frac{M_{cr}}{M_y} = \frac{5.56}{\varepsilon_y}\left(\frac{\sqrt{D_T}}{\ell / r_y}\right)\sqrt{1 + \frac{5.78}{D_T (\ell / r_y)^2}} \tag{P7.1.5}$$

Plot the buckling curve ($1000 M_{cr} \varepsilon_y / M_y$ vs. ℓ / r_r) for W27 × 94 using Eqs. (P7.1.4). Limit $\ell / r_r < 500$.

7.2 Using the Rayleigh-Ritz method, determine the critical uniformly distributed load for a prismatic simply supported beam. The load is applied at the centroid. Assume $u = A \sin \pi z / \ell$, $\phi = \mathrm{B} \sin \pi z / \ell$. Use

$$\Pi = \frac{1}{2}\int_0^\ell \left[EI_y (u'')^2 + EI_w (\phi'')^2 + GK_T (\phi')^2 - 2M_x u'' \phi \right] dz \text{ with}$$

$$M_x = -\frac{w}{2}(\ell z - z^2)$$

What is the coefficient α in association with Eq. (7.7.6)?

7.3 Using the system of differential equations, Eqs. (7.2.22)–(7.2.24), derive the total potential energy expression of Problem 7.2.

7.4 Determine the critical moment of a simply supported prismatic wide flange beam under one end moment only by

(a) Eq. (7.4.2),

(b) the Rayleigh-Ritz method, assuming
$u = A \sin \pi z/\ell,\ \phi = B \sin \pi z/\ell$
For the Rayleigh-Ritz method, use
$\Pi = \frac{1}{2}\int_0^{\ell}[EI_y(u'')^2 + EI_w(\phi'')^2 + GK_T(\phi')^2 - 2M_0\,(z/\ell)\,u''\phi]dz$

7.5 Prove the following relationship regarding the loss of potential energy of the externally applied uniform bending moment in a prismatic simply supported beam:

$$\int_0^{\ell} M_0 u''\phi dz = -\int_0^{\ell} M_0 u'\phi' dz$$

7.6 From the geometry of Fig. 7-1, it is obvious that

$$y_s = v \cos\phi + u \sin\phi$$

Show that $y_s'' = v'' \cos\phi + u'' \sin\phi$

7.7 Ends are simply supported ($\phi = \phi'' = u = u'' = v = v'' = 0$) and $u = \phi = 0$ at the load points. The loads are assumed to apply at the centroid. The W27 × 94 beam is made of A36 steel and is 40 feet long.

(a) Determine the ultimate load, P_u by the 2005 AISC Specifications.

(b) Determine the critical load, P_{cr}, by any refined method and assess the effect of the continuity condition at the load point on the buckling strength of uniform bending in the middle.

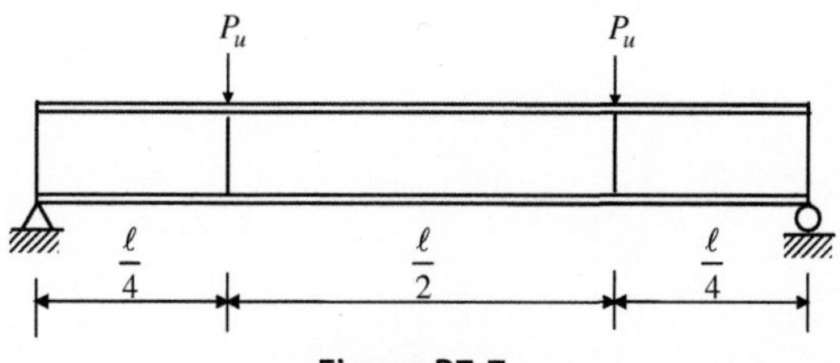

Figure P7-7

CHAPTER 8

Buckling of Plate Elements

Contents

8.1. INTRODUCTION

Equilibrium and stability equations of one-dimensional elements such as beams, columns, and framed members have been treated in the preceding chapters. The analysis of these members is relatively simple as bending, the essential characteristic of buckling, can be assumed to take place in one plane only. The buckling of a plate, however, involves bending in two planes and is therefore much more complicated. From a mathematical point of view, the main difference between framed members and plates is that quantities such as

Stability of Structures
ISBN 978-0-12-385122-2, doi:10.1016/B978-0-12-385122-2.10008-9

deflections and bending moments, which are functions of a single independent variable in framed members, become functions of two independent variables in plates. Consequently, the behavior of plates is governed by partial differential equations, which increases the complexity of analysis.

There is another significant difference in the buckling characteristics of framed members and plates. For a framed member, buckling terminates the member's ability to resist any further load, and the critical load is thus the failure load or the ultimate load. The same is not necessarily true for plates. A plate element may carry additional loading beyond the critical load. This reserve strength is called the postbuckling strength. The relative magnitude of the postbuckling strength to the buckling load depends on various parameters such as dimensional properties, boundary conditions, types of loading, and the ratio of buckling stress to yield stress. Plate buckling is usually referred to as local buckling. Structural shapes composed of plate elements may not necessarily terminate their load-carrying capacity at the instance of local buckling of individual plate elements. Such an additional strength of structural members is attributable not only to the postbuckling strength of the plate elements but also to possible stress redistribution in the member after failure of individual plate elements.

The earliest solution of a simply supported flat-plate stability problem apparently was given by Bryan (1891), almost 150 years after Euler presented the first accurate stability analysis of a column. At the beginning of the twentieth century, plate buckling was again investigated by Timoshenko, who studied not only the simply supported case, but many other combinations of boundary conditions as well. Many of the solutions he obtained are given in Timoshenko and Gere (1961). Treatments of flat-plate stability analysis that are much more extensive than those given here may be found in Bleich (1952) and Timoshenko and Gere (1961). Other references such as Allen and Bulson (1980), Brush and Almroth (1975), Chajes (1974), and Szilard (1974) are reflected for their modern treatments of the subject.

8.2. DIFFERENTIAL EQUATION OF PLATE BUCKLING

8.2.1. Plate Bending Theory

The classical theory of flat plates presented here follows the part of materials leading to the von Kármán equations entailed by Langhaar (1962) excluding the energy that results from heating. Consider an isolated free body of a plate element in the deformed configuration (necessary for stability problems examining equilibrium in the deformed configuration, neighboring equilibrium). The plate material is assumed to be isotropic and homogeneous and to obey Hooke's law. The plate is assumed to be prismatic, and forces expressed per-unit width of the plate are assumed constant along the length direction. The plate is referred to rectangular Cartesian coordinates x, y, z, where x and y lie in the middle plane of the plate and z is measured from the middle plane. The objective of thin-plate theory is to reduce a three-dimensional (complex) problem to an approximate (practical) one based on the following simplifying assumptions:

1. Normals to the undeformed middle plane are assumed to remain normal to the deflected middle plane and inextensional during deformations, so that transverse normal and shearing strains may be ignored in deriving the plate kinematic relations.
2. Transverse normal stresses are assumed to be small compared with the other normal stresses, so that they may be ignored.

Novozhilov (1953) referred to these approximations as the Kirchhoff assumptions. The first approximation is tantamount to the typical plane strain assumption, and the second is part of plane stress assumption.

Internal forces (generalized) acting on the edges of a plate element $dxdy$ are related to the internal stresses by the equations

$$N_x = \int_{-h/2}^{h/2} \bar{\sigma}_x dz \quad N_y = \int_{-h/2}^{h/2} \bar{\sigma}_y dz \quad N_{xy} = \int_{-h/2}^{h/2} \bar{\tau}_{xy} dz$$

$$N_{yx} = \int_{-h/2}^{h/2} \bar{\tau}_{yx} dz \quad Q_x = \int_{-h/2}^{h/2} \bar{\tau}_{xz} dz \quad Q_y = \int_{-h/2}^{h/2} \bar{\tau}_{yz} dz$$

$$M_x = \int_{-h/2}^{h/2} \bar{\sigma}_x z dz \quad M_y = \int_{-h/2}^{h/2} \bar{\sigma}_y z dz \quad M_{xy} = \int_{-h/2}^{h/2} \bar{\tau}_{xy} z dz$$

$$M_{yx} = \int_{-h/2}^{h/2} \bar{\tau}_{yx} z dz \tag{8.2.1}$$

where

N_x, N_y, N_{xy}, N_{yx} = in-plane normal and shearing forces, Q_x, Q_y = transverse shearing forces, M_x, M_y = bending moments, M_{xy}, M_{yx} = twisting moments.

The barred quantities, $\bar{\sigma}_x$, $\bar{\tau}_{xy}$, etc., stand for stress components at any point through the thickness, as distinguished from σ_x, τ_{xy}, etc., which denote corresponding quantities on the middle plane ($z = 0$). The positive in-plane normal and shearing forces, transverse (also referred to as bending) shearing forces, bending moments, and twisting moments are given in Figs. 8-1, 8-2, and 8-3, respectively.

Since $\tau_{xy} = \tau_{yx}$, Eqs. (8.2.1) reveal that $N_{xy} = N_{yx}$ and $M_{xy} = M_{yx}$. The directions of the positive moments given in Fig. 8-3 result in positive stresses at the positive end of the z-axis. As a result of the Kirchhoff's first approximation, the displacement components at any point in the plate, $\bar{u}$, $\bar{v}$, $\bar{w}$, can be expressed in terms of the corresponding middle-plane quantities, u, v, w, by the relations

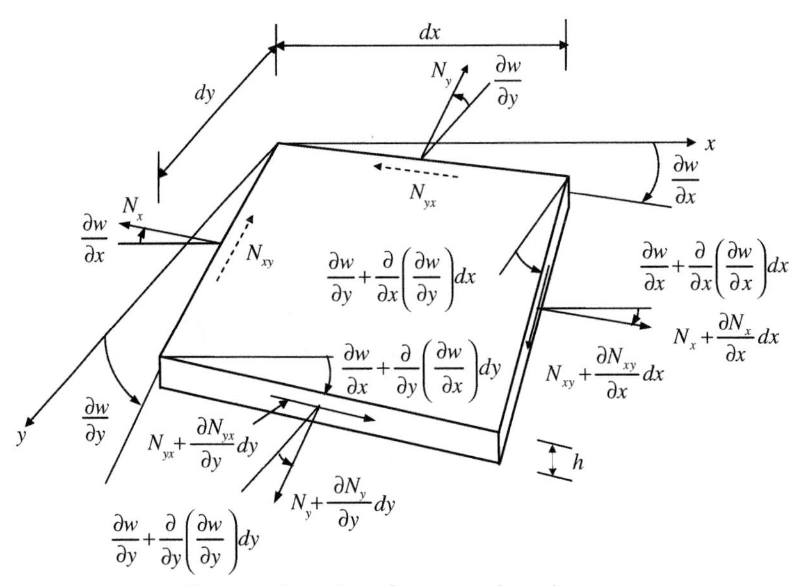

Figure 8-1 In-plane forces on plate element

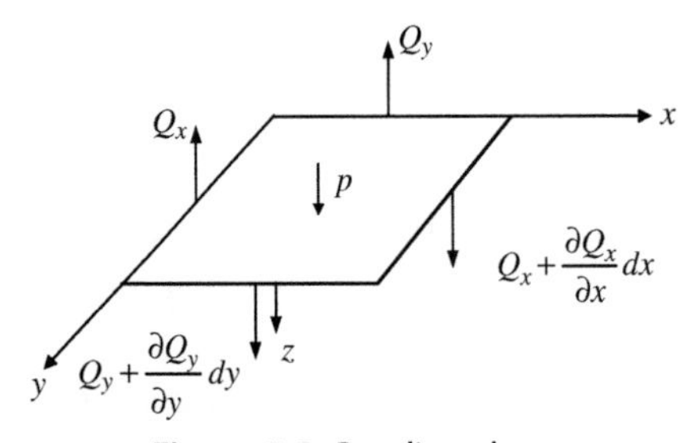

Figure 8-2 Bending shear

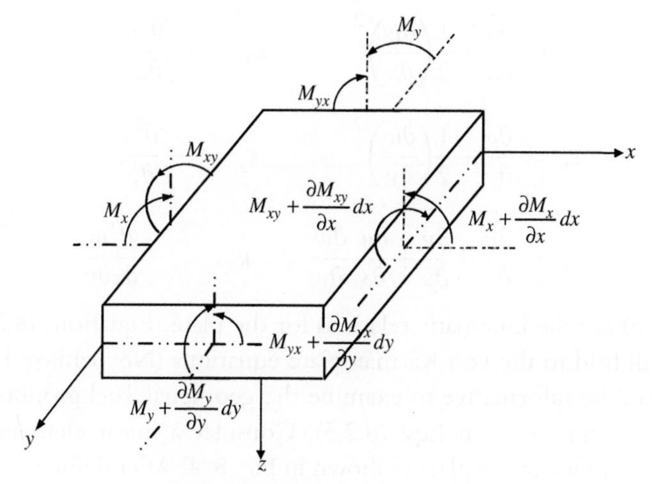

Figure 8-3 Moment components

$$\bar{u} = u - z\frac{\partial w}{\partial x} \quad \bar{v} = v - z\frac{\partial w}{\partial y} \quad \bar{w} = w \tag{8.2.2}$$

where positive rotations are shown in Fig. 8-1.

Ignoring negligibly small higher order terms, the components of the strain-displacement relations for a three-dimensional body as given by Novozhilov (1953) are

$$\bar{\varepsilon}_x = \frac{\partial \bar{u}}{\partial x} + \frac{1}{2}\left(\frac{\partial \bar{w}}{\partial x}\right)^2 \quad \bar{\varepsilon}_y = \frac{\partial \bar{v}}{\partial y} + \frac{1}{2}\left(\frac{\partial \bar{w}}{\partial y}\right)^2 \quad \bar{\gamma}_{xy} = \frac{\partial \bar{u}}{\partial y} + \frac{\partial \bar{v}}{\partial y} + \frac{\partial \bar{w}}{\partial x}\frac{\partial \bar{w}}{\partial y} \tag{8.2.3}$$

The strain components in Eqs. (8.2.3) are the Green-Lagrange strain (Sokolnikoff 1956; Bathe 1996), which is suited for the incremental analysis based on the total Lagrangian formulation. Substituting Eq. (8.2.2) into Eq. (8.2.3) yields

$$\bar{\varepsilon}_x = \varepsilon_x - z\frac{\partial^2 w}{\partial x^2} = \varepsilon_x + z\kappa_x \quad \bar{\varepsilon}_y = \varepsilon_y - z\frac{\partial^2 w}{\partial y^2} = \varepsilon_y + z\kappa_y$$

$$\bar{\gamma}_{xy} = \gamma_{xy} - 2z\frac{\partial^2 w}{\partial x \partial y} = \gamma_{xy} + 2z\kappa_{xy} \tag{8.2.4}$$

where $\bar{\varepsilon}_x$, $\bar{\varepsilon}_y$, $\bar{\gamma}_{xy}$ are normal and shear strain components at any point through the plate thickness and ε_x, ε_y, γ_{xy} are corresponding quantities at points on the middle plane, and where

$$
\begin{aligned}
\varepsilon_x &= \frac{\partial u}{\partial x} + \frac{1}{2}\left(\frac{\partial w}{\partial x}\right)^2 & \kappa_x &= -\frac{\partial^2 w}{\partial x^2} \\
\varepsilon_y &= \frac{\partial v}{\partial y} + \frac{1}{2}\left(\frac{\partial w}{\partial y}\right)^2 & \kappa_y &= -\frac{\partial^2 w}{\partial y^2} \\
\gamma_{xy} &= \frac{\partial u}{\partial y} + \frac{\partial v}{\partial y} + \frac{\partial w}{\partial x}\frac{\partial w}{\partial y} & \kappa_{xy} &= -\frac{\partial^2 w}{\partial x \partial y}
\end{aligned} \tag{8.2.5}
$$

Eqs. (8.2.5) are the kinematic relations for the plate. Equations (8.2.4) and (8.2.5) will lead to the von Kármán plate equations (Novozhilov 1953).

It would be informative to examine the geometric background for the large strain expressions in Eqs. (8.2.5). Consider a linear element AB of the middle surface of the plate as shown in Fig. 8-4. After deformations, the element assumes the new position $A'B'$. The length of the element is changed due to the in-plane deformation in the x direction u and due to the transverse displacement w in the z direction. As a result of the u displacement, the elongation of the element is

$$
\frac{\partial u}{\partial x}dx
$$

The length $A'B'$ due to the transverse displacement alone is computed from the Pythagorean theorem as (after a binomial expansion)

$$
\begin{aligned}
A'B' &= \left[dx^2 + \left(\frac{\partial w}{\partial x}dx\right)^2\right]^{1/2} \\
&= \left[1 + \left(\frac{\partial w}{\partial x}\right)^2\right]^{1/2} dx \simeq \left[1 + \frac{1}{2}\left(\frac{\partial w}{\partial x}\right)^2\right] dx
\end{aligned}
$$

The elongation due to the transverse displacement is

$$
\frac{1}{2}\left(\frac{\partial w}{\partial x}\right)^2 dx
$$

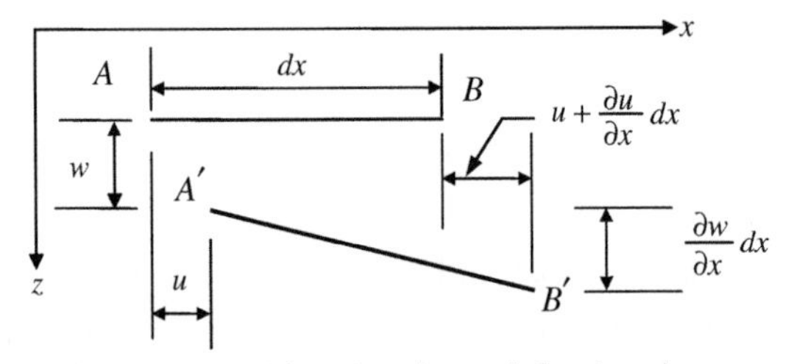

Figure 8-4 Axial strain—large deflection theory

and the total elongation is the sum of the two

$$\frac{\partial u}{\partial x}dx + \frac{1}{2}\left(\frac{\partial w}{\partial x}\right)^2 dx$$

The axial strain is equal to the total elongation divided by the original length of the element dx. Hence,

$$\varepsilon_x = \frac{\partial u}{\partial x} + \frac{1}{2}\left(\frac{\partial w}{\partial x}\right)^2$$

Likewise,

$$\varepsilon_y = \frac{\partial v}{\partial y} + \frac{1}{2}\left(\frac{\partial w}{\partial y}\right)^2$$

The shear strain (angular change) may consist of the in-plane contribution and the bending (vertical) contribution as illustrated in Figs. 8-5 and 8-6.

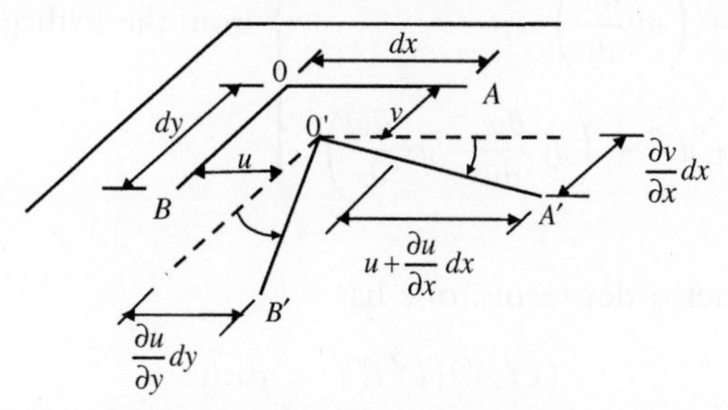

Figure 8-5 In-plane angle change (after Chajes, *Principles of Structural Stability Theory*. Englewood Cliffs, NJ: Prentice-Hall, 1974). Reproduced by permission from the author.

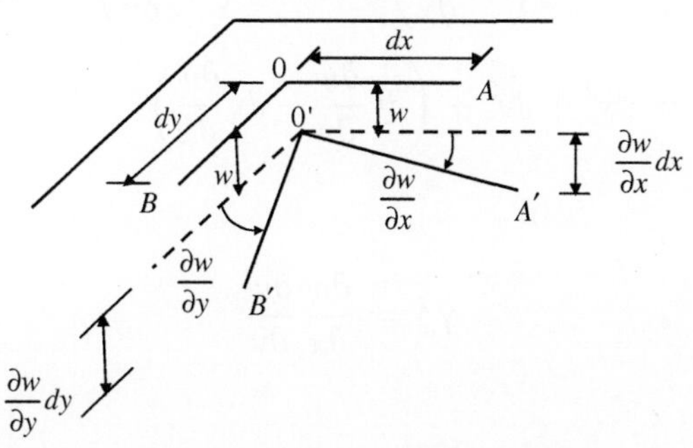

Figure 8-6 Out-of-plane angle change (after Chajes, *Principles of Structural Stability Theory*. Englewood Cliffs, NJ: Prentice-Hall, 1974). Reproduced by permission from the author.

The in-plane contribution is

$$\frac{\partial u}{\partial y} + \frac{\partial v}{\partial x} = \gamma_n$$

The bending contribution is

$$\gamma_w = \angle BOA - \angle B'O'A' = \frac{\pi}{2} - \left(\frac{\pi}{2} - \gamma_w\right)$$

From the elementary geometry (law of cosine),

$$(A'B')^2 = (O'A')^2 + (O'B')^2 - 2(O'A')(O'B')\cos\left(\frac{\pi}{2} - \gamma_w\right)$$

where

$$\left.\begin{aligned}(O'A')^2 &= dx^2 + \left(dx\frac{\partial w}{\partial x}\right)^2 \\ (O'B')^2 &= dy^2 + \left(dy\frac{\partial w}{\partial y}\right)^2 \\ (A'B')^2 &= dx^2 + dy^2 + \left(dy\frac{\partial w}{\partial y} - dx\frac{\partial w}{\partial x}\right)^2\end{aligned}\right\} \text{from the Pythagorean theorem}$$

Neglecting higher order terms, one has

$$(O'A')(O'B') = dxdy$$

Recognizing that $\cos((\pi/2) - \gamma_w) = \gamma_w$ for small angles, then

$$\begin{aligned}(A'B')^2 &= dx^2 + \left(dx\frac{\partial w}{\partial x}\right)^2 + dy^2 + \left(dy\frac{\partial w}{\partial y}\right)^2 - 2\gamma_w dxdy \\ &= dx^2 + dy^2 + \left(dy\frac{\partial w}{\partial y} - dx\frac{\partial w}{\partial x}\right)^2\end{aligned}$$

which leads to

$$\gamma_w = \frac{\partial w}{\partial x}\frac{\partial w}{\partial y}$$

Hence,

$$\gamma = \frac{\partial u}{\partial y} + \frac{\partial v}{\partial x} + \frac{\partial w}{\partial x}\frac{\partial w}{\partial y}$$

Modifying the generalized Hooke's law in a three-dimensional isotropic medium with the Kirchhoff's assumptions leads to the following stress-strain relations:

$$\bar{\sigma}_x = \frac{E}{1-\mu^2}(\bar{\varepsilon}_x + \mu\bar{\varepsilon}_y) \quad \bar{\sigma}_y = \frac{E}{1-\mu^2}(\bar{\varepsilon}_y + \mu\bar{\varepsilon}_x) \quad \bar{\tau}_{xy} = \frac{E}{2(1+\mu)}\bar{\gamma}_{xy} \tag{8.2.6}$$

Substituting Eqs. (8.2.6) and (8.2.4) into Eq. (8.2.1) and integrating the result gives

$$\begin{aligned}
N_x &= C(\varepsilon_x + \mu\varepsilon_y) \quad N_y = C(\varepsilon_y + \mu\varepsilon_x) \quad N_{xy} = C(1-\mu)\gamma_{xy}/2 \\
M_x &= -D\left(\frac{\partial^2 w}{\partial x^2} + \mu\frac{\partial^2 w}{\partial y^2}\right) \quad M_y = -D\left(\frac{\partial^2 w}{\partial y^2} + \mu\frac{\partial^2 w}{\partial x^2}\right) \\
M_{xy} &= -D(1-\mu)\frac{\partial^2 w}{\partial x \partial y}
\end{aligned} \tag{8.2.7}$$

with

$$C = \frac{Eh}{1-\mu^2} \quad \text{and} \quad D = \frac{Eh^3}{12(1-\mu^2)} \tag{8.2.8}$$

The coefficients C and D are axial and bending rigidities of the plate per unit width shown in Fig. 8-7, respectively.

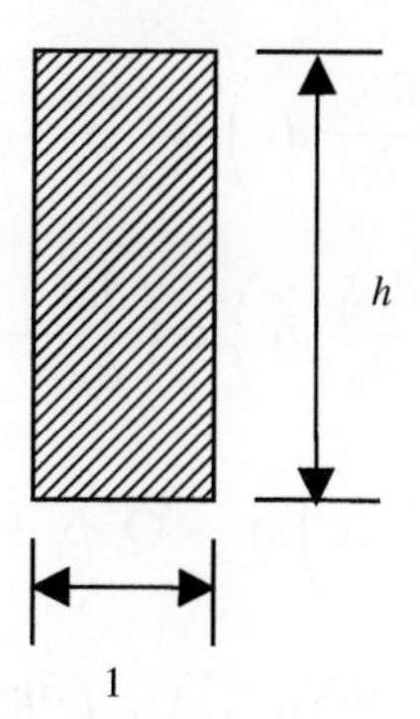

Figure 8-7 Plate cross section

8.2.2. Equilibrium Equations

In order to account for the interaction between forces and deformations, the equations representing equilibrium must be derived in a slightly deformed configuration (neighboring equilibrium), as shown in Fig. 8-1. The forces and deformations are assumed to vary across the plate element.

To simplify the diagrams, bending shearing forces and moment intensities are shown in Figs. 8-2 and 8-3 in their positive directions, respectively. The angles of rotations $\partial w/\partial x$ and $\partial w/\partial y$ are small, and sines and cosines of the angles are replaced by the angles and unity, respectively. Quadratic or higher order terms are assumed to be negligibly small and are ignored.

Summing the forces in Fig. 8-1 in the x direction gives

$$-N_x dy + \left(N_x + \frac{\partial N_x}{\partial x}dx\right)dy - N_{yx}dx + \left(N_{yx} + \frac{\partial N_x}{\partial y}dy\right)dx = 0 \tag{8.2.9}$$

Canceling out the quantity $dx\ dy$ in Eq. (8.2.9) results in

$$\frac{\partial N_x}{\partial x} + \frac{\partial N_{yx}}{\partial y} = 0 \tag{8.2.10}$$

Likewise, summing the forces in the y direction yields

$$\frac{\partial N_{xy}}{\partial x} + \frac{\partial N_y}{\partial y} = 0 \tag{8.2.11}$$

Summation of the forces in the z direction is somewhat more involved. From Figs. 8-1 and 8-2, one obtains

$$\begin{aligned}
&- N_x dy \frac{\partial w}{\partial x} + \left(N_x + \frac{\partial N_x}{\partial x}dx\right)dy\left(\frac{\partial w}{\partial x} + \frac{\partial^2 w}{\partial x^2}dx\right)\\
&- N_y dx \frac{\partial w}{\partial y} + \left(N_y + \frac{\partial N_y}{\partial y}dy\right)dx\left(\frac{\partial w}{\partial y} + \frac{\partial^2 w}{\partial y^2}dy\right)\\
&- Q_x dy + \left(Q_x + \frac{\partial Q_x}{\partial x}dx\right)dy - Q_y dx + \left(Q_y + \frac{\partial Q_y}{\partial y}dy\right)dx\\
&- N_{xy} dy \frac{\partial w}{\partial y} + \left(N_{xy} + \frac{\partial N_{xy}}{\partial x}dx\right)dy\left(\frac{\partial w}{\partial y} + \frac{\partial^2 w}{\partial x \partial y}dx\right)\\
&- N_{yx} dx \frac{\partial w}{\partial x} + \left(N_{yx} + \frac{\partial N_{yx}}{\partial y}dy\right)dx\left(\frac{\partial w}{\partial x} + \frac{\partial^2 w}{\partial x \partial y}dy\right) + p dx\ dy = 0
\end{aligned} \tag{8.2.12}$$

Neglecting higher order terms and regrouping terms in Eq. (8.2.12) gives

$$\left(\frac{\partial N_x}{\partial x}+\frac{\partial N_{yx}}{\partial y}\right)\frac{\partial w}{\partial x}+\left(\frac{\partial N_{xy}}{\partial x}+\frac{\partial N_y}{\partial y}\right)\frac{\partial w}{\partial y}+N_x\frac{\partial^2 w}{\partial x^2}+N_{xy}\frac{\partial^2 w}{\partial x\partial y}$$

$$+N_{yx}\frac{\partial^2 w}{\partial x\partial y}+N_y\frac{\partial^2 w}{\partial y^2}+\frac{\partial Q_x}{\partial x}+\frac{\partial Q_y}{\partial y}+p=0 \tag{8.2.13}$$

As a consequence of Eqs. (8.2.10) and (8.2.11), the quantities inside the parentheses in Eq. (8.2.13) are zero. Since $N_{xy}=N_{yx}$ (this can be readily proved from Eq. (8.2.1) by noting that $\tau_{xy}=\tau_{yx}$), it follows

$$\frac{\partial Q_x}{\partial x}+\frac{\partial Q_y}{\partial y}+N_x\frac{\partial^2 w}{\partial x^2}+N_y\frac{\partial^2 w}{\partial y^2}+2N_{xy}\frac{\partial^2 w}{\partial x\partial y}+p=0 \tag{8.2.14}$$

The condition that the sum of moments about the x-axis must vanish yields.

$$-\frac{\partial M_y}{\partial y}-\frac{\partial M_{xy}}{\partial x}+Q_y=0 \tag{8.2.15}$$

Similarly, moments about the y-axis lead to

$$\frac{\partial M_x}{\partial x}+\frac{\partial M_{yx}}{\partial y}-Q_x=0 \tag{8.2.16}$$

Differentiating Eqs. (8.2.15) and (8.2.16) and substituting the results into Eq. (8.2.14) yields

$$\frac{\partial^2 M_x}{\partial x^2}+2\frac{\partial^2 M_{xy}}{\partial x\partial y}+\frac{\partial^2 M_y}{\partial y^2}+N_x\frac{\partial^2 w}{\partial x^2}+N_y\frac{\partial^2 w}{\partial y^2}+2N_{xy}\frac{\partial^2 w}{\partial x\partial y}+p=0 \tag{8.2.17}$$

If one considers (at least temporarily) N_x, N_y, and N_{xy} are known, then Eq. (8.2.17) contains four unknowns M_x, M_y, M_{xy}, and w. In order to determine these quantities uniquely, one needs three additional relationships. These three additional equations may be obtained from the kinematic and constitutive conditions, Eqs. (8.2.7).

Substituting Eqs. (8.2.7) into Eq. (8.2.17) gives Eq. (8.2.18c)

$$\frac{\partial N_x}{\partial x}+\frac{\partial N_{yx}}{\partial y}=0 \tag{8.2.18a}$$

$$\frac{\partial N_{xy}}{\partial x} + \frac{\partial N_y}{\partial y} = 0 \tag{8.2.18b}$$

$$D\left(\frac{\partial^4 w}{\partial x^4} + 2\frac{\partial^4 w}{\partial x^2 \partial y^2} + \frac{\partial^4 w}{\partial y^4}\right) = N_x \frac{\partial^2 w}{\partial x^2} + N_y \frac{\partial^2 w}{\partial y^2} + 2N_{xy}\frac{\partial^2 w}{\partial x \partial y} + p \tag{8.2.18c}$$

Equations (8.2.18) are a form of von Kármán plate equations, and they are the nonlinear equilibrium equations for all flat and slightly deformed configurations of the plate within the scope of the intermediate class of deformations.

8.2.3. Stationary Potential Energy

It would be interesting to rederive the above nonlinear equilibrium equations on the basis of the principle of minimum potential energy. A loaded plate is in equilibrium if its total potential energy Π is stationary (minimum), and Π is stationary if the integrand in the potential energy functional satisfies the Euler-Lagrange equations of the calculus of variations.

The total potential energy of a plate subjected to transverse loads $p(x,y)$ and in-plane loading is the sum of the strain energy U and the potential energy of the applied load V

$$\Pi = U + V \tag{8.2.19}$$

The strain energy U for a three-dimensional isotropic medium is given by

$$U = \frac{1}{2}\int_v \bar{\sigma}^T \bar{\varepsilon} dv$$

Omission of $\bar{\gamma}_{xz}$, $\bar{\gamma}_{yz}$ (the resulting error would be negligible if the plate lateral dimensions are at least greater than 10 times the plate thickness h) and $\bar{\sigma}_z$ in accordance with Kirchhoff's approximation of thin-plate theory along with Eqs. (8.2.6) leads to

$$U = \frac{E}{2(1-\mu^2)}\iiint\left(\bar{\varepsilon}_x^2 + \bar{\varepsilon}_y^2 + 2\mu\bar{\varepsilon}_x\bar{\varepsilon}_y + \frac{1-\mu}{2}\bar{\gamma}_{xy}^2\right) dxdydz$$

Introducing Eqs. (8.2.4) into the above equation and integrating with respect to z leads to

$$U = U_m + U_b \tag{8.2.20}$$

where

$$U_m = \frac{C}{2} \iint \left(\varepsilon_x^2 + \varepsilon_y^2 + 2\mu\varepsilon_x\varepsilon_y + \frac{1-\mu}{2}\gamma_{xy}^2 \right) dxdy \tag{8.2.21}$$

and

$$U_b = \frac{D}{2} \iint \left[\left(\frac{\partial^2 w}{\partial x^2} \right)^2 + \left(\frac{\partial^2 w}{\partial y^2} \right)^2 + 2\mu \frac{\partial^2 w}{\partial x^2} \frac{\partial^2 w}{\partial y^2} + 2(1-\mu) \left(\frac{\partial^2 w}{\partial x \partial y} \right)^2 \right] dxdy \tag{8.2.22}$$

The quantities in Eqs. (8.2.21) and (8.2.22) are referred to as the membrane strain energy and the bending strain energy of the plate, respectively.

The potential energy of the applied loads for a conservative force system is the negative of the work done by the loads. Hence, for the transverse load p,

$$V = -\iint pwdxdy$$

Consider as an example an in-plane compressive edge load P_x as shown in Fig. 8-8. For such a load the potential energy may be written

$$V = -P_x[u(a) - u(0)]$$

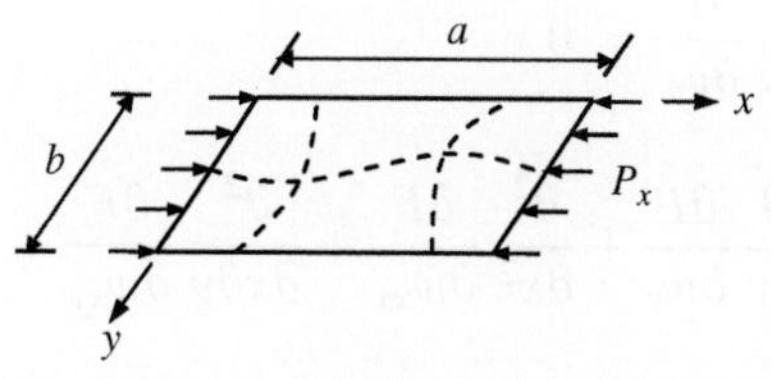

Figure 8-8 Simply supported plate subjected to uniaxial force

or

$$V = -\frac{P_x}{b}\int_0^b \int_0^a \frac{\partial u}{\partial x}\,dxdy$$

Hence, for the transverse and edge loads together, the potential energy of the applied loads is

$$V = \iint \left(\frac{P_x}{b}\frac{\partial u}{\partial x} - pw\right)dxdy \tag{8.2.23}$$

Symbolically, the total potential energy functional is

$$\Pi = \iint F dxdy \tag{8.2.24}$$

where

$$\begin{aligned} F &= \frac{C}{2}\left(\varepsilon_x^2 + \varepsilon_y^2 + 2\mu\varepsilon_x\varepsilon_y + \frac{1-\mu}{2}\gamma_{xy}^2\right) \\ &+ \frac{D}{2}\left[\left(\frac{\partial^2 w}{\partial x^2}\right)^2 + \left(\frac{\partial^2 w}{\partial y^2}\right)^2 + 2\mu\frac{\partial^2 w}{\partial x^2}\frac{\partial^2 w}{\partial y^2} + 2(1-\mu)\left(\frac{\partial^2 w}{\partial x\partial y}\right)^2\right] \\ &+ \left(\frac{P_x}{b}\frac{\partial u}{\partial x} - pw\right) \end{aligned} \tag{8.2.25}$$

For equilibrium the potential energy Π must be stationary (minimum); that is, its first variation $\delta\Pi$ must vanish. Accordingly, the integrand F must satisfy the Euler-Lagrange equations of the calculus of variations. The Euler-Lagrange equations are as follows (Bleich 1952):

$$\begin{aligned} &\frac{\partial F}{\partial u} - \frac{\partial}{\partial x}\frac{\partial F}{\partial u_x} - \frac{\partial}{\partial y}\frac{\partial F}{\partial u_y} = 0 \\ &\frac{\partial F}{\partial v} - \frac{\partial}{\partial x}\frac{\partial F}{\partial v_x} - \frac{\partial}{\partial y}\frac{\partial F}{\partial v_y} = 0 \\ &\frac{\partial F}{\partial w} - \frac{\partial}{\partial x}\frac{\partial F}{\partial w_x} - \frac{\partial}{\partial y}\frac{\partial F}{\partial w_y} + \frac{\partial^2}{\partial x^2}\frac{\partial F}{\partial w_{xx}} + \frac{\partial^2}{\partial x\partial y}\frac{\partial F}{\partial w_{xy}} + \frac{\partial^2}{\partial y^2}\frac{\partial F}{\partial w_{yy}} = 0 \end{aligned} \tag{a}$$

From Eq. (8.2.25)

$$\frac{\partial F}{\partial u} = 0 \quad \frac{\partial F}{\partial u_x} = \frac{C}{2}(2\varepsilon_x + 2\mu\varepsilon_y) + \frac{P_x}{b} \quad \frac{\partial F}{\partial u_y} = \frac{C}{2}(1-\mu)\gamma_{xy}$$

$$\frac{\partial F}{\partial v} = 0 \quad \frac{\partial F}{\partial v_x} = \frac{C}{2}(1-\mu)\gamma_{xy} \quad \frac{\partial F}{\partial v_y} = C(\varepsilon_y + \mu\varepsilon_x)$$

$$\frac{\partial F}{\partial w} = -p \quad \frac{\partial F}{\partial w_x} = \frac{C}{2}\left[(2\varepsilon_x + 2\mu\varepsilon_y)w_x + (1-\mu)\gamma_{xy}w_y\right]$$

$$\frac{\partial F}{\partial w_y} = \frac{C}{2}\left[(2\varepsilon_y + 2\mu\varepsilon_x)w_y + (1-\mu)\gamma_{xy}w_x\right] \quad \frac{\partial F}{\partial w_{xx}} = -D\left(\frac{\partial^2 w}{\partial x^2} + \mu\frac{\partial^2 w}{\partial y^2}\right)$$

$$\frac{\partial F}{\partial w_{xy}} = -2D(1-\mu)\frac{\partial^2 w}{\partial x \partial y} \quad \frac{\partial F}{\partial w_{xx}} = -D\left(\frac{\partial^2 w}{\partial y^2} + \mu\frac{\partial^2 w}{\partial x^2}\right)$$

(b)

Substituting these derivatives Eqs. (b) into the Euler-Lagrange differential equations Eqs. (a) and simplifying gives

$$C\left[\frac{\partial(\varepsilon_x + \mu\varepsilon_y)}{\partial x} + \frac{1-\mu}{2}\frac{\partial\gamma_{xy}}{\partial y}\right] = 0$$

$$C\left[\frac{\partial(\varepsilon_y + \mu\varepsilon_x)}{\partial y} + \frac{1-\mu}{2}\frac{\partial\gamma_{xy}}{\partial x}\right] = 0$$

$$D\left[\frac{\partial^2}{\partial x^2}\left(\frac{\partial^2 w}{\partial x^2} + \mu\frac{\partial^2 w}{\partial y^2}\right) + \frac{\partial^2}{\partial y^2}\left(\frac{\partial^2 w}{\partial y^2} + \mu\frac{\partial^2 w}{\partial x^2}\right) + 2(1-\mu)\frac{\partial^2}{\partial x \partial y}\frac{\partial^2 w}{\partial x \partial y}\right]$$

$$-C\frac{\partial}{\partial x}\left[(\varepsilon_x + \mu\varepsilon_y)\frac{\partial w}{\partial x} + \frac{1-\mu}{2}\gamma_{xy}\frac{\partial w}{\partial y}\right]$$

$$-C\frac{\partial}{\partial y}\left[(\varepsilon_y + \mu\varepsilon_x)\frac{\partial w}{\partial y} + \frac{1-\mu}{2}\gamma_{xy}\frac{\partial w}{\partial x}\right] = p$$

(c)

Substituting the plate constitutive relations from Eqs. (8.2.7) into the above Eqs. (c) yields

$$\frac{\partial N_x}{\partial x} + \frac{\partial N_{xy}}{\partial y} = 0 \quad \frac{\partial N_y}{\partial y} + \frac{\partial N_{xy}}{\partial x} = 0$$

$$D\nabla^4 w - \left(N_x\frac{\partial^2 w}{\partial x^2} + 2N_{xy}\frac{\partial^2 w}{\partial x \partial y} + N_y\frac{\partial^2 w}{\partial y^2}\right) = p \quad \text{(d)}$$

It is noted that the first two relations are reflected in the simplification process to obtain the third equation above. The term containing the edge load P_x disappears in the equilibrium equation. It reenters in the analysis as the boundary condition is $N_x = -P_x/b$ at $x = 0, a$. These equations of equilibrium Eqs. (d) are identical to Eqs. (8.2.18) as expected.

These nonlinear equations of equilibrium contain four unknowns N_x, N_y, N_{xy}, and w. Three equations in three variables u, v, w can be obtained by introducing the kinematic and constitutive relations of the plate (Eqs. (8.2.5) and Eqs. (8.2.7), respectively). The results are

$$\frac{\partial}{\partial x}\left\{\left(\frac{\partial u}{\partial x}\right)+\frac{1}{2}\left(\frac{\partial w}{\partial x}\right)^2+\mu\left[\left(\frac{\partial v}{\partial y}\right)+\frac{1}{2}\left(\frac{\partial w}{\partial y}\right)^2\right]\right\}$$
$$+\frac{1-\mu}{2}\frac{\partial}{\partial y}\left(\frac{\partial u}{\partial y}+\frac{\partial v}{\partial x}+\frac{\partial w}{\partial x}\frac{\partial w}{\partial y}\right)=0$$

$$\frac{\partial}{\partial y}\left\{\left(\frac{\partial v}{\partial y}\right)+\frac{1}{2}\left(\frac{\partial w}{\partial y}\right)^2+\mu\left[\left(\frac{\partial u}{\partial x}\right)+\frac{1}{2}\left(\frac{\partial w}{\partial x}\right)^2\right]\right\}$$
$$+\frac{1-\mu}{2}\frac{\partial}{\partial x}\left(\frac{\partial u}{\partial y}+\frac{\partial v}{\partial x}+\frac{\partial w}{\partial x}\frac{\partial w}{\partial y}\right)=0$$

$$D\nabla^4 w - C\left\{\left(\frac{\partial u}{\partial x}\right)+\frac{1}{2}\left(\frac{\partial w}{\partial x}\right)^2+\mu\left[\left(\frac{\partial v}{\partial y}\right)+\frac{1}{2}\left(\frac{\partial w}{\partial y}\right)^2\right]\right\}\frac{\partial^2 w}{\partial x^2}$$
$$- C\left\{\left(\frac{\partial v}{\partial y}\right)+\frac{1}{2}\left(\frac{\partial w}{\partial y}\right)^2+\mu\left[\left(\frac{\partial u}{\partial x}\right)+\frac{1}{2}\left(\frac{\partial w}{\partial x}\right)^2\right]\right\}\frac{\partial^2 w}{\partial y^2}$$
$$-(1-\mu)C\left(\frac{\partial u}{\partial y}+\frac{\partial v}{\partial x}+\frac{\partial w}{\partial x}\frac{\partial w}{\partial y}\right)\frac{\partial^2 w}{\partial x \partial y}=p \tag{8.2.26}$$

Equations (8.2.26) may be considerably simplified if one introduces a stress function f defined by the following relations (Timoshenko and Woinowsky-Krieger 1959):

$$N_x = h\frac{\partial^2 f}{\partial y^2} \quad N_y = h\frac{\partial^2 f}{\partial x^2} \quad N_{xy} = -h\frac{\partial^2 f}{\partial x \partial y} \tag{8.2.27}$$

where $f = f(x,y)$. These equations satisfy Eqs. (8.2.10) and (8.2.11) automatically. Substituting these equations into Eq. (8.2.18c) gives

$$D\nabla^4 w - h\left(\frac{\partial^2 f}{\partial y^2}\frac{\partial^2 w}{\partial x^2} - 2\frac{\partial^2 f}{\partial x\partial y}\frac{\partial^2 w}{\partial x\partial y} + \frac{\partial^2 f}{\partial x^2}\frac{\partial^2 w}{\partial y^2}\right) = p \tag{8.2.28}$$

From kinematic compatibility from Eqs. (8.2.5), it is seen that

$$\frac{\partial^2 \varepsilon_x}{\partial y^2} + \frac{\partial^2 \varepsilon_y}{\partial x^2} - \frac{\partial^2 \gamma_{xy}}{\partial x\partial y} = \left(\frac{\partial^2 w}{\partial x\partial y}\right)^2 - \frac{\partial^2 w}{\partial x^2}\frac{\partial^2 w}{\partial y^2} \tag{8.2.29}$$

Equation (8.2.29) is a deformation compatibility equation. It follows from Eqs. (8.2.7) that

$$\varepsilon_x = \frac{1}{E}\left(\frac{\partial^2 f}{\partial y^2} - \mu\frac{\partial^2 f}{\partial x^2}\right) \qquad \varepsilon_y = \frac{1}{E}\left(\frac{\partial^2 f}{\partial x^2} - \mu\frac{\partial^2 f}{\partial y^2}\right)$$

$$\gamma_{xy} = -\frac{2(1+\mu)}{E}\frac{\partial^2 f}{\partial x\partial y} \tag{8.2.30}$$

Substituting Eqs. (8.2.30) into Eq. (8.2.29) yields

$$\nabla^4 f - E\left[\left(\frac{\partial^2 w}{\partial x\partial y}\right)^2 - \frac{\partial^2 w}{\partial x^2}\frac{\partial^2 w}{\partial y^2}\right] = 0 \tag{8.2.31}$$

Equations (8.2.28) and (8.2.31) form two equations for the two variables w and f. They were first derived by von Kármán (1910), Love (1944), and Timoshenko (1983), and they are accordingly referred to as von Kármán large-deflection plate equations. They are called the equilibrium and compatibility equations, respectively. These equations, though very useful, are not the only set of equations that can be used to describe the large-deflection behavior of plates. When digital computers were not available and it was necessary to keep the equations as compact as possible, the von Kármán equations were used almost exclusively, as the solution of these equations basically relied on the iterative procedures. This is, however, no longer the case. Plate equations, other than the von Kármán equations, are now in general use, as the availability of the computer makes it possible to work effectively with any set of equations (Chajes 1974).

To obtain the equilibrium equations of linear small-displacement plate theory, it is only necessary to omit higher order terms (quadratic and cubic terms) in the displacement components. The linear equations corresponding to Eqs. (8.2.18) are found to be

$$\frac{\partial N_x}{\partial x} + \frac{\partial N_{xy}}{\partial y} = 0 \qquad \frac{\partial N_y}{\partial y} + \frac{\partial N_{xy}}{\partial x} = 0 \qquad D\nabla^4 w = p \tag{8.2.32}$$

where the in-plane forces are defined the way they were in Eqs. (8.2.7); however, the strain components now take only the elastic parts as

$$\varepsilon_x = \frac{\partial u}{\partial x} \quad \varepsilon_y = \frac{\partial v}{\partial y} \quad \gamma_{xy} = \frac{\partial u}{\partial y} + \frac{\partial v}{\partial x} \tag{8.2.33}$$

It is noted that the third equation of Eqs. (8.2.32) is not coupled. Much of the relative simplicity of classical linear or linearized thin-plate theory is a consequence of this uncoupling.

8.3. LINEAR EQUATIONS

Equations (8.2.26) govern all linear and nonlinear equilibrium conditions of the plate within the confinement of the intermediate class of deformations. The equations include linear, quadratic, and cubic terms of variables u, v, and w, and therefore are nonlinear. Consider a particular example shown in Fig. 8-8. An approximate solution of the nonlinear equations can be obtained (Chajes 1974) based on an assumed displacement function. It is now a fairly simple task to obtain a very good iterative numerical solution by a well-established finite element code. A load-displacement curve based on such solutions for a plate subject to the edge load P_x is shown in Fig. 8-9. The symmetry of Fig. 8-9 indicates that the plate may buckle in either direction. The linear equilibrium equations, Eqs. (8.2.32), govern the primary (static) equilibrium path OA. The nonlinear equations, Eqs. (8.2.26), govern both the primary path and the secondary path AB.

The equilibrium paths determined by solution of the equilibrium equations, Eqs. (8.2.26), show the bifurcation point and the corresponding critical load. Hence, a separate solution for the critical load is not necessary. However, the solution of Eqs. (8.2.26) demands a fairly complicated

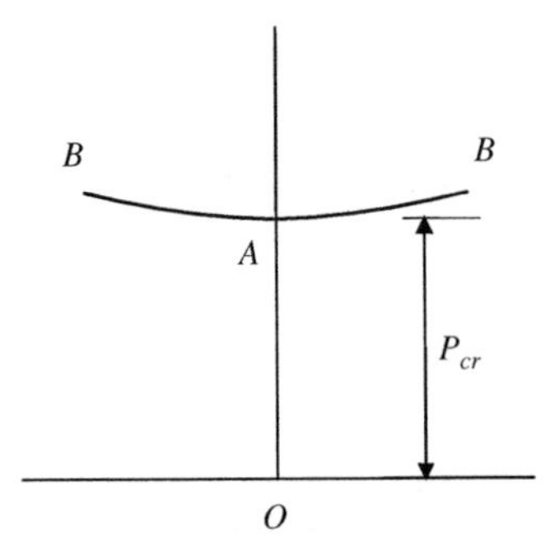

Figure 8-9 Load-deflection curve

numerical procedure. The purpose of stability analysis to be presented herein is to permit determination of the critical load by solution of linear differential equations.

The linear differential equations for the determination of the critical load of a rectangular plate subjected to in-plane loading are derived by applying the adjacent-equilibrium criterion. The same equations are rederived based on the minimum potential energy criterion as was done in the previous section.

8.3.1. Adjacent-Equilibrium Criterion

Adjacent equilibrium configurations are examined using the procedure outlined by Brush and Almroth (1975). Consider the equilibrium configuration at the bifurcation point. Then, the equilibrium configuration is perturbed by the small incremental displacement. The incremental displacement in u_1, v_1, w_1 is arbitrary and tentative. Variables in the two adjacent configurations before and after the increment are represented by u_0, v_0, w_0 and u, v, w. Let

$$
\begin{aligned}
u &\rightarrow u_0 + u_1 \\
v &\rightarrow v_0 + v_1 \\
w &\rightarrow w_0 + w_1
\end{aligned}
\tag{8.3.1}
$$

where the arrow is read "be replaced by." Substitution of Eqs. (8.3.1) into Eqs. (8.2.26) results in equations containing terms that are linear, quadratic, and cubic in u_0, v_0, w_0 and u_1, v_1, w_1 displacement components. In the new equation thus obtained, the terms containing u_0, v_0, w_0 alone are equal to zero as u_0, v_0, w_0 constitute an equilibrium configuration, and terms that are quadratic and cubic in u_1, v_1, w_1 may be ignored because of the smallness of the incremental displacement. Therefore, the resulting equations are homogeneous and linear in u_1, v_1, w_1 with variable coefficients. The coefficients in u_0, v_0, w_0 are governed by the original nonlinear equations. It will simplify the procedure greatly by simply limiting the range of applicability of the linearized equations by requiring that u_0, v_0, w_0 be limited to configurations that are governed by the linear equations, Eqs. (8.2.32). This limitation has the additional advantage of w_0 and its derivatives being equal to zero for in-plane loading (there is no lateral displacement in the primary path as shown in Fig. 8-9).

Equations (8.2.18) will be used instead as it will shorten the derivation. The increment in u, v, w causes a concomitant change in the internal force such as

$$N_x \rightarrow N_{x0} + \Delta N_x$$
$$N_y \rightarrow N_{y0} + \Delta N_y \tag{8.3.2}$$
$$N_{xy} \rightarrow N_{xy0} + \Delta N_{xy}$$

where terms with subscript 0 correspond to the u_0, v_0, w_0 displacement, and ΔN_x, ΔN_y, ΔN_{xy} are increments corresponding to u_1, v_1, w_1. Let also N_{x1}, N_{y1}, N_{xy1} represent parts of ΔN_x, ΔN_y, ΔN_{xy}, respectively, that are linear in u_1, v_1, w_1. For example, from Eqs. (8.2.5) and (8.2.7),

$$N_x = C\left\{\frac{\partial u}{\partial x} + \frac{1}{2}\left(\frac{\partial w}{\partial x}\right)^2 + \mu\left[\frac{\partial v}{\partial y} + \frac{1}{2}\left(\frac{\partial w}{\partial y}\right)^2\right]\right\}$$

As w_0 and its derivatives are equal to zero for in-plane loading, one may write

$$N_{x0} + \Delta N_x = C\left\{\frac{\partial u_0}{\partial x} + \frac{\partial u_1}{\partial x} + \frac{1}{2}\left(\frac{\partial w_1}{\partial x}\right)^2 + \mu\left[\frac{\partial v_0}{\partial y} + \frac{\partial v_1}{\partial y} + \frac{1}{2}\left(\frac{\partial w_1}{\partial y}\right)^2\right]\right\}$$

From which

$$N_{x0} = C\left(\frac{\partial u_0}{\partial x} + \mu\frac{\partial v_0}{\partial y}\right)$$

$$\Delta N_x = C\left\{\frac{\partial u_1}{\partial x} + \frac{1}{2}\left(\frac{\partial w_1}{\partial x}\right)^2 + \mu\left[\frac{\partial v_1}{\partial y} + \frac{1}{2}\left(\frac{\partial w_1}{\partial y}\right)^2\right]\right\}$$

$$N_{x1} = C\left(\frac{\partial u_1}{\partial x} + \mu\frac{\partial v_1}{\partial y}\right)$$

Substituting these into Eqs. (8.2.18) gives

$$\frac{\partial N_{x1}}{\partial x} + \frac{\partial N_{xy1}}{\partial y} = 0 \tag{8.3.3a}$$

$$\frac{\partial N_{y1}}{\partial y} + \frac{\partial N_{xy1}}{\partial x} = 0 \tag{8.3.3b}$$

$$D\nabla^4 w_1 - \left(N_{x0} \frac{\partial^2 w_1}{\partial x^2} + 2N_{xy0} \frac{\partial^2 w_1}{\partial x \partial y} + N_{y0} \frac{\partial^2 w_1}{\partial y^2} \right) = 0 \qquad (8.3.3c)$$

where

$$N_{x0} = C\left(\frac{\partial u_0}{\partial x} + \mu \frac{\partial v_0}{\partial y} \right) \quad N_{x1} = C\left(\frac{\partial u_1}{\partial x} + \mu \frac{\partial v_1}{\partial y} \right)$$
$$N_{y0} = C\left(\frac{\partial v_0}{\partial y} + \mu \frac{\partial u_0}{\partial x} \right) \quad N_{y1} = C\left(\frac{\partial v_1}{\partial y} + \mu \frac{\partial u_1}{\partial x} \right) \qquad (8.3.4)$$
$$N_{xy0} = C\frac{1-\mu}{2}\left(\frac{\partial u_0}{\partial y} + \frac{\partial v_0}{\partial x} \right) \quad N_{xy1} = C\frac{1-\mu}{2}\left(\frac{\partial u_1}{\partial y} + \frac{\partial v_1}{\partial x} \right)$$

Equations (8.3.3) are the stability equations for the plate subjected to in-plane edge loading. As in the case of linear equilibrium equations, Eq. (8.3.3c) is uncoupled from the other two equations. Equation (8.3.3c) is a homogeneous linear equation in w_1 with variable coefficients in N_{x0}, N_{y0}, N_{xy0}, depending on the edge conditions of the plate, which are determined by the other two linear equations (8.3.3a) and (8.3.3b). It is an eigenvalue problem. As such, it has solutions for discrete values of the applied load. At each solution point or bifurcation point, two adjacent equilibrium configurations exist—an undeformed one on the primary equilibrium path and a slightly deformed one on a secondary equilibrium path.

8.3.2. Minimum Potential Energy Criterion

The plate stability equations (8.3.3) will be rederived by applying the minimum potential energy criterion. The equilibrium changes from stable to neutral when the total potential energy functional Π ceases to be a relative minimum. The criterion for the loss of stability is that the integrand in the expression for the second variation of Π satisfies the Euler-Lagrange equations, which is known as the Trefftz criterion according to Langhaar (1962).[1]

Symbolically, the total potential energy increment may be written in the form

$$\Delta\Pi = \delta\Pi + \frac{1}{2!}\delta^2\Pi + \frac{1}{3!}\delta^3\Pi + \ldots \qquad (8.3.5)$$

Each nonzero term in Eq. (8.3.5) is much larger than the sum of the succeeding terms. Since $\delta\Pi$ vanishes by virtue of the principle of minimum

[1] See page 211.

potential energy, the sign of $\Delta\Pi$ is governed by the second variation. For sufficiently small values of the applied load, the second variation is positive definite (condition for Π to be relative minimum). The critical load is defined as the smallest load for which the second variation no longer is positive definite (it is positive semidefinite). According to the Trefftz criterion, the equations for the critical load are given by the Euler-Lagrange equations for the integrand in the second variation. Since the expression for the second variation is a homogeneous quadratic functional, its variational derivatives (Euler-Lagrange equations) are necessarily linear homogeneous differential equations. In order to obtain the second variation, Eqs. (8.3.1) are used again

$$\begin{aligned} u &\rightarrow u_0 + u_1 \\ v &\rightarrow v_0 + v_1 \\ w &\rightarrow w_0 + w_1 \end{aligned} \tag{8.3.1}$$

where (u_0, v_0, w_0) is a configuration on the primary path, including the bifurcation point, and (u_1, v_1, w_1) is a virtual displacement. The total potential energy in a Taylor series expansion is

$$\begin{aligned} \Pi(u_0+u_1, v_0+v_1, w_0+w_1) &= \Pi(u_0, v_0, w_0) + \frac{\partial \Pi}{\partial u_0}u_1 + \frac{\partial \Pi}{\partial v_0}v_1 + \frac{\partial \Pi}{\partial w_0}w_1 \\ &+ \frac{1}{2!}\left[\frac{\partial^2 \Pi}{\partial u_0^2}(u_1)^2 + \frac{\partial^2 \Pi}{\partial v_0^2}(v_1)^2 + \frac{\partial^2 \Pi}{\partial w_0^2}(w_1)^2 + 2\frac{\partial^2 \Pi}{\partial u_0 \partial v_0}u_1 v_1 \right. \\ &\left. + 2\frac{\partial^2 \Pi}{\partial u_0 \partial w_0}u_1 w_1 + 2\frac{\partial^2 \Pi}{\partial v_0 \partial w_0}v_1 w_1\right] + \ldots \end{aligned}$$

The change in potential energy $\Delta\Pi = \Pi(u_0+u_1, v_0+v_1, w_0+w_1) - \Pi(u_0, v_0, w_0)$ can be written as

$$\Delta\Pi = \delta\Pi + \frac{1}{2!}\delta^2\Pi + \ldots$$

where the first variation is equal to zero by virtue of the principle of minimum potential energy and the second variation is defined as

$$\begin{aligned} \delta^2\Pi &= \frac{\partial^2 \Pi}{\partial u_0^2}(u_1)^2 + \frac{\partial^2 \Pi}{\partial v_0^2}(v_1)^2 + \frac{\partial^2 \Pi}{\partial w_0^2}(w_1)^2 + 2\frac{\partial^2 \Pi}{\partial u_0 \partial v_0}u_1 v_1 \\ &+ 2\frac{\partial^2 \Pi}{\partial u_0 \partial w_0}u_1 w_1 + 2\frac{\partial^2 \Pi}{\partial v_0 \partial w_0}v_1 w_1 \end{aligned}$$

The total potential energy functional and the integrand are given by Eq. (8.2.24) and Eq. (8.2.25), respectively. Eq. (8.2.25) will become extremely large when it is expanded according to Eqs. (8.3.1) after each strain term is replaced by Eqs. (8.2.5) that are expanded by Eqs. (8.3.1). Therefore, it would be rather manageable to proceed with the derivation of the second variation term by term. In the derivation of the second variation, it is important to reflect Eqs. (8.3.1) in the strain expression and collect second-order terms in u_1, v_1, w_1, for w_0 (and its derivatives) $= 0$. Membrane strain terms are

$$\varepsilon_x = \frac{\partial u}{\partial x} + \frac{1}{2}\left(\frac{\partial w}{\partial x}\right)^2, \quad \varepsilon_x^2 = \left(\frac{\partial u}{\partial x}\right)^2 + \frac{\partial u}{\partial x}\left(\frac{\partial w}{\partial x}\right)^2 + \frac{1}{4}\left(\frac{\partial w}{\partial x}\right)^4$$

$$\varepsilon_y = \frac{\partial v}{\partial y} + \frac{1}{2}\left(\frac{\partial w}{\partial y}\right)^2, \quad \varepsilon_y^2 = \left(\frac{\partial v}{\partial y}\right)^2 + \frac{\partial v}{\partial y}\left(\frac{\partial w}{\partial y}\right)^2 + \frac{1}{4}\left(\frac{\partial w}{\partial y}\right)^4$$

$$\varepsilon_x\varepsilon_y = \frac{\partial u}{\partial x}\frac{\partial v}{\partial y} + \frac{1}{2}\frac{\partial u}{\partial x}\left(\frac{\partial w}{\partial y}\right)^2 + \frac{1}{2}\frac{\partial v}{\partial y}\left(\frac{\partial w}{\partial x}\right)^2 + \frac{1}{4}\left(\frac{\partial w}{\partial x}\right)^2\left(\frac{\partial w}{\partial y}\right)^2$$

$$\gamma_{xy} = \frac{\partial u}{\partial y} + \frac{\partial v}{\partial x} + \frac{\partial w}{\partial x}\frac{\partial w}{\partial y},$$

$$\gamma_{xy}^2 = \left(\frac{\partial u}{\partial y}\right)^2 + \left(\frac{\partial v}{\partial x}\right)^2 + \left(\frac{\partial w}{\partial x}\right)^2\left(\frac{\partial w}{\partial y}\right)^2 + 2\frac{\partial u}{\partial y}\frac{\partial v}{\partial x} + 2\frac{\partial u}{\partial y}\frac{\partial w}{\partial x}\frac{\partial w}{\partial y} + 2\frac{\partial v}{\partial x}\frac{\partial w}{\partial x}\frac{\partial w}{\partial y}$$

Introducing Eqs. (8.3.1) and carrying out the variations with w_0 (and its derivatives) $= 0$ yields

$$\Pi\left(\varepsilon_x^2\right) = \iint\left[\left(\frac{\partial u}{\partial x}\right)^2 + \frac{\partial u}{\partial x}\left(\frac{\partial w}{\partial x}\right)^2 + \frac{1}{4}\left(\frac{\partial w}{\partial x}\right)^4\right]dxdy$$

$$\begin{aligned}\delta^2\Pi\left(\varepsilon_x^2\right) &= \frac{\partial^2\Pi}{\partial {u_{0,x}}^2}{u_{1,x}}^2 + 2\frac{\partial^2\Pi}{\partial u_{0,x}\partial w_{0,x}}u_{1,x}w_{1,x} + \frac{\partial^2\Pi}{\partial {w_{0,x}}^2}{w_{1,x}}^2\\ &= \iint\left[2\left(\frac{\partial u_1}{\partial x}\right)^2 + 2\frac{\partial u_0}{\partial x}\left(\frac{\partial w_1}{\partial x}\right)^2\right]dxdy\end{aligned}$$

Likewise,

$$\Pi\left(\varepsilon_y^2\right) = \iint\left[\left(\frac{\partial v}{\partial y}\right)^2 + \frac{\partial v}{\partial y}\left(\frac{\partial w}{\partial y}\right)^2 + \frac{1}{4}\left(\frac{\partial w}{\partial y}\right)^4\right]dxdy$$

$$\begin{aligned}\delta^2\Pi\left(\varepsilon_y^2\right) &= \frac{\partial^2\Pi}{\partial {v_{0,y}}^2}{v_{1,y}}^2 + 2\frac{\partial^2\Pi}{\partial v_{0,y}\partial w_{0,y}}v_{1,x}w_{1,y} + \frac{\partial^2\Pi}{\partial {w_{0,y}}^2}{w_{1,y}}^2 \\ &= \iint\left[2\left(\frac{\partial v_1}{\partial y}\right)^2 + 2\frac{\partial v_0}{\partial y}\left(\frac{\partial w_1}{\partial y}\right)^2\right]dxdy\end{aligned}$$

$$\begin{aligned}\Pi\left(\varepsilon_x\varepsilon_y\right) = \iint &\left[\frac{\partial u}{\partial x}\frac{\partial v}{\partial y} + \frac{1}{2}\frac{\partial u}{\partial x}\left(\frac{\partial w}{\partial y}\right)^2 + \frac{1}{2}\frac{\partial v}{\partial y}\left(\frac{\partial w}{\partial x}\right)^2 + \frac{1}{4}\left(\frac{\partial w}{\partial x}\right)^2\right. \\ &\left.\left(\frac{\partial w}{\partial y}\right)^2\right]dxdy\end{aligned}$$

$$\begin{aligned}\delta^2\Pi\left(\varepsilon_x\varepsilon_y\right) &= \frac{\partial^2\Pi}{\partial {u_{0,x}}^2}{u_{1,x}}^2 + \frac{\partial^2\Pi}{\partial {v_{0,y}}^2}{v_{1,y}}^2 + \frac{\partial^2\Pi}{\partial {w_{0,x}}^2}{w_{1,x}}^2 + \frac{\partial^2\Pi}{\partial {w_{0,y}}^2}{w_{1,y}}^2 \\ &+ 2\,\frac{\partial^2\Pi}{\partial u_{0,x}v_{0,y}}u_{1,x}v_{1,y} + 2\frac{\partial^2\Pi}{\partial u_{0,x}\partial w_{0,x}}u_{1,y}w_{1,y} \\ &+ 2\frac{\partial^2\Pi}{\partial u_{0,x}\partial w_{0,y}}u_{1,x}w_{1,y} + 2\frac{\partial^2\Pi}{\partial v_{0,y}\partial w_{0,x}}v_{1,y}w_{1,x} \\ &+ 2\frac{\partial^2\Pi}{\partial v_{0,y}\partial w_{0,y}}\,v_{1,y}w_{1,y} + 2\frac{\partial^2\Pi}{\partial w_{0,x}\partial w_{0,y}}w_{1,x}w_{1,y} \\ &= \iint\left[2\frac{\partial u_1}{\partial x}\frac{\partial v_1}{\partial y} + \frac{\partial v_0}{\partial y}\left(\frac{\partial w_1}{\partial x}\right)^2 + \frac{\partial u_0}{\partial x}\left(\frac{\partial w_1}{\partial y}\right)^2\right]dxdy\end{aligned}$$

$$\begin{aligned}\Pi\left(\gamma_{xy}^2\right) &= \iint\left[\left(\frac{\partial u}{\partial y}\right)^2 + \left(\frac{\partial v}{\partial x}\right)^2 + \left(\frac{\partial w}{\partial x}\right)^2\left(\frac{\partial w}{\partial y}\right)^2 + 2\frac{\partial u}{\partial y}\frac{\partial v}{\partial x}\right. \\ &\left.+ 2\frac{\partial u}{\partial y}\frac{\partial w}{\partial x}\frac{\partial w}{\partial y} + 2\frac{\partial v}{\partial x}\frac{\partial w}{\partial x}\frac{\partial w}{\partial y}\right]dxdy\end{aligned}$$

$$\delta^2\Pi\left(\gamma_{xy}^2\right) = \frac{\partial^2\Pi}{\partial u_{0,y}{}^2}u_{1,y}{}^2 + \frac{\partial^2\Pi}{\partial v_{0,x}{}^2}v_{1,x}{}^2 + \frac{\partial^2\Pi}{\partial w_{0,x}{}^2}w_{1,x}{}^2 + \frac{\partial^2\Pi}{\partial w_{0,y}{}^2}w_{1,y}{}^2$$

$$+ 2\frac{\partial^2\Pi}{\partial u_{0,y}\partial v_{0,x}}u_{1,y}v_{1,x} + 2\frac{\partial^2\Pi}{\partial u_{0,y}\partial w_{0,x}}u_{1,y}w_{1,x}$$

$$+ 2\frac{\partial^2\Pi}{\partial u_{0,y}\partial w_{0,y}}u_{1,y}w_{1,y} + 2\frac{\partial^2\Pi}{\partial v_{0,x}\partial w_{0,x}}v_{1,x}w_{1,x}$$

$$+ 2\frac{\partial^2\Pi}{\partial v_{0,x}\partial w_{0,y}}v_{1,x}w_{1,y} + 2\frac{\partial^2\Pi}{\partial w_{0,x}\partial w_{0,y}}w_{1,x}w_{1,y}$$

$$= \iint \left[2\left(\frac{\partial u_1}{\partial y}\right)^2 + 2\left(\frac{\partial v_1}{\partial x}\right)^2 + 4\frac{\partial u_1}{\partial y}\frac{\partial v_1}{\partial x}\right.$$

$$\left. + 4\frac{\partial u_0}{\partial y}\frac{\partial w_1}{\partial x}\frac{\partial w_1}{\partial y} + 4\frac{\partial v_0}{\partial x}\frac{\partial w_1}{\partial x}\frac{\partial w_1}{\partial y}\right]dxdy$$

The second variations of the bending strain energy terms are

$$\Pi\left[\left(\frac{\partial^2 w}{\partial x^2}\right)^2\right] = \iint \left[\left(\frac{\partial^2 w}{\partial x^2}\right)^2\right]dxdy$$

$$\delta^2\Pi\left[\left(\frac{\partial^2 w}{\partial x^2}\right)^2\right] = \frac{\partial^2\Pi}{\partial w_{0,xx}{}^2}w_{1,xx}{}^2 = \iint 2\left(\frac{\partial^2 w_1}{\partial x^2}\right)^2 dxdy$$

$$\Pi\left[\left(\frac{\partial^2 w}{\partial y^2}\right)^2\right] = \iint \left[\left(\frac{\partial^2 w}{\partial y^2}\right)^2\right]dxdy$$

$$\delta^2\Pi\left[\left(\frac{\partial^2 w}{\partial y^2}\right)^2\right] = \frac{\partial^2\Pi}{\partial w_{0,yy}{}^2}w_{1,yy}{}^2 = \iint 2\left(\frac{\partial^2 w_1}{\partial y^2}\right)^2 dxdy$$

$$\Pi\left(\frac{\partial^2 w}{\partial x^2}\frac{\partial^2 w}{\partial y^2}\right) = \iint \left(\frac{\partial^2 w}{\partial x^2}\frac{\partial^2 w}{\partial y^2}\right)dxdy$$

$$\delta^2 \Pi\left(\frac{\partial^2 w}{\partial x^2}\frac{\partial^2 w}{\partial y^2}\right) = \frac{\partial^2 \Pi}{\partial w_{0,xx}{}^2} w_{1,xx}{}^2 + \frac{\partial^2 \Pi}{\partial w_{0,yy}{}^2} w_{1,yy}{}^2 + 2\frac{\partial^2 \Pi}{\partial w_{0,xx}\partial w_{0,yy}} w_{1,xx}w_{1,yy}$$

$$\delta^2 \Pi\left(\frac{\partial^2 w}{\partial x^2}\frac{\partial^2 w}{\partial y^2}\right) = \iint 2\frac{\partial^2 w_1}{\partial x^2}\frac{\partial^2 w_1}{\partial y^2}\,dxdy$$

$$\Pi\left[\left(\frac{\partial^2 w}{\partial x \partial y}\right)^2\right] = \iint \left[\left(\frac{\partial^2 w}{\partial x \partial y}\right)^2\right] dxdy$$

$$\delta^2 \Pi\left[\left(\frac{\partial^2 w}{\partial x \partial y}\right)^2\right] = \frac{\partial^2 \Pi}{\partial w_{0,xy}{}^2} w_{1,xy}{}^2 = \iint \left[2\left(\frac{\partial^2 w_1}{\partial x \partial y}\right)^2\right] dxdy$$

Hence, the second variation of the membrane strain energy is

$$\frac{1}{2}\delta^2 U_m = \frac{C}{2}\iint \times \left\{\begin{array}{l} [u_{1,x}{}^2 + v_{1,y}{}^2 + 2\mu u_{1,x}v_{1,y}] + \dfrac{1-\mu}{2}(u_{1,y}+v_{1,x})^2 \\ + [(u_{0,x}+\mu v_{0,y})w_{1,x}^2 + (v_{0,y}+\mu u_{0,x})w_{1,y}^2] \\ + (1-\mu)(u_{0,y}+v_{0,x})w_{1,x}w_{1,y} \end{array}\right\} dxdy$$

Substituting Eqs. (8.3.4) into Eq. (8.3.6) yields

$$\frac{1}{2}\delta^2 U_m = \frac{C}{2}\iint \left[u_{1,x}{}^2 + v_{1,y}{}^2 + 2\mu u_{1,x}v_{1,y} + \frac{1-\mu}{2}(u_{1,y}+v_{1,x})^2\right] dxdy + \frac{1}{2}\left(N_{x0}w_{1,x}{}^2 + N_{y0}w_{1,y}{}^2 + 2N_{xy0}w_{1,x}w_{1,y}\right) dxdy \tag{8.3.7}$$

Likewise, the second variation of the bending strain energy is

$$\frac{1}{2}\delta^2 U_b = \frac{D}{2}\iint \left[\left(\frac{\partial^2 w_1}{\partial x^2}\right)^2 + \left(\frac{\partial^2 w_1}{\partial y^2}\right)^2 + 2\mu\frac{\partial^2 w_1}{\partial x^2}\frac{\partial^2 w_1}{\partial y^2} + 2(1-\mu)\left(\frac{\partial^2 w_1}{\partial x \partial y}\right)^2\right] dxdy \tag{8.3.8}$$

Equation (8.2.23) shows no quadratic term or higher order terms in the displacements; therefore it is concluded that $\delta^2 V = 0$. Hence,

$$\delta^2 \Pi = \delta^2 U_m + \delta^2 U_b \tag{8.3.9}$$

and

$$\delta^2 \Pi = \iint F dx dy \tag{8.3.10}$$

where

$$\begin{aligned} F = {} & C\left[{u_{1,x}}^2 + {v_{1,y}}^2 + 2\mu u_{1,x} v_{1,y} + \frac{1-\mu}{2}\left(u_{1,y} + v_{1,x}\right)^2 \right] \\ & + \left(N_{x0} {w_{1,x}}^2 + N_{y0} {w_{1,y}}^2 + 2N_{xy0} w_{1,x} w_{1,y} \right) \\ & + D\left[\left(\frac{\partial^2 w_1}{\partial x^2}\right)^2 + \left(\frac{\partial^2 w_1}{\partial y^2}\right)^2 + 2\mu \frac{\partial^2 w_1}{\partial x^2}\frac{\partial^2 w_1}{\partial y^2} + 2(1-\mu)\left(\frac{\partial^2 w_1}{\partial x \partial y}\right)^2 \right] \end{aligned} \tag{8.3.11}$$

The Euler-Lagrange equations according to the Trefftz criterion are

$$\begin{aligned} & \frac{\partial F}{\partial u_1} - \frac{\partial}{\partial x}\frac{\partial F}{\partial u_{1,x}} - \frac{\partial}{\partial y}\frac{\partial F}{\partial u_{1,y}} = 0 \\ & \frac{\partial F}{\partial v_1} - \frac{\partial}{\partial x}\frac{\partial F}{\partial v_{1,x}} - \frac{\partial}{\partial y}\frac{\partial F}{\partial v_{1,y}} = 0 \\ & \frac{\partial F}{\partial w_1} - \frac{\partial}{\partial x}\frac{\partial F}{\partial w_{1,x}} - \frac{\partial}{\partial y}\frac{\partial F}{\partial w_{1,y}} + \frac{\partial^2}{\partial x^2}\frac{\partial F}{\partial w_{1,xx}} \\ & \quad + \frac{\partial^2}{\partial y^2}\frac{\partial F}{\partial w_{1,yy}} + \frac{\partial^2}{\partial x \partial y}\frac{\partial F}{\partial w_{1,xy}} = 0 \end{aligned} \tag{8.3.12}$$

Substituting the followings into the second equation

$$\frac{\partial F}{\partial u_1} = 0$$

$$\frac{\partial}{\partial x}\frac{\partial F}{\partial u_{1,x}} = 2\left(u_{1,x} + \mu v_{1,y}\right)_{,x}$$

$$\frac{\partial}{\partial y}\frac{\partial F}{\partial u_{1,y}} = 2\frac{(1-\mu)}{2}\left(u_{1,y} + v_{1,x}\right)_{,y}$$

yields

$$\left(u_{1,x}+\mu v_{1,y}\right)_{,x}+\frac{(1-\mu)}{2}\left(u_{1,y}+v_{1,x}\right)_{,y}=0 \tag{8.3.13a}$$

Substituting the followings into the second equation

$$\frac{\partial F}{\partial v_1}=0$$

$$\frac{\partial}{\partial x}\frac{\partial F}{\partial v_{1,x}}=2\frac{(1-\mu)}{2}\left(u_{1,y}+v_{1,x}\right)_{,x}$$

$$\frac{\partial}{\partial y}\frac{\partial F}{\partial v_{1,y}}=2\left(u_{1,x}+\mu v_{1,y}\right)_{,y}$$

yields

$$\left(u_{1,x}+\mu v_{1,y}\right)_{,y}+\frac{(1-\mu)}{2}\left(u_{1,y}+v_{1,x}\right)_{,x}=0 \tag{8.3.13b}$$

Substituting the followings into the third equation

$$\frac{\partial F}{\partial w_1}=0$$

$$\frac{\partial}{\partial x}\frac{\partial F}{\partial w_{1,x}}=2N_{x0}w_{1,xx}+2N_{xy0}w_{1,xy}$$

$$\frac{\partial}{\partial y}\frac{\partial F}{\partial w_{1,y}}=2N_{x0}w_{1,yy}+2N_{xy0}w_{1,xy}$$

$$\frac{\partial^2}{\partial x^2}\frac{\partial F}{\partial w_{1,xx}}=D\left(2\frac{\partial^4 w_1}{\partial x^4}+2\mu\frac{\partial^4 w_1}{\partial x^2\partial y^2}\right)$$

$$\frac{\partial^2}{\partial y^2}\frac{\partial F}{\partial w_{1,xx}}=D\left(2\frac{\partial^4 w_1}{\partial y^4}+2\mu\frac{\partial^4 w_1}{\partial x^2\partial y^2}\right)$$

$$\frac{\partial^2}{\partial x\partial y}\frac{\partial F}{\partial w_{1,xy}}=4D(1-\mu)\frac{\partial^4 w_1}{\partial x^2\partial y^2}$$

yields

$$D\nabla^4 w_1-\left(N_{x0}\frac{\partial^2 w_1}{\partial x^2}+N_{y0}\frac{\partial^2 w_1}{\partial y^2}+2N_{xy0}\frac{\partial^2 w_1}{\partial x\partial y}\right)=0 \tag{8.3.13c}$$

Equations (8.3.13a) and (8.3.13b) can be rewritten as

$$(N_{x1})_{,x}+(N_{xy1})_{,y}=0 \quad (N_{y1})_{,y}+(N_{xy1})_{,x}=0$$

As expected, these equations are identical to Eqs. (8.3.3).

8.4. APPLICATION OF PLATE STABILITY EQUATION

Equation (8.3.3c) governs the buckling problem of a plate subjected to in-plane loads. For a properly posed buckling problem of a plate that is prismatic, homogeneous, and isotropic, N_{xO}, N_{yO}, and N_{xyO} can be functions of the coordinate variables x and y. The demonstrative examples presented here are limited to cases in which these coefficients are constants. For simplicity of notation, the subscript "1" is omitted in the examples.

8.4.1. Plate Simply Supported on Four Edges

Consider a plate simply supported on four edges and subjected to compressive load P_x uniformly distributed at the edges $x = 0$, a as shown in Fig. 8-8. From an equilibrium analysis, the in-plane forces are

$$N_{x0} = -\frac{P_x}{b} = -p_x \quad \text{and} \quad N_y = N_{xy} = 0$$

For all casual analyses of the critical load of a simply supported plate, a typical boundary condition of pin-roller arrangements in two orthogonal directions may be satisfactory, as such boundary conditions are on the conservative side. If a pinned boundary is defined as a support condition that only allows rotation along the edge with constraints for translations in the x, y, z directions intact, then N_{yO} and N_{xyO} are no longer equal to zero. In order to maintain the simplifying assumption of N_{yO} and N_{xyO} to be equal to zero, all in-plane constrains are removed except at a corner point where constraints are provided to eliminate the rigid body motion in a finite element analysis in which constraints can be assigned at each nodal point.

Substituting the simplified analysis results into Eq. (8.3.3c) gives

$$D\nabla^4 w + p_x w_{,xx} = 0 \tag{8.4.1}$$

Since the plate is simply supported on four edges,

$$w = w_{,xx} = 0 \text{ at } x = 0,\ a \quad w = w_{,yy} = 0 \text{ at } y = 0,\ b \tag{8.4.2}$$

Assume the solution to be of a form

$$w_n(x,y) = Y_n(y)\sin\frac{n\pi x}{a} \quad \text{with } n = 1, 2, 3 \ldots \tag{8.4.3}$$

This is a standard procedure of separating variables to transform a partial differential equation into ordinary differential equation, which will reduce the computational efforts significantly. $Y_n(y)$ is a function of the independent variable y only.

Taking appropriate derivatives and substituting into the governing equation above gives

$$\left\{Y_n^{iv} - 2\left(\frac{n\pi}{a}\right)^2 Y_n'' + \left[\left(\frac{n\pi}{a}\right)^4 - \frac{p_x}{D}\left(\frac{n\pi}{a}\right)^2\right] Y_n\right\} \sin\frac{n\pi x}{a} = 0 \quad (8.4.4)$$

Since $\sin(n\pi x/a) \neq 0$ for all values of x, the expression inside the brace must vanish.

Let $u^2 = \frac{p_x}{D}\left(\frac{a}{n\pi}\right)^2$, then

$$Y_n^{iv} - 2\left(\frac{n\pi}{a}\right)^2 Y_n'' + \left(\frac{n\pi}{a}\right)^4 (1 - u^2) Y_n = 0 \quad (8.4.5)$$

Assume the homogeneous solution of Eq. (8.4.5) to be of a form $Y_n = ce^{my}$. Taking successive derivatives, substituting back to Eq. (8.4.5), and solving the resulting characteristic equation gives

$$Y_n = c_1 \cosh k_1 y + c_2 \sinh k_1 y + c_3 \cos k_2 y + c_4 \sin k_2 y$$

$$k_1 = \left(\frac{n\pi}{a}\right)\sqrt{u+1} \quad \text{and} \quad k_2 = \left(\frac{n\pi}{a}\right)\sqrt{u-1}$$

Assume that the rectangular plate shown in Fig. 8-10 is simply supported at $x = \pm a/2$ and elastically restrained at $y = \pm b/2$. Then, the buckling deflection corresponding to the smallest p_x is a symmetric function of y based on the coordinate system given. Hence, Y_n must be an even function and $c_2 = c_4 = 0$. The deflection surface becomes

$$w(x,y) = (c_1 \cosh k_1 y + c_3 \cos k_2 y)\cos\frac{n\pi x}{a} \quad (8.4.6)$$

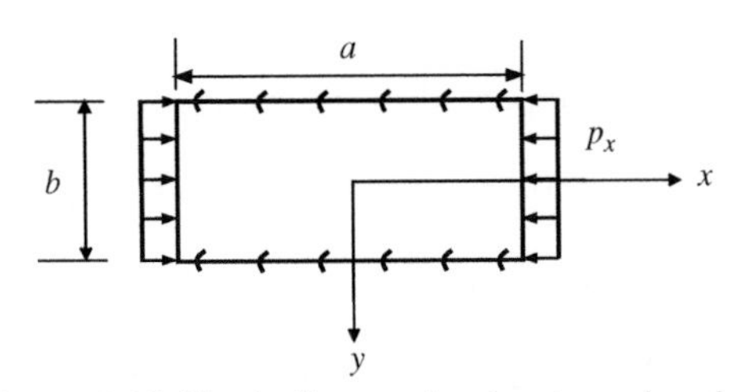

Figure 8-10 Elastically restrained rectangular plate

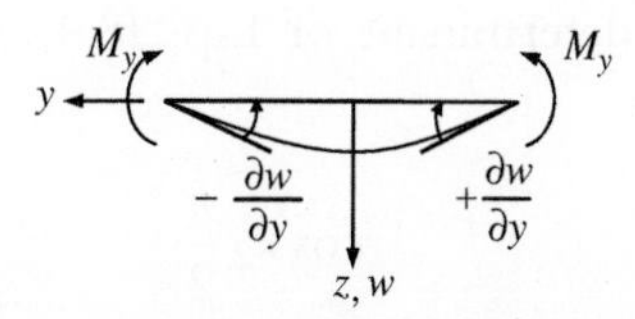

Figure 8-11 Elastically restrained boundary condition

The elastically restrained boundary conditions shown in Fig. 8-11 are

$$M_y = -k\frac{\partial w}{\partial y} \text{ at } y = b/2 \quad \text{and} \quad M_y = +k\frac{\partial w}{\partial y} \text{ at } y = -b/2$$

where k is rotational spring constant per unit width
and since $w = w_{,xx} = 0$ at $y = \pm b/2$

$$M_y\big|_{y=\pm b/2} = -D(w_{,yy} + \mu w_{,xx}) = -Dw_{,yy} = -DY_n'' \cos\frac{n\pi x}{a}$$

From $w = 0$ at $y = \pm b/2$

$$c_1 \cosh k_1\frac{b}{2} + c_3 \cos k_2\frac{b}{2} = 0 \tag{8.4.7}$$

From $M_y = -k\frac{\partial w}{\partial y}$ at $y = +b/2$

$$-D\left(c_1 k_1^2 \cosh k_1\frac{b}{2} - c_3 k_2^2 \cos k_2\frac{b}{2}\right) = -k\left(c_1 k_1 \sinh k_1\frac{b}{2} - c_3 k_2 \sin k_2\frac{b}{2}\right) \tag{8.4.8}$$

From $M_y = +k\frac{\partial w}{\partial y}$ at $y = -b/2$

$$-D\left(c_1 k_1^2 \cosh k_1\frac{b}{2} - c_3 k_2^2 \cos k_2\frac{b}{2}\right) = -k\left(c_1 k_1 \sinh k_1\frac{b}{2} - c_3 k_2 \sin k_2\frac{b}{2}\right) \tag{8.4.9}$$

It is noted that Eq. (8.4.9) is identical to Eq. (8.4.8). Let $\rho = 2D/bk$. Then Eq. (8.4.8) becomes

$$\frac{D}{k}\left(c_1 k_1^2 \cosh k_1\frac{b}{2} - c_3 k_2^2 \cos k_2\frac{b}{2}\right) = \left(c_1 k_1 \sinh k_1\frac{b}{2} - c_3 k_2 \sin k_2\frac{b}{2}\right)$$

$$c_1\left(k_1 \sinh k_1\frac{b}{2} - k_1^2\frac{b}{2}\rho \cosh k_1\frac{b}{2}\right) - c_3\left(k_2 \sin k_2\frac{b}{2} - k_2^2\frac{b}{2}\rho \cos k_2\frac{b}{2}\right) = 0 \tag{8.4.10}$$

Setting the coefficient determinant of Eqs. (8.4.7) and (8.4.10) for the constants c_1 and c_3 yields

$$\begin{vmatrix} \cosh k_1 \dfrac{b}{2} & \cos k_2 \dfrac{b}{2} \\ k_1 \sinh k_1 \dfrac{b}{2} - k_1^2 \dfrac{b}{2} \rho \cosh k_1 \dfrac{b}{2} & -k_2 \sin k_2 \dfrac{b}{2} + k_2^2 \dfrac{b}{2} \rho \cos k_2 \dfrac{b}{2} \end{vmatrix} = 0$$

Expanding the above determinant gives

$$k_1 \tanh \frac{k_1 b}{2} + k_2 \tan \frac{k_2 b}{2} - \frac{1}{2} b\rho\left(k_1^2 + k_2^2\right) = 0 \tag{8.4.11}$$

Let $\alpha = a/b$ be the aspect ratio of the rectangular plate. Then

$$k_1 \frac{b}{2} = \frac{n\pi}{2\alpha}\sqrt{u+1} \quad \text{and} \quad k_2 \frac{b}{2} = \frac{n\pi}{2\alpha}\sqrt{u-1}$$

$$\sqrt{u+1}\tanh\left(\frac{\pi}{2}\sqrt{u+1}\,\frac{n}{\alpha}\right) + \sqrt{u-1}\tan\left(\frac{\pi}{2}\sqrt{u-1}\,\frac{n}{\alpha}\right) - \pi\rho u\left(\frac{n}{\alpha}\right) = 0 \tag{8.4.12}$$

Equation (8.4.12) is the general buckling condition equation.

If the plate is simply supported along the boundary at $y = \pm b/2$, then $k = 0$ and $\rho = \infty$. Therefore, Eq. (8.4.2) becomes

$$\sqrt{u-1}\tan\left(\frac{\pi}{2}\sqrt{u-1}\frac{n}{\alpha}\right) = \infty$$

since $\sqrt{u+1}\tanh\left(\frac{\pi}{2}\sqrt{u+1}\frac{n}{\alpha}\right)$ is a finite value.

Hence

$$\sqrt{u-1}\frac{n}{\alpha} = 1$$

from which

$$u^2 = \left[\left(\frac{\alpha}{n}\right)^2 + 1\right]^2 = \frac{p_x}{D}\left(\frac{a}{n\pi}\right)^2$$

Then

$$p_{xcr} = D\left[\left(\frac{\alpha}{n}\right)^2 + 1\right]^2\left(\frac{n\pi}{a}\right)^2 = \frac{D\pi^2}{b^2}\left(\frac{\alpha}{n} + \frac{n}{\alpha}\right)^2$$

$$\sigma_{cr} = \frac{p_{xcr}}{h} = \left(\frac{h}{b}\right)^2 \frac{E\pi^2}{12(1-\mu^2)}\left(\frac{\alpha}{n} + \frac{n}{\alpha}\right)^2$$

Let

$$k' = \left(\frac{n}{\alpha} + \frac{\alpha}{n}\right)^2$$

which is called the buckling coefficient.
Then

$$\sigma_{cr} = k' \frac{\pi^2 E}{12(1-\mu^2)\left(\frac{b}{h}\right)^2} \tag{8.4.13}$$

It is known that Bryan (1891) derived Eq. (8.4.12) for the first time.

For the smallest p_{xcr},

$$\frac{dp_{xcr}}{d\alpha} = \frac{2D\pi^2}{b^2}\left(\frac{\alpha}{n} + \frac{n}{\alpha}\right)\left(\frac{1}{n} - \frac{n}{\alpha^2}\right) = 0$$

which leads to $n^2 = \alpha^2$.
If $n = 1$, then $\alpha = 1$ and $k' = 4$. The plot of buckling coefficient for $n = 1$ is given in Fig. 8-12. In a similar manner, the curves for $n = 2, 3, 4\ldots$ can be obtained. The solid curves represent lowest critical values, and the dotted lines higher critical values, for given plate aspect ratios. The buckling coefficient k' for plates with other boundary conditions are given by Gerard and Becker (1957).

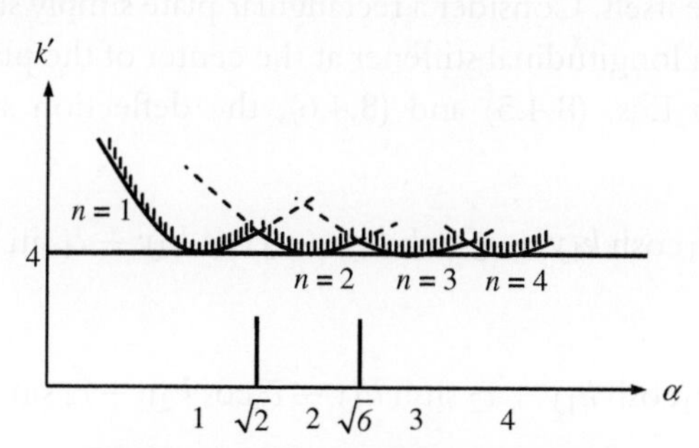

Figure 8-12 Plate buckling coefficient

8.4.2. Longitudinally Stiffened Plates

Longitudinally stiffened compression plates are believed to have been used from the quite early days of steel structures. They render an effective utilization of materials and thus offer a lightweight structure as in the case of box girder bridges, bridge decks, ship hulls, offshore drilling platforms, storage tanks, and so on. Although fragmented research efforts were made on the subject, including those of Barbre (1939), Seide and Stein (1949),

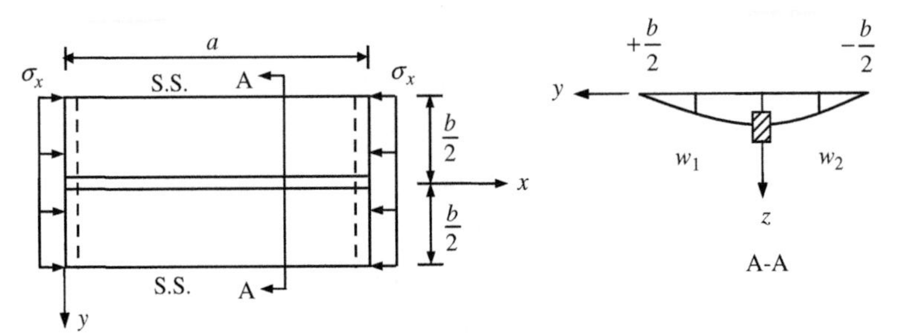

Figure 8-13 Longitudinally stiffened plate strip

Bleich (1952), Timoshenko and Gere (1961), and Sharp (1966), no organized research efforts were initiated until the early 1970s, when an urgent research program was launched in the UK as a result of a series of tragic collapses that occurred during the erection of bridges over the Danube, Milford Haven Bridge in Wales, West Gate Bridge in Australia, and Koblenze Bridge in Germany.

Following is a brief review of what Bleich (1952)[2] presents. Although Timoshenko and Gere (1961)[3] use the energy method of computing the critical stress and the plate buckling coefficient k'as compared to Bleich's approach of solving the differential equation, they use the same parameters: the aspect ratio of the plate, the bending rigidity ratio, and the area ratio of the stiffened plate itself. Consider a rectangular plate simply supported on all four edges with a longitudinal stiffener at the center of the plate as shown in Fig. 8-13. From Eqs. (8.4.5) and (8.4.6), the deflection surfaces can be written as

$$w_1 = \sin\frac{n\pi x}{a}\left(c_1\cosh k_1 y + c_2 \sinh k_1 y + c_3 \cos k_2 y + c_4 \sin k_2 y\right) \text{ for } y \geq 0$$

$$w_2 = \sin\frac{n\pi x}{a}\left(\bar{c}_1\cosh k_1 y + \bar{c}_2 \sinh k_1 y + \bar{c}_3 \cos k_2 y + \bar{c}_4 \sin k_2 y\right) \text{ for } y < 0$$

Boundary conditions (8 bc's) to determine $c_1 - \bar{c}_4$ are

$$w_1 = w_2 \quad \text{at } y = 0 \tag{a}$$

$$\frac{\partial w_1}{\partial y} = \frac{\partial w_2}{\partial y} \quad \text{at } y = 0 \tag{b}$$

[2] See page 360.
[3] See page 394.

$$\frac{\partial^2 w_1}{\partial y^2} = 0 \quad \text{at } y = b/2 \tag{c}$$

$$w_1 = 0 \quad \text{at } y = b/2 \tag{d}$$

$$\frac{\partial^2 w_2}{\partial y^2} = 0 \quad \text{at } y = -b/2 \tag{e}$$

$$w_2 = 0 \quad \text{at } y = -b/2 \tag{f}$$

One needs two additional conditions. Consider the juncture where the stiffener and the plate meet as shown in Fig. 8-14. Consider then the isolated stiffener alone.

The behavior of the stiffener can be described by a beam equation with $\sigma A = N_x$ and $w_1 = w_2 = w$ at $y = 0$

$$EI\frac{\partial^4 w}{\partial x^4} + \sigma A\frac{\partial^2 w}{\partial x^2} = q \;\left(\text{bm: } EIy^{iv} + Py'' = q\right) \tag{g}$$

From the theory of plates

$$Q_1 - Q_2 = q$$

$$= -D\frac{\partial}{\partial y}\left[\frac{\partial^2 w_1}{\partial y^2} + (2-\mu)\frac{\partial^2 w_1}{\partial x^2} - \frac{\partial^2 w_2}{\partial y^2} - (2-\mu)\frac{\partial^2 w_2}{\partial x^2}\right]_{y=0}$$

The distributed torque on the stiffener is

$$M^T = M_{y2} - M_{y1} = GK_T\frac{\partial \theta}{\partial x} - EI_w\frac{\partial^3 \theta}{\partial x^3} \tag{h}$$

where counterclockwise torque is positive and GK_T and EI_w are properties of the stiffener.

$$M_{y2} - M_{y1} = \left[-D\left(\frac{\partial^2 w_2}{\partial y^2} + \mu\frac{\partial^2 w_2}{\partial x^2}\right) + D\left(\frac{\partial^2 w_1}{\partial y^2} + \mu\frac{\partial^2 w_1}{\partial x^2}\right)\right] = M^T$$

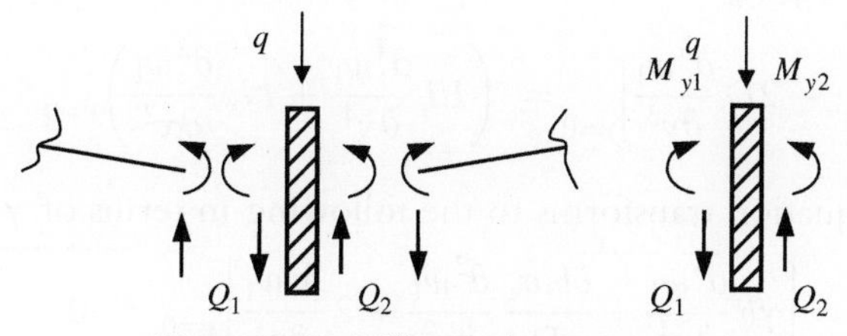

Figure 8-14 Stiffener-plate juncture

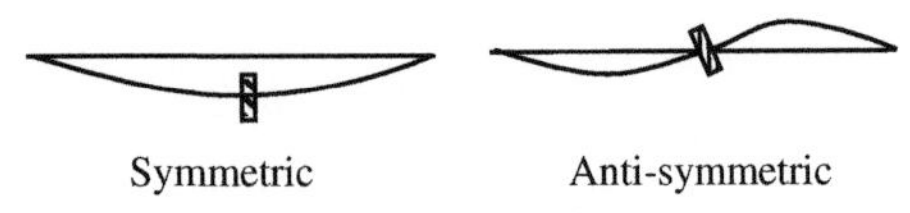

Figure 8-15 Definition of mode shapes

If GK_T and EI_w are assumed negligibly small, though it is not convincing, then

$$M_{y1} = M_{y2} \quad \text{or } M^T = 0 \text{ at } y = 0 \tag{i}$$

Bleich showed the derivation of the critical stress for the symmetric buckling only because the critical stress for antisymmetric buckling was equal to the critical stress for a simply supported plate of width b/2. Refer to Fig. 8-15 for the definition of the terminology.
Three parameters were introduced: the aspect ratio of the plate, the bending rigidity ratio, and the area ratio of the stiffener and the plate, respectively. They are

$$\alpha = \frac{a}{b}$$

$$\gamma = \frac{12(1-\mu^2)I}{bh^3} = \frac{EI}{Db}$$

$$\delta = \frac{A}{bh}$$

where Db = bending rigidity of a plate of width b and A = area of the stiffener. It is evident from symmetry shown in Fig. 8-14 that

$$\left(\frac{\partial^2 w_1}{\partial x^2} = \frac{\partial^2 w_2}{\partial x^2}\right)_{y=0} \quad \text{and} \quad \left(\frac{\partial^3 w_1}{\partial y^3} = -\frac{\partial^3 w_2}{\partial y^3}\right)_{y=0}$$

Therefore

$$q = Q_1 - Q_2 = -2D\left.\frac{\partial^3 w_1}{\partial y^3}\right|_{y=0} \tag{j}$$

Hence, from Eq. (g)

$$-2D\left.\frac{\partial^3 w_1}{\partial y^3}\right|_{y=0} = \left(EI\frac{\partial^4 w_1}{\partial x^4} + N_x\frac{\partial^2 w_1}{\partial x^2}\right)_{y=0}$$

The above equation transforms to the following in terms of γ and δ:

$$\left[\gamma b\frac{\partial^4 w_1}{\partial x^4} + \frac{\delta bt\sigma_c}{D}\frac{\partial^2 w_1}{\partial x^2} + 2\frac{\partial^3 w_1}{\partial y^3}\right]_{y=0} = 0 \tag{k}$$

The available boundary conditions for the four unknowns $c_1 - c_4$ are (c) and (d), $(\partial w_1/\partial y)|_{y=0} = 0$ (due to symmetry), and (k). Applying these four boundary conditions yields four homogeneous equations for the integral constants, $c_1 - c_4$. Setting the coefficient determinant of these simultaneous equations equal to zero gives the stability condition for the symmetric mode of buckling.

Boundary condition (c) gives

$$k_1^2 \cosh \frac{k_1 b}{2} c_1 + k_1^2 \sinh \frac{k_1 b}{2} c_2 - k_2^2 \cos \frac{k_2 b}{2} c_3 - k_2^2 \sin \frac{k_2 b}{2} c_4 = 0 \quad \text{(l)}$$

Boundary condition (d) gives

$$\cosh \frac{k_1 b}{2} c_1 + \sinh \frac{k_1 b}{2} c_2 + \cos \frac{k_2 b}{2} c_3 + \sin \frac{k_2 b}{2} c_4 = 0 \quad \text{(m)}$$

From $\left.\frac{\partial w_1}{\partial y}\right|_{y=0} = 0$ (symmetric condition), one obtains

$$k_1 c_2 + k_2 c_4 = 0 \quad \text{(n)}$$

Boundary condition (k) gives

$$\frac{\gamma}{b^3} \frac{n^4 \pi^4}{\alpha^4} (c_1 + c_3) - \frac{\delta t \sigma_c}{Db} \frac{n^2 \pi^2}{\alpha^2} (c_1 + c_3) + 2\left(k_1^3 c_2 - k_2^3 c_4\right) = 0$$

or

$$\left(\frac{\gamma}{b^3} \frac{n^4 \pi^4}{\alpha^4} - \frac{\delta t \sigma_c}{Db} \frac{n^2 \pi^2}{\alpha^2}\right) c_1 + 2\, k_1^3 c_2 + \left(\frac{\gamma}{b^3} \frac{n^4 \pi^4}{\alpha^4} - \frac{\delta t \sigma_c}{Db} \frac{n^2 \pi^2}{\alpha^2}\right) c_3 - 2 k_2^3 c_4 = 0 \quad \text{(o)}$$

The coefficient determinant is

$$\begin{vmatrix} k_1^2 \cos h \frac{k_1 b}{2} & k_1^2 \sin h \frac{k_1 b}{2} & -k_2^2 \cos \frac{k_2 b}{2} & -k_2^2 \sin \frac{k_2 b}{2} \\ \cos h \frac{k_1 b}{2} & \sin h \frac{k_1 b}{2} & \cos \frac{k_2 b}{2} & \sin \frac{k_2 b}{2} \\ 0 & k_1 & 0 & k_2 \\ \frac{\gamma}{b^3} \frac{n^4 \pi^4}{\alpha^4} - \frac{\delta t \sigma_c}{Db} \frac{n^2 \pi^2}{\alpha^2} & 2k_1^3 & \frac{\gamma}{b^3} \frac{n^4 \pi^4}{\alpha^4} - \frac{\delta t \sigma_c}{Db} \frac{n^2 \pi^2}{\alpha^2} & -2k_2^3 \end{vmatrix}$$

Let

$$big = \frac{\gamma}{b^3}\frac{n^4\pi^4}{\alpha^4} - \frac{\delta t\sigma_c}{Db}\frac{n^2\pi^2}{\alpha^2}.$$

Then, the determinant becomes

$$\begin{vmatrix} k_1^2 \cos h\dfrac{k_1 b}{2} & k_1^2 \sin h\dfrac{k_1 b}{2} & -k_2^2 \cos\dfrac{k_2 b}{2} & -k_2^2 \sin\dfrac{k_2 b}{2} \\ \cos h\dfrac{k_1 b}{2} & \sin h\dfrac{k_1 b}{2} & \cos\dfrac{k_2 b}{2} & \sin\dfrac{k_2 b}{2} \\ 0 & k_1 & 0 & k_2 \\ big & 2k_1^3 & big & -2k_2^3 \end{vmatrix}$$

Expanding the determinant gives

$$\det = 2k_1k_2\left(k_1^2+k_2^2\right)^2 ch\cos + big\left[k_1\left(k_1^2+k_2^2\right)ch\sin - k_2\left(k_1^2+k_2^2\right)sh\cos\right]$$

where

$$ch = \cosh\left(\frac{k_1 b}{2}\right),\ sh = \sinh\left(\frac{k_1 b}{2}\right),\ \cos = \cos\left(\frac{k_2 b}{2}\right),\ \sin = \sin\left(\frac{k_2 b}{2}\right)$$

Letting the determinant equal to zero for the stability condition yields

$$2k_1k_2\left(k_1^2+k_2^2\right)ch\cos + big\left[k_1 ch\sin - k_2 sh\cos\right] = 0$$

Dividing both sides by $-k_1k_2 ch\cos$ gives

$$\left(\frac{1}{k_1}\tanh\frac{k_1 b}{2} - \frac{1}{k_2}\tan\frac{k_2 b}{2}\right)\left(\frac{\gamma}{b^3}\frac{n^4\pi^4}{\alpha^4} - \frac{\delta t\sigma_c}{Db}\frac{n^2\pi^2}{\alpha^2}\right) - 2\left(k_1^2+k_2^2\right) = 0 \tag{8.4.14}$$

Equation (8.4.14) gives the relationship between the stiffener rigidity versus the compressive stress, σ_c, at the instance of symmetric buckling. Bleich then lists the case of two stiffeners subdividing the plate into three equal panels without showing the derivation process for the critical stress. Bleich simply shows a plot of the limiting value of the rigidity ratio γ obtained for the case by Barbre (1939).

It will be informative to review briefly the early development of the design rules applicable to longitudinally stiffened compression panels. A literature search (Choi 2002) reveals that the early design guides were BSI (1982), DIN 4114 (1978), ECCS (1976), and AASHO (1965). According to Wolchuk and Mayrbaurl (1980), the British design specification (BSI, 1982) is influenced to a large degree by the general design philosophy of the "Interim design and workmanship rules" ("the Merrison Rules") (Inquiry

1974). The Merrison Rules method is essentially the culmination of the urgent research program in response to the series of collapses. The method considers the individual stiffener strut separately, which consists of a stiffener with a corresponding width of the flange plate. The strength of the entire stiffened plate is then obtained by multiplying the ultimate stress of the strut by the total area of the plate. This is referred to as the "column behavior" theory, which prevails in European countries.

Highly theoretical and extremely complex analytical research on compression panels stiffened by one or two stiffeners has been carried out by Barbre (1939), Bleich (1952), and Timoshenko and Gere (1961). It appears that their research results on the antisymmetric buckling mode, which might be classified as the "plate behavior" theory, are not currently in use in any national design specifications. Mattock et al. (1967) prepared the "Commentary on criteria for design of steel-concrete composite box girder highway bridges" in August 1967. These criteria were intended to supplement the provisions of Division I of the Standard Specifications for Highway Bridges of the AASHO (1965). An overly conservative approach appears to have been adopted during the course of simplifying and extrapolating the limited research results (some of which appear questionable) to incorporate the case where the number of longitudinal stiffeners was greater than two. Although the equations in the AASHO (1965) give a reasonable value for the minimum required moment of inertia of the stiffener when the number of stiffeners is less than or equal to two, the equations require unreasonably large value for the moment of inertia when the number of stiffeners becomes large. It was found that an old bridge (curved box girder approach spans to the Fort Duquesne Bridge in Pittsburg) designed and built before the enactment of the criteria did not rate well, despite having served safely for many years. After this incident, the latest AASHTO (2007) specifications limit the maximum number of stiffeners to two as a stopgap measure.

In a series of numerical researches at Auburn University, Yoo and his colleagues (Yoo 2001; Yoo et al. 2001), extracted a regression formula for the minimum required moment of inertia for the longitudinal stiffener to assure an antisymmetric buckling mode. The coefficient of correlation R was found to be greater than 0.95.

$$I_s = 0.3\eta\alpha^2\sqrt{n}wh^3 \tag{8.4.15}$$

where α = aspect ratio of subpanel; n = number of stiffeners; h = thickness of plate; w = width of stiffened subpanel; η = ratio of the postbuckling stress to the elastic buckling stress. The elastic buckling stress is to be computed by Eq. (8.4.13) with a value of the buckling coefficient k' equal to four. The ratio of the postbuckling stress to the elastic buckling stress η should be set equal to one when the postbuckling strength is not recognized for reasons other than the strength or the analysis is carried out in the inelastic zone. Choi and Yoo (2005) showed that Eq. (8.4.15) works well for horizontally curved box girder compression flanges too, and its validity has been verified by an experimental study (Choi et al. 2009).

It is reassuring to note that Eq. (8.4.15) includes the length of the member (indirectly by the aspect ratio α). The longitudinal stiffener is, after all, a compression member whether it is examined in the "column behavior" theory or in the "plate behavior" theory. As such, the length of the compression member must be a prominent variable in determining the strength. In order to control the length of the longitudinal stiffener (the aspect ratio α, shall not exceed, say 7), transverse stiffeners are to be used. Choi, Kang, and Yoo (2007) furnish a design guide for transverse stiffeners. Mittelstedt (2008) demonstrates the superiority of the "column behavior" theory by an explicit elastic analysis of longitudinally stiffened plates for buckling loads and the minimum stiffener requirements.

Compression members in general can be classified into three groups: compact, noncompact, and slender. Yielding, inelastic buckling, and elastic buckling, respectively, control the ultimate strength of the members in each group. Geometric imperfections appear to affect the inelastic buckling strength of the members belonging to the noncompact group. Residual stresses are particularly onerous to the postbuckling strength of the slender members and affect the inelastic buckling strength to a much smaller degree. The ultimate strengths of the stocky members in the compact group are not affected by the presence of either initial imperfections or residual stresses. The current AASHTO (2007) provisions for the limiting value of the width-to-thickness ratio classifying the subpanels into these three groups appear reasonable. However, it seems reasonable to classify the zones into just two—the elastic buckling zone and the inelastic buckling zone—as is being done in AISC (2005).

Based on the observations made during the series of investigations by Yoo and his coworkers, a new simple formula is proposed for the ultimate stress in the inelastic buckling zone.

$$\sigma_{cr} = \sigma_y - \frac{\sigma_r}{C_c^2}\left(\frac{w}{h}\right)^2 = \sigma_y\left[1 - \frac{\sigma_r}{\sigma_y C_c^2}\left(\frac{w}{h}\right)^2\right] \tag{8.4.16}$$

where $\sigma r =$ maximum compressive residual stress and $C_c =$ threshold value of the width-to-thickness ratio dividing the elastic buckling and inelastic buckling of the subpanel, which is given by

$$C_c = \sqrt{\frac{\pi^2 E}{3(1-\mu^2)(\sigma_y - \sigma_r)}} \tag{8.4.17}$$

where $\mu =$ Poisson's ratio.

If the intensity of the residual stress σ_r is arbitrarily taken to be 0.5 σ_y, Eq. (8.4.16) reduces to

$$\sigma_{cr} = \sigma_y\left[1 - \frac{1}{2C_c^2}\left(\frac{w}{h}\right)^2\right] \tag{8.4.18}$$

It seems apparent that AISC adopted a residual stress measurement at Lehigh University in the early 1960s conducted on A7 ($\sigma_y = 33$ ksi) steel specimens, in which a maximum residual stress value of 16.5 ksi was reported. Taking the intensity of the residual stress σ_r equal to 0.5 σ_y ensures that the inelastic buckling stress curve given by Eq. (8.4.180 and the elastic buckling stress curve, Eq. (8.4.13), have a common tangent, as shown in Fig. 8-16. AISC (1989) retained the residual stress value of $\sigma_y/2$ up to its ninth edition of the *Steel Construction Manual*. Although AISC (2005) does not use the term *residual stress*, it would seem that the idea remains unchanged as the maximum elastic buckling stress (F_e) is limited to 0.44 σ_y. Limited test results indicate that the intensity of the residual stress in high-strength steels is considerably less than 0.5 σ_y (Choi et al., 2009).

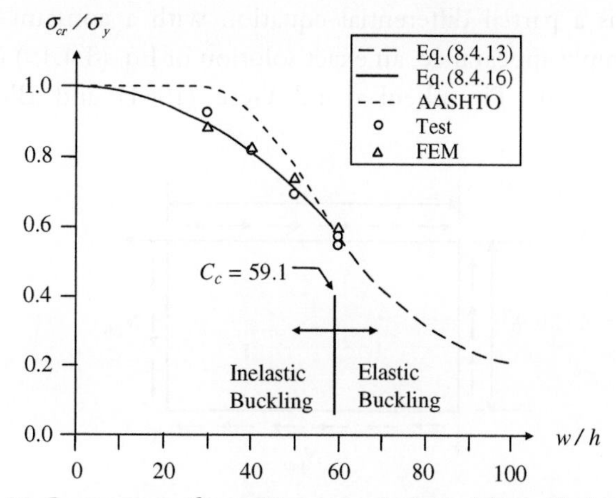

Figure 8-16 Comparison of transition curve (adopted from Choi et al. 2009)

Numerical values presented in Fig. 8-16 were generated assuming the intensity of the residual stress equal to $0.4\sigma_y$ as is currently used in the AASHTO (2007).

The nonlinear iterative finite element analysis reflected the residual stress as well as an initial geometric imperfection Δ of $w/100$, a maximum value allowed by the AWS (2008). The mill-specified yield stress of the test specimens is 50 ksi, yielding the threshold value of the width-to-thickness ratio C_c equal to 59.1 as per Eq. (8.4.17). If an initial imperfection Δ of $w/1000$ simulating a flat plate and zero residual stress are incorporated in the finite element analysis model, AASHTO curves are better represented. However, those are unconservative assumptions that do not reflect realistic construction conditions. It would seem appropriate to replace the outdated AASHTO (2007) provisions for the minimum required stiffness of the longitudinal stiffener with Eq. (8.4.15) and the strength predictor equations with Eq. (8.4.16). It should be remembered that Eq. (8.4.15) is valid for inelastic buckling and is applicable to horizontally curved box girders, as well as ship hulls.

8.4.3. Shear Loading

For a plate subjected to uniformly distributed shear loading as shown in Fig. 8-17, Eq. (8.3.3c) reduces to ($N_{xy0} = N_{yx0}$)

$$D\nabla^4 w - 2N_{xy0}\frac{\partial^2 w}{\partial x \partial y} = 0 \tag{8.4.19}$$

Equation (8.4.19), similar to the case of uniform compression loading in Eq. (8.4.1), is a partial differential equation with a constant coefficient. Despite its simple appearance, an exact solution of Eq. (8.4.19) is extremely difficult to obtain. Timoshenko and Gere (1961) and Bleich (1952)

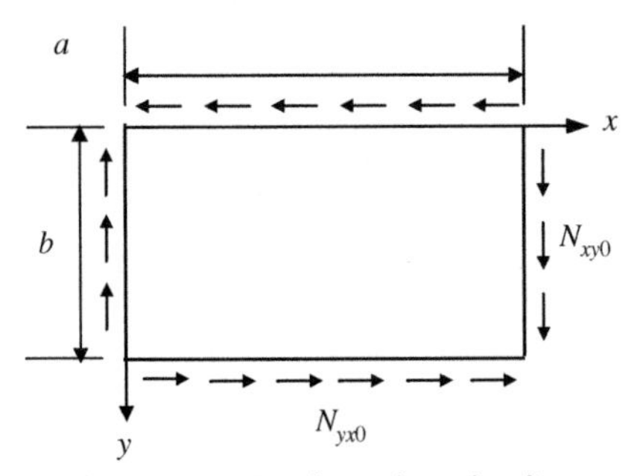

Figure 8-17 In-plane shear loading

assumed the deflected surface of the buckled plate in the form of the double series

$$w = \sum_{m=1}^{\infty}\sum_{n=1}^{\infty} a_{mn} \sin\frac{m\pi x}{a} \sin\frac{n\pi y}{b} \tag{8.4.20}$$

and then applied to the principle of minimum potential energy. Although four terms were used ($m = 1\text{-}2$, $n = 1\text{-}2$), the critical stress was 15% higher than the correct one for the square plate and the differences were even greater than 15% for long narrow rectangular plates. Southwell and Skan (1924) assumed the deflection function in the form

$$w(x,y) = f(x)g(y) = ce^{mx}e^{ny} \tag{8.4.21}$$

After transforming the partial differential equation into an ordinary differential equation, a procedure of the characteristic polynomial can be applied. Exact solutions of Eq. (8.4.19) are available only for the case of an infinitely long strip (Brush and Almroth 1975). Such a solution is available in Southwell and Skan (1924). Their results may be expressed in the form

$$N_{xy0} = k_s\frac{\pi^2 D}{b^2} \tag{8.4.22}$$

In this notation their results for infinitely long simply supported and clamped strips are $k_s = 5.34$ and $k_s = 8.98$, respectively.

For plates of finite dimensions, available numerical solutions by numerous researchers are summarized in Gerard and Becker (1957). Bleich introduces k_s values for simply supported and clamped square plates 9.34 and 14.71, respectively. Empirical formulas for k_s, along with source information given in Galambos (1998), are as follows:

Plate Simply Supported on Four Edges

$$k_s = 4.00 + \frac{5.34}{\alpha^2} \quad \text{for } \alpha \leq 1 \tag{8.4.23a}$$

$$k_s = 5.34 + \frac{4.00}{\alpha^2} \quad \text{for } \alpha \geq 1 \tag{8.4.23b}$$

Plate Clamped on Four Edges

$$k_s = 5.60 + \frac{8.98}{\alpha^2} \quad \text{for } \alpha \leq 1 \tag{8.4.24a}$$

$$k_s = 8.98 + \frac{5.60}{\alpha^2} \quad \text{for } \alpha \geq 1 \tag{8.4.24b}$$

Plate Clamped on Two Opposite Edges and Simply Supported on the Other Two Edges

Long edges clamped:

$$k_s = \frac{8.98}{\alpha^2} + 5.61 - 1.99\alpha \quad \text{for } \alpha \leq 1 \tag{8.4.25a}$$

$$k_s = 8.98 + \frac{5.61}{\alpha^2} - \frac{1.99}{\alpha^3} \quad \text{for } \alpha \geq 1 \tag{8.4.25b}$$

Short edges clamped:

$$k_s = \frac{5.34}{\alpha^2} + \frac{2.31}{\alpha} - 3.44 + 8.39\alpha \quad \text{for } \alpha \leq 1 \tag{8.4.26a}$$

$$k_s = 5.34 + \frac{2.31}{\alpha} - \frac{3.44}{\alpha^2} + \frac{8.39}{\alpha^3} \quad \text{for } \alpha \geq 1 \tag{8.4.26b}$$

One can very well appreciate scientists and engineers' struggles in the bygone era in solving such a straightforward equation as Eq. (8.4.19) simply because they lacked the analytical tools that are currently available. Perhaps the single most important application of the elastic buckling strength of thin rectangular panels subjected to shear loading is to the stiffened and/or unstiffened webs of plate- and box-girders. If that is the case, then it would be desirable to reflect the realistic boundary condition of the web panels, particularly at the juncture between the flange and web. It would seem reasonable to assume the boundary condition of the web panel to be simply supported at the intermediate transverse stiffener location, as they are designed to give the nodal line during buckling. However, the boundary condition at the flange and web juncture must be in between a clamped and a simply supported condition. Lee et al. (1996) proposed that the following two equations be used in the determining the shear buckling coefficients for the plate girder web panels:

$$k_s = k_{ss} + \frac{4}{5}\left(k_{sf} - k_{ss}\right)\left[1 - \frac{2}{3}\left(2 - \frac{t_f}{t_w}\right)\right] \quad \text{for } \frac{1}{2} < \frac{t_f}{t_w} < 2 \tag{8.4.27a}$$

$$k_s = k_s + \frac{4}{5}\left(k_{sf} - k_{ss}\right) \quad \text{for } \frac{t_f}{t_w} \geq 2 \tag{8.4.27b}$$

where t_f = flange thickness; t_w = web thickness; k_{ss} = shear buckling coefficient given by Eqs. (8.4.23); k_{sf} = shear buckling coefficient of plate clamped at the flange and web juncture and simply supported at the intermediate transverse stiffener location given by

$$k_s = \frac{5.34}{(a/D)^2} + \frac{2.31}{a/D} - 3.44 + 8.39\frac{a}{D} \quad \text{for} \frac{a}{D} < 1 \qquad (8.4.28a)$$

$$k_s = 8.98 + \frac{5.61}{(a/D)^2} - \frac{1.99}{(a/D)^3} \quad \text{for} \frac{a}{D} \geq 1 \qquad (8.4.28b)$$

where D = web depth; a = transverse stiffener spacing.

Equations (8.4.27) are regression formulas based on three-dimensional finite element analyses of numerous hypothetical plate-girder models encompassing a wide range of practical parameters. The correlation coefficient R of Eqs. (8.2.27) is greater than 0.95, and the validity and accuracy of Eqs. (8.4.27) have been demonstrated in numerous subsequent studies (Lee and Yoo 1998; Lee and Yoo 1999; Lee et al. 2002; Lee et al. 2003).

Shear buckling is a misnomer. The diagonal compressive stress causes the web to buckle. Elastic plate buckling is essentially local buckling. Therefore, there always exists postbuckling reserve strength. Frequently, excessive deformations are required to develop postbuckling strength. Web postbuckling, however, does not require excessive deformations. That is why engineers have reflected the postbuckling reserve strength in the design of thin web panels over the past 50 years. Postbuckling behavior of a web panel is indeed a very complex phenomenon. The nonlinear shear stress and normal stress interaction that takes place from the onset of elastic shear buckling to the ultimate strength state is so complex that any attempt to address this phenomenon using classical closed-form solutions appears to be a futile exercise. Even after codification of the Basler (1961) model and the Rockey or Cardiff model (Porter et al., 1975), there has been an ongoing controversy among researchers as they attempt to adequately explain the physical postbuckling behavior of web panels. The fact that more than a dozen theories and their derivatives have been suggested for explaining the phenomenon testifies to the complexity of tension field action.

Finally, Yoo and Lee (2006) put the postbuckling controversy to rest by discovering that the diagonal compression continuously increases in close proximity to the edges after elastic buckling, thereby producing in the web panel a self-equilibrating force system that does not depend on the flanges and stiffeners. As a result of this discovery, wholesale revisions must be made to the specification provisions, as well as steel design textbooks. The sole function of the intermediate transverse stiffener is to demarcate the web panel by establishing a nodal line in the buckling mode shape. It is not subjected to a resultant compressive force that was assumed to act on the post in a Pratt truss in the Basler model. Hence, there is no area requirement for the stiffener. Since the end panel is also in a self-equilibrating force system, it is certainly capable of developing tension field. The restriction of ignoring any tension field in the end panels, therefore, needs to be revised. Again, the flange anchoring mechanism in the Cardiff model is not needed. An arbitrary limitation of the web panel aspect ratio of three is not required (Lee et al. 2008; Lee et al. 2009a; Lee et al. 2009b).

8.5. ENERGY METHODS

8.5.1. Strain Energy of a Plate Element

For thin-walled plates where the thickness h is not greater than, say, one-tenth of the plate side dimensions, the constitutive relationship becomes a plane stress problem: that is, $\bar{\sigma}_z = \bar{\gamma}_{xz} = \bar{\gamma}_{yz} = 0$.

Although general expressions for the strain energy of a flat-plate element have been derived in Section 8.2, it would be interesting to examine the contribution of each stress component to total potential energy. Consider the plate element shown in Fig. 8-18 subjected first to $\bar{\sigma}_x$ only. Then, the force $P = \bar{\sigma}_x dA = \bar{\sigma}_x dzdy$ moves a distance equal to $\Delta_x = \bar{\varepsilon}_{xx} dx = \bar{\sigma}_x dx/E$. Hence,

$$dU_1 = \frac{1}{2}\frac{1}{E}\bar{\sigma}_x^2\, dxdydz \tag{8.5.1}$$

Then, the element is subjected to $\bar{\sigma}_y$. The strain energy due to $\bar{\sigma}_y$ is

$$dU_2 = \frac{1}{2}\frac{1}{E}\bar{\sigma}_y^2\, dxdydz \tag{8.5.2}$$

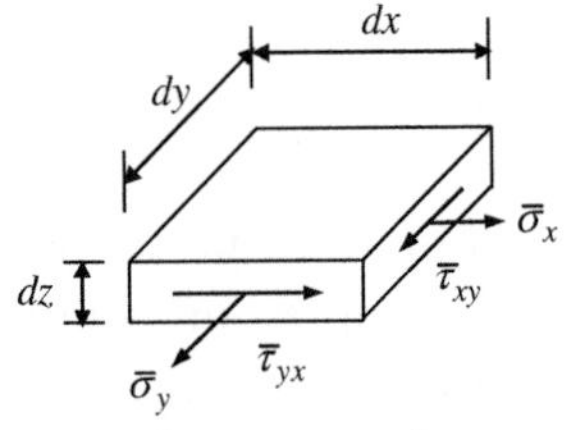

Figure 8-18 Stresses on plate elements

However, this time the force in the x direction rides a distance $= -\mu\,(\bar{\sigma}_y/E)\,dx$. Hence,

$$dU_3 = -\mu \frac{1}{E}\bar{\sigma}_y\bar{\sigma}_x\,dxdydz \tag{8.5.3}$$

Assuming that normal stresses produce no shear stresses and vice versa, it is possible to obtain strain energy of a plate element due to shear independently of the normal forces. Due to a shear stress, there exists a force, $\bar{\tau}_{xy}dxdz$ and a corresponding deformation $\bar{\gamma}_{xy}dy$ as shown in Fig. 8-19. Hence,

$$dU_4 = \frac{1}{2}\bar{\tau}_{xy}dxdz\left(\bar{\gamma}_{xy}dy\right) = \frac{1+\mu}{E}\bar{\tau}_{xy}^2\,dxdydz \tag{8.5.4}$$

The total strain energy is then

$$dU = \frac{1}{2E}[\bar{\sigma}_x^2 + \bar{\sigma}_y^2 - 2\mu\bar{\sigma}_x\bar{\sigma}_y + 2(1+\mu)\bar{\tau}_{xy}^2]dxdydz \tag{8.5.5}$$

For the entire plate of length a, width b, and thickness h, the strain energy becomes

$$U = \int_{-h/2}^{h/2}\int_0^b\int_0^a \frac{1}{2E}\left[\bar{\sigma}_x^2 + \bar{\sigma}_y^2 - 2\mu\bar{\sigma}_x\bar{\sigma}_y + 2(1+\mu)\bar{\tau}_{xy}^2\right]dxdydz \tag{8.5.6}$$

As a consequence of neglecting $\bar{\sigma}_z$, $\bar{\gamma}_{xz}$, $\bar{\gamma}_{yz}$, Eq. (8.5.6) is limited to thin plates only. It is also limited to linearly elastic materials and/or linearized problems but it is not limited to problems of either small displacements or membrane forces only. Substituting Eqs. (8.2.4) into Eq. (8.5.6) and carrying out the integration with respect to z, one obtains:

$$U_m = \frac{C}{2}\iint\left(\varepsilon_x^2 + \varepsilon_y^2 + 2\mu\varepsilon_x\varepsilon_y + \frac{1-\mu}{2}\gamma_{xy}^2\right)dxdy \tag{8.5.7}$$

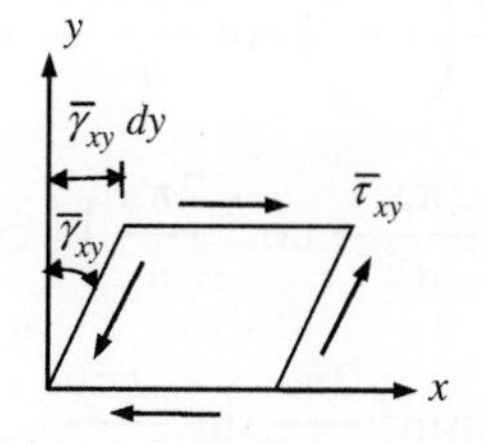

Figure 8-19 Shear strain

$$U_b = \frac{D}{2}\int_0^b \int_0^a \left[\left(\frac{\partial^2 w}{\partial x^2}\right)^2 + \left(\frac{\partial^2 w}{\partial y^2}\right)^2 + 2\mu \frac{\partial^2 w}{\partial x^2}\frac{\partial^2 w}{\partial y^2} + 2(1-\mu)\left(\frac{\partial^2 w}{\partial x \partial y}\right)^2\right] dxdy \tag{8.5.8}$$

8.5.2. Critical Loads of Rectangular Plates by the Energy Method

Herein, the energy method is applied to a square plate fixed on all four edges subjected to uniform compression shown in Fig. 8-20. In a plate buckling problem, the classical boundary condition at a support is applicable to the rotation only with the translation permitted as long as it does not create the rigid body motion.

The geometric boundary conditions are

$$w = \frac{\partial w}{\partial x} = 0 \text{ at } x = 0, a \quad w = \frac{\partial w}{\partial y} = 0 \text{ at } y = 0, a$$

The following displacement function will meet these boundary conditions:

$$w = A\left(1 - \cos\frac{2\pi x}{a}\right)\left(1 - \cos\frac{2\pi y}{a}\right)$$

Taking partial derivatives and substituting them into Eq. (8.5.8) leads to

$$U = \frac{D}{2}\frac{16\pi^4 A^2}{a^4} \times \int_0^a \int_0^a \left[\begin{array}{l} \cos^2\frac{2\pi x}{a}\left(1 - 2\cos\frac{2\pi y}{a} + \cos^2\frac{2\pi y}{a}\right) \\ + \cos^2\frac{2\pi y}{a}\left(1 - 2\cos\frac{2\pi x}{a} + \cos^2\frac{2\pi x}{a}\right) \\ + 2\mu\left(\cos\frac{2\pi x}{a} - \cos^2\frac{2\pi x}{a}\right)\left(\cos\frac{2\pi y}{a} - \cos^2\frac{2\pi y}{a}\right) \\ + 2(1-\mu)\sin^2\frac{2\pi x}{a}\sin^2\frac{2\pi y}{a} \end{array}\right] dxdy$$

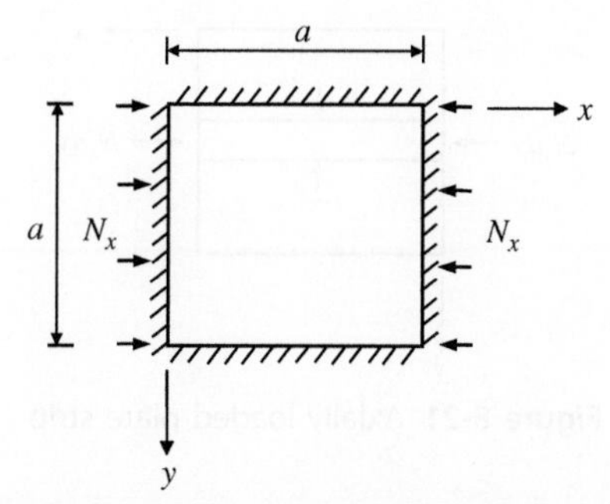

Figure 8-20 Square plate fixed on all four sides

Making use of the following definite integrals

$$\int_0^\alpha \sin^2 \frac{\beta \pi x}{\alpha} dx = \frac{\alpha}{2}, \int_0^\alpha \cos^2 \frac{\beta \pi x}{\alpha} dx = \frac{\alpha}{2}$$

$$\int_0^\alpha \sin \frac{\beta \pi x}{\alpha} dx = 0, \int_0^\alpha \cos \frac{\beta \pi x}{\alpha} dx = 0$$

where α, β are any integer, the strain energy becomes

$$U = \frac{16D\pi^4 A^2}{a^2}$$

The loss of potential energy of externally applied load due to shortening of the plate strip shown in Fig. 8-21 is

$$dV = -(N_x dy)\left[\frac{1}{2}\int_0^a \left(\frac{\partial w}{\partial x}\right)^2 dx\right]$$

Integrating dV gives

$$V = \int_0^a dV$$

$$= -\int_0^a (N_x dy)\left[\frac{1}{2}\int_0^a \left(\frac{\partial w}{\partial x}\right)^2 dx\right]$$

$$= -\frac{3N_x \pi^2 A^2}{2}$$

Then the total potential energy is

$$\Pi = U + V = \frac{16D\pi^4 A^2}{a^2} - \frac{3N_x \pi^2 A^2}{2}$$

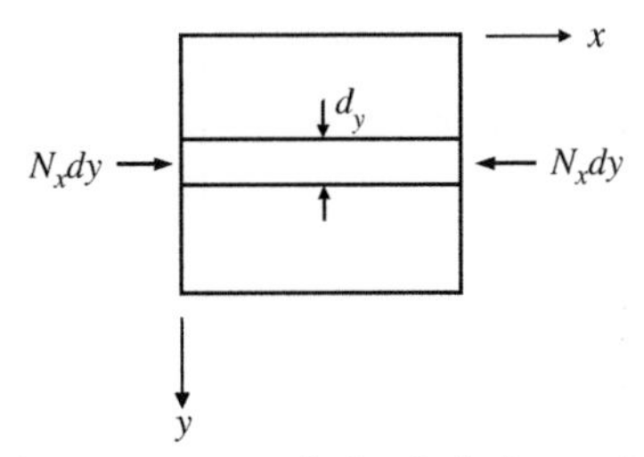

Figure 8-21 Axially loaded plate strip

Taking the first derivative with respect to A and setting it equal to zero gives

$$\frac{d\Pi}{dA} = \frac{32D\pi^4 A}{a^2} - 3N_x\pi^2 A = 0$$

Since A is not zero,

$$N_{xcr} = \frac{32D\pi^2}{3a^2} = \frac{10.67D\pi^2}{a^2}$$

which is upper-bound solution.
Using an infinite series for w, Levy (1942) obtained an exact solution $N_{xcr} = 10.07D\pi^2/a^2$, which is approximately 6% less than the above.

8.5.3. Shear Buckling of a Plate Element by the Galerkin Method

Consider the simply supported square plate (a square plate is chosen here just to simplify the computation effort) shown in Fig. 8-22. The plate is loaded by uniform shearing forces N_{xy} on four edges. To determine the critical load, the Galerkin method will be used. Although the procedure (without accompanying background information) was introduced in Section 1.8, it would be useful to examine the fundamentals of the method. Sokolnikoff (1956)[4] shows that the Galerkin and Rayleigh-Ritz methods are equivalent when applied to variational problems with quadratic functionals. In 1915 Galerkin proposed an approximate solution method that is of much wider scope than the Ritz method.

Sokolnikoff (1956) presents the following background information on the Galerkin method:
Consider a differential equation of the form

$$L(u) = 0 \tag{8.5.9}$$

[4] See page 413.

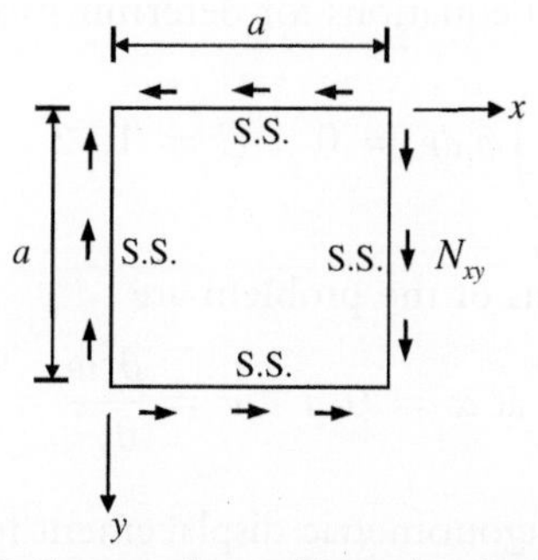

Figure 8-22 Square plate subjected to in-plane shear

where L is the differential operator and u is the displacement field. Suppose an approximate solution is sought to the problem in the form

$$u_n = \sum_{j=1}^{n} a_j\varphi_j \tag{8.5.10}$$

where the φ_j are kinematically admissible functions and the a_j are constants. As the finite sum, Eq. (8.5.10) will not satisfy Eq. (8.5.9), it follows that

$$L(u_n) = \varepsilon_n \text{ and } \varepsilon_n \neq 0 \tag{8.5.11}$$

If Max ε_n is small, then u_n can be considered a satisfactory approximation to u and the task at hand is to select a_j to minimize ε_n.

A reasonable minimization technique is as follows: If one represents u by the series

$$u = \sum_{i=1}^{\infty} a_i\varphi_i$$

with suitable properties and considers the nth partial sum

$$u_n = \sum_{i=1}^{n} a_i\varphi_i$$

then the orthogonality condition

$$\int_v L(u_n)\varphi_i dv = 0 \quad \text{as } n \to \infty \tag{8.5.12}$$

is equivalent to the statement that $L(u) = 0$.

This led Galerkin to impose on the error function a set of orthogonality conditions

$$\int_v L(u_n)\varphi_i dv = 0 \quad (i = 1, 2, \ldots, n) \tag{8.5.13}$$

which yields the set of n equations for determination of the constants a_j

$$\int_v L\left(\sum_{j=1}^{n} a_j\varphi_j\right)\varphi_i dv = 0 \quad (i = 1, 2, \ldots, n) \tag{8.5.14}$$

The boundary conditions of the problem are

$$w = \frac{\partial^2 w}{\partial x^2} = 0 \text{ at } x = 0, a \quad w = \frac{\partial^2 w}{\partial y^2} = 0 \text{ at } y = 0, a$$

Consider a two-term trigonometric displacement function such that

$$w = A_1 \sin\frac{\pi x}{a}\sin\frac{\pi y}{a} + A_2\sin\frac{2\pi x}{a}\sin\frac{2\pi y}{a}$$

The assumed displacement function meets geometric boundary conditions and natural boundary conditions.

The Galerkin equation takes the following form:

$$\int_0^a\int_0^a L(w)\varphi_i(x,y)dxdy = 0 \quad \text{with } i = 1, 2$$

where

$$L(w) = \frac{\partial^4 w}{\partial x^4} + 2\frac{\partial^4 w}{\partial x^2\partial y^2} + \frac{\partial^4 w}{\partial y^4} + \frac{2N_{xy}}{D}\frac{\partial^2 w}{\partial x\partial y}$$

$$\varphi_1(x,y) = \sin\frac{\pi x}{a}\sin\frac{\pi y}{a}$$

$$\varphi_2(x,y) = \sin\frac{2\pi x}{a}\sin\frac{2\pi y}{a}$$

Since there are two terms in the assumed displacement function, two Galerkin equations must be written.

$$\int_0^a\int_0^a L(w)\varphi_1(x,y)dxdy$$

$$= \int_0^a\int_0^a\left[\begin{array}{l}\dfrac{4A_1\pi^4}{a^4}\sin^2\dfrac{\pi x}{a}\sin^2\dfrac{\pi y}{a} + \dfrac{64A_2\pi^4}{a^4}\sin\dfrac{2\pi x}{a}\sin\dfrac{\pi x}{a}\sin\dfrac{2\pi y}{a}\sin\dfrac{\pi y}{a} \\ + \dfrac{2N_{xy}}{D}\left(\begin{array}{l}\dfrac{A_1\pi^2}{a^2}\cos\dfrac{\pi x}{a}\sin\dfrac{\pi x}{a}\cos\dfrac{\pi y}{a}\sin\dfrac{\pi y}{a} \\ + \dfrac{4A_2\pi^2}{a^2}\cos\dfrac{2\pi x}{a}\sin\dfrac{\pi x}{a}\cos\dfrac{2\pi y}{a}\sin\dfrac{\pi y}{a}\end{array}\right)\end{array}\right]dxdy$$

and

$$\int_0^a \int_0^a L(w)\varphi_2(x,y)dxdy$$

$$= \int_0^a \int_0^a \left[\begin{array}{l} \dfrac{4A_1\pi^4}{a^4}\sin\dfrac{\pi x}{a}\sin\dfrac{2\pi x}{a}\sin\dfrac{\pi y}{a}\sin\dfrac{2\pi y}{a} + \dfrac{64A_2\pi^4}{a^4}\sin^2\dfrac{2\pi x}{a}\sin^2\dfrac{2\pi y}{a} \\ + \frac{2N_{xy}}{D}\left(\begin{array}{l} \dfrac{A_1\pi^2}{a^2}\cos\dfrac{\pi x}{a}\sin\dfrac{2\pi x}{a}\cos\dfrac{\pi y}{a}\sin\dfrac{2\pi y}{a} \\ + \dfrac{4A_2\pi^2}{a^2}\cos\dfrac{2\pi x}{a}\sin\dfrac{2\pi x}{a}\cos\dfrac{2\pi y}{a}\sin\dfrac{2\pi y}{a} \end{array} \right) \end{array} \right] dxdy$$

Recalling

$$\int_0^a \sin^2\frac{m\pi x}{a}dx = \frac{a}{2}, \int_0^a \cos^2\frac{m\pi x}{a}dx = \frac{a}{2}$$

$$\int_0^a \sin\frac{m\pi x}{a}\sin\frac{n\pi x}{a} = 0 \text{ and } \int_0^a \cos\frac{m\pi x}{a}\cos\frac{n\pi x}{a} = 0 \text{ if } m \neq n$$

$$\int_0^a \cos\frac{2\pi x}{a}\sin\frac{\pi x}{a}dx = -\frac{2a}{3\pi}, \int_0^a \sin\frac{2\pi x}{a}\cos\frac{\pi x}{a}dx = \frac{4a}{3\pi}$$

the Galerkin equations reduce to

$$\frac{4A_1\pi^4}{a^4}\left(\frac{a}{2}\right)^2 + \frac{2N_{xy}}{D}\frac{4A_2\pi^2}{a^2}\left(-\frac{2a}{3\pi}\right)^2 = 0$$

$$\frac{64A_2\pi^4}{a^4}\left(\frac{a}{2}\right)^2 + \frac{2N_{xy}}{D}\frac{A_1\pi^2}{a^2}\left(\frac{4a}{3\pi}\right)^2 = 0$$

or

$$\frac{\pi^4}{a^2}A_1 + \frac{32N_{xy}}{9D}A_2 = 0$$

$$\frac{32N_{xy}}{9D}A_1 + \frac{16\pi^4}{a^2}A_2 = 0$$

Setting the determinant for A_1 and A_2 equal to zero gives

$$N_{xycr} = 11.1\frac{\pi^2}{a^2}D$$

which is approximately 18.8% greater than the exact value $N_{xycr} = 9.34(\pi^2/a^2)D$ obtained by Stein and Neff (1947). A numerical solution to an accuracy of this level can be obtained from most commercially available general-purpose three-dimensional finite element codes with a discretization of the plate less than 20 nodes per each edge (Lee et al., 1996).

Inelastic plate buckling analysis may be performed using an iterative procedure on commercially available general-purpose three-dimensional finite element packages such as ABAQUS, NASTRAN, or ADINA. Inelastic buckling at the transition zone is fairly sensitively affected by the initial imperfection assumed.

8.5.4. Postbuckling of Plate Elements

Equations (8.2.28) and (8.2.31) are nonlinear coupled partial differential equations. As is the case for all nonlinear equations, there is no closed-form general solution available to these equations. Consider as an example a square plate simply supported on all four edges and subjected to a uniform compressive force N_x as shown in Fig. 8-23. In order to examine the stress pattern in the postbuckling range, the following assumptions are made:

1. All edges remain straight and maintain the original 90 degrees.
2. The shearing forces, N_{xy} (N_{yx}) vanish on all four edges.
3. The edges, $y = 0,\ a$ are free to move in the y direction.

Let $\sigma_{x\ avg}$ be the average value of the applied compressive stress. Then

$$\sigma_{x\ avg} = -\frac{1}{ah}\int_0^a N_x dy \tag{a}$$

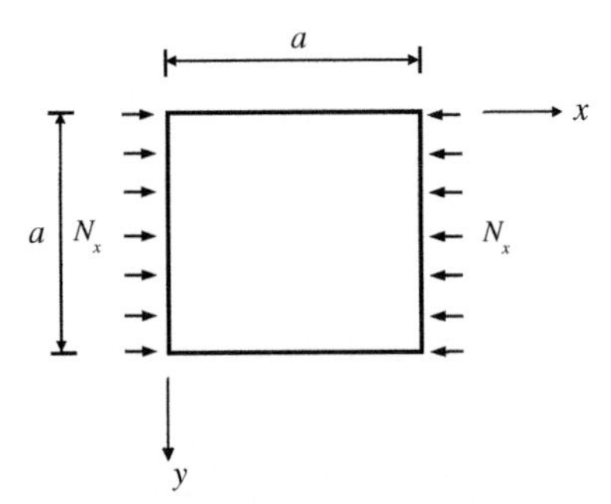

Figure 8-23 Simply supported square plate subjected to uniaxial force

where $\sigma_{x\ avg}$ is positive when N_x is in compression. N_x is assumed to vary internally as a result of large displacement in the postbuckling range. The plate boundary conditions are

$$w = \frac{\partial^2 w}{\partial x^2} = 0 \quad \text{at } x = 0, a \quad \text{and} \quad w = \frac{\partial^2 w}{\partial y^2} = 0 \quad \text{at } y = 0, a$$

Assume

$$w = g \sin \frac{\pi x}{a} \sin \frac{\pi y}{a} \tag{b}$$

Substituting Eq. (b) into Eq. (8.2.31) gives

$$\frac{\partial^4 f}{\partial x^4} + 2\frac{\partial^4 f}{\partial x^2 \partial y^2} + \frac{\partial^4 f}{\partial y^4} = g^2 \frac{E\pi^4}{a^4}\left(\cos^2 \frac{\pi x}{a} \cos^2 \frac{\pi y}{a} - \sin^2 \frac{\pi x}{a} \sin^2 \frac{\pi y}{a}\right)$$

$$= g^2 \frac{E\pi^4}{2a^4}\left(\cos \frac{2\pi x}{a} + \cos \frac{2\pi y}{a}\right) \tag{c}$$

Let the solution of Eq. (c) be $f_t = f_h + f_p$. The implication of the homogeneous solution is that the right-hand side of Eq. (c) is equal to zero. That is, the transverse deflection of the plate is either zero or negligibly small in the state just prior to buckling. At this state, N_x is constant at any point of the plate.

$$N_x = h\frac{\partial^2 f}{\partial y^2} = \text{constant}$$

Hence, the homegeneous solution is

$$f_h = Ay^2$$

Noting that

$$\sigma_{x\ avg} = -\frac{N_x}{h} = \frac{\partial^2 f}{\partial y^2}$$

the homegeneous solution can be rewritten as

$$f_h = -\frac{\sigma_{x\ avg} y^2}{2} \tag{d}$$

Examining the form of the right-hand side of Eq. (c), one may assume the particular solution as

$$f_p = B \cos \frac{2\pi x}{a} + C \cos \frac{2\pi y}{a}$$

Substituting this into Eq. (c) and equating coefficient of the terms yields

$$B\frac{16\pi^4}{a^4}\cos\frac{2\pi x}{a} = g^2\frac{E\pi^4}{2a^4}\cos\frac{2\pi x}{a}$$

$$B = \frac{Eg^2}{32} = C$$

Hence, the total solution is

$$f_t = \frac{Eg^2}{32}\left(\cos\frac{2\pi x}{a} + \cos\frac{2\pi y}{a}\right) - \frac{\sigma_{x\ avg}y^2}{2} \tag{e}$$

To determine the coefficient g, use the Galerkin method

$$\int_0^a\int_0^a L(g)\varphi(x,y)dxdy = 0 \tag{f}$$

where

$$L(g) = \frac{\partial^4 w}{\partial x^4} + 2\frac{\partial^4 w}{\partial x^2\partial y^2} + \frac{\partial^4 w}{\partial y^4} - \frac{h}{D}\left(\frac{\partial^2 f}{\partial y^2}\frac{\partial^2 w}{\partial x^2} + \frac{\partial^2 f}{\partial x^2}\frac{\partial^2 w}{\partial y^2} - 2\frac{\partial^2 f}{\partial x\partial y}\frac{\partial^2 w}{\partial x\partial y}\right)$$

$$\varphi(x,y) = \sin\frac{\pi x}{a}\sin\frac{\pi y}{a}$$

Using (b) for w and (e) for f, one can write $L(g)$ as

$$L(g) = \frac{1}{D}\left[\frac{4gD\pi^4}{a^4} - \frac{Ehg^3\pi^4}{8a^4}\left(\cos\frac{2\pi x}{a} + \cos\frac{2\pi y}{a}\right) - \sigma_{x\ avg}hg\frac{\pi^2}{a^2}\right]\sin\frac{\pi x}{a}\sin\frac{\pi y}{a}$$

Hence, the Galerkin equation takes the following form:

$$\int_0^a\int_0^a L\left(g\right)\varphi\left(x,y\right)dxdy$$
$$= \frac{1}{D}\int_0^a\int_0^a\left[\left(\frac{4gD\pi^4}{a^4} - \sigma_{x\ avg}hg\frac{\pi^2}{a^2}\right)\left(\sin^2\frac{\pi x}{a}\sin^2\frac{\pi y}{a}\right) - \frac{Ehg^3\pi^4}{8a^4}\right.$$
$$\left.\times\left(\cos\frac{2\pi x}{a}\sin^2\frac{\pi x}{a}\sin^2\frac{\pi y}{a} + \cos\frac{2\pi y}{a}\sin^2\frac{\pi x}{a}\sin^2\frac{\pi y}{a}\right)\right]dxdy = 0$$

Recalling

$$\int_0^a \sin^2 \frac{2\pi x}{a} dx = \frac{a}{2}$$

the Galerkin equation reduces to

$$\left(\frac{4gD\pi^4}{a^4} - \sigma_{xavg} hg \frac{\pi^2}{a^2}\right)\frac{a^2}{4} - \frac{Ehg^3\pi^4}{8a^4}\frac{a}{2}\left(\int_0^a \cos\frac{2\pi x}{a}\sin^2\frac{\pi x}{a}dx\right.$$
$$\left. + \int_0^a \cos\frac{2\pi y}{a}\sin^2\frac{\pi y}{a}dy\right) = 0$$

Making use of the following relations

$$\cos\frac{2\pi x}{a}\sin^2\frac{\pi x}{a} = \frac{1}{2}\left(\cos\frac{2\pi x}{a} - \cos^2\frac{2\pi x}{a}\right)$$

$$\int_0^a \cos^2\frac{2\pi x}{a}dx = \frac{a}{2}, \int_0^a \cos\frac{2\pi x}{a}dx = 0$$

The Galerkin Equation can be further simplified as

$$\frac{gD\pi^4}{a^2} - \sigma_{xavg} hg\frac{\pi^2}{4} + \frac{Ehg^3\pi^4}{32a^2} = 0$$

Hence

$$\sigma_{x\ avg} = \frac{4D\pi^2}{ha^2} + \frac{E\pi^2 g^2}{8a^2} = \sigma_{cr} + \frac{E\pi^2 g^2}{8a^2} \tag{g}$$

or

$$g^2 = \frac{8a^2}{E\pi^2}\left(\sigma_{x\ avg} - \sigma_{cr}\right) \tag{h}$$

Figure 8-24 graphically shows the relationship between the average applied stress $\sigma_{x\ avg}$ and the maximum lateral deflection g subsequent to onset of buckling.

In order to understand why the plate is able to develop postbuckling strength, one has to investigate the middle-surface stresses subsequent to buckling.

Recall that the longitudinal stress σ_x is

$$\sigma_x = \frac{N_x}{h} = -\frac{\partial^2 f}{\partial y^2}$$

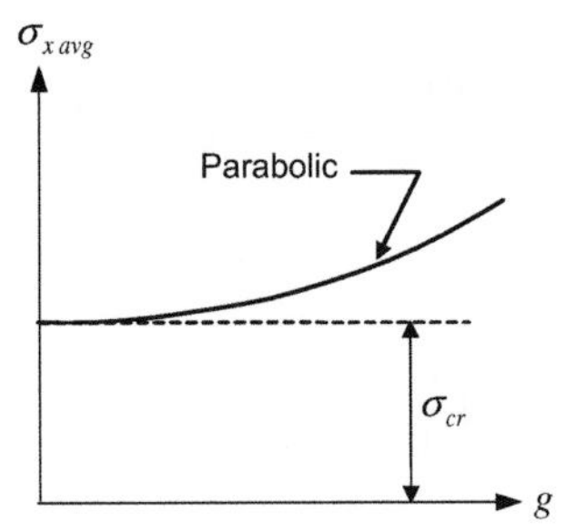

Figure 8-24 Postbuckling stress

Substituting Eq. (e) into the above gives

$$\sigma_x = \frac{E\pi^2 g^2}{8a^2}\cos\frac{2\pi y}{a} + \sigma_{x\ avg} \tag{i}$$

Substituting Eq. (h) into (i)

$$\sigma_x = \sigma_{x\ avg} + \left(\sigma_{xavg} - \sigma_{cr}\right)\cos\frac{2\pi y}{a} \tag{k}$$

In a similar manner, the stress in the transverse direction is

$$\sigma_y = \frac{N_y}{h} = -\frac{\partial^2 f}{\partial x^2} = \frac{E\pi^2 g^2}{8a^2}\cos\frac{2\pi x}{a} = \left(\sigma_{x\ avg} - \sigma_{cr}\right)\cos\frac{2\pi x}{a}$$

Figure 8-25 shows the variation of σ_x and σ_y.

As can be seen in Fig. 8-25, the tensile stress σ_y developed in the middle of the plate is believed to be the source of the postbuckling strength. Also, the degree of the uneven stress distribution of σ_x in the postbuckling stage could be reflected in the determination of the effective width of thin plates in compression.

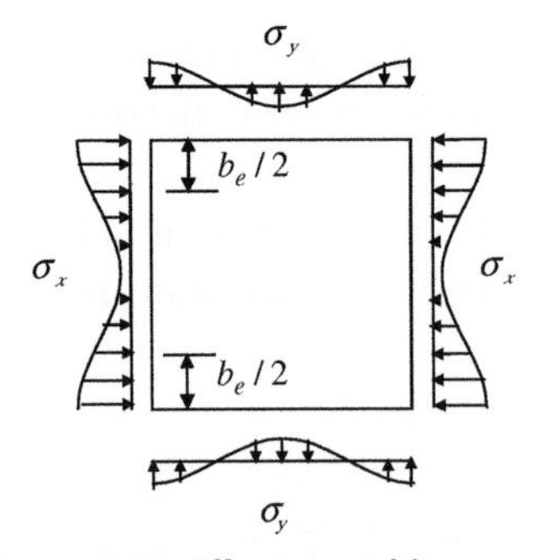

Figure 8-25 Effective width concept

The results shown above are based on an approximate analysis and hence may contain inaccuracies. For example, a refined analysis would show that in-plane shear stresses develop in addition to transverse tensile stresses after initial buckling. Such a refined analysis can readily be performed on any three-dimensional general-purpose finite element codes. However, the refined analysis results would not likely add any significant new information that the simplified analysis was unable to detect.

It should be noted that the postbuckling strength discussed above is due to the geometric nonlinearity. As can be seen in Fig. 8-24, any significant postbuckling (reserve) strength can only be recognized after a considerable deformation. Therefore, in most design specifications dealing with hot-rolled structural plates subjected to in-plane compression, the postbuckling strength is not recognized, whereas it is recognized in the design of cold-formed structures.

8.6. DESIGN PROVISIONS FOR LOCAL BUCKLING OF COMPRESSION ELEMENTS

In proportioning the width-to-thickness ratio of flat-plate elements of hot-rolled structural shapes, it is common practice to design the member so that overall failure occurs prior to local buckling failure. When a shape is produced with the same dimensions for different yield stresses, a section that satisfies the local buckling provision for a lower yield stress may not do so for a higher yield stress. In the AISC Specifications, the local buckling stress is kept above the yield stress for most rolled shapes, thereby making it possible to specify a single provision for beam and column sections.

If local buckling is not to occur at a stress smaller than the yield stress, σ_{cr} must be greater than σ_y. Since the plate buckling stress is given by Eq. (8.4.13), it requires

$$\frac{k'\pi^2 E}{12(1-\mu^2)}\left(\frac{h}{b}\right)^2 > \sigma_y$$

For $\mu = 0.3$, one obtains

$$\frac{b}{h} < 0.95\sqrt{\frac{k'E}{\sigma_y}} \tag{8.6.1}$$

For an unstiffened element of free-simple edge conditions, the plate buckling coefficient k' is 0.425. Substituting this value into Eq. (8.6.1) gives

$$\frac{b}{h} < 0.615\sqrt{\frac{E}{\sigma_y}} \tag{8.6.2}$$

The plate buckling coefficient k' could be as low as 0.35 due to web–flange interactions. Using the lower value of $k' = 0.35$ yields

$$\frac{b}{h} < 0.56\sqrt{\frac{E}{\sigma_y}} \tag{8.6.3}$$

The width-to-thickness ratio is further reduced in the current AISC (2005) and AASHTO (2007) to reflect the initial imperfections and residual stresses for compact sections.

$$\frac{b}{h} < 0.38\sqrt{\frac{E}{\sigma_y}} \tag{8.6.4}$$

For the corresponding width-to-thickness ratio of the stiffened element of a box-girder flange, the plate buckling coefficient k' of 4.0 for simple supports along both unloaded edges is substituted into Eq. (8.6.1) to yield

$$\frac{b}{h} < 1.90\sqrt{\frac{E}{\sigma_y}} \tag{8.6.5}$$

AISC (2005) reduces this further for compact sections to

$$\frac{b}{h} < 1.12\sqrt{\frac{E}{\sigma_y}} \tag{8.6.6}$$

8.7. INELASTIC BUCKLING OF PLATE ELEMENTS

When the applied load is increased beyond the elastic buckling load, the plate structure's response exhibits some form of nonlinear behavior, either geometric or material or a combination of these two. In the past, attempts were made to solve the material nonlinear problems by adjusting the modulus of elasticity either by the tangent-modulus theory or the reduced-modulus theory, or the combination. As these procedures are essentially empirical, their accuracy or success depends largely on the success of accurately extracting necessary data from experiments to determine the proper values for the adjusted modulus of elasticity.

Today, engineers are blessed with the availability of high-power digital computers at their fingertips and the advancement of sophisticated software. It is now just a matter of preparing a good set of input data that will evaluate the effect of complex residual stress distributions and geometric imperfections due to either milling or welding practice. The iterative procedure automatically evaluates the ultimate strength of structures. The embedded postprocessor in most advanced software provides engineers with practically inexhaustible information in graphical and/or tabular forms.

Despite enormous computation power, a computer program cannot design a structure. No computer program has been developed to design a structure automatically. And it is not expected to see one in the near future. Hence, engineers' input will be required in many future years to come. This is one reason why engineers need advanced knowledge of structural behavior.

8.8. FAILURE OF PLATE ELEMENTS

The neighboring equilibrium path in Fig. 8-24 for an initially flat plate subjected to in-plane compression is shown again in Fig. 8-26, along with a corresponding curve of a slightly imperfect plate. Two important observations from Fig. 8-26 are worthy to note: (1) Buckling of real (imperfect) plates is generally so gradual that it is difficult to indicate at precisely what load the buckling takes place. Therefore, it takes an element of judgment call to declare the critical load. (2) Unlike a column, the plate continues to carry additional loads after buckling.

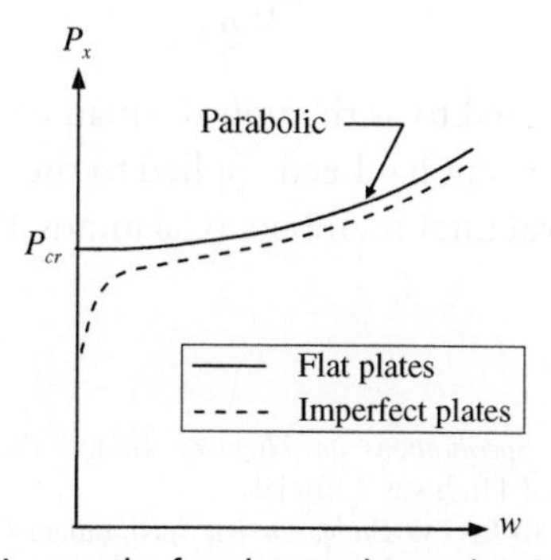

Figure 8-26 Equilibrium paths for plates subjected to in-plane compression

Hence, P_{cr} for the plate is not the ultimate strength. In order to take advantage of the additional load-carrying capacity, the postbuckling strength of plates must be correctly assessed.

In fact, Brush and Almroth (1975) credits Wagner (1929) for establishing a criterion for the postbuckling strength of a shear web. As alluded to in Section 8.4.3, this was the beginning of the long series of attempts to understand the true mechanics behind the tension field action. Nearly 80 years would elapse until Yoo and Lee (2006) could finally explain the true mechanics of the tension field action.

Unlike the shear web, a plate subjected to uniformly distributed in-plane compressive load P_x, it is much simpler to do. The applied load P_x as shown in Fig. 8-23 can be expressed as

$$P_x = h \int_0^b \sigma_x dy$$

where h and b are plate thickness and width, respectively. When $P_x \le P_{cr}$, the stress across the plate is uniform. Then $P_x = hb\sigma_x$. If $P_x = P_{cr}$, then

$$P_{cr} = hb\sigma_x \tag{8.8.1}$$

For $P_x > P_{cr}$, the stress at $y = 0, b$ is greater than that at the center of the plate because of the stiffening effect of the supports as shown in Fig. 8-25. For design purposes, it is customary to express the results of the analysis in terms of an effective width over which the stress is assumed to be uniform. Then

$$P_x = hb_e\sigma_{\max} \tag{8.8.2}$$

where $\sigma_{\max}$ is the maximum stress at the supports $y = 0, b$. An approximate expression for the effective plate width b_e is

$$b_e = b\sqrt{\frac{\sigma_{cr}}{\sigma_{\max}}} \tag{8.8.3}$$

Equation (8.8.3) is referred to as the von Kármán effective-width formula. The effective-width concept has been applied to the design of cold-formed steel and aluminum structural members (Galambos 1998).

REFERENCES

AASHO. (1965). *Standard Specifications for Highway Bridges* (9th ed.). Washington, DC: American Association of Highway Officials.

AASHTO. (2007). AASHTO LRFD *Bridge Design Specifications* (4th ed.). Washington, DC: American Association of Highway Officials.

AISC. (2005). *Specification for Structural Steel Building* (13th ed.). Chicago, IL: American Institute of Steel Construction.
Allen, H. G., & Bulson, P. S. (1980). *Background to Buckling*. London: McGraw-Hill (UK).
AWS. (2008). *Bridge Welding Code* (5th ed.). Miami, FL: AASHTO/AWS D1.5M/ D1.5:2008, A Joint Publication of American Association of State Highway and Transportation Officials, Washington, DC and American Welding Society.
Barbre, F. (1939). Stability of Rectangular Plates with Longitudinal or Transverse Stiffeners Under Uniform Compression. Washington, DC: NACA-TM-904.
Basler, K. (1961). Strength of Plate Girders in Shear. *Journal of the Structural Div., ASCE., Vol. 87*(No. 7), 151–181.
Bathe, K. J. (1996). *Finite Element Procedures*. Englewood Cliffs, NJ: Prentice-Hall.
Bleich, F. (1952). *Buckling Strength of Metal Structures*. New York: McGraw-Hill.
Brush, D. O., & Almroth, B. O. (1975). *Buckling of Bars, Plates, and Shells*. New York: McGraw-Hill.
Bryan, G. H. (1891). On the Stability of a Plane Plate Under Thrusts in Its Own Plane with Applications to the Buckling of the Sides of a Ship. *Proceedings London Mathematical Society, Vol. 22*, pp. 54–67.
BSI. (1982). *Steel, Concrete, and Composite Bridges, BS 5400, Part 3, Code of Practice for Design of Steel Bridges*. London: British Standard Institution.
Chajes, A. (1974). *Principles of Structural Stability Theory*. Englewood Cliffs, NJ: Prentice-Hall.
Choi, B. H. (2002). *Design Requirements for Longitudinally Stiffeners for Horizontally Curved Box Girders*. Auburn, AL: Ph.D. dissertation, Auburn University.
Choi, B. H., & Yoo, C. H. (2005). Strength of Stiffened Flanges in Horizontally Curved Box Girders. *Journal of Engineering Mechanics, ASCE, Vol. 131*(No. 1), 20–27, January.
Choi, B. H., Kang, Y. J., & Yoo, C. H. (2007). Stiffness Requirements for Transverse Stiffeners of Compression Flanges, *Engineering Structures*. Elsevier. *Vol. 29*, No. 9, September, pp. 2087–2096.
Choi, B. H., Hwang, M., Yoon, T., & Yoo, C. H. (2009). Experimental Study on Inelastic Buckling Strength and Stiffness Requirements for Longitudinally Stiffened Compression Panels, *Engineering Structures*. Elsevier. *Vol. 31*, No.5, May, pp. 1141–1153.
DIN 4114 (1978). *Beulsicherheitsnachweise fur Platten*, DASt Richtline 12, Deutscher Ausschuss fur Stahlbau, Draft, Berlin (in German).
ECCS (1976). *Manual on the Stability of Steel Structures, Introductory Report* (2nd. ed.). Tokyo: Second International Colloquium on Stability-ECCS.
Galambos, T. V. (1998). Guide to Stability Design Criteria for Metal Structures. In *Structural Stability Research Council* (5th ed.). New York: John Wiley and Sons.
Gerard, G., & Becker, H. (1957). *Handbook of Structural Stability, Part I, Buckling of Flat Plates*. Washington, DC: NACA, Tech. Note No. 3781.
Inquiry (1974). Inquiry into the Basis of Design and Method of Erection of Steel Box Girder Bridges, Appendix I, Interim Design and Workmanship Rules, Part III: Basis for the Design Rules and for the Design of Special Structures not within the Scope of Part II. Report of the Committee. London: Her Majesty's Stationery Office.
Langhaar, H. L. (1962). *Energy Methods in Applied Mechanics*. New York: John Wiley and Sons.
Lee, S. C., Davidson, J. S., & Yoo, C. H. (1996). Shear Buckling Coefficients of Plate Girder Web Panels. *Computers and Structures, Vol. 59*(No. 5), 789–795.
Lee, S. C., & Yoo, C. H. (1998). Strength of Plate Girder Web Panels Under Pure Shear. *Journal of Structural Engineering, ASCE, Vol. 124*(No. 2), 184–194.
Lee, S. C., & Yoo, C. H. (1999). Strength of Curved I-Girder Web Panels Under Pure Shear. *Journal of Structural Engineering, ASCE, Vol. 125*(No. 8), 847–853.
Lee, S. C., & Yoo, C. H. (1999). Experimental Study on Ultimate Shear Strength of Web Panels. *Journal of Structural Engineering, ASCE, Vol. 125*(No. 8), 838–846.

Lee, S. C., Yoo, C. H., & Yoon, D. Y. (2002). Behavior of Intermediate Transverse Stiffeners Attached on Web Panels. *Journal of Structural Engineering, ASCE, Vol. 128*(No. 3), 337–345.

Lee, S. C., Yoo, C. H., & Yoon, D. Y. (2003). New Design Rule for Intermediate Transverse Stiffeners Attached on Web Panels. *Journal of Structural Engineering, ASCE, Vol. 129*(No. 12), 1607–1614.

Lee, S. C., Lee, D. S., & Yoo, C. H. (2008). Ultimate Shear Strength of Long Web Panels, *Journal of Constructional Steel Research*. Elsevier. *No. 12* (pp. 1357–1365).

Lee, S. C., Lee, D. S., & Yoo, C. H. (2009a). Further Insight into Postbuckling of Web Panels, I: Review of Flange Anchoring Mechanism. *Journal of Structural Engineering, ASCE, Vol. 135*(No.1), 3–10.

Lee, S. C., Lee, D. S., Park, C. S., & Yoo, C. H. (2009b). Further Insight into Postbuckling of Web Panels, II: Experiments and Verification of New Theory. *Journal of Structural Engineering, ASCE, Vol. 135*(No. 1), 11–18.

Levy, S. (1942). Buckling of Rectangular Plates with Built-in Edges. *Journal of Applied Mechanics, ASME, Vol. 9*, A-171–174.

Love, A. E. H. (1944). In *A Treatise on the Mathematical Theory of Elasticity* (4th ed.). New York: Dover Publications.

Mattock, A. H. (1967). *Criteria for Design of Steel Concrete Composite Box Girder Highway Bridges, Appendix B.* Report of the Committee (unpublished).

Mittelstedt, C. (2008). Explicit Analysis and Design Equations for Buckling Loads and Minimum Stiffener Requirements of Orthotropic and Isotropic Plates Under Compressive Load Braced by Longitudinal Stiffeners. *Thin-Walled Structures, 46*(No. 12), 1409–1429.

Novozhilov, V. V. (1953). *Foundations of the Nonlinear Theory of Elasticity*. In F. Bagemihl, H. Komm, & W. Seidel (Eds.). Translated from the 1st (1948) Russian edition. Rochester, NY: Greylock Press.

Porter, D. M., Rockey, K. C., & Evans, H. R. (1975). The Collapse Behavior of Plate Girders Loaded in Shear. *Structural Engineering, Vol. 53*(No. 8), 313–325.

Seide, P., & Stein, M. (1949). In *Compressive Buckling of Simply Supported Plates with Longitudinal Stiffeners*. Washington, DC: NACA Technical Note, No. 1825.

Sharp, M. L. (1966). Longitudinal Stiffeners for Compression Members. *Journal of Structural Div., ASCE, Vol. 92*(No. ST5), 187–212.

Sokolnikoff, I. S. (1956). *Mathematical Theory of Elasticity*. New York: McGraw-Hill.

Southwell, R. V., & Skan, S. W. (1924). On the Stability Under Shearing Forces of a Flat Elastic Strip. *Proc. Roy. Soc. London, Ser. A, Vol. 105*, 582–607.

Stein, M., & Neff, J. (1947). *Buckling Stresses of Simply Supported Rectangular Plates in Shear.* Washington, DC: NACA Technical Note, No. 1222.

Szilard, R. (1974). *Theory and Analysis of Plate: Classical and Numerical Methods*. Englewood Cliffs, NJ: Prentice-Hall.

Timoshenko, S. P. (1983). *History of Strength of Materials*. New York: Dover Publications.

Timoshenko, S. P., & Gere, J. M. (1961). *Theory of Elastic Stability* (2nd ed.). New York: McGraw-Hill.

Timoshenko, S. P., & Woinowsky-Krieger, S. (1959). In *Theory of Plates and Shells* (2nd ed.). New York: McGraw-Hill.

von Kármán, T. (1910). "Festigkeit im Machinenbau," Enzyklopädie der mathematischn Wissenschaften, Vol. 4, pp. 349.

Wagner, H. (1929). "Ebene Blechweandträger mit sehr dünnem Stegblech," Z. Flugtech. Motorluftschiffahrt, Vol. 20, pp. 200, 227, 256, 279, and 306.

Wolchuk, R., & Mayrbaurl, R. M. (1980). Proposed Design Specifications for Steel Box Girder Bridges. *Report No. FHWA-TS-80-205.* Washington, DC: Federal Highway Administration.

Yoo, C. H. (2001). Design of Longitudinal Stiffeners on Box Girder Flanges. *International Journal of Steel Structures, Vol. 1*(No. 1), 12–23, June.
Yoo, C. H., Choi, B. H., & Ford, E. M. (2001). Stiffness Requirements for Longitudinally Stiffened Box Girder Flanges. *Journal of Structural Engineering, ASCE, Vol. 127*(No. 6), 705–711, June.
Yoo, C. H., & Lee, S. C. (2006). Mechanics of Web Panel Postbuckling Behavior in Shear. *Journal of Structural Engineering, ASCE, Vol. 132*(No.10), 1580–1589.

PROBLEMS

8.1 For a thin flat plate that is subjected to a uniform compressive force P_x in the longitudinal direction, the governing differential equation may be written as per Eq. (8.3.3c) as $D\nabla^4 w + (P_x/b)w_{,xx} = 0$. If the loaded edges $x = 0$, a are simply supported, solutions of the form $w = Y(y)\sin(m\pi x/a)$ satisfy the differential equation. The transformed ordinary differential equation formed is

$$Y^{iv} - 2\left(\frac{m\pi}{a}\right)^2 Y'' + \left[\left(\frac{m\pi}{a}\right)^4 - \frac{P_x}{Db}\left(\frac{m\pi}{a}\right)^2\right] Y = 0$$

As this is an ordinary homogeneous differential equation, a solution of the form $Y = e^{\lambda y}$ will satisfy the equation. The characteristic equation is

$$\lambda^4 - 2\left(\frac{m\pi}{a}\right)^2 \lambda^2 + \left[\left(\frac{m\pi}{a}\right)^4 - \frac{P_x}{Db}\left(\frac{m\pi}{a}\right)^2\right] = 0$$

and the roots of this polynomial are

$$\lambda = \pm\left[\frac{m\pi}{a}\left(\frac{m\pi}{a} \pm \sqrt{\frac{P_x}{Db}}\right)\right]^{1/2}$$

Let the four roots λ be α, $-\alpha$, $i\beta$, and $-i\beta$. Then

$$\alpha = \left[\left(\frac{m\pi}{a}\right)^2 + \frac{m\pi}{a}\sqrt{\frac{P_x}{Db}}\right]^{1/2} \quad \beta = \left[-\left(\frac{m\pi}{a}\right)^2 + \frac{m\pi}{a}\sqrt{\frac{P_x}{Db}}\right]^{1/2}$$

Show that the characteristic equation for the critical load is

$$2\alpha\beta + \left(\alpha^2 - \beta^2\right)\sinh \alpha b \sin \beta b - 2\alpha\beta(\cosh \alpha b \cos \beta b) = 0$$

if the plate is clamped at the unloaded edges $y = 0$, b.

8.2 For a plate subjected to a compressive force P_x, Eq. (8.4.13) gives for the average stress $\sigma_x = k'\pi^2 E/12(1-\mu^2)(b/h)^2$.
For an infinitely long plate that is simply supported on one unloaded edge and free on the other, $k' = 0.425$ for $\mu = 0.3$. Then for $E = 29 \times 10^3$ ksi,

$$\sigma_{cr} = 11,000\left(\frac{h}{b}\right)^2$$

Using this information, determine the critical stress for local buckling of one leg of the angle $L6 \times 6 \times 5/16$ if the other leg is assumed to furnish only simple support to the leg. Review your answer with the current AISC local buckling provision, Q_s. Assume the torsional-flexural instability does not control. Neglect the fillet effect.

8.3 For a thin rectangular flat plate that is subjected to a uniform compressive force P_x in the longitudinal direction, the governing differential equation may be written as per Eq. (8.3.3c) as $D\nabla^4 w + (P_x/b)w_{,xx} = 0$. If all four edges are simply supported, a solution of the form $w = C_1 \sin(m\pi x/a)\sin(n\pi y/b)$ $m, n = 1, 2, 3, \ldots$ is seen to be the exact solution. Prove it.

8.4 A square plate of dimension "a" is simply supported on all four boundaries. The plate is subjected to a uniformly distributed compressive load on four sides as shown in Fig. P8-4. Using the differential equation method discussed, determine the critical load.

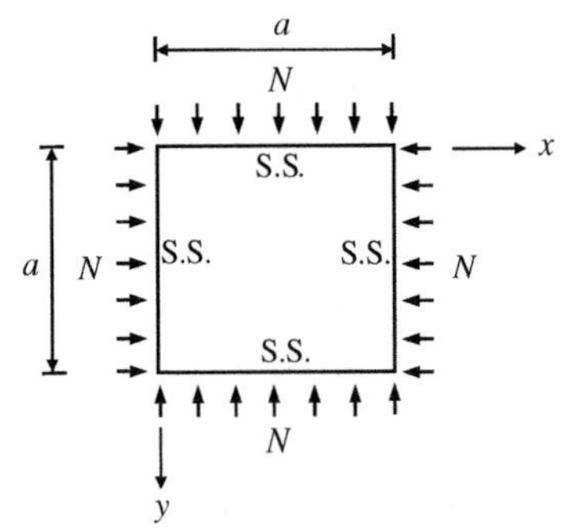

Figure P8-4 Square plate subjected to load on four sides

8.5 Consider a square plate of dimension "a" subjected to N_x. The boundary condition perpendicular to N_x is changed to pinned (immovable). Due to the effect of Poisson's ratio, $\mu = 0.3$, forces are induced in the y direction equal to $N_y = \mu N_x$. Determine the critical load, $N_{x\,cr}$.

8.6 A square plate of dimension "a" is simply supported on all four boundaries. The plate is subjected to a linearly varying compressive load, N_x, as shown in Fig. P8-6. Using the energy method discussed, determine the critical load.

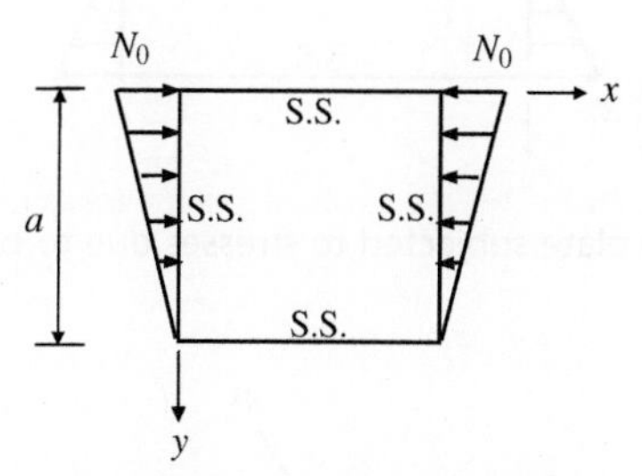

Figure P8-6 Linearly varying load

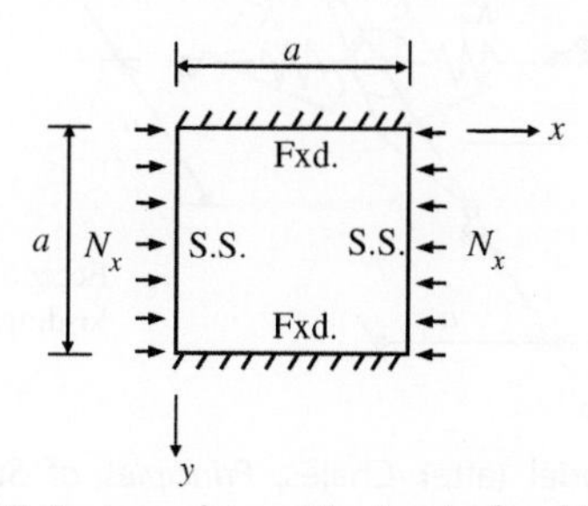

Figure P8-7 Square plate with simple-fixed boundaries

8.7 A square plate of dimension "a" is simply supported on edges parallel to the uniformly distributed load, N_x, and fixed on edges perpendicular to the load. Using the energy method discussed, determine the critical load.

8.8 A square plate of dimension "a" is simply supported on all four boundaries. The plate is subjected to a linearly varying compressive load, N_x, as shown in Fig. P8-8. Using the energy method discussed, determine the critical load.

8.9 Using the energy method, determine the critical load for the one-degree-of-freedom model of a flat plate shown in Fig. P8-9 (Model analysis I). The model consists of four rigid bars pin-connected to each other and to the supports. At the center of the model, two linear rotational springs of stiffness $C = M/\theta$ connect opposite bars to each

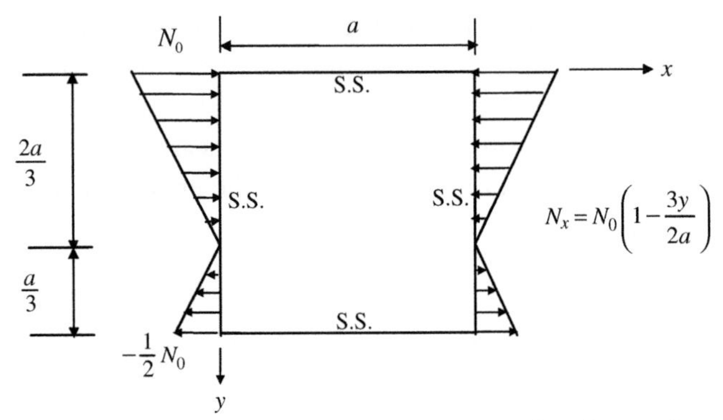

Figure P8-8 Square plate subjected to stresses due to bending and axial force

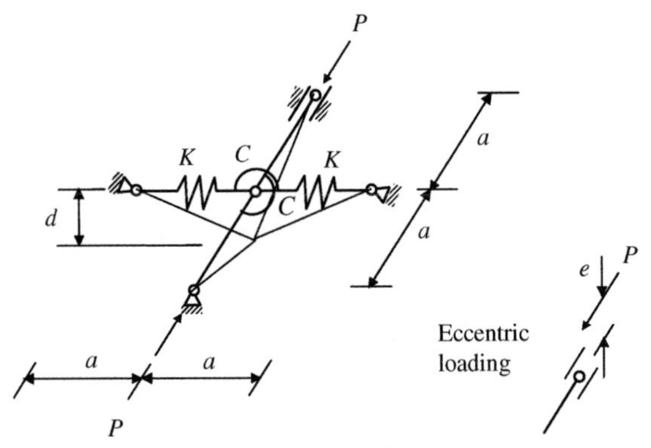

Figure P8-9 Plate model (after Chajes, *Principles of Structural Stability Theory.* Englewood Cliffs, NJ: Prentice-Hall, 1974). Reproduced by permission from the author.

other. Also, each of the two transverse bars contains a linear extensional springs of stiffness K. For small lateral deflections, the energy in the extensional springs can be neglected.

Using the same model, obtain and plot relationships for the load P versus the lateral deflection d when Model analysis II):

(a) The lateral deflection is large.

(b) The lateral deflection is large, and the loads are applied eccentrically to the plane of the undeformed model.

Which fundamental buckling characteristics of an actual plate are demonstrated by these models? (*Note:* For large deflections the energy in the extensional springs must be considered.)

CHAPTER 9

Buckling of Thin Cylindrical Shell Elements

Contents

9.1. INTRODUCTION

Shell buckling has become one of the important areas of interest in structural mechanics in recent times. The difference between a plate element and a shell element is that the plate element has curvatures in the unloaded state, whereas the shell element is assumed to be initially flat. Although the presence of initial curvature is of little consequence for bending, it does affect the membrane action of the element significantly.

Membrane action is caused by in-plane forces. These forces may be the primary forces caused by applied edge loads or edge deformations, or they may be secondary forces resulting from flexural deformations. In a stability analysis, primary in-plane forces must be considered whether or not initial curvature exists. However, the same is not necessarily the case regarding secondary in-plane forces. If the element is initially flat, secondary in-plane forces do not affect membrane action significantly unless the bending deformations are large. It is for this reason that membrane action due to secondary forces is ignored in the small-deflection plate theory, but not in

Stability of Structures
ISBN 978-0-12-385122-2, doi:10.1016/B978-0-12-385122-2.10009-0

the large-deflection plate theory. If the element has initial curvature, on the other hand, membrane action caused by secondary in-plane forces will be significant regardless of the magnitude of the bending deformations. Membrane action resulting from secondary forces therefore must be accounted for in both small- and large-deflection shell theories (Chajes 1974).

In addition to this complication is the fact that in many shell problems the initially buckled form is in a condition of unstable equilibrium and a new position of equilibrium can exist at a much lower buckling load. Thus, the theoretical initial buckling load calculated by the classical theories of stability is rarely attained in experiments. Discussion of shell behavior in the postbuckling range, which is governed to a great extent by the nature of the initial imperfection, is therefore a necessary element of any buckling analysis (Allen and Bulson 1980).

Examination of these problems has resulted in thousands of papers and reports over the years as well as a number of books. It would be impractical to condense the whole of this work into a chapter or two, and the aim here will be to introduce the student to the fundamentals and at the same time indicate selected simple formulas of interest to the practicing engineers. To do this, no attempt has been made at a general analysis, but each practical problem is examined separately.

Development of many governing equations has followed the procedure given by Brush and Almroth (1975) and Chajes (1974).

9.2. LARGE-DEFLECTION EQUATIONS (DONNELL TYPE)

As the reliability and efficiency of the incremental finite element analysis have been well established, much of the work in shell analysis is being carried out on digital computers these days. In such environments, the simplicity of the governing equations is of little importance other than initial programming efforts. As a result, interest the Donnell equations has diminished somewhat. However, the relative simplicity of the equations makes them well suited for this introductory examination of shell buckling.

Consider a differential shell element of thickness h with a radius of curvature R as shown in Fig. 9-1(a). The coordinate system is a pointwise orthogonal rectangular coordinate system with the origin in the middle surface of the shell so that the x-axis is parallel to the axis of the cylinder, the y-axis is tangent to the circular arc, and the z-axis is normal to the middle surface directed toward the center of curvature.

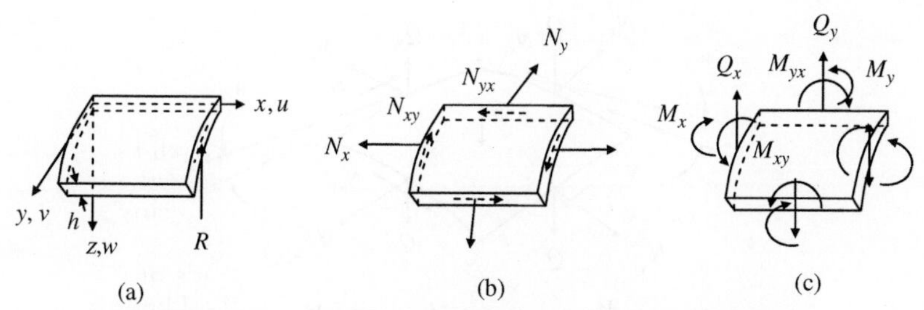

Figure 9-1 Cylindrical shell displacements and forces

As in plate theory, it is convenient in shell theory to express internal forces (generalized) per unit distance along the edge of the shell element as shown in Fig. 9-1(b, c). They are related to the internal stresses by

$$
\begin{aligned}
N_x &= \int_{-h/2}^{h/2} \overline{\sigma}_x \left(1 + \frac{z}{R}\right) dz \qquad N_y = \int_{-h/2}^{h/2} \overline{\sigma}_y dz \\
N_{xy} &= \int_{-h/2}^{h/2} \overline{\tau}_{xy} \left(1 + \frac{z}{R}\right) dz \qquad N_{yx} = \int_{-h/2}^{h/2} \overline{\tau}_{yx} dz
\end{aligned}
\tag{9.2.1a}
$$

$$
Q_x = \int_{-h/2}^{h/2} \overline{\tau}_{xz} \left(1 + \frac{z}{R}\right) dz \qquad Q_y = \int_{-h/2}^{h/2} \overline{\tau}_{zx} dz \tag{9.2.1b}
$$

$$
\begin{aligned}
M_x &= \int_{-h/2}^{h/2} z\overline{\sigma}_x \left(1 + \frac{z}{R}\right) dz \qquad M_y = \int_{-h/2}^{h/2} z\overline{\sigma}_y dz \\
M_{xy} &= \int_{-h/2}^{h/2} z\overline{\tau}_{xy} \left(1 + \frac{z}{R}\right) dz \qquad M_{yx} = \int_{-h/2}^{h/2} z\overline{\tau}_{yx} dz
\end{aligned}
\tag{9.2.1c}
$$

where N_x, N_y, N_{xy}, N_{yx} are in-plane normal and shearing forces; Q_x, Q_y are transverse shearing forces; M_x, M_y are bending moments; and M_{xy}, M_{yx} are twisting moments. As in Chapter 8, the quantities with bar $\overline{\sigma}_x$, $\overline{\tau}_{xy}$ are stresses at any point through the wall thickness, as distinguished from σ_x, τ_{xy}, which refer to corresponding stresses on the middle surface ($z = 0$) only.

The nonlinear equilibrium equations may be obtained by summing the generalized internal forces for a cylindrical shell element in a slightly deformed configuration as shown in Fig. 9-2. The positive directions of moments and in-plane forces are the same as defined in Chapter 8, and their directions are taken to produce positive stresses at the positive ends of the element. The double arrow for moments follows the right-hand screw rule. The internal forces (generalized) and rotations vary across the element, and

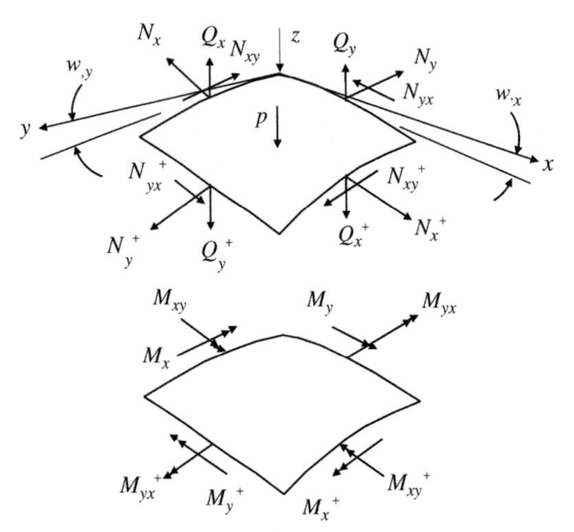

Figure 9-2 Internal forces in cylindrical shell

the notation N_x^+ is used for $N_x + N_{x,x}\, dx$. For the intermediate class of deformations considered herein, the angles of rotation $w_{,x}$ and $w_{,y}$ are assumed to be small so that sines and cosines of the angles can be replaced by the angles themselves and by unity, respectively (micro geometry holds). Furthermore, quadratic terms are assumed to be small.

It is necessary to consider the initial curvature to derive the equation of equilibrium in the z direction. Due to the initial curvature of the shell element, the N_y forces as shown in Fig. 9-3 have a component in the z direction.

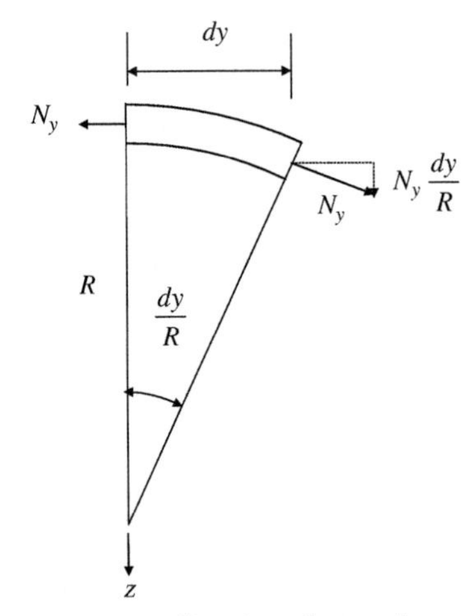

Figure 9-3 Z components of in-plane forces due to initial curvature

None of the other in-plane forces has components in the z direction due to the initial curvature. All in-plane forces, however, have z components due to the curvature produced by bending. These components are identical to the ones that were considered in a plate element (Eq. (8.2.14).

Summation of forces in the x and y directions, respectively yields the following equations:

$$N_{x,x} + N_{yx,y} = 0 \tag{9.2.2}$$

$$N_{xy,x} + N_{y,y} = 0 \tag{9.2.3}$$

Adding the component of force shown in Fig. 9-3 to Eq. (8.2.14) yields

$$\frac{\partial Q_x}{\partial x} + \frac{\partial Q_y}{\partial y} + N_x \frac{\partial^2 w}{\partial x^2} + N_y \left(\frac{\partial^2 w}{\partial y^2} + \frac{1}{R} \right) + 2N_{xy} \frac{\partial^2 w}{\partial x \partial y} + p = 0 \tag{9.2.4}$$

It is noted that another simplifying assumption has been introduced in Eq. (9.2.2c) in that z/R is neglected relative to unity in Eqs. (9.2.1). Then, it follows immediately that $N_{xy} = N_{yx}$ and $M_{xy} = M_{yx}$ as $\overline{\tau}_{xy} = \overline{\tau}_{yx}$.

Since the equations of moment equilibrium about the x- and y-axes are not altered in going from the plate to the shell element, Eqs. (8.2.15) and (8.2.16) are also valid for the shell element. Replacing the first two shear terms in Eq. (9.2.2c) by moments given by Eqs. (8.2.15) and (8.2.16) yields

$$M_{x,xx} + 2\,M_{xy,xy} + M_{y,yy} + N_x w_{,xx} + 2\,N_{xy} w_{,xy} + N_y \left(\frac{1}{R} + w_{,yy} \right) + p = 0 \tag{9.2.5}$$

The constitutive equations for thin-walled isotropic elastic cylinders are the same as those for flat-plate elements in Eqs. (8.2.7), which are

$$\begin{aligned} N_x &= C(\varepsilon_x + \mu \varepsilon_y) \quad N_y = C(\varepsilon_y + \mu \varepsilon_x) \\ N_{xy} &= C(1-\mu)\gamma_{xy}/2 \quad M_x = -D\left(\frac{\partial^2 w}{\partial x^2} + \mu \frac{\partial^2 w}{\partial y^2} \right) \\ M_y &= -D\left(\frac{\partial^2 w}{\partial y^2} + \mu \frac{\partial^2 w}{\partial x^2} \right) \quad M_{xy} = -D(1-\mu)\frac{\partial^2 w}{\partial x \partial y} \end{aligned} \tag{9.2.6}$$

where the coefficients C and D are the same as defined in Eqs. (8.2.8), which are

$$C = \frac{Eh}{1-\mu^2} \quad \text{and} \quad D = \frac{Eh^3}{12(1-\mu^2)} \tag{9.2.7}$$

The kinematic relations at the middle surface on which the Donnell equations are based are identical to those given in Eqs. (8.2.5) with one exception for ε_y (Donnell 1933), which are

$$\begin{aligned} \varepsilon_x &= \frac{\partial u}{\partial x} + \frac{1}{2}\left(\frac{\partial w}{\partial x}\right)^2 & \kappa_x &= -\frac{\partial^2 w}{\partial x^2} \\ \varepsilon_y &= \frac{\partial v}{\partial y} - \frac{w}{R} + \frac{1}{2}\left(\frac{\partial w}{\partial y}\right)^2 & \kappa_y &= -\frac{\partial^2 w}{\partial y^2} \\ \gamma_{xy} &= \frac{\partial u}{\partial y} + \frac{\partial v}{\partial x} + \frac{\partial w}{\partial x}\frac{\partial w}{\partial y} & \kappa_{xy} &= -\frac{\partial^2 w}{\partial x \partial y} \end{aligned} \tag{9.2.8}$$

Substituting the above constitutive and kinematic relations into Eq. (9.2.3) yields a coupled set of three nonlinear differential equations in the three variables u, v, and w.

$$N_{x,x} + N_{xy,y} = 0 \tag{9.2.9a}$$

$$N_{xy,x} + N_{y,y} = 0 \tag{9.2.9b}$$

$$D(w_{,xxxx} + 2\,w_{,xxyy} + w_{,yyyy}) - [N_x w_{,xx} + 2\,N_{xy} w_{,xy} + N_y(1/R + w_{,yy})] = p \tag{9.2.9c}$$

Equations (9.2.9) are nonlinear equilibrium equations for thin cylindrical shells. They have been widely used in the large-deflection analyses of cylindrical shells (von Kármán and Tsien 1941).

9.3. ENERGY METHOD

It would be informative to rederive the nonlinear equilibrium equations in Eqs. (9.2.9) based on the principle of minimum potential energy. The total potential energy Π is the sum of the strain energy U of the cylindrical shell and the loss of the potential energy of the applied load V.

$$\Pi = U + V \tag{9.3.1}$$

The strain energy of a deformed shell can be expressed in two parts: (1) the strain energy due to bending and (2) the strain energy due to the membrane action. Using the general expression of the strain energy of a plate (Eq. (8.5.6)) and using a separate expression for the bending stress and the membrane stress, one obtains

$$U_b = \frac{D}{2}\iint \left[{\kappa_x}^2 + {\kappa_y}^2 + 2\mu\kappa_x\kappa_y + 2(1-\mu){\kappa_{xy}}^2\right] dA \qquad (9.3.2)$$

$$U_m = \frac{C}{2}\iint \left[{\varepsilon_x}^2 + {\varepsilon_y}^2 + 2\mu\varepsilon_x\varepsilon_y + \frac{(1-\mu)}{2}{\gamma_{xy}}^2\right] dA \qquad (9.3.3)$$

where ε_x, ε_y, ..., κ_{xy} are given Eqs. (9.2.6). Hence, the total strain energy is

$$U = U_b + U_m \qquad (9.3.4)$$

For a cylindrical shell subjected to lateral pressure p, the potential energy of the applied pressure is

$$V = -\iint_A pwdxdy \qquad (9.3.5)$$

The Euler-Lagrange differential equations for an integrand of Eq. (9.3.1) are

$$\frac{\partial F}{\partial u} - \frac{\partial}{\partial x}\frac{\partial F}{\partial u_{,x}} - \frac{\partial}{\partial y}\frac{\partial F}{\partial u_{,y}} = 0 \qquad (9.3.6a)$$

$$\frac{\partial F}{\partial v} - \frac{\partial}{\partial x}\frac{\partial F}{\partial v_{,x}} - \frac{\partial}{\partial y}\frac{\partial F}{\partial v_{,y}} = 0 \qquad (9.3.6b)$$

$$\frac{\partial F}{\partial w} - \frac{\partial}{\partial x}\frac{\partial F}{\partial w_{,x}} - \frac{\partial}{\partial y}\frac{\partial F}{\partial w_{,y}} + \frac{\partial^2}{\partial x^2}\frac{\partial F}{\partial w_{,xx}} + \frac{\partial^2}{\partial x \partial y}\frac{\partial F}{\partial w_{,xy}} + \frac{\partial^2}{\partial y^2}\frac{\partial F}{\partial w_{,yy}} = 0 \qquad (9.3.6c)$$

It can be shown that the execution of the Euler-Lagrange differential equations, Eqs. (9.3.6), will lead to the nonlinear equilibrium equations, Eqs. (9.2.9) (see Problem 9.2).

Equations (9.2.9) are nonlinear equilibrium equations for thin cylindrical shells. They are the counterpart for shells of Eqs. (8.2.18). There are four unknowns N_x, N_y, N_{xy}, and w. Three equations in three unknowns u, v, w may be obtained by introducing the constitutive and kinematic relations of Eqs. (9.2.4) and (9.2.6). As was done in Chapter 8, a simpler set of two equations in two variables can be obtained by use of a stress function identical to Eqs. (8.2.27).

$$N_x = f_{,yy},\ N_y = f_{,xx},\ N_{xy} = -f_{,xy} \qquad (9.3.7)$$

Rearranging Eqs. (9.2.4), one obtains the following relations:

$$\varepsilon_x = \frac{1}{Eh}(N_x - \mu N_y) \qquad (9.3.8a)$$

$$\varepsilon_y = \frac{1}{Eh}(N_y - \mu N_x) \tag{9.3.8b}$$

$$\gamma_{xy} = \frac{2}{Eh}(1+\mu)N_{xy} \tag{9.3.8c}$$

Differentiating the in-plane strains of Eqs. (9.2.6), ε_x twice with respect to y, ε_y twice with respect to x, and γ_{xy} with respect to x and y, respectively, one obtains the compatibility relation

$$\frac{\partial^2 \varepsilon_x}{\partial y^2} + \frac{\partial^2 \varepsilon_y}{\partial x^2} - \frac{\partial^2 \gamma_{xy}}{\partial x \partial y} = \left(\frac{\partial^2 w}{\partial x \partial y}\right)^2 - \frac{\partial^2 w}{\partial x^2}\frac{\partial^2 w}{\partial y^2} - \frac{1}{R}\frac{\partial^2 w}{\partial x^2} \tag{9.3.9}$$

Substituting Eqs. (9.3.7) into Eqs. (9.3.8) yields

$$\varepsilon_x = \frac{1}{Eh}\left(\frac{\partial^2 f}{\partial y^2} - \mu\frac{\partial^2 f}{\partial x^2}\right) \tag{9.3.10a}$$

$$\varepsilon_y = \frac{1}{Eh}\left(\frac{\partial^2 f}{\partial x^2} - \mu\frac{\partial^2 f}{\partial y^2}\right) \tag{9.3.10b}$$

$$\gamma_{xy} = -\frac{2(1+\mu)}{Eh}\frac{\partial^2 f}{\partial x \partial y} \tag{9.3.10c}$$

Making use of Eqs. (9.3.7), (9.3.8), and (9.3.10), Eqs. (9.2.9c) and (9.3.9) become

$$D\nabla^4 w - \left[f_{,yy}w_{,xx} - 2f_{,xy}w_{,xy} + f_{,xx}(1/R + w_{,yy})\right] = p \tag{9.3.11a}$$

$$\nabla^4 f = Eh[(w_{,xy})^2 - w_{,xx}w_{,yy} - 1/Rw_{,xx}] \tag{9.3.11b}$$

Equations (9.3.11) were first presented by Donnell (1934) when he combined the strain-displacement relations in the von Kármán large-deflection plate theory with his own linear shell theory. The equations are therefore called the von Kármán-Donnell large-displacement equations.

The linear equilibrium equations corresponding to Eqs. (9.2.9) are obtained by dropping all quadratic and higher order terms in u, v, w from the nonlinear equations. The resulting equations are

$$N_{x,x} + N_{xy,y} = 0 \tag{9.3.12a}$$

$$N_{xy,x} + N_{y,y} = 0 \tag{9.3.12b}$$

$$D\nabla^4 w - N_y/R = p \tag{9.3.12c}$$

where

$$\begin{aligned} N_x &= C(\varepsilon_x + \mu\varepsilon_y), & \varepsilon_x &= u_{,x} \\ N_y &= C(\varepsilon_y + \mu\varepsilon_x), & \varepsilon_y &= v_{,y} - w/R \\ N_{xy} &= C\gamma_{xy}(1-\mu)/2, & \gamma_{xy} &= u_{,y} + v_{,x} \end{aligned} \tag{9.3.13}$$

It is noted that Eq. (9.3.12c) is still coupled to Eq. (9.3.12b) whereas in the case of the plate, the third of Eqs. (8.2.32) is uncoupled from the other equations.

The linear equilibrium equations given by Eqs. (9.3.12) are a coupled set of three equations in four unknowns N_x, N_y, N_{xy}, and w. A set of three equations in three unknowns u, v, and w can be obtained by substituting appropriate constitutive and kinematic relations into Eqs. (9.3.12). The resulting equations are

$$u_{,xx} - \frac{\mu w_{,x}}{R} + \frac{1-\mu}{2}u_{,yy} + \frac{1+\mu}{2}v_{,xy} = 0 \tag{9.3.14a}$$

$$\frac{1-\mu}{2}v_{,xx} + \frac{1+\mu}{2}u_{,xy} + v_{,yy} - \frac{w_{,y}}{R} = 0 \tag{9.3.14b}$$

$$D\nabla^4 w - \frac{C}{R}\left(v_{,y} - \frac{w}{R} + \mu u_{,x}\right) = p \tag{9.3.14c}$$

These equations may be partially uncoupled (Donnell 1933) to give (see Problem 9.3)

$$\nabla^4 u = \frac{w_{,xyy}}{R} - \frac{\mu w_{,xxx}}{R} \tag{9.3.15a}$$

$$\nabla^4 v = \frac{(2+\mu)w_{,xxy}}{R} + \frac{w_{,yyy}}{R} \tag{9.3.15b}$$

$$D\nabla^8 w + \frac{1-\mu^2}{R^2}Cw_{,xxxx} = \nabla^4 p \tag{9.3.15c}$$

It is of interest to note that the linear membrane equations are obtained by setting the bending rigidity of the shell element equal to zero ($D = 0$) in Eqs. (9.3.12). The resulting equations are

$$N_{x,x} + N_{xy,y} = 0 \tag{9.3.16a}$$

$$N_{xy,x} + N_{y,y} = 0 \tag{9.3.16b}$$

$$-N_y/R = p \tag{9.3.16c}$$

These equations are statically determinate; that is, there are three variables in three equations. Equation (9.3.16c) gives the well-known hoop compression ($-N_y$) due to the external pressure p.

9.4. LINEAR STABILITY EQUATIONS (DONNELL TYPE)

Equations (9.2.9) govern all linear and nonlinear equilibrium conditions of the cylindrical shell within the confinement of the intermediate class of deformations. The equations include linear, quadratic, and cubic terms of variables u, v, and w, and therefore are nonlinear. It is now a fairly simple task to obtain a very good iterative numerical solution by a well-established finite element code. A load-displacement curve based on such solutions for a cylinder subject to the edge load is shown in Fig. 9-4. The linear equilibrium equations, Eqs. (9.3.12), govern the primary (static) path. The nonlinear equations, Eqs. (9.2.9), govern both the primary path and the secondary path.

The equilibrium paths determined by solution of the equilibrium equations, Eqs. (9.2.9), show the bifurcation point and the corresponding critical load. Hence, a separate solution for the critical load is not necessary. However, the solution of Eqs. (9.2.9) demands a fairly complicated numerical procedure. The purpose of stability analysis to be presented herein is to permit determination of the critical load by solution of linear differential equations.

The linear differential equations for determination of the critical load of a cylinder subjected to external loading are derived by application of the adjacent-equilibrium criterion. The same equations are rederived based on the minimum potential energy criterion as was done in the previous section.

9.4.1. Adjacent-Equilibrium Criterion

Adjacent (or neighboring) equilibrium configurations are examined using the procedure outlined by Brush and Almroth (1975) as was done in Chapter 8. Consider the equilibrium configuration at the bifurcation point. Then, the equilibrium configuration is perturbed by the small incremental

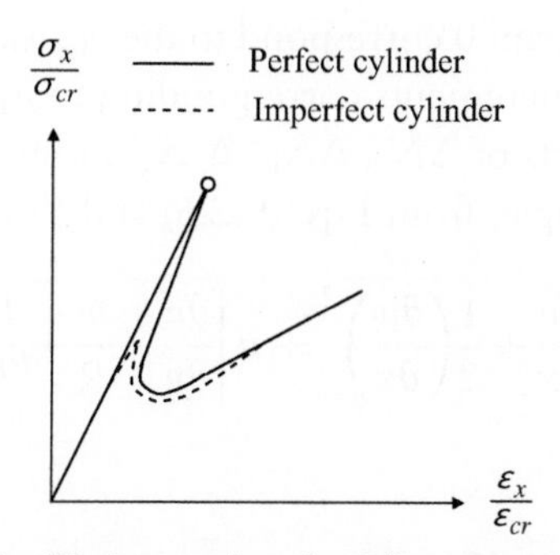

Figure 9-4 Equilibrium paths of axially compressed cylinder

displacement. The incremental displacement in u_1, v_1, w_1 is arbitrary and tentative. Variables in the two adjacent configurations before and after the increment are represented by u_0, v_0, w_0 and u, v, w. Let

$$u \rightarrow u_0 + u_1$$
$$v \rightarrow v_0 + v_1 \tag{9.4.1}$$
$$w \rightarrow w_0 + w_1$$

where the arrow is read "be replaced by." Substitution of Eqs. (9.4.1) into Eqs. (9.2.9) results in equations containing terms that are linear, quadratic, and cubic in u_0, v_0, w_0 and u_1, v_1, w_1 displacement components. In the new equation obtained, the terms containing u_0, v_0, w_0 alone are equal to zero as u_0, v_0, w_0 constitute an equilibrium configuration, and terms that are quadratic and cubic in u_1, v_1, w_1 may be ignored because of the smallness of the incremental displacement. Therefore, the resulting equations are homogeneous and linear in u_1, v_1, w_1 with variable coefficients. The coefficients in u_0, v_0, w_0 are governed by the original nonlinear equations. It will simplify the procedure greatly by simply limiting the range of applicability of the linearized equations by requiring that u_0, v_0, w_0 be limited to configurations that are governed by the linear equations, Eqs. (9.3.12).

The increment in u, v, w causes a concomitant change in the internal force such as

$$N_x \rightarrow N_{x0} + \Delta N_x$$
$$N_y \rightarrow N_{y0} + \Delta N_y \tag{9.4.2}$$
$$N_{xy} \rightarrow N_{xy0} + \Delta N_{xy}$$

where terms with subscript 0 correspond to the u_0, v_0, w_0 displacement, and ΔN_x, ΔN_y, ΔN_{xy} are increments corresponding to u_1, v_1, w_1. Let also N_{x1}, N_{y1}, N_{xy1} represent parts of ΔN_x, ΔN_y, ΔN_{xy}, respectively, that are linear in u_1, v_1, w_1. For example, from Eqs. (9.2.6) and (9.2.8),

$$N_x = C\left\{\frac{\partial u}{\partial x}+\frac{1}{2}\left(\frac{\partial w}{\partial x}\right)^2+\mu\left[\frac{\partial v}{\partial y}-\frac{w}{R}+\frac{1}{2}\left(\frac{\partial w}{\partial y}\right)^2\right]\right\}$$

Then

$$N_{x0}+\Delta N_x = C\left[\begin{array}{l}\dfrac{\partial u_0}{\partial x}+\dfrac{\partial u_1}{\partial x}+\dfrac{1}{2}\dfrac{\partial w_0}{\partial x}^2+\dfrac{\partial w_0}{\partial x}\dfrac{\partial w_1}{\partial x}+\dfrac{1}{2}\dfrac{\partial w_1}{\partial x}^2\\ +\mu\left(\dfrac{\partial v_0}{\partial y}-\dfrac{w_0}{R}+\dfrac{\partial v_1}{\partial y}-\dfrac{w_1}{R}+\dfrac{1}{2}\dfrac{\partial w_0}{\partial y}^2+\dfrac{\partial w_0}{\partial y}\dfrac{\partial w_1}{\partial y}+\dfrac{1}{2}\dfrac{\partial w_1}{\partial y}^2\right)\end{array}\right]$$

From which

$$N_{x0} = C\left[\frac{\partial u_0}{\partial x}+\frac{1}{2}\frac{\partial w_0}{\partial x}^2+\mu\left(\frac{\partial v_0}{\partial y}-\frac{w_0}{R}+\frac{1}{2}\frac{\partial w_0}{\partial y}^2\right)\right]$$

$$\Delta N_x = C\left\{\frac{\partial u_1}{\partial x}+\frac{\partial w_0}{\partial x}\frac{\partial w_1}{\partial x}+\frac{1}{2}\left(\frac{\partial w_1}{\partial x}\right)^2 +\mu\left[\frac{\partial v_1}{\partial y}-\frac{w_1}{R}+\frac{\partial w_0}{\partial y}\frac{\partial w_1}{\partial y}+\frac{1}{2}\left(\frac{\partial w_1}{\partial y}\right)^2\right]\right\}$$

$$N_{x1} = C\left[\frac{\partial u_1}{\partial x}+\frac{\partial w_0}{\partial x}\frac{\partial w_1}{\partial x}+\mu\left(\frac{\partial v_1}{\partial y}-\frac{w_1}{R}+\frac{\partial w_0}{\partial y}\frac{\partial w_1}{\partial y}\right)\right]$$

Expressions for N_{y1} and N_{xy1} are determined following the similar procedure shown above.

Substituting these into Eqs. (9.2.9) gives

$$\frac{\partial N_{x1}}{\partial x}+\frac{\partial N_{xy1}}{\partial y} = 0 \tag{9.4.3a}$$

$$\frac{\partial N_{y1}}{\partial y}+\frac{\partial N_{xy1}}{\partial x} = 0 \tag{9.4.3b}$$

$$D\nabla^4 w_1 - \frac{N_{y1}}{R} - \left(N_{x0}\frac{\partial^2 w_1}{\partial x^2} + N_{x1}\frac{\partial^2 w_0}{\partial x^2} + 2N_{xy0}\frac{\partial^2 w_1}{\partial x \partial y} + 2N_{xy1}\frac{\partial^2 w_0}{\partial x \partial y} + N_{y0}\frac{\partial^2 w_1}{\partial y^2} + N_{y1}\frac{\partial^2 w_0}{\partial y^2} \right) = 0 \tag{9.4.3c}$$

where

$$\begin{aligned} N_{x0} &= C(\varepsilon_{x0} + \mu\varepsilon_{y0}) \quad N_{x1} = C(\varepsilon_{x1} + \mu\varepsilon_{y1}) \\ N_{y0} &= C(\varepsilon_{y0} + \mu\varepsilon_{x0}) \quad N_{x1} = C(\varepsilon_{y1} + \mu\varepsilon_{x1}) \\ N_{xy0} &= C\frac{1-\mu}{2}\gamma_{xy0} \qquad N_{xy1} = C\frac{1-\mu}{2}\gamma_{xy1} \end{aligned} \tag{9.4.4}$$

and

$$\begin{aligned} \varepsilon_{x0} &= u_{0,x} + \frac{1}{2}w_{0,x}{}^2 \qquad \varepsilon_{x1} = u_{1,x} + w_{0,x}w_{1,x} \\ \varepsilon_{y0} &= v_{0,y} - \frac{w_0}{R} + \frac{1}{2}w_{0,y}{}^2 \qquad \varepsilon_{y1} = v_{1,x} - \frac{w_1}{R} + w_{0,y}w_{1,y} \\ \gamma_{xy0} &= v_{0,x} + u_{0,y} + w_{0,x}w_{1,y} \quad \gamma_{xy1} = v_{1,x} + u_{1,y} + w_{0,x}w_{1,y} + w_{0,y}w_{1,x} \end{aligned} \tag{9.4.5}$$

Equations (9.4.3) correspond to Eqs. (8.3.3).

In the stability analysis, the displacement (u_0, v_0, w_0) is referred to as the prebuckling deformation, and (u_1, v_1, w_1) is called the buckling mode. Equations (9.4.3) to (9.4.5) include $w_{0,x}$ and $w_{0,y}$ representing prebuckling rotations. The presence of these prebuckling rotations in the stability equations introduces a substantial complication. Fortunately, though, the influence of prebuckling rotations is negligibly small; hence, they are omitted in the remainder of this chapter.

The resulting equations are

$$\frac{\partial N_{x1}}{\partial x} + \frac{\partial N_{xy1}}{\partial y} = 0 \tag{9.4.6a}$$

$$\frac{\partial N_{y1}}{\partial y} + \frac{\partial N_{xy1}}{\partial x} = 0 \tag{9.4.6b}$$

$$D\nabla^4 w_1 - \frac{N_{y1}}{R} - \left(N_{x0}\frac{\partial^2 w_1}{\partial x^2} + 2N_{xy0}\frac{\partial^2 w_1}{\partial x \partial y} + N_{y0}\frac{\partial^2 w_1}{\partial y^2} \right) = 0 \tag{9.4.6c}$$

Similarly, neglecting the prebuckling rotation terms in the kinematic relations Eqs. (9.4.5) yields

$$\begin{aligned} \varepsilon_{x0} &= u_{0,x} & \varepsilon_{x1} &= u_{1,x} \\ \varepsilon_{y0} &= v_{0,y} - \frac{w_0}{R} & \varepsilon_{y1} &= v_{1,x} - \frac{w_1}{R} \\ \gamma_{xy0} &= v_{0,x} + u_{0,y} & \gamma_{xy1} &= v_{1,x} + u_{1,y} \end{aligned} \tag{9.4.7}$$

Equations (9.4.6) are the stability equations for the cylinder. As in the case of linear equilibrium equations, Eq. (9.4.6c) is uncoupled from the other two equations. Equation (9.4.6c) is a homogeneous linear equation in w_1 with variable coefficients in N_{x0}, N_{y0}, N_{xy0}, depending on the edge conditions of the cylinder, which are determined by the other two linear equations (9.4.6a) and (9.4.6b).

9.4.2. Trefftz Criterion

The stability equations of the cylindrical shell Eqs. (9.4.6) will be rederived using the Trefftz criterion. The criterion for the loss of stability is that the integrand in the expression for the second variation of the total potential energy functional satisfies the Euler-Lagrange equations, which is known as the Trefftz criterion.

An expression for the total potential energy of a circular cylindrical shell is given by Eqs. (9.3.1) to (9.3.5). In order to obtain the corresponding expression for the second variation of the total potential energy, the deformations are replaced by the sum of the deformations in the primary path and the incremental virtual deformations in the adjacent equilibrium path as

$$u \rightarrow u_0 + u_1$$

$$v \rightarrow v_0 + v_1$$

$$w \rightarrow w_0 + w_1$$

Then, one collects all terms in the resulting expression that are quadratic in the virtual deformations u_1, v_1, w_1. Since the potential energy of the applied load Eq. (9.3.5) is a linear functional, $\delta^2 V = 0$. Hence, the second variation is found to be (see Problem 9.4)

$$\begin{aligned}\frac{1}{2}\delta^2 \Pi &= \frac{C}{2}\iint\left(\varepsilon_{x1}{}^2 + \varepsilon_{y1}{}^2 + 2\mu\varepsilon_{x1}\varepsilon_{y1} + \frac{1-\mu}{2}\gamma_{xy1}{}^2\right)dxdy \\ &+ \frac{1}{2}\iint\left(N_{x0}w_{1,x}{}^2 + N_{y0}w_{1,y}{}^2 + 2N_{xy0}w_{1,x}w_{1,y}\right)dxdy \\ &+ \frac{D}{2}\iint\left[w_{1,xx}{}^2 + w_{1,yy}{}^2 + 2\mu w_{1,xx}w_{1,yy} + 2(1-\mu)w_{1,yy}{}^2\right]dxdy\end{aligned} \tag{9.4.8}$$

Applying the Euler-Lagrange differential equations Eqs. (9.3.6) to Eq. (9.4.8) yields the linear stability equations Eqs. (9.4.6) (see Problem 9.5).

9.5. APPLICATIONS OF LINEAR BUCKLING EQUATIONS

Applications of Donnell-type linear stability equations are given in this section. For notational simplicity the subscripts 1 are omitted from the incremental quantities, and quantities with subscripts 0 are treated as constants.

9.5.1. Uniform External Pressure

Consider a circular cylindrical shell that is simply supported at its ends and is subjected to external lateral pressure p_e, in pounds per square inch. The prebuckling static deformation is axisymmetric, as shown in Fig. 9-5.

The simply supported end condition, as in most classical analyses of multidimensional entities, implies that there will be no moment developed at the boundary. However, the boundary condition does allow longitudinal and radial translations within the limitation of preventing any rigid body motion. As a consequence $N_{x0} = 0$, and $N_{xy} = 0$ if no torsional load is applied.

Assuming the coefficient N_{y0} is governed by the membrane action Eq. (9.3.1), then

$$N_{y0} = -p_e R \tag{9.5.1}$$

Incorporating this into Eq. (9.3.15c) gives

$$D\nabla^8 w + \frac{1-\mu^2}{R^2}Cw_{,xxxx} + p_e R\nabla^4 w_{,yy} = 0 \tag{9.5.2}$$

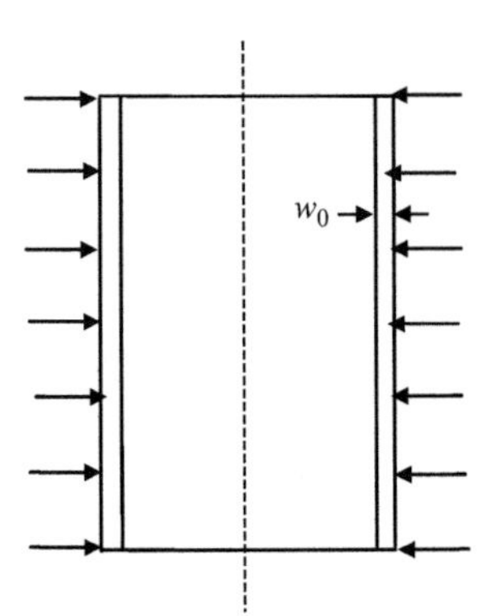

Figure 9-5 Cylinder subjected to external pressure

The boundary conditions corresponding to simply supported ends are

$$w = w_{,xx} = 0 \quad \text{at } x = 0, \ell \tag{9.5.3}$$

and w is to be a periodic function of y. Both the governing differential equation and the boundary conditions are satisfied if the lateral displacement is of the form

$$w = a \sin\frac{m\pi x}{\ell} \sin\frac{n\pi y}{\pi R} \tag{9.5.4}$$

where m is the number of half-waves in the longitudinal direction and n is the number of half-waves in the circumferential direction. Introducing a variable β such that

$$\beta = \frac{n\ell}{\pi R} \tag{9.5.5}$$

then, Eq. (9.5.4) becomes

$$w = a \sin\frac{m\pi x}{\ell} \sin\frac{\beta\pi y}{\ell} \tag{9.5.6}$$

Substituting Eq. (9.5.6) into Eq. (9.5.2) gives

$$D\left(\frac{\pi}{\ell}\right)^8 (m^2 + \beta^2)^4 + \frac{(1-\mu^2)C}{R^2} m^4 \left(\frac{\pi}{\ell}\right)^4 - p_e R \left(\frac{\pi}{\ell}\right)^6 \beta^2 (m^2 + \beta^2)^2 = 0 \tag{9.5.7}$$

Dividing Eq. (9.5.7) by $(\pi/\ell)^6$ and solving for $p_e R$, one obtains

$$p_e R = \frac{D\left(m^2 + \beta^2\right)^2}{\beta^2} \left(\frac{\pi}{\ell}\right)^2 + \frac{\left(1-\mu^2\right) C m^4}{R^2 \beta^2 \left(m^2 + \beta^2\right)^2} \left(\frac{\ell}{\pi}\right)^2 \tag{9.5.8}$$

Substituting $C = Eh/(1-\mu^2)$ and $D = Eh^3/[12(1-\mu^2)]$ and rearranging gives

$$\frac{p_e R}{Eh} = \frac{h^2(m^2+\beta^2)^2}{12(1-\mu^2)\beta^2}\left(\frac{\pi}{\ell}\right)^2 + \frac{m^4}{R^2\beta^2(m^2+\beta^2)^2}\left(\frac{\ell}{\pi}\right)^2 \tag{9.5.9}$$

For particular values of ℓ/R and R/h, the m and n corresponding to the smallest eigenvalue may be calculated by trial and error.

As an example, calculate p_{cr} for a cylinder with $R = 20$ in., $\ell = 20$ in., $h = 0.2$ in., $E = 10 \times 10^6$ psi, and $\mu = 0.3$. It is found from executing Eq. (9.5.9) by **Maple**® that for $m = 1$ and $n = 7$, 8, and 9, respectively, $p_e = 122.13$, 105.97, and 107.92 psi and that p_e is higher for all other values of m and n. Therefore, p_{cr} is taken to be 105.97 psi. The values for the sine-wave length parameters m and n indicate that the shell has one-half sine wave in the axial direction and eight full sine waves in the circumferential direction in the eigen mode shape. It would seem intuitively clear that m must be equal to one, otherwise p_e would become larger than that for $m = 1$. Then, Eq. (9.5.9) may be rewritten as

$$p_e = \frac{Eh}{R}\left[\frac{h^2(1+\beta^2)^2}{12(1-\mu^2)\beta^2}\left(\frac{\pi}{\ell}\right)^2 + \frac{1}{R^2\beta^2(1+\beta^2)^2}\left(\frac{\ell}{\pi}\right)^2\right] \tag{9.5.10}$$

If ℓ/R approaches infinity, Eq. (9.5.10) reduces to the following as β also approaches infinity:

$$p_e = n^2\frac{D}{R^3} \tag{9.5.11}$$

Equation (9.5.11) agrees with the Donnell analysis of the circular ring (Brush and Almroth 1975).[1] For $n = 2$, this value is 33% higher (taking the classical value as the base) than the classical eigenvalue for the ring given in Timoshenko and Gere (1961).[2] The error occurs because of the approximations. Langhaar (1962) believes that Donnell's equation gives more accurate results when multiple wave patterns occur in the buckled form. When $\ell = 2000$ in. in the above example with $n = 2$, Eq. (9.5.10) gives

[1] See page 139.
[2] See page 291.

$p_e = 3.666$ psi, while Eq. (9.5.11) yields $p_e = 3.663$ psi. Although Eq. (9.5.10) gives much lower p_e for $n = 1$, Timoshenko and Gere (1961)[3] show that the smallest possible value of n must be equal to 2 considering an initial ellipticity and the inextensibility of the member in the circumferential direction.

For external lateral pressure p, the hoop compressive stress σ_y is related by $\sigma_y = pR/h$, although the validity of this relationship is questionable at the ends of the cylinder where simply supported boundary conditions and corresponding displacement function Eq. (9.5.6) are assumed. Hence, $\sigma_{cr} = p_{cr}R/h$. As the cylinder radius approaches infinity, the critical stress σ_{cr} approaches the value given in Fig. 8-12 for long flat plates; that is, $\sigma_{cr} = 4\pi^2 D/\ell^2 h$. In the above example, when $R = 500$ in. and $\ell = 1$ in. with $n = 1570$, Eq. (9.5.10) gives $p_e = 578.4389$ psi, while the simple hoop compression relation yields $p_e = 578.4383$ psi.

A good overview of the historical development on the subject is given in Allen and Bulson (1980), which includes the contribution of von Mises (1914), Southwell (1914), Donnell (1933), Batdorf (1947), Kraus (1967), and Brush and Almroth (1975).

9.5.2. Axially Loaded Cylinders

Consider a circular cylinder of length ℓ and radius R that is simply supported at its ends and subjected to a uniformly distributed axial compressive load P.

Under the action of the load, the cylinder shortens and except at the ends, increases its diameter. The prebuckling static deformation is axisymmetric, and the critical load P_{cr} is the lowest load at which equilibrium in the axisymmetric form ceases to be stable. Although the lateral displacement w_0 is likely to be a function of x, it is assumed, for simplicity, uniform as shown in Fig. 9-6 and the prebuckling deformation may be determined by the linear membrane equations. Under these simplifying assumptions, the critical load can be determined by solving Donnell equation Eq. (9.3.15c) in the manner outlined by Batdorf (1947).

From a membrane analysis of the unbuckled cylinder

$$N_{x0} = -\frac{P}{2\pi R} \quad \text{and} \quad N_{xy0} = N_{y0} = 0$$

Substituting these values into Donnell equation Eq. (9.3.15c) gives

$$D\nabla^8 w + \frac{1-\mu^2}{R^2} C w_{,xxxx} + \frac{P}{2\pi R}\nabla^4 w_{,xx} = 0 \tag{9.5.12}$$

[3] See page 295.

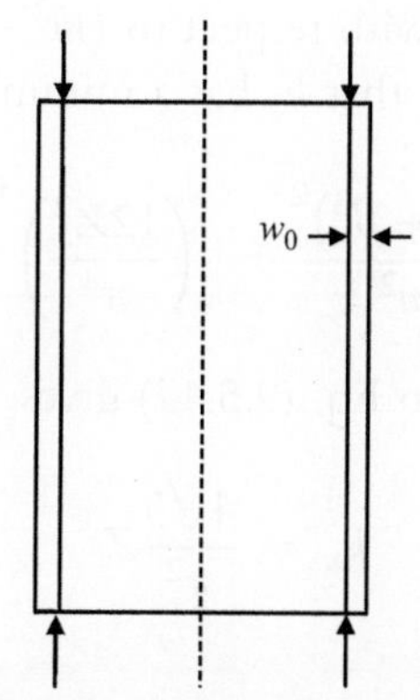

Figure 9-6 Cylinder subjected to axial compression

Equation (9.5.12) is a linear partial differential equation with constant coefficients. The boundary conditions and solution form are the same as for the previous example. Substituting Eq. (9.5.6) into Eq. (9.5.12) gives

$$D\left(\frac{\pi}{\ell}\right)^8 \left(m^2+\beta^2\right)^4 + \frac{Eh}{R^2}m^4\left(\frac{\pi}{\ell}\right)^4 - \sigma_x h\left(\frac{\pi}{\ell}\right)^6 m^2\left(m^2+\beta^2\right)^2 = 0 \quad (9.5.13)$$

Dividing Eq. (9.5.13) by $D(\pi/\ell)^8$ and introducing two new variables gives

$$\left(m^2+\beta^2\right)^4 + \frac{12m^4Z^2}{\pi^4} - k_x m^2\left(m^2+\beta^2\right)^2 = 0 \quad (9.5.14)$$

where

$$Z = \frac{\ell^2}{Rh}\left(1-\mu^2\right)^{1/2} \quad (9.5.15)$$

$$k_x = \frac{\sigma_x h \ell^2}{D\pi^2} \quad (9.5.16)$$

The nondimensionalized variable Z is known as the Batdorf parameter useful for distinguishing short and long cylinders and k_x is a buckling stress parameter similar to the one that appears in the plate buckling equation Eq. (8.4.13).

Solving Eq. (9.5.14) for k_x yields

$$k_x = \frac{\left(m^2+\beta^2\right)^2}{m^2} + \frac{12Z^2m^2}{\pi^4\left(m^2+\beta^2\right)^2} \quad (9.5.17)$$

Differentiating Eq. (9.5.17) with respect to $(m^2+\beta^2)^2/m^2$ and setting the result equal to zero indicates that k_x has a minimum value when

$$\frac{(m^2+\beta^2)^2}{m^2} = \left(\frac{12Z^2}{\pi^4}\right)^{1/2} \tag{9.5.18}$$

Substituting Eq. (9.5.18) into Eq. (9.5.17) gives

$$k_x = \frac{4\sqrt{3}}{\pi^2}Z \tag{9.5.19}$$

from which

$$\sigma_{cr} = \frac{1}{\sqrt{3(1-\mu^2)}}\frac{Eh}{R} \tag{9.5.20}$$

Equation (9.5.20) is considered to be the classical solution for axially compressed cylinders. It is noted that Eq. (9.5.20) is independent from the length of the cylinder, indicating that the critical stress is for the case of local buckling. It is also of interest to note that Eq. (9.5.20) is the solution for axisymmetric as well as asymmetric modes (see Problem 9.6). For $\mu = 0.3$, Eq. (9.5.20) becomes

$$\sigma_{cr} = 0.605\frac{Eh}{R} \tag{9.5.21}$$

Equation (9.5.18) indicates that the cylindrical shell subjected to axial compression has a large number of instability modes corresponding to a single bifurcation point. Since m and n are positive integers, it is impossible to satisfy Eq. (9.5.18) for short cylinders. Such a difficulty arises for values of the Batdorf parameter Z less than 2.85 (see Problem 9.7). In such cases, Eq. (9.5.13) and the trial-and-error procedure may be used to determine the critical load.

If $Z < 2.85$, the critical stress coefficient k_x is determined by setting $m = 1$ and $\beta = 0$ (as ℓ approaches to zero) in Eq. (9.5.17). This leads to

$$k_x = 1 + \frac{12Z^2}{\pi^4} \tag{9.5.22}$$

As the cylinder radius approaches infinity (or the cylinder length approaches zero), the coefficient k_x approaches 1. Then,

$$N_{x0_{cr}} = \sigma_{cr}h = \frac{\pi^2 D}{\ell^2} \tag{9.5.23}$$

This is the equation for the critical load intensity, in pounds per inch, of a wide column, that is, a flat plate that is simply supported on the loaded ends and free on the unloaded edges.

A very long cylinder can buckle as a column with undeformed cross section ($m = n = 1$). The present Donnell formulation does not yield the correct result for this case as compared to that given by Timoshenko and Gere (1961).[4] More accurate values than those given by Eq. (9.5.21) are given by, for example, Timoshenko and Gere (1961).[5] Figure 9-6 shows possible built-in eccentricities due to the expansion of the cylinder wall during loading. Inclusion of such eccentricities will likely lower the critical load.

9.5.3. Torsional Load

Consider a circular cylinder of length ℓ and radius R that is simply supported at its ends and subjected to a twisting moment. Assume, for simplicity, that a linear membrane analysis is adequate for the prebuckling deformation. Then, N_{xt0} is constant, and the Donnell equation Eq. (9.3.15c) may be rewittten in the form

$$D\nabla^8 w + \frac{1-\mu^2}{R^2} C w_{,xxxx} - \frac{2}{R} N_{xy0} \nabla^4 w_{,xy} = 0 \tag{9.5.24}$$

Equation (9.5.24) has odd-ordered derivatives with respect to each of the coordinate variables in one term $((2/R)N_{xy0}\nabla^4 w_{,xy})$, and even-ordered derivatives in the other two terms. Therefore, a deflection function of the form of Eq. (9.5.6) will not work. Under torsional loading, the buckling deformation of a cylinder consists of a number of circumferential waves that spiral around the tube from one end to the other. Such waves can be represented by a deflection function of the form

$$w = a \sin\left(\frac{m\pi x}{\ell} - \frac{\beta\pi y}{\ell}\right) \tag{9.5.25}$$

where β is defined by Eq. (9.5.5). Equation (9.5.25) satisfies the differential equation and the requirement of periodicity in the circumferential direction. But it does not satisfy any of the commonly used boundary conditions at the cylinder ends. Therefore, this simple deflection function may be used for only long cylinders whose end conditions have little effect in the critical load.

[4] See page 466.
[5] See page 464, Eq. (i).

For such cylinders, substituting Eq. (9.5.25) into Eq. (9.5.24) yields

$$D\left(\frac{\pi}{\ell}\right)^8\left(m^2+\beta^2\right)^4+\frac{Eh}{R^2}m^4\left(\frac{\pi}{\ell}\right)^4-\frac{2}{R}N_{xy0}\left(\frac{\pi}{\ell}\right)^6 m\beta\left(m^2+\beta^2\right)^2=0 \tag{9.5.26}$$

Dividing Eq. (9.5.26) by $D(\pi/\ell)^6$ and solving for N_{xy0} gives

$$N_{xy0}=\frac{DR\left(m^2+\beta^2\right)^2}{2m\beta}\left(\frac{\pi}{\ell}\right)^2+\frac{Ehm^3}{R\beta\left(m^2+\beta^2\right)^2}\left(\frac{\ell}{\pi}\right)^2 \tag{9.5.27}$$

A distinct eigenvalue corresponding to each pair of m and n can be determined by trial-and-error. For long tubes, the smallest values of N_{xy0} correspond to $n = 2$ (Donnell 1933).

For a sufficiently long cylinder, Brush and Almroth (1975)[6] give the critical shear stress

$$\tau_{cr}=\frac{0.271E}{\left(1-\mu^2\right)^{3/4}}\left(\frac{h}{R}\right)^{3/2} \tag{9.5.28}$$

which is again 15% higher than a value given by Timoshenko and Gere (1961)[7] with a coefficient of 0.236 instead of 0.271.

9.5.4. Combined Axial Compression and External Pressure

Consider a cylindrical shell of length ℓ and radius R subjected to an axial compressive force P and uniform external lateral pressure p_e. If a linear membrane analysis is assumed satisfactory for the axisymmetric prebuckling deformation, Eq. (9.3.15c) becomes

$$D\nabla^8 w+\frac{1-\mu^2}{R^2}Cw_{,xxxx}+\nabla^4\left(\frac{P}{2\pi R}w_{,xx}+Rp_e w_{,yy}\right)=0 \tag{9.5.29}$$

Equation (9.5.29) may be simplified by letting

$$\frac{P}{2\pi R}=Fp_eR \tag{9.5.30}$$

where F is a dimensionless constant.

[6] See page 171.
[7] See page 504.

Substituting Eq. (9.5.30) into Eq. (9.5.29) yields

$$D\nabla^8 w + \frac{1-\mu^2}{R^2} C w_{,xxxx} + Rp_e \nabla^4 (F w_{,xx} + w_{,yy}) = 0 \qquad (9.5.31)$$

Substituting Eq. (9.5.6) into Eq. (9.5.31) gives

$$D\left(\frac{\pi}{\ell}\right)^8 (m^2+\beta^2)^4 + \frac{(1-\mu^2)C}{R^2} m^4 \left(\frac{\pi}{\ell}\right)^4 - p_e R \left(\frac{\pi}{\ell}\right)^6 \left(Fm^2+\beta^2\right)\left(m^2+\beta^2\right)^2 = 0$$

Hence

$$p_e R = \frac{D\left(m^2+\beta^2\right)^2}{\left(Fm^2+\beta^2\right)} \left(\frac{\pi}{\ell}\right)^2 + \frac{(1-\mu^2)Cm^4}{R^2\left(Fm^2+\beta^2\right)\left(m^2+\beta^2\right)^2} \left(\frac{\ell}{\pi}\right)^2 \qquad (9.5.32)$$

A distinct eigenvalue corresponding to each pair of m and n can be determined by trial and error. A case of particular interest is when $F = 1/2$. For that value the cylinder is subjected to the same pressure p_e on both its lateral and end surfaces. Such load is termed *hydrostatic pressure loading.*

9.5.5. Effect of Boundary Conditions

The uncoupled Donnell equations Eqs. (9.3.15) are not suitable for a general analysis, as shown below. The assumed solution function, Eq. (9.5.6) for the boundary conditions $w = w_{,xx} = 0$ at $x = 0$, ℓ is of the form

$$w = c_1 \sin\frac{m\pi x}{\ell} \sin\frac{\beta\pi y}{\ell}$$

Substituting this into Eqs. (9.3.15a) and (9.3.15b) reveals that the corresponding expressions for u and v, respectively, must be of the forms

$$u = a_1 \cos\frac{m\pi x}{\ell} \sin\frac{\beta\pi y}{\ell}$$
$$v = b_1 \sin\frac{m\pi x}{\ell} \cos\frac{\beta\pi y}{\ell} \qquad (9.5.33)$$

These assumed displacement functions are suitable only for boundary conditions $u_{,x} = v = 0$ at $x = 0$, ℓ. The common boundary condition $w = w_{,xx} = 0$ at $x = 0$, ℓ, for example, is excluded.

For a general analysis of cylinder end conditions, the coupled form of the Donnell equation Eqs. (9.3.14) may be used. Equations (9.3.14) are of second order in u and v and fourth order in w. Therefore, each set of boundary conditions consists of four boundary conditions at each end of the cylinder. The conditions need not be the same at the two ends of the cylinder and there may be many combinations of eight boundary conditions for a cylinder.

As an example, consider a cylindrical tank subjected to external hydrostatic pressure p_e. From a linear static analysis, one obtains

$$N_{x0} = -\frac{1}{2}p_e R \quad N_{xy0} = 0 \quad N_{y0} = -p_e R \tag{9.5.34}$$

Substituting Eqs. (9.5.34) into Eqs. (9.3.14) yields

$$u_{,xx} + \frac{1-\mu}{2}u_{,yy} + \frac{1+\mu}{2}v_{,xy} - \frac{\mu w_{,x}}{R} = 0 \tag{9.5.35a}$$

$$\frac{1-\mu}{2}v_{,xx} + \frac{1+\mu}{2}u_{,xy} + v_{,yy} - \frac{w_{,y}}{R} = 0 \tag{9.5.35b}$$

$$D\nabla^4 w - \frac{C}{R}\left(v_{,y} - \frac{w}{R} + \mu u_{,x}\right) + p_e R\left(\frac{1}{2}w_{,xx} + w_{,yy}\right) = 0 \tag{9.5.35c}$$

The following displacement functions will satisfy the differential equations:

$$\begin{aligned} u &= u_n(x)\cos\frac{\beta\pi y}{\ell} \\ v &= v_n(x)\sin\frac{\beta\pi y}{\ell} \\ w &= w_n(x)\cos\frac{\beta\pi y}{\ell} \end{aligned} \tag{9.5.36}$$

Substituting Eqs. (9.5.36) into Eqs. (9.5.35) gives

$$u_n'' - \frac{1-\mu}{2}u\left(\frac{\pi}{\ell}\right)^2\beta^2 + \frac{1+\mu}{2}v'\left(\frac{\pi}{\ell}\right)\beta - \frac{\mu}{R}w' = 0 \tag{9.5.37a}$$

$$-\frac{1+\mu}{2}u_n'\left(\frac{\pi}{\ell}\right)\beta + \frac{1-\mu}{2}v_n'' + v_n\left(\frac{\pi}{\ell}\right)^2\beta^2 + \frac{1}{R}\left(\frac{\pi}{\ell}\right)\beta w_n = 0 \tag{9.5.37b}$$

$$D\left[w_n{}^{\text{iv}} - 2\left(\frac{\pi}{\ell}\right)^2 \beta^2 w_n{}'' + \left(\frac{\pi}{\ell}\right)^4 \beta^4 w_n\right]$$
$$-\frac{C}{R}\left(\frac{\pi}{\ell}\beta v_n - \frac{w_n}{R} + \mu u_n'\right) + p_e R\left[\frac{1}{2}w_n{}'' - \left(\frac{\pi}{\ell}\right)^2 \beta^2 w_n\right] = 0$$
(9.5.37c)

where primes denote differentiation with respect to x. A general solution may be obtained by setting the determinant for the unknown arbitrary coefficients Eqs. (9.5.33) equal to zero. Such an analysis has been carried out by Sobel (1964). Even for the case of the constant coefficient shown here, the amount of labor involved in the algebra is formidable. It does not look surprising that modern engineers rely more and more on computer solutions for moderately complex problems.

9.6. FAILURE OF CYLINDRICAL SHELLS

The classical solution to the buckling problem of axially loaded cylinders was obtained by Lorenz (1908). It was later independently arrived at by Timoshenko (1910), von Mises (1914), and Southwell (1914) in a slightly modified form. The equilibrium paths of an initially perfect cylinder and a slightly imperfect cylinder subjected to axial compression are shown in Fig. 9-4. Three distinct characteristics may be observed from the figure: (1) The buckling load represents the ultimate strength of the cylinder. (2) The buckling load of the imperfect shell could be substantially lower than that given by the classical theory. Buckling loads as small as 30% of the load given by the classical solution were not unusual. (3) For shell specimens that are nominally alike, the buckling loads may vary widely due to unintentional differences in the initial shape of the shell. In fact, the test results exhibited an unusually large degree of scatter.

The first progress toward solving this troublesome problem was achieved by Donnell (1934) when he proposed that a nonlinear finite-deflection theory was required. Donnell added the same terms that von Kármán had used in formulating the nonlinear plate equation to his small deflection equations. His analysis, however, did not lead to satisfactory results due to oversimplification.

Using essentially the same large-deflection equations as Donnell used and employing a better function that adequately represented the buckling pattern of the shell, von Kármán and Tsien (1941) were able to obtain the first meaningful solution to the problem. Although their work was far from complete, it proved to be a significant milestone of the large-deflection theory of axially loaded cylindrical shells. In the years that followed, several researchers improved the solution by adding more relevant terms; more and more accurate postbuckling curves were realized.

The next significant progress in the study of axially loaded cylindrical shells was made by Donnell and Wan (1950) when they introduced initial imperfections into the analysis. As a result of this work, it is now generally believed that initial imperfections are the main reason for the discrepancy between the classical buckling solution and experimentally observed values. Lately this conclusion has been verified by carefully planned and executed experimental investigation (Tennyson 1964; Stein 1968).

As there are significant discrepancies between the test data and the classical theoretical buckling loads, particularly for cylindrical shells subjected to axial compressive loads, the design of cylindrical shells is based on the theoretical critical load modified by empirical reduction, or *knockdown*, factors for each kind of loading. The magnitude of the reduction factor in each case depends on both the average difference between theoretical and experimental values for the critical load and the severity of scatter of the test data. Comprehensive collections of test data and design recommendations for cylindrical shells and curved plates are available in Gerard and Becker (1957) and Baker et al. (1972). It is of interest to note that the minimum width-to-thickness ratio of a tubular section specified in AISC (2005) to be classified for a noncompact section ($2R/h < 0.11E/\sigma_y$) is such that the local buckling of a tubular section is effectively eliminated in structural steel buildings.

9.7. POSTBUCKLING OF CYLINDRICAL SHELLS

Equations (9.3.11) were derived for large-deflection nonlinear analysis of cylindrical shells. In order to make these equations valid for a shell with initial imperfections, a few modifications must be made. First, assume that the lateral deflection consists of an initial distortion w_i in addition to the deflection w induced by the applied loads.

$$D\left(w_{,xxxx}+2w_{,xxyy}+w_{,yyyy}\right)-\left[f_{,yy}w_{,xx}-2f_{,xy}w_{,xy}+f_{,xx}\left(1/R+w_{,yy}\right)\right]=p \tag{9.3.11a}$$

$$f_{,xxxx}+2f_{,xxyy}+f_{,yyyy}=Eh\left[\left(w_{,xy}\right)^2-w_{,xx}w_{,yy}-1/Rw_{,xx}\right] \tag{9.3.11b}$$

Equation (9.3.11a) is an equation of equilibrium in the radial direction. The first three terms in Eq. (9.3.11a) are related to transverse shear forces. The initial distortion does not affect them. The remaining terms are components of middle surface forces obtained by multiplying the forces by the surface curvature. Since the total curvature applies here, w must be replaced with $w + w_i$. The equation of equilibrium for the initially imperfect shell thus takes the form

$$\begin{aligned} D(\nabla^4 w) - \big[f_{,yy}(w_{,xx} + w_{i,xx}) - 2f_{,xy}(w_{,xy} + w_{i,xy}) \\ + f_{,xx}(1/R + w_{,yy} + w_{i,yy}) \big] = p \end{aligned} \tag{9.7.1}$$

In order to obtain the compatibility equation for an initially distorted shell, the strain-displacement relations must be modified in order to reflect the effects of an initial imperfection. Excluding quadratic terms of the initial imperfection, it can be shown that the modified strain-displacement relations after replacing w with $w + w_i$ take the form [see Eqs. (9.4.5)]

$$\varepsilon_x = u_{,x} + \frac{1}{2}w_{,x}{}^2 + w_{,x}w_{i,x} \tag{9.7.2a}$$

$$\varepsilon_y = v_{,y} - \frac{w}{R} + \frac{1}{2}w_{,y}{}^2 + w_{,y}w_{i,y} \tag{9.7.2b}$$

$$\gamma_{xy} = u_{,y} + v_{,x} + w_{,x}w_{,y} + w_{i,x}w_{,y} + w_{,x}w_{i,y} \tag{9.7.2c}$$

Differentiating Eqs. (9.7.2) yields

$$\begin{aligned} \varepsilon_{x,yy} + \varepsilon_{y,xx} - \gamma_{xy,xy} = {} & w_{xy}{}^2 + 2w_{i,xy}w_{,xy} - w_{,xx}w_{,yy} \\ & - w_{i,xx}w_{,yy} - w_{i,yy}w_{,xx} - \frac{1}{R}w_{,xx} \end{aligned} \tag{9.7.3}$$

Substituting stress function Eq. (9.3.7) into Eq. (9.7.3) gives

$$\nabla^4 f = Eh\left(w_{xy}{}^2 + 2w_{i,xy}w_{,xy} - w_{,xx}w_{,yy} - w_{i,xx}w_{,yy} - w_{,xx}w_{i,yy} - \frac{1}{R}w_{,xx} \right) \tag{9.7.4}$$

Equations (9.7.1) and (9.7.4) are the governing large-deflection equations for an initially imperfect cylindrical shell.

Consider a rectangular cylindrical panel whose postbuckling behavior is very similar to that of an entire cylinder. Thus, the consideration is limited to such a panel avoiding lengthy computational efforts. The analysis presented herein follows the general outline of that given by Volmir (1967). A cylindrical panel is a section of an entire cylindrical shell bounded by two generators and two circular arcs. The radius of the shell is R, its thickness h. The length of each edge of the panel is a. The panel is subjected to a uniform axial compression stress p_x as shown in Fig. 9-7. The x-axis is in the direction of the cylinder, and y is in the circumferential direction.

The assumed boundary conditions of the panel are that (1) the edges are simply supported, (2) the shear force N_{xy} vanishes along each edge, (3) the edges at $y = 0$, a are free to move in the y direction, and (4) the panel retains its original rectangular shape. These conditions are satisfied if the displacement function is taken as

$$w = c \sin \frac{\pi x}{a} \sin \frac{\pi y}{a} \tag{9.7.5}$$

It is assumed that the initial distortion can also be given by

$$w_i = c_i \sin \frac{\pi x}{a} \sin \frac{\pi y}{a} \tag{9.7.6}$$

The first step is to evaluate the stress function f in terms of the assumed deformation functions. Substituting Eqs. (9.7.5) and (9.7.6) into Eq. (9.7.4) yields

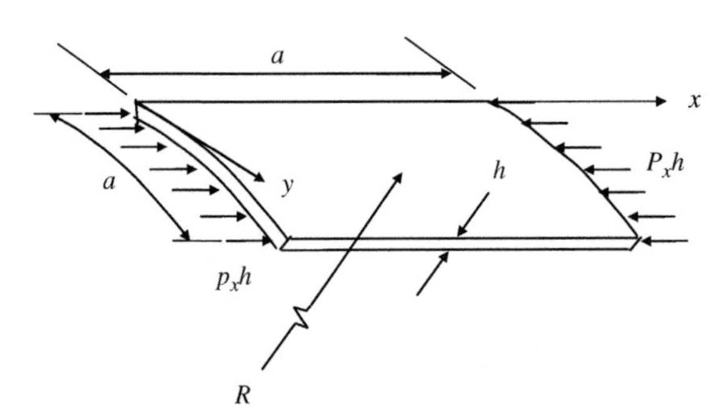

Figure 9-7 Cylindrical panel subjected to axial compression

$$\nabla^4 f = Eh\left[\left(c^2 + 2cc_i\right)\frac{\pi^4}{2a^4}\left(\cos\frac{2\pi x}{a} + \cos\frac{2\pi y}{a}\right)\right.$$
$$\left. + \frac{c}{R}\frac{\pi^2}{a^2}\left(\sin\frac{\pi x}{a}\sin\frac{\pi y}{a}\right)\right] \tag{9.7.7}$$

A particular solution to this equation, obtained by using the method of undetermined coefficients, is

$$f_p = \frac{Eh(c^2 + 2cc_i)}{32}\left(\cos\frac{2\pi x}{a} + \cos\frac{2\pi y}{a}\right) + \frac{Ehca^2}{4R\pi^2}\sin\frac{\pi x}{a}\sin\frac{\pi y}{a} \tag{9.7.8}$$

The homogeneous solution is obtained considering the prebuckling stress p_x in the x direction and $N_y = N_{xy} = 0$. Noting that $N_x = -p_x h$ and recalling $N_x = f_{,xx}$ of Eqs. (9.3.7), one obtains the homogeneous solution

$$f_h = -\frac{hp_x}{2}y^2 \tag{9.7.9}$$

Hence, the total solution of Eq. (9.7.7) is

$$f = \frac{Eh(c^2 + 2cc_i)}{32}\left(\cos\frac{2\pi x}{a} + \cos\frac{2\pi y}{a}\right) + \frac{Ehca^2}{4R\pi^2}\sin\frac{\pi x}{a}\sin\frac{\pi y}{a} - \frac{hp_x}{2}y^2 \tag{9.7.10}$$

A relation between c, c_i, and p_x will be determined from Eq. (9.7.1) by means of the Galerkin method. For the problem at hand, the Galerkin equation takes the form

$$\int_0^a \int_0^a L(c)g(x,y)\,dxdy = 0 \tag{9.7.11}$$

where $L(c)$ is the left-hand side of Eq. (9.7.1)

$$L(c) = D\nabla^4 w - \left[f_{,yy}\left(w_{,xx} + w_{i,xx}\right) - 2f_{,xy}\left(w_{,xy} + w_{i,xy}\right)\right.$$
$$\left. + f_{,xx}\left(1/R + w_{,yy} + w_{i,yy}\right)\right] \tag{9.7.12}$$

and $g(x,y) = \sin\,(\pi x/a)\,\sin(\pi y/a)$. Substituting Eqs. (9.7.5), (9.7.6), and (9.7.10) for w, w_i, and f, respectively, Eq. (9.7.12) becomes

$$L(c) = 4Dc\left(\frac{\pi}{a}\right)^4 \sin\frac{\pi x}{a}\sin\frac{\pi y}{a} - \left[\frac{\pi^2}{8a^2}Eh\left(c^2+2cc_i\right)\cos\frac{2\pi y}{a}\right.$$
$$\left. + \frac{1}{4}Ehc\sin\frac{\pi x}{a}\sin\frac{\pi y}{a} - hp_x\right]\left[\left(\frac{\pi}{a}\right)^2\left(c+c_i\right)\sin\frac{\pi x}{a}\sin\frac{\pi y}{a}\right]$$
$$+\frac{1}{2R}\left(\frac{\pi}{a}\right)^2 Ehc\cos^3\frac{\pi x}{a}\cos^3\frac{\pi y}{a}\left(c+c_i\right) - \left[\frac{\pi^2}{8a^2}Eh\left(c^2+2cc_i\right)\cos\frac{2\pi x}{a}\right.$$
$$\left.+\frac{1}{4}Ehc\sin\frac{\pi x}{a}\sin\frac{\pi y}{a}\right]\left[-\frac{1}{R}+\left(\frac{\pi}{a}\right)^2\left(c+c_i\right)\sin\frac{\pi x}{a}\sin\frac{\pi y}{a}\right]$$

Substituting L (c) and g (x,y) into Eq. (9.7.11) and carrying out integration **Maple®** gives

$$\frac{Dc\pi^4}{a^2} - \frac{p_x h\pi^2}{4}\left(c+c_i\right) + \frac{Eha^2c}{16R^2} - \frac{Eh}{R}\left(\frac{5}{6}c^2+cc_i\right) + \frac{Eh\pi^4}{32a^2}\left(c^3+3c^2c_i+2cc_i^{\,2}\right) = 0 \tag{9.7.13}$$

from which

$$p_x = \left[\frac{4D\pi^2}{a^2h} + \frac{Ea^2}{4\pi^2R^2} + \frac{E\pi^2}{8a^2}\left(c^2+3cc_i+2c_i^2\right) - \frac{4E}{\pi^2R}\left(\frac{5}{6}c+c_i\right)\right]\frac{c}{c+c_i} \tag{9.7.14}$$

Introduce the following nondimensional parameters into Eq. (9.7.14) to characterize the influence of each parameter to the postbuckling behavior of the panel:

$$\bar{p}_x = \frac{p_x a^2}{Eh^2}$$
$$k = \frac{a^2}{hR}$$
$$\delta = \frac{c}{h}$$
$$\delta_i = \frac{c_i}{h}$$

The parameters are measures of the loading ($\bar{p}_x$), of the curvature of the panel (k), and of the deflections (δ and δ_i), respectively. Substituting the parameters into Eq. (9.7.14) yields

$$\bar{p}_x = \left[\frac{\pi^2}{3(1-\mu^2)} + \frac{k^2}{4\pi^2} + \frac{\pi^2}{8}\left(\delta^2 + 3\delta\delta_i + 2\delta_i^2\right) - \frac{4k}{\pi^2}\left(\frac{5}{6}\delta + \delta_i\right)\right]\frac{\delta}{\delta + \delta_i} \tag{9.7.15}$$

for $\mu = 0.3$,

$$\bar{p}_x = \left[3.615 + \frac{k^2}{39.48} + 1.2337\left(\delta^2 + 3\delta\delta_i + 2\delta_i^2\right) - 0.4053k\left(0.8333\delta + \delta_i\right)\right]\frac{\delta}{\delta + \delta_i} \tag{9.7.16}$$

The load-deflection relationship of Eq. (9.7.16) is plotted in Figs. 9-8 and 9-9. Figure 9-8 (see Problem 8.9) is for a panel with $k = 0$ ($R = \infty$) depicting a flat plate, and Fig. 9-9 (see Problem 9.8) is for a cylindrical panel with $k = 24$. The curves in each figure show the variation of the load parameter $\bar{p}_x$ with the total lateral deflection parameter $\delta + \delta_i$. The distinct characteristics of the curves in Fig. 9-8 are that bending of an initially deformed plate begins as soon as the load is applied, deflections increase slowly at first and then more rapidly in the neighborhood of the critical load, and, as the deflections increase in magnitude, the curves of the initially deformed plates approach that of the perfect plate. Thus, the elastic critical load does not represent the maximum carrying capacity of the panel. Koiter (1945) termed this phenomenon imperfection-insensitive.

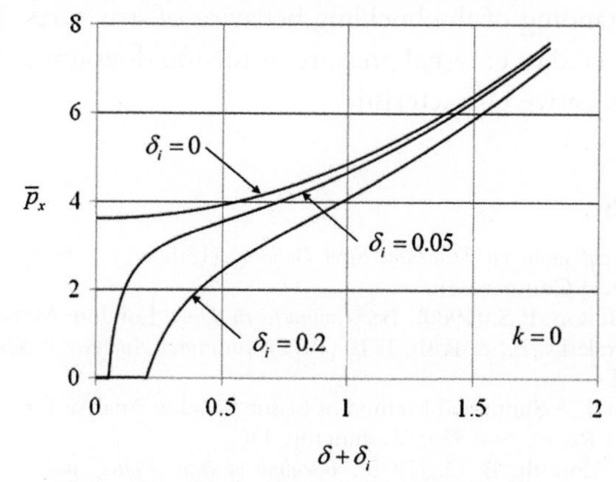

Figure 9-8 Postbuckling curves for flat plates

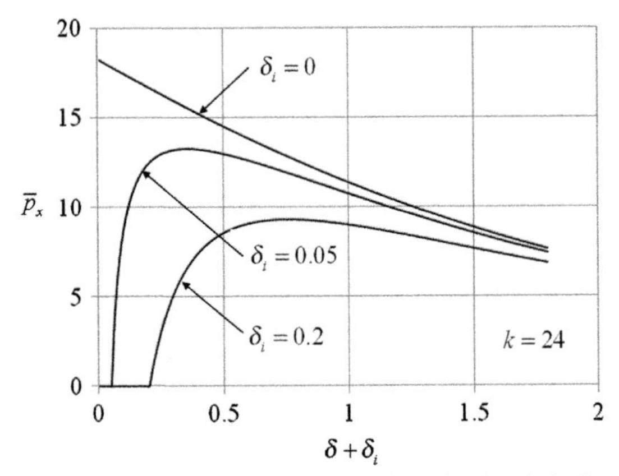

Figure 9-9 Postbuckling curves for cylindrical shells

With regard to the curved panel with initial imperfection, the following observations can be made. When the applied axial load is small, bending increases slowly with an increase in the load. Then, at a certain load depending on the size of the initial imperfections, bending suddenly grows rapidly and the load begins to drop. As the deflection continues to increase, the curve of the imperfect panel approaches that of the perfect panel. The important conclusion that can be drawn form this observation is that the maximum load that an initially imperfect panel can support is significantly less than the critical load given by the classical theory, a dangerous phenomenon no one should ignore. This is an imperfection-sensitive case.

Brush and Almroth (1975) praise Koiter's (1945) initial-postbuckling theory as one of the most important contributions in recent years to the general understanding of the buckling behavior of structures. Interestingly, a cylinder subjected to external pressure or torsion does not exhibit such an imperfection-sensitive characteristic.

REFERENCES

AISC. (2005). *Specification for Structural Steel Building* (13th ed.). Chicago, IL: American Institute of Steel Construction.

Allen, H. G., & Bulson, P. S. (1980). *Background to Buckling*. London: McGraw-Hill (UK).

Baker, E. H., Kovalesky, L., & Rish, F. L. (1972). *Structural Analysis of Shells*. New York: McGraw-Hill.

Batdorf, S. B. (1947). A Simplified Method of Elastic Stability Analysis for Thin Cylindrical Shells. *NACA Report*. No. 874, Washington, DC.

Brush, D. O., & Almroth, B. O. (1975). *Buckling of Bars, Plates, and Shells*. New York: McGraw-Hill.

Chajes, A. (1974). *Principles of Structural Stability Theory*. Englewood Cliffs, NJ: Prentice-Hall.
Donnell, L. H. (1933). Stability of Thin-Walled Tubes under Torsion. *NACA Technical Report*. No. 479, Washington, DC.
Donnell, L. H. (1934). New Theory for Buckling of Thin Cylinders Under Axial Compression and Bending. *Transactions, ASME, Vol. 56*, 795–806.
Donnell, L. H., & Wan, C. C. (1950). Effect of Imperfections on Buckling of Thin Cylinders and Columns Under Axial Compression. *Journal of Applied Mechanics, Vol. 17* (No. 1), 73–88.
Gerard, G., & Becker, H. (1957). Handbook of Structural Stability, Part III, Buckling of Curved Plates and Shells. *NACA Technical Note*. No. 3787, Washington, DC.
Koiter, W. T. (1945). On the Stability of Elastic Equilibrium (in Dutch with English summary), thesis, Delft, H.J. Paris, Amsterdam. English translation, Air Force Flight Dynamics Laboratory, Technical Report, AFFDL-TR-70-25, February 1970.
Kraus, H. (1967). *Thin Elastic Shells*. New York: John Wiley and Sons.
Langhaar, H. L. (1962). *Energy Methods in Applied Mechanics*. New York: John Wiley and Sons.
Lorenz, R. (1908). Achsensymmetrische Verzerrungen in dηnnwandigen Hohlzylindern, *Zeitschrift des Vereines Deutscher Ingeniere, Vol. 52*(No. 43), 1766.
Sobel, L. H. (1964). Effects of Boundary Conditions on the Stability of Cylinders Subjected to Lateral and Axial Pressure. *AIAA J., Vol. 2*, 1437–1440.
Southwell, R. V. (1914). On the General Theory of Elastic Stability. *Phil. Trans. Roy. Soc. London, Series A, 213*, 187.
Stein, M. (1968). Some Recent Advances in the Investigation of Shell Buckling. *AIAA Journal, Vol. 6*(No. 12), 2339–2345.
Tennyson, R. C. (1964). *An Experimental Investigation of the Buckling of Circular Cylindrical Shells in Axial Compression Using the Photoelastic Technique*. Institute of Aerospace Science, University of Toronto. Report No. 102.
Timoshenko, S. P. (1910). Einige Stabilit≅tsprobleme der Elastizit≅ts-theorie. *Zeitschrift für Mathematik und Physik, Vol. 58*(No.4), 378.
Timoshenko, S. P., & Gere, J. M. (1961). *Theory of Elastic Stability* (2nd ed.). New York: McGraw-Hill.
Volmir, A. S. (1967). A Translation of Flexible Plates and Shells. Air Force Flight Dynamics Laboratory, Technical Report No. 66-216, Wright-
von Kármán, T., & Tsien, H. S. (1941). The Buckling of Thin Cylindrical Shells under Axial Compression. *Journal of the Aeronautical Sciences, Vol. 8*(No. 8), 303–312.
von Mises, R. (1914). *Z. Ver. deut. Ingr., Vol. 58*, 750.

PROBLEMS

9.1 Show that Eqs. (9.2.9) may be derived from those in Eqs. (9.2.2), (9.2.3), and (9.2.5) by introducing appropriate constitutive and kinematic relations.

9.2 Show that the application of the Euler-Lagrange differential equations Eqs. (9.3.6) to the integrand of Eq. (9.3.1) leads to the equilibrium equations given in Eqs. (9.2.9).

9.3 Show that Eqs. (9.3.14a) and (9.3.14b) can be partially uncoupled to obtain Eqs. (9.3.15a) and (9.3.15b). Show that the u and v may be eliminated from Eq. (9.3.14c) by applying the operator ∇^4 and use of Eqs. (9.3.15a) and (9.3.15b). In this way, derive Eq. (9.3.15c).

9.4 Derive Eq. (9.4.8) following the procedure outlined in Section 8.3.

9.5 Apply the Euler-Lagrange differential equations Eqs. (9.3.6) to Eq. (9.4.8) and derive Eqs. (9.4.6).

9.6 Equation (9.5.20) has been derived for the asymmetric buckling mode for cylindrical shells subjected to axial compression. Show that it is also the correct eigenvalue for the axisymmetric buckling mode.

9.7 Show why Eq. (9.5.18) cannot be satisfied for cylinders whose Batdorf parameter Z is less than 2.85.

9.8 Using the energy method, examine the behavior of the one-degree-of-freedom model of a curved plate shown in Fig. P9-8. The model consists of four rigid bars pin-connected to each other and to the supports. At the center of the model two linear rotational springs of stiffness $C = M/\theta$ connect opposite bars to each other. Also, each of the two transverse bars contains a linear extensional spring of stiffness K. Determine the load-deflection relation for finite deflections when the load P is applied

(a) concentric with the axis of the longitudinal bars,

(b) eccentric to the axis of the longitudinal bars.

Discuss the problem.

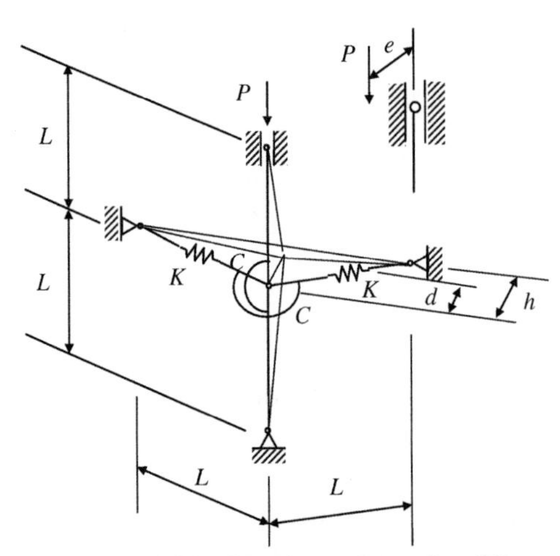

Figure P9-8 Cylindrical shell model (after Chajes, *Principles of Structural Stability Theory*. Englewood Cliffs, NJ: Prentice-Hall, 1974). Reproduced by permission from the author.

CHAPTER 10

Buckling of General Shell Elements

Contents

10.1. INTRODUCTION

A large number of publications on the equilibrium equations of general thin elastic shells have appeared since the first useful shell theory was presented by Love in 1888 (Love 1944). Naghdi (1963) points out that many of the theories presented, including Love's theory, contain some inconsistencies. Practical results can be obtained only with the aid of approximations, yet the subject has proved to be very sensitive in this respect, particularly in problems of buckling (Langhaar 1962). The continuing effort in the field is motivated by a desire to define a theory that is characterized by simplicity, consistency, and clarity.

All shell theories available today are based on the assumption that the strains in the shell are small enough to be discarded in comparison with unity. It is also assumed that the shell is thin enough that quantities such as the thickness/radius ratio may be discarded in comparison with unity

In addition to the assumption of small strains and small thickness/radius ratios, Love used the approximations previously applied by Kirchhoff in

Stability of Structures
ISBN 978-0-12-385122-2, doi:10.1016/B978-0-12-385122-2.10010-7

thin-plate analysis. That is, Love assumed that (1) normals to the reference surface remain normal during deformation, and (2) the transverse normal stress is negligibly small. The assumption that normals remain normal to the deformed surface implies that the resistance to the deformation under transverse shear is infinite.

For the derivation of nonlinear shell theory, different levels of assumptions have been employed. The nonlinear equilibrium and linear stability equations presented in this chapter based on the energy criterion are based on analyses by Brush and Amroth (1975). Their approach followed analyses presented in Koiter (1967) and Sanders (1963). The same nonlinear equilibrium equations based on the concept of equilibrium of forces and couples are derived following the procedure outlined by Novozhilov (1964), Klaus (1967), and Gould (1988). Yet more physical approaches relying mainly on the free-body diagrams and trigonometry (Timoshenko and Gere, 1961; Flügge, 1973) are available for the derivation of governing equations. The present equations are limited to shell coordinates that coincide with the lines of principal curvature, and equations for only the intermediate class of deformations are considered.

10.2. NONLINEAR EQUILIBRIUM EQUATIONS

In shell theory, a special type of curvilinear coordinate system is usually employed. The middle surface of the shell is defined by $X = X(x,y), Y = Y(x,y)$, and $Z = Z(x,y)$, where X, Y, Z are rectangular coordinates and x, y are surface coordinates, as shown in Fig. 10-1. The normal distance from the middle surface in the thickness direction is denoted by $\pm z$. Positive z is measured in the sense of the positive normal n of the middle surface. To any

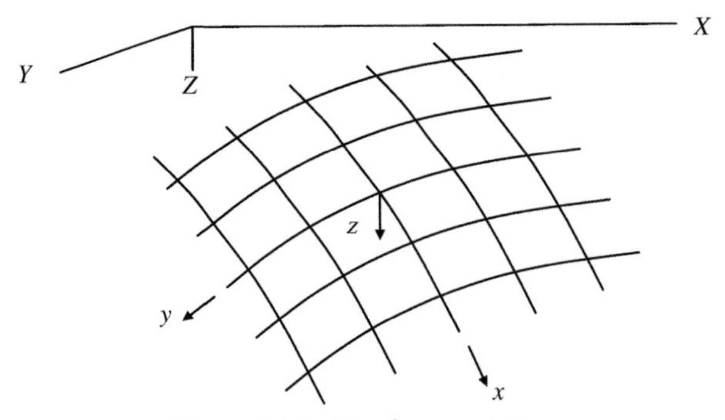

Figure 10-1 Coordinate systems

set of x,y,z, there corresponds a point in the shell. Hence x,y,z are curvilinear space coordinates, and they will be called shell coordinates. If the shell coordinates are orthogonal, the coordinate lines on the middle surface must be lines of principal curvature. Conversely, if the coordinate lines on the middle surface are lines of principal curvature, the shell coordinates are orthogonal. The proof of this geometric property is given in Langhaar (1962), Novozhilov (1964), and Gould (1988). The exterior surfaces of the shell are represented by $z = \pm h/2$, where h is thickness of the shell. If h is constant, the exterior surfaces are coordinate surfaces. The principal radii of curvature are denoted by R_x and R_y, respectively. Distances ds_x and ds_y along the coordinate lines are given by the relations

$$ds_x = A\,dx \qquad ds_y = B\,dy \tag{10.2.1}$$

where A, B are given by

$$\begin{aligned} A &= \left[\left(\frac{\partial X}{\partial x}\right)^2 + \left(\frac{\partial Y}{\partial x}\right)^2 + \left(\frac{\partial Z}{\partial x}\right)^2\right]^{1/2} \\ B &= \left[\left(\frac{\partial X}{\partial y}\right)^2 + \left(\frac{\partial Y}{\partial y}\right)^2 + \left(\frac{\partial Z}{\partial y}\right)^2\right]^{1/2} \end{aligned} \tag{10.2.2}$$

The proof of Eq. (10.2.2) can be accomplished by a simple vector analysis (see Problem 10.1). The area of any part of the surface is evidently determined for the orthogonal surface coordinates by

$$\text{area} = \iint AB\,dxdy \tag{10.2.3}$$

Additional geometric relations are presented herein without proofs as they can be found in texts on differential geometry.

For orthogonal surface coordinates, the magnitude of vectors $\mathbf{r}_{,x}$ and $\mathbf{r}_{,y}$ are A and B, respectively. Therefore, the unit vector normal to the surface is (see Problem 10.2)

$$\mathbf{n} = \frac{\mathbf{r}_{,x} \times \mathbf{r}_{,y}}{AB} \tag{10.2.4}$$

If the lines of principal curvature are coordinate lines (that is, $\mathbf{r}_{,x} \cdot \mathbf{r}_{,y} = 0$), a theorem of Rodrigues is expressed (see Problem 10.3)[1]

$$\frac{\partial \mathbf{n}}{\partial x} = \frac{1}{R_x}\frac{\partial \mathbf{r}}{\partial x} \qquad \frac{\partial \mathbf{n}}{\partial y} = \frac{1}{R_y}\frac{\partial \mathbf{r}}{\partial y} \tag{10.2.5}$$

[1] See Novozhilov (1964), page 10.

It should be noted that when the positive normal vector is directing toward the concave side (inward) of the shell, the sign of principal radii in Eq. (10.2.5) is reversed.

The product $K = 1/(R_x R_y)$ is known as the Gauss curvature of the surface. If the surface coordinates are orthogonal, K satisfies the following differential equation of Gauss:

$$-KAB = \frac{\partial}{\partial x}\left(\frac{B_{,x}}{A}\right) + \frac{\partial}{\partial y}\left(\frac{A_{,x}}{B}\right) \tag{10.2.6}$$

The functions A, B, R_x, R_y satisfy two differential equations of Codazzi. If the coordinate lines coincide with the lines of principal curvature, the Codazzi differential equations take the form (see Problem 10.4)

$$\frac{\partial}{\partial y}\left(\frac{A}{R_x}\right) = \frac{1}{R_y}\frac{\partial A}{\partial y} \quad \frac{\partial}{\partial x}\left(\frac{B}{R_y}\right) = \frac{1}{R_x}\frac{\partial B}{\partial x} \tag{10.2.7}$$

Figure 10-2 represents a portion of a cross section of a shell. The position vector of a point on the middle surface is **r**, and the position vector of the corresponding point at distance z from the middle surface is **Q**. From Fig. 10-2

$$\mathbf{Q} = \mathbf{r} + \mathbf{n}z \tag{a}$$

Taking partial derivatives with respect to the middle surface coordinators x and y yields

$$\mathbf{Q}_{,x} = \mathbf{r}_{,x} + \mathbf{n}_{,x}z \quad \mathbf{Q}_{,y} = \mathbf{r}_{,y} + \mathbf{n}_{,y}z \tag{b}$$

Since the coordinate lines are lines of principal curvature, the Rodrigues formulas Eq. (10.2.5) (with the reversed sign of the principal radii) apply. Hence,

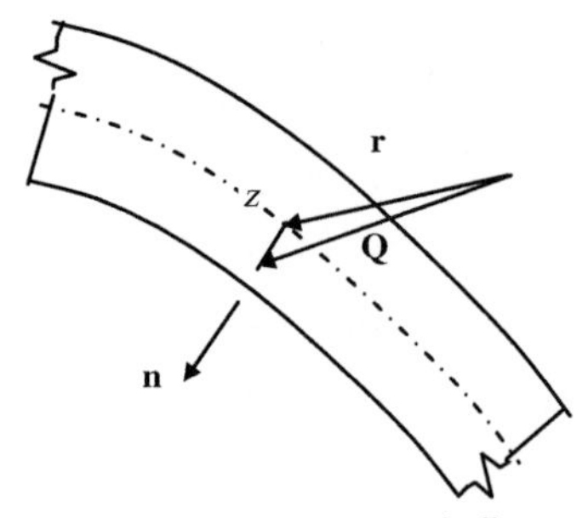

Figure 10-2 Position vectors on shell cross section

$$\mathbf{Q}_{,x} = \left(1 - \frac{z}{R_x}\right)\mathbf{r}_{,x} \quad \mathbf{Q}_{,y} = \left(1 - \frac{z}{R_y}\right)\mathbf{r}_{,y} \quad \mathbf{Q}_{,z} = \mathbf{n} \qquad \text{(c)}$$

By Eq. (10.2.2)

$$\mathbf{r}_{,x}\cdot\mathbf{r}_{,x} = A^2 \text{ and } \mathbf{r}_{,y}\cdot\mathbf{r}_{,y} = B^2$$

Recall

$$\mathbf{r}_{,x}\cdot\mathbf{n} = \mathbf{r}_{,y}\cdot\mathbf{n} = \mathbf{r}_{,x}\cdot\mathbf{r}_{,y} = 0$$

Consequently, since $ds^2 = d\mathbf{Q}\cdot d\mathbf{Q} = (\mathbf{Q}_{,x}dx + \mathbf{Q}_{,y}dy + \mathbf{Q}_{,z}dz)^2$, Eq. (c) gives

$$ds^2 = \alpha^2 dx^2 + \beta^2 dy^2 + \gamma^2 dz^2 \qquad (10.2.8)$$

where

$$\alpha = A\left(1 - \frac{z}{R_x}\right) \quad \beta = B\left(1 - \frac{z}{R_y}\right) \quad \gamma = 1 \qquad (10.2.9)$$

The factors α, β, γ are called the Lamé coefficients.[2] Note that $\alpha = A$ and $\beta = B$, if $z = 0$, that is, on the middle surface. Eq. (c) shows that $\mathbf{Q}_{,x}$ and $\mathbf{Q}_{,y}$ are parallel to the vectors $\mathbf{r}_{,x}$ and $\mathbf{r}_{,y}$.

10.2.1. Strain Energy of General Shells

Koiter (1960) theory is based on a strain energy expression derived in terms of the following three simplifying assumptions:

1. The shell is thin, that is, $h/R <<< 1$, where R is the smallest principal radius of curvature of the undeformed middle surface of the shell.
2. The strains are small compared with unity, and the strain energy density function is given by the quadratic function of the strain components for an isotropic elastic material.
3. The state of stress is approximately plane; the effects of transverse shearing and normal stresses to the middle surface may be neglected in the strain energy density function.

The third of Koiter's assumptions is equivalent to the Kirchhoff approximation used in Chapter 8. Under these assumptions, Koiter (1960) presents the following equations for the strain energy of a thin elastic shell:

$$U = U_m + U_b \qquad (10.2.10a)$$

[2] See Langhaar (1962), page 181.

$$U_m = \frac{C}{2}\iint\left(\varepsilon_x^2 + \varepsilon_y^2 + 2\,\mu\varepsilon_x\varepsilon_y + \frac{1-\mu}{2}\,\gamma_{xy^2}\right)AB\,dxdy \qquad (10.2.10b)$$

$$U_b = \frac{D}{2}\iint\left[\kappa_x^2 + \kappa_y^2 + 2\,\mu\kappa_x\kappa_y + 2(1-\mu)\kappa_{xy}{}^2\right]AB\,dxdy \qquad (10.2.10c)$$

where ε_x, ε_y, and γ_{xy} are middle-surface normal and shear strain components, and κ_x, κ_y, and κ_{xy} are middle-surface curvature changes and rate of twist.

The total potential energy Π of a loaded shell is the sum of the strain energy U and the loss of the potential energy V of the external loads:

$$\Pi = U + V \qquad (10.2.11)$$

Let p_x, p_y, and p_z denote the x, y, and z components, respectively, of the uniformly distributed load over the surface of the shell element, and let u, v, w be the corresponding components of the displacements of a point on the shell middle surface. Then, the loss of potential energy is

$$V = -\iint\left(p_x u + p_y v + p_z w\right)AB\,dxdy \qquad (10.2.12)$$

The equilibrium and stability equations presented here are based on nonlinear middle-surface kinematic relations of the simple form

$$\begin{aligned}
\varepsilon_x &= e_{xx} + \frac{1}{2}\beta_x^2 & \kappa_x &= \chi_{xx} \\
\varepsilon_y &= e_{yy} + \frac{1}{2}\beta_y^2 & \kappa_y &= \chi_{yy} \\
\gamma_{xy} &= e_{xy} + \beta_x\beta_y & \kappa_{xy} &= \chi_{xy}
\end{aligned} \qquad (10.2.13)$$

where e_{ij}, β_i, and χ_{ij} are linear functions of the middle-surface displacement components u, v, w. Sanders kinematic relations (Sanders 1963) at the middle surface may be rewritten

$$\begin{aligned}
e_{xx} &= \frac{u_{,x}}{A} + \frac{A_{,y}v}{AB} - \frac{w}{R_x} \\
e_{yy} &= \frac{v_{,y}}{B} + \frac{B_{,x}u}{AB} - \frac{w}{R_y} \\
e_{xy} &= \frac{v_{,x}}{A} + \frac{u_{,y}}{B} - \frac{B_{,x}v + A_{,y}u}{AB}
\end{aligned} \qquad (10.2.14a)$$

$$\beta_x = -\frac{w_{,x}}{A} - \frac{u}{R_x}$$
$$\beta_y = -\frac{w_{,y}}{B} - \frac{v}{R_y} \tag{10.2.14b}$$

$$\chi_{xx} = \frac{\beta_{x,x}}{A} + \frac{A_{,y}\beta_y}{AB}$$
$$\chi_{yy} = \frac{\beta_{y,y}}{B} + \frac{B_{,x}\beta_x}{AB} \tag{10.2.14c}$$
$$2\,\chi_{xy} = \frac{\beta_{y,x}}{A} + \frac{\beta_{x,y}}{B} - \frac{A_{,y}\beta_x + B_{,x}\beta_y}{AB}$$

Experience with numerical solutions has shown that the terms containing u and v in Eqs. (10.2.14b) are of negligibly small influence for shell segments that are almost flat and for shells whose displacements are rapidly varying functions of shell coordinates, that is, consisting of many relatively small buckles of which bases are significantly smaller than the radius. Such a shell is called "quasi-shallow" even when the shell as a whole is not shallow. If terms containing u and v are discarded, Eqs. (10.2.14b) can be simplified as Eq. (10.2.15). The ramification of ignoring these terms is detailed in Sanders (1963), Koiter (1960), Novozhilov (1964), and Brush and Almroth (1975).

$$\beta_x = -\frac{w_{,x}}{A}$$
$$\beta_y = -\frac{w_{,y}}{B} \tag{10.2.15}$$

Substituting Eqs. (10.2.15) into Eqs. (10.2.14c) and simplifying gives

$$\chi_{xx} = -\frac{w_{,xx}}{A^2} + \frac{A_{,x}w_{,x}}{A^3} - \frac{A_{,y}w_{,y}}{AB^2}$$
$$\chi_{yy} = -\frac{w_{,yy}}{B^2} + \frac{B_{,y}w_{,y}}{B^3} - \frac{B_{,x}w_{,x}}{A^2B} \tag{10.2.16}$$
$$\chi_{xy} = -\frac{w_{,xy}}{AB} + \frac{A_{,y}w_{,x}}{A^2B} + \frac{B_{,x}w_{,y}}{AB^2}$$

Equations (10.2.16) are given by Novozhilov (1964) and Brush and Almroth (1975). The simplified expressions in Eqs. (10.2.14a), (10.2.15), and (10.2.16) are the kinematic relations underlying the Donnell-Mushtari-Vlasov (DMV) equations for a quasi-shallow shell.

The strain components are defined as for rectangular coordinates by Eqs. (8.2.3). The derivation of the expressions for strains in other coordinates is a routine problem of tensor calculus. Most derivations of strains for general shell elements are done by such tensor analyses. General expressions for three-dimensional strains in orthogonal curvilinear coordinates are also derived without the use of tensors by Novozhilov (1953). Equations (10.2.14) may also be obtained by taking only the strains in a plane stress problem and neglecting higher order terms from the general strains derived by Novozhilov.

Nonlinear equilibrium equations may be derived from Eqs. (10.2.10), along with Eqs. (10.2.13), (10.2.14a), (10.2.15), and (10.2.16), by applying the Euler-Lagrange differential equations. The combination of Eqs. (10.2.15) and (10.2.14c) will also work as Eq. (10.2.16) is derived from Eq. (10.2.14c) by way of Eq. (10.2.15). However, the combination of Eqs. (10.2.14b) and (10.2.14c) fails to lead to a relatively simple system of nonlinear equilibrium equations. The resulting equilibrium equations are found to be (see Problem 10.5)

$$\begin{aligned}
&(BN_x)_{,x}+(AN_{xy})_{,y}-B_{,x}N_y+A_{,y}N_{xy}=-ABp_x\\
&(AN_y)_{,y}+(BN_{xy})_{,x}-A_{,y}N_x+B_{,x}N_{xy}=-ABp_y\\
&\left[\frac{1}{A}(BM_x)_{,x}\right]_{,x}-\left(\frac{A_{,y}}{B}M_x\right)_{,y}+\left[\frac{1}{B}(AM_y)_{,y}\right]_{,y}-\left(\frac{B_{,x}}{A}M_y\right)_{,x}\\
&\quad+2\left[M_{xy,xy}+\left(\frac{A_{,y}}{A}M_{xy}\right)_{,x}+\left(\frac{B_{,x}}{B}M_{xy}\right)_{,y}\right]+AB\left(\frac{N_x}{R_x}+\frac{N_y}{R_y}\right)\\
&\quad-\left[\left(BN_x\beta_x+BN_{xy}\beta_y\right)_{,x}+\left(AN_y\beta_y+AN_{xy}\beta_x\right)_{,y}\right]=-ABp_z
\end{aligned}\tag{10.2.17}$$

where

$$\begin{aligned}
N_x&=C(\varepsilon_x+\mu\varepsilon_y) & M_x&=D\left(\kappa_x+\mu\kappa_y\right)\\
N_y&=C(\varepsilon_y+\mu\varepsilon_x) & M_y&=D\left(\kappa_y+\mu\kappa_x\right)\\
N_{xy}&=C\frac{1-\mu}{2}\gamma_{xy} & M_{xy}&=D\,(1-\mu)\kappa_{xy}
\end{aligned}\tag{10.2.18}$$

Equations (10.2.17) can be specialized for a circular cylindrical shell Eqs. (9.2.9) by setting $A=1$, $B=R_y=R$, and $1/R_x=0$. Equations (10.2.17) can also be converted to the nonlinear equilibrium equations for a rectangular flat plate (von Kármán plate equations) Eqs. (8.2.18) by setting $A=B=1$ and $1/R_x=1/R_y=0$. Similarly, Eqs. (10.2.17) can also be

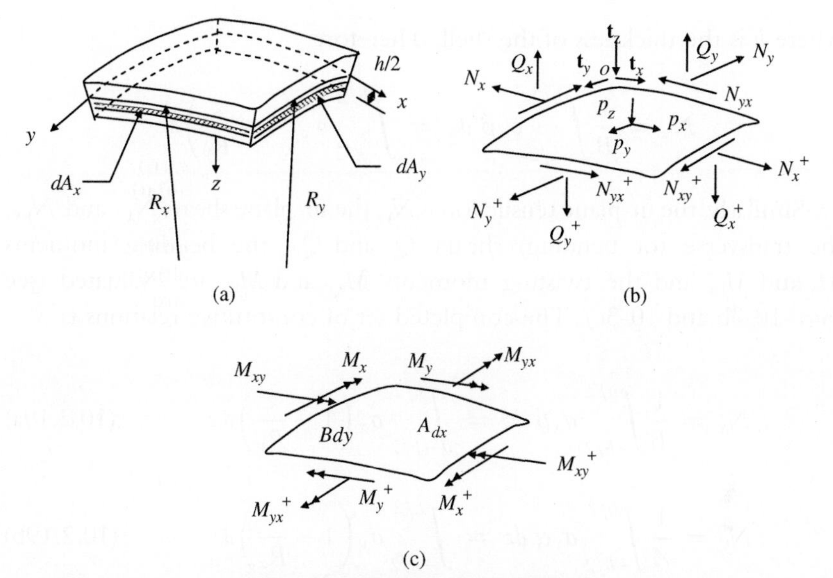

Figure 10-3 Positive internal forces

converted to the nonlinear equilibrium equation for a column Eq. (1.7.14) (see Problem 10.3).

10.2.2. Equilibrium of Shells

Consider a differential element of a general shell, cut out by surfaces $x =$ constant and $y =$ constant as shown in Fig. 10-3. The variables (x, y, z) are orthogonal shell coordinates. Hence, the coordinate lines on the middle surface are lines of principal curvature. By Eqs. (10.2.9), the differential areas of the cross sections shown by the hatched lines in Fig. 10-3a are

$$dA_x = \alpha\, dxdz = A\left(1 - \frac{z}{R_x}\right) dxdz$$

$$dA_y = \beta\, dydz = B\left(1 - \frac{z}{R_y}\right) dydz$$

where R_x and R_y are the principal radii of curvature of the middle surface.

Let N_x be the tensile in-plane force on a cross section per unit length of a y-coordinate line as shown in Fig. 10-3b. Then the total tensile force on the differential element in the x direction is $N_x Bdy$. Hence

$$N_x Bdy = \int \sigma_x dA_y = dy \int_{-h/2}^{h/2} \sigma_x \beta\, dz$$

where h is the thickness of the shell. Therefore

$$N_x = \frac{1}{B}\int_{-h/2}^{h/2} \sigma_x \beta \, dz = \int_{-h/2}^{h/2} \sigma_x \left(1 - \frac{z}{R_y}\right) dz$$

Similarly, the in-plane tensile force N_y, the in-plane shears N_{xy} and N_{yx}, the transverse (or bending) shears Q_x and Q_y, the bending moments M_x and M_y, and the twisting moments M_{xy} and M_{yx} are evaluated (see Figs. 10-3b and 10-3c). The completed set of constitutive relations is

$$N_x = \frac{1}{B}\int_{-h/2}^{h/2} \sigma_x \beta \, dz = \int_{-h/2}^{h/2} \sigma_x \left(1 - \frac{z}{R_y}\right) dz \tag{10.2.19a}$$

$$N_y = \frac{1}{A}\int_{-h/2}^{h/2} \sigma_y \alpha \, dz = \int_{-h/2}^{h/2} \sigma_y \left(1 - \frac{z}{R_x}\right) dz \tag{10.2.19b}$$

$$N_{xy} = \frac{1}{B}\int_{-h/2}^{h/2} \tau_{xy} \beta \, dz = \int_{-h/2}^{h/2} \tau_{xy} \left(1 - \frac{z}{R_y}\right) dz \tag{10.2.19c}$$

$$N_{yx} = \frac{1}{A}\int_{-h/2}^{h/2} \tau_{yx} \alpha \, dz = \int_{-h/2}^{h/2} \tau_{yx} \left(1 - \frac{z}{R_x}\right) dz \tag{10.2.19d}$$

$$Q_x = \frac{1}{B}\int_{-h/2}^{h/2} \tau_{xz} \beta \, dz = \int_{-h/2}^{h/2} \tau_{xz} \left(1 - \frac{z}{R_y}\right) dz \tag{10.2.19e}$$

$$Q_y = \frac{1}{A}\int_{-h/2}^{h/2} \tau_{yz} \alpha \, dz = \int_{-h/2}^{h/2} \tau_{yz} \left(1 - \frac{z}{R_x}\right) dz \tag{10.2.19f}$$

$$M_x = \frac{1}{B}\int_{-h/2}^{h/2} z\sigma_x \beta \, dz = \int_{-h/2}^{h/2} z\sigma_x \left(1 - \frac{z}{R_y}\right) dz \tag{10.2.19g}$$

$$M_y = \frac{1}{A}\int_{-h/2}^{h/2} z\sigma_y \alpha \, dz = \int_{-h/2}^{h/2} z\sigma_y \left(1 - \frac{z}{R_x}\right) dz \tag{10.2.19h}$$

$$M_{xy} = \frac{1}{B}\int_{-h/2}^{h/2} z\tau_{xy} \beta \, dz = \int_{-h/2}^{h/2} z\tau_{xy} \left(1 - \frac{z}{R_y}\right) dz$$

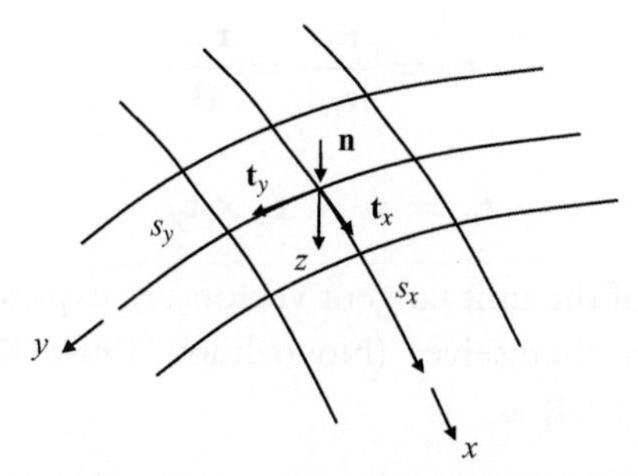

Figure 10-4 Unit tangent and normal vectors

$$M_{yx} = \frac{1}{A}\int_{-h/2}^{h/2} z\tau_{yx}\alpha\, dz = \int_{-h/2}^{h/2} z\tau_{yx}\left(1 - \frac{z}{R_x}\right) dz \qquad (10.2.19i)$$

The positive senses of forces and moments are shown in Figs. 10-3. Equations (10.2.19) are also valid for flat plates, with $1/R_x = 1/R_y = 0$. Likewise they are valid for circular cylindrical shells, with $1/R_x = 0$ and $A = 1$ and $B = R_y = R$.

The nonlinear equilibrium equations may be derived by summing forces and moments for a general shell shown in Fig. 10-3. For an intermediate class of deformations, the angles of rotation β_x and β_y are assumed to be small, and sines and cosines of the angles are replaced by the angles themselves and by unity, respectively. Furthermore, quadratic terms representing nonlinear interaction between the small transverse shearing forces and the rotations are assumed to be negligibly small.

Novozhilov (1964)[3] states that the thicknesses of shells for a large number of applications lie in the range of $1/1000 \leq h/R \leq 1/50$, that is, in the range of thin shells. It appears appropriate to neglect z/R relative to unity in the DMV form of the equations. Then $N_{xy} = N_{yx}$ and $M_{xy} = M_{yx}$.

Define unit tangent vectors as shown in Fig. 10-4. When the curvilinear coordinate lines are defined by a position vector **r** emanating from the origin of the rectangular orthogonal coordinate system, the derivatives of **r** with respect to the curvilinear coordinates $\partial\mathbf{r}/\partial x = \mathbf{r}_{,x}$ and $\partial\mathbf{r}/\partial y = \mathbf{r}_{,y}$ are vectors that are tangent to the s_x and s_y coordinate lines, and the corresponding unit tangent vectors are given by

$$\mathbf{t}_x = \frac{\mathbf{r}_{,x}}{|\mathbf{r}_{,x}|} = \frac{\mathbf{r}_{,x}}{A} \qquad (10.2.20a)$$

[3] See page 2.

$$\mathbf{t}_y = \frac{\mathbf{r}_{,y}}{|\mathbf{r}_{,y}|} = \frac{\mathbf{r}_{,y}}{B} \tag{10.2.20b}$$

$$\mathbf{t}_n = \mathbf{n} = \mathbf{t}_x \times \mathbf{t}_y \tag{10.2.20c}$$

The derivatives of the unit tangent vectors are expressed in terms of the unit tangent vectors themselves (Novozhilov 1964; Klaus 1967; Gould 1988) (see Problem 10.4) as

$$\begin{Bmatrix} \mathbf{t}_{x,x} \\ \mathbf{t}_{x,y} \\ \mathbf{t}_{y,x} \\ \mathbf{t}_{y,y} \\ \mathbf{t}_{n,x} \\ \mathbf{t}_{n,y} \end{Bmatrix} = \begin{bmatrix} 0 & \dfrac{-A_{,y}}{B} & \dfrac{A}{R_x} \\ 0 & \dfrac{B_{,x}}{A} & 0 \\ \dfrac{A_{,y}}{B} & 0 & 0 \\ \dfrac{-B_{,x}}{A} & 0 & \dfrac{B}{R_y} \\ \dfrac{-A}{R_x} & 0 & 0 \\ 0 & \dfrac{-B}{R_y} & 0 \end{bmatrix} \begin{Bmatrix} \mathbf{t}_x \\ \mathbf{t}_y \\ \mathbf{t}_n \end{Bmatrix} \tag{10.2.21}$$

In terms of the stress resultants and moments, the resulting vectors are

$$\mathbf{F}_x = \left(N_x\mathbf{t}_x + N_{xy}\mathbf{t}_y + Q_x\mathbf{t}_n\right)B\,dy \tag{10.2.22a}$$

$$\mathbf{F}_y = \left(N_y\mathbf{t}_y + N_{yx}\mathbf{t}_x + Q_y\mathbf{t}_n\right)A\,dx \tag{10.2.22b}$$

$$\mathbf{C}_x = \left(-M_{xy}\mathbf{t}_x + M_x\mathbf{t}_y\right)B\,dy \tag{10.2.22c}$$

$$\mathbf{C}_y = \left(-M_y\mathbf{t}_x + M_{yx}\mathbf{t}_y\right)A\,dx \tag{10.2.22d}$$

and the load vector is

$$\mathbf{p}AB\,dxdy = \left(p_x\mathbf{t}_x + p_y\mathbf{t}_y + p_z\mathbf{t}_n\right)AB\,dxdy \tag{10.2.22e}$$

Applying the equations of static equilibrium yields

$$\sum \mathbf{F} = 0 \tag{10.2.23a}$$

$$\sum \mathbf{C} = 0 \tag{10.2.23b}$$

Equation (10.2.23a) becomes

$$\left(\mathbf{F}_x + \mathbf{F}_{x,x}\, dx - \mathbf{F}_x\right) + \left(\mathbf{F}_y + \mathbf{F}_{y,y} dx - \mathbf{F}_y\right) + \mathbf{p}AB\, dxdy = 0$$

or

$$\mathbf{F}_{x,x}\, dx + \mathbf{F}_{y,y}\, dy + \mathbf{p}AB\, dxdy = 0 \tag{10.2.24}$$

Substituting Eqs. (10.2.22a), (10.2.22b), and (10.2.22e) into Eq. (10.2.24) and dividing through by $dxdy$ gives

$$\left[(N_x \mathbf{t}_x + N_{xy}\mathbf{t}_y + Q_x \mathbf{t}_n)B\right]_{,x} + \left[(N_{yx}\mathbf{t}_x + N_y \mathbf{t}_y + Q_y \mathbf{t}_n)A\right]_{,y}$$

$$+ \left(\mathbf{p}_x \mathbf{t}_x + \mathbf{p}_y \mathbf{t}_y + \mathbf{p}_z \mathbf{t}_n\right)AB = 0 \tag{10.2.25}$$

The differentiation indicated in Eq. (10.2.25) can be evaluated using the identities given by Eq. (10.2.21). The resulting vector equations may be factored into

$$F_x \mathbf{t}_x + F_y \mathbf{t}_y + F_z \mathbf{t}_n = 0 \tag{10.2.26}$$

Since the unit tangent vectors are independent, Eq. (10.2.26) can only be satisfied if

$$F_x = 0; \quad F_y = 0; \quad F_z = 0 \tag{10.2.27}$$

which yields the three scalar equations of force equilibrium. They are (see Problem 10.7):

$$F_x = \left[(BN_x)_{,x} + \left(AN_{yx}\right)_{,y} + A_{,y}N_{xy} - B_{,x}N_y\right] - Q_x \frac{AB}{R_x} + p_x AB = 0 \tag{10.2.28a}$$

$$F_y = \left[\left(BN_{xy}\right)_{,x} + \left(AN_y\right)_{,y} + B_{,x}N_{yx} - A_{,y}N_x\right] - Q_y \frac{AB}{R_y} + p_y AB = 0 \tag{10.2.28b}$$

$$F_z = \left[(BQ_x)_{,x} + \left(AQ_y\right)_{,y}\right] + N_x \frac{AB}{R_x} + N_y \frac{AB}{R_y} + p_z AB = 0 \tag{10.2.28c}$$

Next, the moment equilibrium equation Eq. (10.2.23b) is evaluated about axes through point "o" in Fig. 10-3(b). Summing the moments given in Eqs. (10.22c) and (10.2.22d) yields

$$\sum \mathbf{C} = \mathbf{C}_x + \mathbf{C}_{x,x}\, dx - \mathbf{C}_x + \mathbf{C}_y + \mathbf{C}_{y,y}\, dy - \mathbf{C}_y \qquad (10.2.29a)$$

Substituting Eqs. (10.2.22c) and (10.2.22d) into Eq. (10.2.29a) gives:

$$\sum \mathbf{C} = \left[\left(-M_{xy}\mathbf{t}_x + M_x\mathbf{t}_y\right)B\right]_{,x} dxdy + \left[\left(-M_y\mathbf{t}_x + M_{yx}\mathbf{t}_y\right)A\right]_{,y} dxdy \qquad (10.2.29b)$$

Additional stress vectors contributing to the moment equilibrium include

$$\left[Q_y\mathbf{t}_x - Q_x\mathbf{t}_y + \left(N_{xy} - N_{yx}\right)\mathbf{t}_z\right]AB\, dxdy \qquad (10.2.30)$$

Expanding Eq. (10.2.29b) in accordance with Eqs. (10.2.21) and combining the results with Eq. (10.2.30), one obtains

$$\mathbf{G}_x\mathbf{t}_x + \mathbf{G}_y\mathbf{t}_y + \mathbf{G}_z\mathbf{t}_z = 0 \qquad (10.2.31)$$

As the unit tangent vectors are independent, Eq. (10.2.31) can only be satisfied if

$$\mathbf{G}_x = 0; \quad \mathbf{G}_y = 0; \quad \mathbf{G}_z = 0 \qquad (10.2.32)$$

Equations (10.2.32) lead to three scalar moment-equilibrium equations. As the third moment-equilibrium equation, however, is identically satisfied if the symmetry of the stress tensor $\sigma_{ij} = \sigma_{ji}$ is invoked (Novozhilov 1964), only two moment-equilibrium equations will be evaluated. There exist modified shell theories, such as Klaus (1967), which attempt to redefine the stress resultants and couples so that $\mathbf{G}_z = 0$ can be satisfied. However, none of these attempts succeed in satisfying the so-called sixth equation unconditionally.

They are (see Problem 10.8):

$$\mathbf{G}_x = -\left(BM_{xy}\right)_{,x} - \left(AM_y\right)_{,y} - B_{,x}M_{yx} + A_{,y}M_x + Q_yAB = 0 \qquad (10.2.33a)$$

$$\mathbf{G}_y = \left(BM_x\right)_{,x} + \left(AM_{yx}\right)_{,y} - B_{,x}M_y + A_{,y}M_{xy} - Q_xAB = 0 \qquad (10.2.33b)$$

Rearranging Eqs. (10.2.33a) and (10.2.33b) gives

$$Q_yA = \frac{1}{B}\left[\left(BM_{xy}\right)_{,x} + \left(AM_y\right)_{,y} + B_{,x}M_{yx} - A_{,y}M_x\right] \qquad (10.2.34a)$$

$$BQ_x = \frac{1}{A}\left[(BM_x)_{,x} + (AM_{yx})_{,y} - B_{,x}M_y + A_{,y}M_{xy}\right] \quad (10.2.34b)$$

Substituting the appropriate derivatives of Eqs. (10.2.34) into Eq. (10.2.28c) yields

$$\begin{aligned} F_z = & \left\{\frac{1}{A}\left[(BM_x)_{,x} + (AM_{yx})_{,y} - B_{,x}M_y + A_{,y}M_{xy}\right]\right\}_{,x} \\ & + \left\{\frac{1}{B}\left[(BM_{xy})_{,x} + (AM_y)_{,y} + B_{,x}M_{yx} - A_{,y}M_x\right]\right\}_{,y} \\ & + N_x\frac{AB}{R_x} + N_y\frac{AB}{R_y} + p_zAB = 0 \end{aligned} \quad (10.2.35)$$

Although all stress resultants necessary for the small displacement theory of general shells have been accounted for by Eqs. (10.2.28a), (10.2.28b), and (10.2.35), there are yet other force components to be added to Eq. (10.2.35) in order to account for the effect of large displacements that will lead to the derivation of the equation for elastic buckling of shells. They are

$$\begin{aligned} & N_xB\,dy\beta_x - \left[N_xB\,dy\beta_x + (N_xB\,dy\beta_x)_{,x}dx\right] + N_yA\,dx\beta_y \\ & - \left[N_yA\,dx\beta_y + (N_yA\,dx\beta_y)_{,y}dy\right] + N_{xy}B\,dy\beta_y \\ & - \left[N_{xy}B\,dy\beta_y + (N_{xy}B\,dy\beta_y)_{,x}dx\right] + N_{yx}A\,dx\beta_x \\ & - \left[N_{yx}A\,dx\beta_x + (N_{yx}A\,dx\beta_x)_{,y}dy\right] \\ & = -(BN_x\beta_x + BN_{xy}\beta_y)_{,x}\,dxdy - (AN_y\beta_y + AN_{yx}\beta_x)_{,y}\,dxdy \end{aligned} \quad (10.2.36)$$

Dividing Eq. (10.2.36) by $dxdy$ for consistent dimensions and adding to Eq. (10.2.35) gives

$$\begin{aligned} F_z = & \left\{\frac{1}{A}\left[(BM_x)_{,x} + (AM_{yx})_{,y} - B_{,x}M_y + A_{,y}M_{xy}\right]\right\}_{,x} \\ & + \left\{\frac{1}{B}\left[(BM_{xy})_{,x} + (AM_y)_{,y} + B_{,x}M_{yx} - A_{,y}M_x\right]\right\}_{,y} \quad (10.2.37) \\ & - (BN_x\beta_x + BN_{xy}\beta_y)_{,x} - (AN_y\beta_y + AN_{yx}\beta_x)_{,y} \\ & + N_x\frac{AB}{R_x} + N_y\frac{AB}{R_y} + p_zAB = 0 \end{aligned}$$

In the DMV form of equilibrium equations, terms involving the vertical shear Q_x or Q_y divided by the radius of curvature in Eqs. (10.2.28) are ignored for being small for thin shells as in Sanders (1963) and Brush and Almroth (1975). When they are ignored, Eqs. (10.2.28a) and (10.2.28b), and (10.2.37) are identical to Eqs. (10.2.17) (see Problem 10.9) derived by the principle of minimum total potential energy by way of the calculus of variations.

It should be noted that the use of the equilibrium method of deriving a governing differential equation based on an isolated free-body diagram is much easier for a simple structure, such as a column, but for complex three-dimensional structures, such as a general shell, the energy method is much more straightforward.

10.3. LINEAR STABILITY EQUATIONS (DONNELL TYPE)

As was done in Chapter 9, the linear differential equations for the determination of the critical load of a general shell subjected to external loading are derived by application of the adjacent-equilibrium criterion. The same equations are then rederived for loss of stability by application of Trefftz criterion in terms of linear displacement parameters e_{ij}, β_i, χ_{ij} of Eqs. (10.2.13). Equations (10.2.17) govern all linear and nonlinear equilibrium conditions of the general shell within the confinement of the intermediate class of deformations. The equations include linear, quadratic, and cubic terms of variables u, v, and w, and therefore are nonlinear. It is now a fairly simple task to obtain a very good iterative numerical solution by a well-established finite element code.

10.3.1. Adjacent-Equilibrium Criterion

Adjacent (or neighboring) equilibrium configurations are examined using the procedure outlined by Brush and Almroth (1975), as was done in Chapters 8 and 9. Consider the equilibrium configuration at the bifurcation point. Then, the equilibrium configuration is perturbed by the small incremental displacement. The incremental displacement in u_1, v_1, w_1 is arbitrary and tentative. Variables in the two adjacent configurations before and after the increment are represented by u_0, v_0, w_0 and u, v, w. Let

$$
\begin{aligned}
u &\rightarrow u_0 + u_1 \\
v &\rightarrow v_0 + v_1 \\
w &\rightarrow w_0 + w_1
\end{aligned}
\tag{10.3.1}
$$

where the arrow is read as "be replaced by." Substitution of Eqs. (10.31) into Eqs. (10.2.17) results in equations containing terms that are linear, quadratic, and cubic in $u_0,\ v_0,\ w_0$ and $u_1,\ v_1,\ w_1$ displacement components. In the new equation obtained, the terms containing $u_0,\ v_0,\ w_0$ alone are equal to zero as $u_0,\ v_0,\ w_0$ constitute an equilibrium configuration, and terms that are quadratic and cubic in $u_1,\ v_1,\ w_1$ may be ignored because of the smallness of the incremental displacement. Therefore, the resulting equations are homogeneous and linear in $u_1,\ v_1,\ w_1$ with variable coefficients. The increment in $u,\ v,\ w$ causes a concomitant change in the internal force such as

$$\begin{aligned} N_x &\rightarrow N_{x0} + \Delta N_x \\ N_y &\rightarrow N_{y0} + \Delta N_y \\ N_{xy} &\rightarrow N_{xy0} + \Delta N_{xy} \end{aligned} \tag{10.3.2}$$

where terms with subscript 0 correspond to the $u_0,\ v_0,\ w_0$ displacement, and $\Delta N_x,\ \Delta N_y,\ \Delta N_{xy}$ are increments corresponding to $u_1,\ v_1,\ w_1$. Let also $N_{x1},\ N_{y1},\ N_{xy1}$ represent parts of $\Delta N_x, \Delta N_y,\ \Delta N_{xy}$, respectively, that are linear in $u_1,\ v_1,\ w_1$. For example, from Eqs. (10.2.13), (10.2.14a), (10.2.15), (10.2.16), and (10.2.18),

$$\begin{aligned} N_x = C\Bigg\{&\frac{1}{A}\frac{\partial u}{\partial x} + \frac{A_{,y}}{AB}v - \frac{w}{R_x} + \frac{1}{2}\left(\frac{\partial w}{\partial x}\right)^2 \\ &+ \mu\left[\frac{1}{B}\frac{\partial v}{\partial y} + \frac{B_{,x}}{AB}u - \frac{w}{R_y} + \frac{1}{2}\left(\frac{\partial w}{\partial y}\right)^2\right]\Bigg\} \end{aligned}$$

Then

$$N_{x0} + \Delta N_x = C\left\{\begin{aligned} &\frac{1}{A}\left(\frac{\partial u_0}{\partial x} + \frac{\partial u_1}{\partial x}\right) + \frac{A_{,y}}{AB}(v_0 + v_1) - \frac{1}{R_x}(w_0 + w_1) \\ &+ \frac{1}{2}\frac{\partial w_0}{\partial x}^2 + \frac{\partial w_0}{\partial x}\frac{\partial w_1}{\partial x} + \frac{1}{2}\frac{\partial w_1}{\partial x}^2 \\ &+ \mu\left[\frac{1}{B}\left(\frac{\partial v_0}{\partial y} + \frac{\partial v_1}{\partial y}\right) + \frac{B_{,x}}{AB}(u_0 + u_1)\right. \\ &\left. - \frac{1}{R_y}(w_0 + w_1) + \frac{1}{2}\frac{\partial w_0}{\partial y}^2 + \frac{\partial w_0}{\partial y}\frac{\partial w_1}{\partial y} + \frac{1}{2}\frac{\partial w_1}{\partial y}^2\right] \end{aligned}\right\}$$

From which

$$N_{x0} = C\left[\frac{1}{A}\frac{\partial u_0}{\partial x} + \frac{A_{,y}}{AB}v_0 - \frac{1}{R_x}w_0 + \frac{1}{2}\frac{\partial w_0}{\partial x}^2 + \mu\left(\frac{1}{B}\frac{\partial v_0}{\partial y} + \frac{B_{,x}}{AB}u_0 - \frac{1}{R_y}w_0 + \frac{1}{2}\frac{\partial w_0}{\partial y}^2\right)\right]$$

$$\Delta N_x = C\left[\begin{array}{l}\frac{1}{A}\frac{\partial u_1}{\partial x} + \frac{A_{,y}}{AB}v_1 - \frac{1}{R_x}w_1 + \frac{\partial w_0}{\partial x}\frac{\partial w_1}{\partial x} + \frac{1}{2}\frac{\partial w_0}{\partial x}^2 \\ + \mu\left(\frac{1}{B}\frac{\partial v_1}{\partial y} + \frac{B_{,x}}{AB}u_1 - \frac{1}{R_y}w_1 + \frac{\partial w_0}{\partial y}\frac{\partial w_1}{\partial y} + \frac{1}{2}\frac{\partial w_1}{\partial y}^2\right)\end{array}\right]$$

$$N_{x1} = C\left[\frac{1}{A}\frac{\partial u_1}{\partial x} + \frac{A_{,y}}{AB}v_1 - \frac{1}{R_x}w_1 + \frac{\partial w_0}{\partial x}\frac{\partial w_1}{\partial x} + \mu\left(\frac{1}{B}\frac{\partial v_1}{\partial y} + \frac{B_{,x}}{AB}u_1 - \frac{1}{R_y}w_1 + \frac{\partial w_0}{\partial y}\frac{\partial w_1}{\partial y}\right)\right] \quad (10.3.3a)$$

Similarly,

$$N_y = C\left\{\frac{1}{B}\frac{\partial v}{\partial y} + \frac{B_{,x}}{AB}u - \frac{w}{R_y} + \frac{1}{2}\left(\frac{\partial w}{\partial y}\right)^2 + \mu\left[\frac{1}{A}\frac{\partial u}{\partial x} + \frac{A_{y}}{AB}v - \frac{w}{R_x} + \frac{1}{2}\left(\frac{\partial w}{\partial x}\right)^2\right]\right\}$$

Then

$$N_{y0} + \Delta N_y = C\left\{\begin{array}{l}\frac{1}{B}\left(\frac{\partial v_0}{\partial y} + \frac{\partial v_1}{\partial y}\right) + \frac{B_{,x}}{AB}(u_0 + u_1) - \frac{1}{R_y}(w_0 + w_1) \\ + \frac{1}{2}\frac{\partial w_0}{\partial y}^2 + \frac{\partial w_0}{\partial y}\frac{\partial w_1}{\partial y} + \frac{1}{2}\frac{\partial w_1}{\partial y}^2 \\ + \mu\left[\frac{1}{A}\left(\frac{\partial u_0}{\partial x} + \frac{\partial u_1}{\partial x}\right) + \frac{A_y}{AB}(v_0 + v_1)\right. \\ \left. - \frac{1}{R_x}(w_0 + w_1) + \frac{1}{2}\frac{\partial w_0}{\partial x}^2 + \frac{\partial w_0}{\partial x}\frac{\partial w_1}{\partial x} + \frac{1}{2}\frac{\partial w_1}{\partial x}^2\right]\end{array}\right\}$$

From which

$$N_{y0} = C\left[\frac{1}{B}\frac{\partial v_0}{\partial y} + \frac{B_{,x}}{AB}u_0 - \frac{1}{R_y}w_0 + \frac{1}{2}\frac{\partial w_0}{\partial y}^2 + \mu\left(\frac{1}{A}\frac{\partial u_0}{\partial x} + \frac{A_{,y}}{AB}v_0 - \frac{1}{R_x}w_0 + \frac{1}{2}\frac{\partial w_0}{\partial x}^2\right)\right]$$

$$\Delta N_y = C\left[\begin{array}{l} \dfrac{1}{B}\dfrac{\partial v_1}{\partial y} + \dfrac{B_{,x}}{AB}u_1 - \dfrac{1}{R_y}w_1 + \dfrac{\partial w_0}{\partial y}\dfrac{\partial w_1}{\partial y} + \dfrac{1}{2}\dfrac{\partial w_0}{\partial y}^2 \\ +\mu\left(\dfrac{1}{A}\dfrac{\partial u_1}{\partial x} + \dfrac{A_{,y}}{AB}v_1 - \dfrac{1}{R_x}w_1 + \dfrac{\partial w_0}{\partial x}\dfrac{\partial w_1}{\partial x} + \dfrac{1}{2}\dfrac{\partial w_1}{\partial x}^2\right) \end{array}\right]$$

$$N_{y1} = C\left[\frac{1}{B}\frac{\partial v_1}{\partial y} + \frac{B_{,x}}{AB}u_1 - \frac{1}{R_y}w_1 + \frac{\partial w_0}{\partial y}\frac{\partial w_1}{\partial y} + \mu\left(\frac{1}{A}\frac{\partial u_1}{\partial x} + \frac{A_{,y}}{AB}v_1 - \frac{1}{R_x}w_1 + \frac{\partial w_0}{\partial x}\frac{\partial w_1}{\partial x}\right)\right] \tag{10.3.3b}$$

$$N_{xy} = C\frac{1-\mu}{2}\left(\frac{1}{A}\frac{\partial v}{\partial x} + \frac{\partial u}{B} - \frac{B_{,x}v + A_{,y}u}{AB} + \frac{\partial w}{\partial x}\frac{\partial w}{\partial y}\right)$$

Then

$$N_{xy0} + \Delta N_{xy} = C\frac{1-\mu}{2}\left[\begin{array}{c} \dfrac{1}{A}\left(\dfrac{\partial v_0}{\partial x} + \dfrac{\partial v_1}{\partial x}\right) + \dfrac{1}{B}(u_0 + u_1) \\ -\dfrac{B_{,x}(v_0 + v_1) + A_{,y}(u_0 + u_1)}{AB} \\ +\dfrac{\partial(w_0 + w_1)}{\partial x}\dfrac{\partial(w_0 + w_1)}{\partial y} \end{array}\right]$$

From which

$$N_{xy0} = C\frac{1-\mu}{2}\left[\frac{1}{A}\left(\frac{\partial v_0}{\partial x}\right) + \frac{1}{B}(u_0) - \frac{B_{,x}(v_0) + A_{,y}(u_0)}{AB} + \frac{\partial(w_0)}{\partial x}\frac{\partial(w_0)}{\partial y}\right]$$

$$\Delta N_{xy} = C\frac{1-\mu}{2}\left[\frac{1}{A}\left(\frac{\partial v_1}{\partial x}\right) + \frac{1}{B}(u_1) - \frac{B_{,x}(v_1) + A_{,y}(u_1)}{AB} + w_{0,x}w_{1,y} + w_{1,x}w_{0,y} + w_{1,x}w_{1,y}\right]$$

$$N_{xy1} = C\frac{1-\mu}{2}\left[\frac{1}{A}\left(\frac{\partial v_1}{\partial x}\right) + \frac{1}{B}(u_1) - \frac{B_{,x}(v_1) + A_{,y}(u_1)}{AB} + w_{0,x}w_{1,y} + w_{1,x}w_{0,y}\right] \quad (10.3.3c)$$

A similar procedure is taken for the expression of bending moments.

$$M_{x1} = -D\left[\left(\frac{\beta_{x1,x}}{A} + \frac{A_{,y}\beta_{y1}}{AB}\right) + \mu\left(\frac{\beta_{y1,y}}{B} + \frac{B_{,x}\beta_{x1}}{AB}\right)\right] \quad (10.3.3d)$$

$$M_{y1} = -D\left[\left(\frac{\beta_{y1,y}}{B} + \frac{B_{,x}\beta_{x1}}{AB}\right) + \mu\left(\frac{\beta_{x1,x}}{A} + \frac{A_{,y}\beta_{y1}}{AB}\right)\right] \quad (10.3.3e)$$

$$M_{xy1} = D\frac{1-\mu}{2}\left(\frac{\beta_{y1,x}}{A} + \frac{\beta_{x1,y}}{B} - \frac{A_{,y}\beta_{x1} + B_{,x}\beta_{y1}}{AB}\right) \quad (10.3.3f)$$

Substituting Eqs. (10.3.3) into Eqs. (10.2.17) gives

$$(BN_{x1})_{,x} + (AN_{xy1})_{,y} - B_{,x}N_{y1} + A_{,y}N_{xy1} = 0$$

$$(AN_{y1})_{,y} + (BN_{xy1})_{,x} - A_{,y}N_{x1} + B_{,x}N_{xy1} = 0$$

$$\left[\frac{1}{A}(BM_{x1})_{,x}\right]_{,x} - \left(\frac{A_{,y}}{B}M_{x1}\right)_{,y} + \left[\frac{1}{B}(AM_{y1})_{,y}\right]_{,y} - \left(\frac{B_{,x}}{A}M_{y1}\right)_{,x}$$
$$+ 2\left[M_{xy1,xy} + \left(\frac{A_{,y}}{A}M_{xy1}\right)_{,x} + \left(\frac{B_{,x}}{B}M_{xy1}\right)_{,y}\right] + AB\left(\frac{N_{x1}}{R_x} + \frac{N_{y1}}{R_y}\right)$$
$$-\left[\begin{array}{l}(BN_{x0}\beta_{x1} + BN_{xy0}\beta_{y1})_{,x} + (BN_{x1}\beta_{x1} + BN_{xy1}\beta_{y1})_{,x} \\ + (AN_{y0}\beta_{y} + AN_{xy0}\beta_{x})_{,y} + (AN_{y1}\beta_{y} + AN_{xy1}\beta_{x})_{,y}\end{array}\right] = 0 \quad (10.3.4)$$

where

$$e_{xx1} = \frac{u_{1,x}}{A} + \frac{A_{,y}v_1}{AB} - \frac{w_1}{R_x} \quad (10.3.5a)$$

$$e_{yy1} = \frac{v_{1,y}}{B} + \frac{B_{,x}u_1}{AB} - \frac{w_1}{R_y} \tag{10.3.5b}$$

$$e_{xy1} = \frac{v_{1,x}}{A} + \frac{u_{1,y}}{B} - \frac{B_{,x}v_1 + A_{,y}u_1}{AB} \tag{10.3.5c}$$

$$\beta_{x1} = -\frac{w_{1,x}}{A} \tag{10.3.5d}$$

$$\beta_{y1} = -\frac{w_{1,y}}{B} \tag{10.3.5e}$$

Substituting Eqs. (10.3.3) and (10.3.5) into Eqs. (10.3.4) yields a set of three linear homogeneous equations in u_1, v_1, w_1 with variable coefficients in N_{x0}, N_{y0}, N_{xy0}, β_{x0}, and β_{y0}. These coefficients are evaluated by Eqs. (10.2.17). Eqs. (10.3.3), (10.3.4), and (10.3.5) are linear stability equations for the quasi-shallow shell of general shape under the DMV approximations.

10.3.2. The Trefftz Criterion

Equations (10.3.4) are rederived on the basis of the minimum potential energy criterion. Equations (10.2.10) to (10.2.13) represent a general expression for the potential energy in terms of parameters e_{ij}, β_i, χ_{ij} that are linear functions of the middle surface displacement components u, v, w. To obtain an expression for the second variation of the total potential energy, the displacement components are again disturbed

$$\begin{aligned} u &\rightarrow u_0 + u_1 \\ v &\rightarrow v_0 + v_1 \\ w &\rightarrow w_0 + w_1 \end{aligned} \tag{10.3.1}$$

Then one collects all terms in the resulting expression that are quadratic in the virtual deformations u_1, v_1, w_1. Consequently, e_{ij} is replaced by $e_{ij0} + e_{ij1}$, etc., and terms that are quadratic in the quantities with subscript 1 are collected. Since the potential energy of the applied load Eq. (10.2.12) is a linear functional of the displacement components and makes no contribution to the second variation, $\delta^2 V = 0$. Therefore

$$\delta^2 \Pi = \delta^2 U$$

or

$$\delta^2 \Pi = \delta^2 U_m + \delta^2 U_b \tag{10.3.6}$$

Hence, the expressions for the second variation of the membrane and bending strain energy are found to be (see Problem 10.10):

$$\frac{1}{2}\delta^2 U_m = \frac{C}{2}\iint \left\{ \begin{aligned} &\left[(\delta\varepsilon_x)^2 + \left(\delta\varepsilon_y\right)^2 + 2\mu(\delta\varepsilon_x)\left(\delta\varepsilon_y\right) \right. \\ &\left. + \frac{1-\mu}{2}\left(\delta\gamma_{xy}\right)^2\right] + \left[\left(\varepsilon_{x0} + \mu\varepsilon_{y0}\right)\left(\delta^2\varepsilon_x\right) \right. \\ &\left. + \left(\varepsilon_{y0} + \mu\varepsilon_{x0}\right)\left(\delta^2\varepsilon_y\right) + \frac{1-\mu}{2}\gamma_{xy0}\left(\delta^2\gamma_{xy}\right)^2\right] \end{aligned} \right\} ABdxdy \tag{10.3.7a}$$

and

$$\begin{aligned} \frac{1}{2}\delta^2 U_b &= \frac{D}{2}\iint \left[(\delta\kappa_x)^2 + \left(\delta\kappa_y\right)^2 + 2\mu(\delta\kappa_x)\left(\delta\kappa_y\right)\right. \\ &\quad \left. + 2(1-\mu)\left(\delta\kappa_{xy}\right)^2\right] AB\, dxdy \end{aligned} \tag{10.3.7b}$$

From Eqs. (10.2.13),

$$\begin{aligned} \delta\varepsilon_x &= e_{xx1} + \beta_{x0}\beta_{x1} & \delta^2\varepsilon_x &= \beta_{x1} & \delta\kappa_x &= \chi_{xx1} \\ \delta\varepsilon_y &= e_{yy1} + \beta_{y0}\beta_{y1} & \delta^2\varepsilon_y &= \beta_{y1} & \delta\kappa_y &= \chi_{yy1} \\ \delta\gamma_{xy} &= e_{y1} + \beta_{y0}\beta_1 + \beta_0\beta_{y1} & \delta^2\gamma_{xy} &= 2\beta_{x1}\beta_{y1} & \delta\kappa_{xy} &= \chi_{xy1} \end{aligned} \tag{10.3.8}$$

From Eq. (10.2.18),

$$\begin{aligned} N_{x0} &= C\left(\varepsilon_{x0} + \mu\varepsilon_{y0}\right) \\ N_{x0} &= C\left(\varepsilon_{y0} + \mu\varepsilon_{x0}\right) \\ N_{xy0} &= C(1-\mu)\gamma_{xy0}/2 \end{aligned} \tag{10.3.9}$$

Therefore the expression for the second variation of the total potential energy is

$$\begin{aligned} \frac{1}{2}\delta^2\Pi &= \frac{C}{2}\iint \left[\begin{aligned} &\left(e_{xx1} + \beta_{x0}\beta_{x1}\right)^2 + \left(e_{yy1} + \beta_{y0}\beta_{y1}\right)^2 \\ &+ 2\mu\left(e_{xx1} + \beta_{x0}\beta_{x1}\right)\left(e_{yy1} + \beta_{y0}\beta_{y1}\right) \\ &+ \frac{1-\mu}{2}\left(e_{xy1} + \beta_{x0}\beta_{y1} + \beta_{y0}\beta_{x1}\right) \end{aligned} \right] AB\, dxdy \\ &+ \frac{1}{2}\iint \left(N_{x0}{\beta_{x1}}^2 + N_{y0}{\beta_{y1}}^2 + 2N_{xy0}\beta_{x1}\beta_{y1}\right) AB\, dxdy \\ &+ \frac{D}{2}\iint \left[{\chi_{xx1}}^2 + {\chi_{yy1}}^2 + 2\mu\chi_{xx1}\chi_{yy1} + 2(1-\mu){\chi_{xy1}}^2\right] AB\, dxdy \end{aligned} \tag{10.3.10}$$

Equation (10.3.10) is a general expression for the second variation of the total potential energy of a thin shell of general shape (shallow or nonshallow) as no simplifying expressions are adopted in the derivation.

For quasi-shallow shells, the incremental deformation parameters given by the DMV approximations are, from Eqs. (10.2.14a), (10.2.15), and (10.2.16),

$$
\begin{aligned}
e_{xx1} &= \frac{u_{1,x}}{A} + \frac{A_{,y}v_1}{AB} - \frac{w_1}{R_x} \\
e_{yy1} &= \frac{v_{1,y}}{B} + \frac{B_{,x}u_1}{AB} - \frac{w_1}{R_y} \\
e_{xy1} &= \frac{v_{1,x}}{A} + \frac{u_{1,y}}{B} - \frac{B_{,x}v_1 + A_{,y}u_1}{AB} \\
\beta_{x1} &= -\frac{w_{1,x}}{A} \\
\beta_{y1} &= -\frac{w_{1,y}}{B} \\
\chi_{xx1} &= -\frac{w_{,xx}}{A^2} + \frac{A_{,x}w_{,x}}{A^3} - \frac{A_{,y}w_{,y}}{AB^2} \\
\chi_{yy1} &= -\frac{w_{1,yy}}{B^2} + \frac{B_{,y}w_{1,y}}{B^3} - \frac{B_{,x}w_{1,x}}{A^2B} \\
\chi_{xy1} &= -\frac{w_{1,xy}}{AB} + \frac{A_{,y}w_{1,x}}{A^2B} + \frac{B_{,x}w_{1,y}}{AB^2}
\end{aligned}
\qquad (10.3.11)
$$

Substituting Eqs. (10.3.9) and (10.3.11) into Eq. (10.3.10) and applying the Euler-Lagrange differential equation given by Eqs. (8.3.12) yields (see Problem 10.11)

$$(BN_{x1})_{,x} + (AN_{xy1})_{,y} - B_{,x}N_{y1} + A_{,y}N_{xy1} = 0 \qquad (10.3.12a)$$

$$(AN_{y1})_{,y} + (BN_{xy1})_{,x} - A_{,y}N_{x1} + B_{,x}N_{xy1} = 0 \qquad (10.3.12b)$$

$$\left[\frac{1}{A}(BM_{x1})_{,x}\right]_{,x}-\left(\frac{A_{,y}}{B}M_{x1}\right)_{,y}+\left[\frac{1}{B}(AM_{y1})_{,y}\right]_{,y}-\left(\frac{B_{,x}}{A}M_{y1}\right)_{,x}$$
$$+2\left[M_{xy1,xy}+\left(\frac{A_{,y}}{A}M_{xy1}\right)_{,x}+\left(\frac{B_{,x}}{B}M_{xy1}\right)_{,y}\right]+AB\left(\frac{N_{x1}}{R_x}+\frac{N_{y1}}{R_y}\right)$$
$$-\begin{bmatrix}(BN_{xo}\beta_{x1}+BN_{xy0}\beta_{y1})_{,x}+(BN_{x1}\beta_{x0}+BN_{xy1}\beta_{y0})_{,x}\\+(AN_{y0}\beta_{y1}+AN_{xy0}\beta_{x1})_{,y}+(AN_{y1}\beta_{y0}+AN_{xy1}\beta_{x0})_{,y}\end{bmatrix}=0$$
(10.3.12c)

where

$$N_{x1}=C\left[(e_{xx1}+\beta_{x0}\beta_{x1})+\mu(e_{yy1}+\beta_{y0}\beta_{y1})\right] \quad (10.3.13a)$$

$$N_{y1}=C\left[(e_{yy1}+\beta_{y0}\beta_{y1})+\mu(e_{xx1}+\beta_{x0}\beta_{x1})\right] \quad (10.3.13b)$$

$$N_{xy1}=C\frac{1-\mu}{2}(e_{xy1}+\beta_{x0}\beta_{y1}+\beta_{y0}\beta_{x1}) \quad (10.3.13c)$$

$$M_{x1}=D\left[\left(-\frac{w_{,xx}}{A^2}+\frac{A_{,x}w_{,x}}{A^3}-\frac{A_{,y}w_{,y}}{AB^2}\right)+\mu\left(-\frac{w_{1,yy}}{B^2}+\frac{B_{,y}w_{1,y}}{B^3}-\frac{B_{,x}w_{1,x}}{A^2B}\right)\right]$$
(10.3.13d)

$$M_{y1}=D\left[\left(-\frac{w_{1,yy}}{B^2}+\frac{B_{,y}w_{1,y}}{B^3}-\frac{B_{,x}w_{1,x}}{A^2B}\right)+\mu\left(-\frac{w_{,xx}}{A^2}+\frac{A_{,x}w_{,x}}{A^3}-\frac{A_{,y}w_{,y}}{AB^2}\right)\right]$$
(10.3.13e)

$$M_{xy1}=D\frac{1-\mu}{2}\left(-\frac{w_{1,xy}}{AB}+\frac{A_{,y}w_{1,x}}{A^2B}+\frac{B_{,x}w_{1,y}}{AB^2}\right) \quad (10.3.13f)$$

Equations (10.3.12) and (10.3.13) are the linear stability equations for the shell of general shape, under the DMV approximations.

10.4. APPLICATIONS

10.4.1. Shells of Revolution

Structural shells often take the shapes of shells of revolution. The middle surface of a shell of revolution is formed by rotating a plane curve (generator) with respect to an axis in the plane of the curve as shown in Fig. 10-5. The lines of principal curvature are called the meridians (surface curves

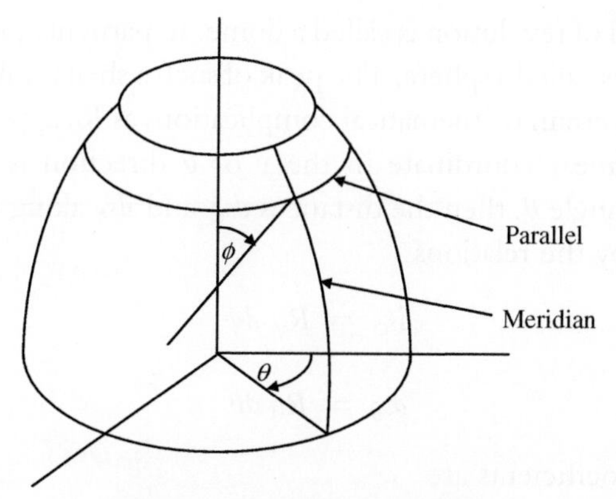

Figure 10-5 Shell of revolution

intersected by planes containing the axis of rotation) and parallels (surface curves intersected by planes perpendicular to the axis of rotation). The parallels and meridians are the same as the latitudes and the longitudes in a glove. In Fig. 10-6 the meridian of a shell of revolution of positive Gaussian curvature is illustrated. Points on the middle surface may be referred to coordinates ϕ and θ, where ϕ denotes the angle between the axis of rotation and a normal to the middle surface, and θ is a circumferential coordinate as shown in Fig. 10-5. The principal radii of curvature of the surface in the ϕ and θ directions may be denoted by R_ϕ and R_θ, respectively. It is convenient to define an additional variable R_0 defined by the relation

$$R_0 = R_\theta \sin \phi \tag{10.4.1}$$

Note that R_0 is not a principal radius of curvature as it is not normal to the surface. Rather, it is a projection of R_θ on the horizontal plane.

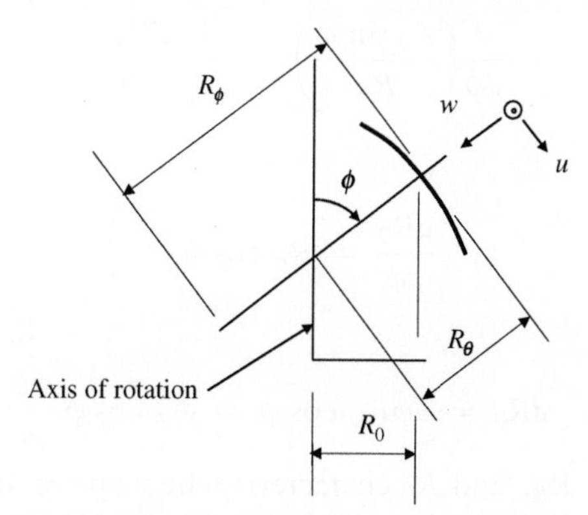

Figure 10-6 Meridian of shell of revolution

A closed shell of revolution is called a dome. In particular, if the generator is a half circle, it is called a sphere. The peak of such a shell is called the pole. A pole introduces certain mathematical complications as R_0 approaches to zero.

If the curvilinear coordinate in the y or θ direction is chosen as the circumferential angle θ, then the distances ds_ϕ and ds_θ along the coordinate lines are given by the relations

$$ds_\phi = R_\phi \, d\phi \tag{10.4.2a}$$

$$ds_\theta = R_0 \, d\theta \tag{10.4.2b}$$

and the Lamé coefficients are

$$A = R_\phi \tag{10.4.3a}$$

$$B = R_0 \tag{10.4.3b}$$

Furthermore,

$$R_x = R_\phi \tag{10.4.4a}$$

$$R_y = R_\theta \tag{10.4.4b}$$

Recall the second Codazzi equations given in Eq. (10.2.7)

$$\frac{\partial}{\partial x}\left(\frac{B}{R_y}\right) = \frac{1}{R_x}\frac{\partial B}{\partial x}$$

Making use of Eqs. (10.4.1), (10.4.3) and (10.4.4), the second equation becomes

$$\frac{d}{d\phi}\left(\frac{R_\theta \sin \phi}{R_\theta}\right) = \frac{1}{R_\phi}\frac{dR_o}{d\phi} \tag{10.4.5}$$

From which

$$\frac{dR_0}{d\phi} = R_\phi \cos \phi \tag{10.4.6a}$$

or

$$dR_0 = d\phi R_\phi \cos \phi = ds_\phi \cos \phi \tag{10.4.6b}$$

The variables R_ϕ, R_θ, and R_0 characterize the shape of the middle surface of the undeformed shell and are a function of ϕ only. Variables u, v, w denote middle-surface displacement components in the ϕ, θ, and normal directions, respectively. The displacement components, in general, are functions of both ϕ and θ.

10.4.2. Stability Equations

Specializing Eqs. (10.3.12), (10.3.11), and (10.3.13) for a shell of revolution by neglecting prebuckling rotation terms ($\beta_{\phi o}$ and $\beta_{\theta 0}$) yields (Brush and Almroth 1975[4]):

$$\left(R_0 N_{\phi 1}\right)_{,\phi} + R_\phi N_{\phi\theta 1,\theta} - R_\phi N_{\phi 1}\cos\phi = 0$$

$$\left(R_0 N_{\phi\theta 1}\right)_{,\phi} + R_\phi N_{\theta 1,\theta} + R_\phi N_{\phi\theta 1}\cos\phi = 0$$

$$\left[\frac{1}{R_\phi}\left(R_0 M_{\phi 1}\right)_{,\phi}\right]_{,\phi} + \left[\frac{R_\phi}{R_0}M_{\theta 1,\theta\theta} - \left(M_{\theta 1}\cos\phi\right)_{,\phi}\right]$$

$$+2\left(M_{\phi\theta 1,\phi\theta} + \frac{R_\phi}{R_0}M_{\phi\theta 1,\theta}\cos\phi\right) - \left(R_0 N_{\phi 1} + R_\phi N_{\theta 1}\sin\phi\right)$$

$$-\left[\left(R_0 N_{\phi 0}\beta_{\phi 1} + R_0 N_{\phi\theta 0}\beta_{\theta 1}\right)_{,\phi} + \left(R_\phi N_{\theta 0}\beta_{\theta 1} + R_\phi N_{\phi\theta 0}\beta_{\phi 1}\right)_{,\theta}\right] = 0 \quad (10.4.7)$$

where

$$N_{\phi 1} = C\left[\frac{1}{R_\phi}\left(u_{1,\phi} - w_1\right) + \frac{\mu}{R_0}\left(v_{1,\theta} + u_1\cos\phi - w_1\sin\phi\right)\right]$$
$$= C\left(e_{\phi\phi 1} + \mu e_{\theta\theta 1}\right) \quad (10.4.8a)$$

$$N_{\theta 1} = C\left[\frac{1}{R_0}\left(v_{1,\theta} + u_1\cos\phi - w_1\sin\phi\right) + \frac{\mu}{R_\phi}\left(u_{1,\phi} - w_1\right)\right]$$
$$= C\left(e_{\theta\theta 1} + \mu e_{\phi\phi 1}\right) \quad (10.4.8b)$$

$$N_{\phi\theta 1} = C\frac{1-\mu}{2}\left(\frac{R_0}{R_\phi}\left(\frac{v_1}{R_0}\right)_{,\phi} + \frac{u_{1,\theta}}{R_0}\right) = C\frac{1-\mu}{2}e_{\phi\theta 1} \quad (10.4.8c)$$

$$M_{\phi 1} = D\left[\frac{\beta_{\phi 1,\phi}}{R_\phi} + \frac{\mu}{R_0}\left(\beta_{\theta 1,\theta} + \beta_{\phi 1}\cos\phi\right)\right] \quad (10.4.8d)$$

[4] See page 206.

$$M_{\theta 1} = D\left[\frac{1}{R_o}(\beta_{\theta 1,\theta} + \beta_{\phi 1}\cos\phi) + \frac{\mu\beta_{\phi 1,\phi}}{R_\phi}\right] \tag{10.4.8e}$$

$$M_{\phi\theta 1} = D\frac{1-\mu}{2}\left[\frac{R_0}{R_\phi}\left(\frac{\beta_{\theta 1}}{R_0}\right)_{,\phi} + \frac{\beta_{\phi 1,\theta}}{R_0}\right] \tag{10.4.8f}$$

$$\beta_{\phi 1} = -\frac{w_{1,\phi}}{R_\phi} \tag{10.4.8g}$$

$$\beta_{\theta 1} = -\frac{w_{1,\theta}}{R_0} \tag{10.4.8h}$$

Equations (10.4.7) are the essence of DMV theory of the symmetrically loaded quasi-shallow shell of revolution. According to Novozhilov (1964),[5] Donnell (1933) in the United States and Mushtari (1938) in the Soviet Union apparently derived the theory independently. Later, Vlasov (1964) improved and generalized the theory significantly.

The coefficients $N_{\phi o}$, $N_{\phi\theta o}$, $N_{\theta o}$ in Eqs. (10.4.7) are determined by the linear equilibrium equations obtained from the specialization of Eqs. (10.2.17) for axisymmetric deformation of a shell of revolution. They are

$$\begin{aligned}
&\frac{d}{d\phi}(R_0N_\phi) - R_\phi N_\theta\cos\phi = -R_0R_\phi p_\phi\\
&\frac{d}{d\phi}(R_0N_{\phi\theta}) + R_\phi N_{\phi\theta}\cos\phi = -R_0R_\phi p_\theta\\
&\frac{d}{d\phi}\left[\frac{1}{R_\phi}\frac{d}{d\phi}(R_0M_\phi)\right] - \frac{d}{d\phi}(M_\theta\cos\phi) - (R_0N_\phi + R_\phi N_\theta\sin\phi) = -R_0R_\phi p_z
\end{aligned} \tag{10.4.9}$$

where the constitutive and kinematic relations are given by Eqs. (10.4.13).

As a simplifying approximation in the determination of the coefficients in the stability equations, the linear bending equation is frequently replaced by the corresponding linear membrane equation. The first and second terms in the third equations of Eqs. (10.4.9) are considered to be small compared to the remaining terms. Hence, they are frequently neglected. Then Eqs. (10.4.9) become

[5] See pages 88–94.

$$\begin{aligned}
&\frac{d}{d\phi}(R_0N_\phi) - R_\phi N_\theta \cos\phi = -R_0R_\phi p_\phi \\
&\frac{d}{d\phi}(R_0N_{\phi\theta}) + R_\phi N_{\phi\theta} \cos\phi = -R_0R_\phi p_\theta \qquad (10.4.10)\\
&R_0N_\phi + R_\phi N_\theta \sin\phi = R_0R_\phi p_z
\end{aligned}$$

Equations (10.4.10) are statically determinate. Hence, solutions can be determined without constitutive and kinematic relations given by Eqs. (10.4.13).

If the shell is not subjected to torsional loading, the coefficient $N_{\phi\theta 0}$ becomes zero in Eqs. (10.4.7). In such cases the stability equations obtained by substitution of Eqs. (10.4.13) into (10.4.7) may be reduced to ordinary differential equation by selection of solutions of the form

$$\begin{aligned}
u_1 &= u_n(\phi)\cos n\theta \\
v_1 &= v_n(\phi)\sin n\theta \qquad (10.4.11)\\
w_1 &= w_n(\phi)\cos n\theta
\end{aligned}$$

To sum up, stability equations for shells of revolution are given in Eqs. (10.4.7) in which prebuckling rotation terms are omitted. Linear equilibrium equations for symmetrically loaded shells of revolution are given in Eqs. (10.4.9), and corresponding linear membrane equations are given in Eqs. (10.4.10).

10.4.3. Circular Flat Plates

The middle plane of a circular flat plate may be described by a polar coordinate system of, r and θ as shown in Fig. 10-7. In specialization of equations of the shell of revolution for circular flat plates, it is required that R_ϕ and R_θ go to infinity, the angle ϕ goes to zero, and by virtue of Eq. (10.4.6b).

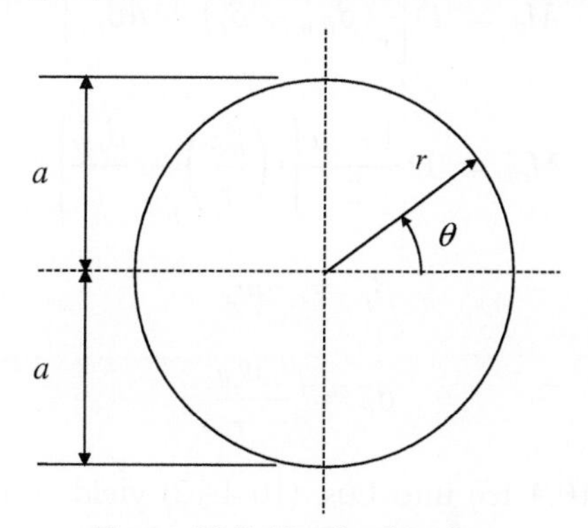

Figure 10-7 Circular flat plate

$$\lim_{R_\phi \to \infty} (R_\phi d\phi) = dR_0 = dr$$

Then $\sin\phi = 0$ and $\cos\phi = 1$. Substituting these values into Eqs. (10.4.7) gives

$$(rN_r)_{,r} + N_{r\theta,\theta} - N_\theta = 0 \quad (10.4.12a)$$

$$(rN_{r\theta})_{,r} + N_{\theta,\theta} + N_{r\theta} = 0 \quad (10.4.12b)$$

$$\begin{aligned} &(rM_r)_{,rr} + 2\left(M_{r\theta,r\theta} + \frac{1}{r}M_{r\theta,\theta}\right) + \left(\frac{1}{r}M_{\theta,\theta\theta} - M_{\theta,r}\right) \\ &- \left[(rN_{ro}\beta_r + rN_{r\theta o}\beta_\theta)_{,r} + (N_{r\theta o}\beta_r + N_{\theta o}\beta_\theta)_{,\theta}\right] = 0 \end{aligned} \quad (10.4.12c)$$

where the subscript ϕ has been replaced by r. The corresponding constitutive and kinematic relations to Eqs. (10.4.8) are

$$N_r = C\left[u_{,r} + \frac{\mu}{r}(v_{,\theta} + u)\right] = C(e_{rr} + \mu e_{\theta\theta}) \quad (10.4.13a)$$

$$N_\theta = C\left[\frac{1}{r}(v_{,\theta} + u) + \mu u_{,r}\right] = C(e_{\theta\theta} + \mu e_{rr}) \quad (10.4.13b)$$

$$N_{r\theta} = C\frac{1-\mu}{2}\left[r\left(\frac{v}{r}\right)_{,r} + \frac{u_{,\theta}}{r}\right] = C\frac{1-\mu}{2}e_{r\theta} \quad (10.4.13c)$$

$$M_r = D\left[\beta_{r,r} + \frac{\mu}{r}(\beta_{\theta,\theta} + \beta_r)\right] \quad (10.4.13d)$$

$$M_\theta = D\left[\frac{1}{r}(\beta_{\theta,\theta} + \beta_r) + \mu\beta_{r,r}\right] \quad (10.4.13e)$$

$$M_{r\theta} = D\frac{1-\mu}{2}\left[r\left(\frac{\beta_\theta}{r}\right)_{,r} + \frac{\beta_{r,\theta}}{r}\right] \quad (10.4.13f)$$

$$\beta_r = -w_{,r} \quad (10.4.13g)$$

$$\beta_\theta = -\frac{w_{,\theta}}{r} \quad (10.4.13h)$$

Substituting Eqs. (10.4.13) into Eqs. (10.4.12) yields a set of three homogeneous equations in u, v, w in which the third equation is uncoupled from the first two as in the case for a rectangular plate. The moment expressions in Eqs. (10.4.13) are identical to those given by Szilard (1974).

As a specific example, consider the axisymmetric buckling of a plate subjected to a uniform compressive force around the circumference $N_{ro} = -N$ lb/in. Then $\beta_\theta = N_{r\theta} = 0$. Let $\beta_r = \beta$ to simplify the notation. Equation (10.4.12c) specializes to

$$\frac{d^2}{dr^2}(rM_r) - \frac{d}{dr}M_\theta + \frac{d}{dr}(rN\beta) = 0 \tag{10.4.14}$$

where

$$M_r = D\left[\frac{d\beta}{dr} + \frac{\mu}{r}\beta\right]$$

$$M_\theta = D\left[\frac{1}{r}\beta + \mu\frac{d\beta}{dr}\right]$$

Integrating Eq. (10.4.14) gives

$$\frac{d}{dr}(rM_r) - M_\theta + rN\beta = c_1$$

where c_1 is an integral constant. As $M_r = M_\theta = 0$ for $N = 0$, c_1 must be equal to zero. Substituting the expressions for M_r and M_θ into the above equation and rearranging gives

$$r^2\frac{d^2\beta}{dr^2} + r\frac{d\beta}{dr} - \left(1 - \frac{N}{D}r^2\right)\beta = 0 \tag{10.4.15}$$

Equation (10.4.15) is the same as Eqs. (a) and (b) in Timoshenko and Gere (1961).[6] As a homogeneous equation, it has nontrivial solutions only for discrete values of the applied load N. The smallest solution is the critical load N_{cr}.

Following the procedure given in Timoshenko and Gere (1961),[7] the general solution is readily obtained. Let $\alpha^2 = N/D$ and $u = \alpha r$. With these new variables, Eq. (10.4.15) may be rewritten as

$$u^2\frac{d\beta^2}{dr^2} + u\frac{d\beta}{dr} + \left(u^2 - 1\right)\beta = 0 \tag{10.4.16}$$

[6] See page 389.
[7] See page 390.

The general solution of this equation is given by Grossman and Derrick (1988)[8]

$$\beta = A_1 J_1(u) + A_2 Y_1(u) \tag{10.4.17}$$

where $J_1(u)$ and $Y_1(u)$ are Bessel functions of the first and second kinds of order one, respectively. At the center of the plate ($r = u = 0$), β must be equal to zero in order to satisfy the condition of symmetry. Since $Y_1(0) \rightarrow \infty$, A_2 must be equal to zero[9] and

$$\beta = A_1 J_1(u)$$

Solutions for two boundary conditions are given by Timoshenko and Gere (1961). For the clamped edge, $\beta = 0$ at $r = a$ and therefore $J_1(\alpha a) = 0$. **Maple®** gives the smallest nontrivial solution $\alpha a = 3.8317$. Hence the critical load is

$$N_{cr} = \frac{3.8317^2 D}{a^2} = \frac{14.68D}{a^2} \tag{10.4.17}$$

For the simply supported plate

$$(M_r)_{r=a} = D\left(\frac{d\beta}{dr} + \frac{\mu\beta}{r}\right)_{r=a} = 0$$

Therefore

$$\left[\frac{dJ_1(u)}{dr} + \mu\frac{J_1(u)}{r}\right]_{r=0} = 0$$

or

$$\left[u\frac{dJ_1(u)}{du} + \mu J_1(u)\right]_{u=\alpha a} = 0$$

Applying the derivative formula $dJ_1(u)/du = J_0 - J_1/u$, where J_0 is the Bessel function of the first kind of order zero from Grossman and Derrick (1988) [10] to the above equation gives

$$\alpha a J_0(\alpha a) - (1 - \mu)J_1(\alpha a) = 0$$

For $\mu = 0.3$,

$$\alpha a J_0(\alpha a) - 0.7J_1(\alpha a) = 0$$

[8] See page 276.
[9] See Grossman and Derrick (1988), page 277.
[10] See page 278.

The smallest nonzero value of αa satisfying the above equation obtained from **Maple**® is $\alpha a = 2.04885$, say $\alpha a = 2.049$. Hence, the critical load is

$$N_{cr} = \frac{2.049^2 D}{a^2} = \frac{4.2D}{a^2} \tag{10.4.18}$$

10.4.4. Conical Shells

As an example of shells of revolution, consider a truncated conical shell with a vertex angle of 2α shown in Fig. 10-8. The longitudinal coordinate s and a circumferential coordinate θ are chosen as the orthogonal curvilinear coordinates. Of course, the axial coordinate η can be chosen as the other curvilinear coordinate instead of s, if so desired.

In the equations for shells of revolution Eqs. (10.4.7), R_ϕ approaches to infinity for a cone, and hence

$$\lim_{R_\phi \to \infty} \left(R_\phi \, d\phi\right) = ds \tag{10.4.19}$$

Furthermore, from Fig. 10-8, it is evident that the meridian angle $\phi = (\pi/2) - \alpha =$ constant and $R_0 = s \sin \alpha$. Also $\sin \phi = \cos \alpha$ and $\cos \phi = \sin \alpha$. Substituting these values into Eqs. (10.4.7) and rearrangement yields

$$(sN_s)_{,s} + \frac{1}{\sin \alpha} N_{s\theta,\theta} - N_\theta = 0 \tag{10.4.20a}$$

$$\frac{1}{\sin \alpha} N_{\theta,\theta} + \frac{1}{s}\left(s^2 N_{s\theta}\right)_{,s} = 0 \tag{10.4.20b}$$

$$\begin{aligned}(sM_s)_{,ss} &+ \frac{2}{\sin \alpha}\left(M_{s\theta,s\theta} + \frac{1}{s} M_{s\theta,\theta}\right) + \frac{1}{s \sin^2 \alpha} M_{\theta,\theta\theta} - M_{\theta,s} - N_\theta \cot \alpha \\ &- \left[\left(sN_{s0}\beta_s + sN_{s\theta 0}\beta_\theta\right)_{,s} + \frac{1}{\sin \alpha}\left(N_{s\theta 0}\beta_s + N_{\theta 0}\beta_\theta\right)_{,\theta}\right] = 0\end{aligned} \tag{10.4.20c}$$

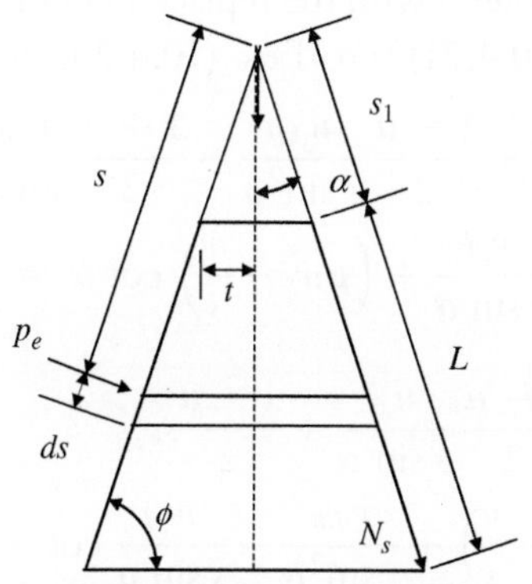

Figure 10-8 Conical shell

where the subscript ϕ has been replaced by s using the relationship given in Eq. (10.4.19). Converting directly from Eqs. (10.4.8), the constitutive relations for conical shells are

$$N_s = C\left[u_{,s} + \frac{\mu}{s}\left(\frac{v_{,\theta}}{\sin\alpha} + u + w\cot\alpha\right)\right] \tag{10.4.21a}$$

$$N_\theta = C\left[\frac{1}{s}\left(\frac{v_{,\theta}}{\sin\alpha} + u + w\cot\alpha\right) + \mu u_{,s}\right] \tag{10.4.21b}$$

$$N_{s\theta} = C\frac{1-\mu}{2}\left(v_{,s} - \frac{v}{s} + \frac{u_{,\theta}}{s\sin\alpha}\right) \tag{10.4.21c}$$

$$M_s = D\left[\beta_{s,s} + \frac{\mu}{s}\left(\frac{\beta_{\theta,\theta}}{\sin\alpha} + \beta_s\right)\right] \tag{10.4.21d}$$

$$M_\theta = D\left[\frac{1}{s}\left(\frac{\beta_{\theta,\theta}}{\sin\alpha} + \beta_s\right) + \mu\beta_{s,s}\right] \tag{10.4.21e}$$

$$M_{s\theta} = D\frac{1-\mu}{2}\left[s\left(\frac{\beta_\theta}{s}\right)_{,s} + \frac{\beta_\theta}{s\sin\alpha}\right] \tag{10.4.21f}$$

$$\beta_s = -w_{,s} \tag{10.4.21g}$$

$$\beta_\theta = -\frac{w_{,\theta}}{s\sin\alpha} \tag{10.4.21h}$$

It should be of interest to note that at one extreme, when $\alpha = \pi/2$, these equations reduce to the equations for flat circular plates and at the other extreme, for $\alpha = 0$, these equations correspond to the expressions for cylindrical shells in Chapter 9 with the replacement of $s\sin\alpha$ by the radius R.

Substituting Eqs. (10.4.21) into Eqs. (10.4.20) gives

$$\begin{aligned} &su_{ss} + u_{,s} - \frac{u}{s} + \frac{1-\mu}{2}\frac{u_{,\theta\theta}}{s\sin^2\alpha} + \frac{1+\mu}{2}\frac{v_{,s\theta}}{\sin\alpha} \\ &-\frac{3-\mu}{2}\frac{v_{,\theta}}{s\sin\alpha} + \left(\mu w_s - \frac{w}{s}\right)\cot\alpha = 0 \end{aligned} \tag{10.4.22a}$$

$$\begin{aligned} &\frac{1+\mu}{2}\frac{u_{,s\theta}}{\sin\alpha} + \frac{3-\mu}{2}\frac{u_{,\theta}}{s\sin\alpha} + \frac{1-\mu}{2}sv_{,ss} \\ &+\frac{1-\mu}{2}\left(v_{,s} - \frac{v}{s}\right) + \frac{v_{,\theta\theta}}{s\sin^2\alpha} + \frac{w_{,\theta}}{s\sin\alpha}\cot\alpha = 0 \end{aligned} \tag{10.4.22b}$$

$$Ds\left(w_{,ssss}+2\frac{w_{,sss}}{s}-\frac{w_{,ss}}{s^2}+\frac{w_{,s}}{s^3}-2\frac{w_{,s\theta\theta}}{s^3\sin^2\alpha}+2\frac{w_{,ss\theta\theta}}{s^2\sin^2\alpha}+4\frac{w_{,\theta\theta}}{s^4\sin^2\alpha}\right.$$
$$\left.+\frac{w_{,\theta\theta\theta\theta}}{s^4\sin^4\alpha}\right)+C\left(\frac{v_{,\theta}}{s\sin\alpha}+\frac{u}{s}+\frac{w\cot\alpha}{s}+\mu u_{,s}\right)\cot\alpha$$
$$-\left[\left(N_{s0}sw_{,s}+N_{s\theta0}\frac{w_{,\theta}}{\sin\alpha}\right)_{,s}+\frac{1}{\sin\alpha}\left(N_{s\theta0}w_{,s}+N_{\theta0}\frac{w_{,\theta}}{s\sin\alpha}\right)_{,\theta}\right]=0$$
$$\text{(10.4.22c)}$$

Equations (10.4.22) give a coupled set of three homogeneous differential equations in u, v, and w.

Consider, as an example, a conical shell subjected to uniform external hydrostatic pressure or internal suction p_e in pounds per square inch and an axial compressive force P in pounds. Suppose that a membrane analysis is adequate for the prebuckling static deformation; then the coefficients are computed by the simple static relations

$$N_{s0}=-\frac{1}{2}p_es\tan\alpha-\frac{P}{2\pi s\sin\alpha},\quad N_{\theta0}=-p_es\tan\alpha,\quad N_{s\theta0}=0 \quad (10.4.23)$$

Substituting these values into Eqs. (10.4.22) reveals that Eq. (10.4.22c) is a stability equation with variable coefficients. In general, a solution for critical values of the applied load needs to rely on numerical methods. An excellent numerical analysis is reported by Baruch, Harari, and Singer (1967). They obtained extensive numerical results based on the Galerkin procedure for hydrostatic-pressure loading, with a fairly wide range of parameters.

10.4.5. Shallow Spherical Caps

A cross section of a spherical cap is shown in Fig. 10-8. The middle surface is described by curvilinear coordinates r and θ. The rise H of the shell is much smaller than the base chord ($2a$).

From Fig. 10-9, $r_\phi = R$, a constant, and $\sin\phi = r/R$. For the shallow shell, approximately, $\cos\phi = 1$ and $r_\phi d\phi = dr$. Substituting these approximations into Eq. (10.4.7) yields

$$(rN_r)_{,r}+N_{r\theta,\theta}-N_\theta=0 \quad (10.4.24a)$$

$$(rN_{r\theta})_{,r}+N_{\theta,\theta}+N_{r\theta}=0 \quad (10.4.24b)$$

$$(rM_r)_{,rr}+2\left(M_{r\theta,r\theta}+\frac{1}{r}M_{r\theta,\theta}\right)+\left(\frac{1}{r}M_{\theta,\theta\theta}-M_{\theta,r}\right)+\frac{r}{R}(N_r+N_\theta)$$
$$-\left[(rN_{ro}\beta_r+rN_{r\theta o}\beta_\theta)_{,r}+(N_{r\theta o}\beta_r+N_{\theta o}\beta_\theta)_{,\theta}\right]=0 \quad (10.4.24c)$$

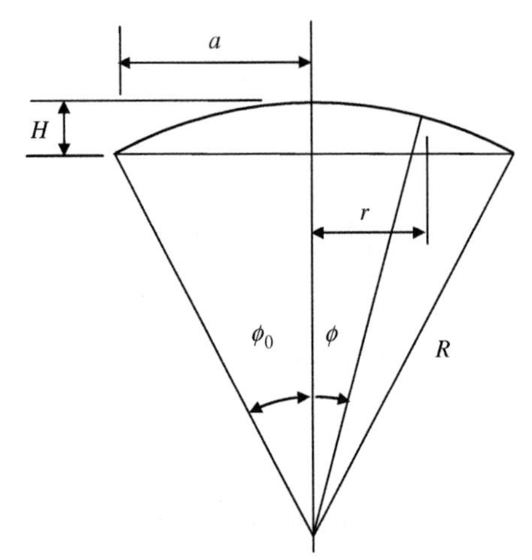

Figure 10-9 Shallow spherical cap (after Brush and Almroth, *Buckling of Bars, Plates, and Shells*. McGraw-Hill, 1975). Reproduced by permission.

where

$$N_r = C(e_{rr} + \mu e_{\theta\theta}) \tag{10.4.25a}$$

$$N_\theta = C(e_{\theta\theta} + \mu e_{rr}) \tag{10.4.25b}$$

$$N_{r\theta} = C\frac{1-\mu}{2}e_{r\theta} \tag{10.4.25c}$$

$$e_{rr} = u_{,r} - \frac{w}{R} \tag{10.4.25d}$$

$$e_{\theta\theta} = \frac{v_{,\theta} + u}{r} - \frac{w}{R} \tag{10.4.25e}$$

$$e_{r\theta} = r\left(\frac{v}{r}\right)_{,r} + \frac{u_{,\theta}}{r} \tag{10.4.25f}$$

$$M_r = D\left[\beta_{r,r} + \frac{\mu}{r}(\beta_{\theta,\theta} + \beta_r)\right] \tag{10.4.26a}$$

$$M_\theta = D\left[\frac{1}{r}(\beta_{\theta,\theta} + \beta_r) + \mu\beta_{r,r}\right] \tag{10.4.26b}$$

$$M_{r\theta} = D\frac{1-\mu}{2}\left[r\left(\frac{\beta_\theta}{r}\right)_{,r} + \frac{\beta_{r,\theta}}{r}\right] \tag{10.4.26c}$$

$$\beta_r = -w_{,r} \tag{10.4.26d}$$

$$\beta_\theta = -\frac{w_{,\theta}}{r} \tag{10.4.26e}$$

For $R \to \infty$, Eqs. (10.4.24) and (10.4.25) reduce to Eqs. (10.4.12) and (10.4.13) for the case of a circular flat plate.

It is noted that substitution of constitutive and kinematic relations of Eqs. (10.4.25) into Eq. (10.4.24c) and considerable algebraic operations and rearrangements (see Problem 10.12) give

$$D\nabla^4 w - \frac{N_r + N_\theta}{R} - \frac{1}{r}\left[\left(rN_{r0}w_{,r} + N_{r\theta 0}w_{,\theta}\right)_{,r} + \left(N_{r\theta 0}w_{,r} + N_{\theta 0}\frac{w_{,\theta}}{r}\right)_{,\theta}\right] = 0 \tag{10.4.27}$$

where

$$\nabla^2(\,) = \left[(\,)_{,rr} + \frac{1}{r}(\,)_{,r} + \frac{1}{r^2}(\,)_{,\theta\theta}\right]$$

$$\nabla^4(\,) = \nabla^2\nabla^2(\,)$$

Equations (10.4.24a), (10.4.24b), (10.4.25), and (10.4.27) give a coupled set of three homogeneous equations in u, v, *and* w.

Suppose the spherical cap is subjected to a uniform external pressure p_e normal to the middle plane and that the prebuckling state may be approximated by a membrane analysis. Then $N_{r0} = N_{\theta 0} = -p_e R/2$ and $N_{r\theta} = 0$. Substituting these values into Eq. (10.4.27) and rearranging yields

$$D\nabla^4 w - \frac{N_r + N_\theta}{R} + \frac{1}{2}p_e R\nabla^2 w = 0 \tag{10.4.28}$$

The set of three equations in u, v, w mentioned above can be reduced to two equations in w and a stress function f. According to Novozhilov (1964),[11] Vlasov (1944) first introduced an arbitrary function known as the Airy stress function f defined by

$$N_r = \frac{1}{r}f_{,r} + \frac{1}{r^2}f_{,\theta\theta} \qquad N_\theta = f_{,rr} \quad N_{r\theta} = -\left(\frac{f_{,\theta}}{r}\right)_{,r} \tag{10.4.29}$$

[11] See page 90.

Equation (10.4.27) can be written

$$D\nabla^4 w - \frac{1}{R}\nabla^2 f + \frac{1}{2} p_e R \nabla^2 w = 0 \tag{10.4.30}$$

However, it is found (see Problem 10.13) from Eqs. (10.4.25) that

$$-\frac{1}{R}\nabla^2 w = \frac{1}{r^2} e_{rr,\theta\theta} - \frac{1}{r} e_{rr,r} + \frac{1}{r^2}\left(r^2 e_{\theta\theta,r}\right)_{,r} - \frac{1}{r^2}\left(r e_{r\theta}\right)_{,r\theta} \tag{10.4.31a}$$

$$e_{rr} = \frac{1}{Eh}(N_r - \mu N_\theta) \quad e_{\theta\theta} = \frac{1}{Eh}(N_\theta - \mu N_r) \quad e_{r\theta} = \frac{2(1+\mu)}{Eh} N_{r\theta} \tag{10.4.31b}$$

Hence, the stress function f must satisfy the compatibility condition (see Problem 10.14)

$$\nabla^4 f = -\frac{Eh}{R}\nabla^2 w \tag{10.4.32}$$

Equations (10.4.30) and (10.4.32) reduce the problem to a set of two coupled homogeneous differential equations in f and w. These equations have nontrivial solutions only for discrete values of p_e, which may be termed periodic eigenvalues. The smallest eigenvalue is called p_{cr}.

Hutchinson (1967) gives a simple solution. Let $x = r\cos\theta$ and $y = r\sin\theta$. As the Laplacian, in general terms, is given by

$$\nabla^2(\,) = \frac{1}{AB}\left\{\left[\frac{B}{A}(\,)_{,\alpha}\right]_{,\alpha} + \left[\frac{A}{B}(\,)_{,\beta}\right]_{,\beta}\right\} \tag{10.4.33}$$

for the case of shells of revolution, it becomes

$$\nabla^2(\,) = \frac{1}{r}\left[r(\,)_{,rr} + (\,)_{,r}\right] + \frac{1}{r^2}(\,)_{,\theta\theta}$$

$$= (\,)_{,rr} + \frac{1}{r}(\,)_{,r} + \frac{1}{r^2}(\,)_{,\theta\theta}$$

For the case of Cartesian coordinate system, it is

$$\nabla^2(\,) = (\,)_{,xx} + (\,)_{,yy}$$

Equations (10.4.30) and (10.4.32) are satisfied by products of sinusoidal functions of the form

$$w = \cos\left(k_x \frac{x}{R}\right)\cos\left(k_y \frac{y}{R}\right) \quad f = C_1 \cos\left(k_x \frac{x}{R}\right)\cos\left(k_y \frac{y}{R}\right) \tag{10.4.34}$$

where k_x and k_y are mode shape parameters and C_1 is a constant. Substituting Eq. (10.4.34) into Eq. (10.4.32) gives

$$C_1 = EhR\sqrt{\left(k_x^{\ 2} + k_y^{\ 2}\right)} \tag{10.4.35}$$

Substituting Eq. (10.4.35) and $D = Eh^3/[12(1-\mu^2)]$ into Eq. (10.4.30) yields

$$P_e = \frac{2Eh}{R}\left[\sqrt{\left(k_x^{\ 2} + k_y^{\ 2}\right)} + \frac{1}{12(1-\mu^2)}\left(k_x^{\ 2} + k_y^{\ 2}\right)\left(\frac{h}{r}\right)^2\right] \tag{10.4.36}$$

The classical buckling pressure is found by minimizing Eq. (10.4.36) with respect to $k_x^{\ 2} + k_y^{\ 2}$. The smallest is P_e found for

$$k_x^{\ 2} + k_y^{\ 2} = 2\sqrt{3(1-\mu^2)}\,\frac{R}{h}$$

Substituting this value to Eq. (10.4.36) gives

$$P_{cr} = \frac{2E}{\sqrt{3(1-\mu^2)}}\left(\frac{h}{R}\right)^2 \tag{10.4.37}$$

This is the same result as given by Hutchinson (1967).

Equation (10.4.37) is the same as that given for a complete spherical shell subjected to hydrostatic pressure by Timoshenko and Gere (1961) based on Legendre functions. It is interesting to note that Gould (1988) introduces Vlasov's effort of investigating the stability of pressurized shells without even considering the buckling mode shape functions. The solution functions in Eqs. (10.4.34) do not satisfy the boundary conditions at the edge of a spherical cap, and for a full sphere, the edge on which a combination of boundary conditions can be assessed is not well defined. Therefore, the present simplified buckling analysis is limited to buckling-mode wavelengths that are sufficiently small compared with the radius of the shell. Even under such limitations, the critical pressure predicted by Eq. (10.4.37) is in very poor agreement with test results. It is now firmly believed that the source of such discrepancy is due to two factors: the neglect of nonlinearity in the prebuckling static analysis and the influence of initial imperfections. A means for introducing further refinements into the analysis, such as finite deformation analysis of shallow shells and postbuckling and imperfection sensitivity analysis, may be realized by well-established modern-day finite element codes. However, many design procedures are based on the elastic critical load, reduced by a "knockdown factor" of five or even more. This is perhaps the reason why Miller[12] of CB&I (Chicago Bridge and Iron) relied heavily on experimental investigations for the company's new form of shell structures to build until the late 1980s, when reliable finite element codes were made available.

The subject of this book is the buckling behavior of structural members that are subjected to loading that induces compressive stresses in the body. Buckling is essentially flexural behavior. As such, it has been necessary to investigate the flexural behavior of each structural element covered in the book. However, quite a few structural members can carry the applied load primarily or dominantly through membrane actions. In such cases, the static analysis for membrane action is considerably less complicated than the analysis for combined membrane and flexural actions. When compressive stresses are developed in the body, an elastic buckling strength check is necessary, but for loading cases that produce no, or low, compressive stress, a simplified membrane analysis may suffice.

REFERENCES

Baruch, M., Harari, O., & Singer, J. (1967). Influence of In-Plane Boundary Conditions on the Stability of Conical Shells Under Hydrostatic Pressure. *Israel Journal of Technology, 5*(1–2), 12–24.

Brush, D. O., & Almroth, B. O. (1975). *Buckling of Bars, Plates, and Shells*. New York, NY: McGraw-Hill.

Donnell, L. H. (1933). *Stability of Thin-Walled Tubes Under Torsion*. Washington, DC: NACA Technical Report, No. 479.

Flügge, W. (1973). *Stresses in Shells* (2nd ed.). Berlin: Springer-Verlag.

Gould, P. L. (1988). *Analysis of Shells and Plates*. New York: Springer-Verlag.

Grossman, S. I., & Derrick, W. R. (1988). *Advanced Engineering Mathematics*. New York: Harper and Row.

Hutchinson, J. W. (1967). Imperfection Sensitivity of Externally Pressurized Spherical Shells. *Journal of Applied Mechanics, Vol. 34*, 49–56.

Klaus, H. (1967). *Thin Elastic Shells*. New York: John Wiley and Sons.

Koiter, W. T. (1960). A Consistent First Approximation in the General Theory of Thin Elastic Shells. In *The Theory of Thin Elastic Shells* (pp. 12–33). Armsterdam: North-Holland.

Koiter, W. T. (1967). General Equations of Elastic Stability for Thin Shells. *Proc. Symp. Theory of Shells to Honor Lloyd Hamilton Donnell*. Houston, TX: University of Houston. 187–223.

Langhaar, H. L. (1962). *Energy Methods in Applied Mechanics*. New York: John Wiley and Sons.

Love, A. E. H. (1944). *A Treatise on the Mathematical Theory of Elasticity* (4th ed.). New York: Dover.

[12] Private communication.

Mushtari, K. M. (1938). On the Stability of Cylindrical Shells Subjected to Torsion. Trudy Kaz. aviats. in-ta., 2 (in Russian).
Naghdi, P. M. (1963). Foundations of Elastic Shell Theory. In *Progress in Solid Mechanics, Vol 4*. New York: John Wiley and Sons.
Novozhilov, V. V. (1953). Foundations of the Nonlinear Theory of Elasticity. In F. Bagemihl, H. Komm, & W. Seidel (Eds.), *Translated from the 1st (1948) Russian*. Rochester, NY: Graylock Press.
Novozhilov, V. V. (1964). *Thin Shell Theory*. In: Lowe, P.G. and Radok, J.R. (eds), Translated from the 2nd Russian ed. The Netherlands: Noordhoff, Groningen.
Sanders, J. L. (1963). Nonlinear Theories for Thin Shells. *Q. Appl. Math., 21*(1), 21–36.
Timoshenko, S. P., & Gere, J. M. (1961). Theory of Elastic Stability (2nd ed.). New York: McGraw-Hill.
Szilard, R. (1974). *Theory and Analysis of Plates, Classical and Numerical Methods*. Englewood Cliffs, NJ: Prentice-Hall.
Vlasov, V. Z. (1944). The fundamental differential equations of the general theory of elastic shells, *Prikl. Mat. Mekh.*, Akademiya, Nauk. SSSR, Vol. VIII, No. 2, also Basic Differential Equations in the General Theory of Elastic Shells, NACA TM 1241, February 1951.
Vlasov, V. Z. (1964). *General Theory of Shells and Its Application in Engineering*. Washington, DC: NASA TTF-99 National Aeronautics and Space Administration.

PROBLEMS

10.1 Derive the Lamé coefficients Eq. (10.2.2) from Fig. 10-1.

10.2 Derive Eq. (10.2.4).

10.3 Derive Eq. (10.2.5).

10.4 Derive Eq. (10.2.7).

10.5 Show that the application of Euler-Lagrange differential equations to the energy equations Eqs. (10.2.10) to (10.2.12) along with Eqs. (10.2.14a), (10.2.15), and (10.2.16) yields the nonlinear equilibrium equations for the shell of the general shape in Eqs. (10.2.17).

10.6 Show that (a) Equations (10.2.17) can be specialized for a circular cylindrical shell Eqs. (9.2.9) by setting $A = 1$, $B = R_y = R$, and $1/R_x = 0$, (b) Equations (10.2.17) can also be converted to the nonlinear equilibrium equations for a rectangular flat plate (von Kármán plate equations) Eqs. (8.2.18) by setting $A = B = 1$ and $1/R_x = 1/R_y = 0$, and (c) Similarly, Eqs. (10.2.17) can also be converted to the nonlinear equilibrium equation for a column Eq. (1.7.14).

10.7 Verify Eqs. (10.2.28).

10.8 Verify Eqs. (10.2.33).

10.9 Verify that Eqs. (10.2.28a) and (10.2.28b), and (10.2.37) are identical to Eqs. (10.2.17).

10.10 Derive Eq. (10.3.7) following the procedure outlined in Section 8.3.

10.11 Derive Eqs. (10.3.12) by applying Eqs. (8.3.12) and (10.3.11) on Eq. (10.3.10).

10.12 Derive Eq. (10.4.27) by substituting Eqs. (10.4.26) into Eq. (10.4.24c).

10.13 Derive Eq. (10.4.31a) from Eqs. (10.4.25a)–10.4.25f).

10.14 Derive Eq. (10.4.32) relating the stress function f and the displacement component w for a shallow spherical cap.

AUTHOR INDEX

R

S

T

V

W

Y

Z

SUBJECT INDEX

Printed and bound by CPI Group (UK) Ltd, Croydon, CR0 4YY
08/05/2025
01864884-0001